Langlands Correspondence for Loop Groups

The Langlands Program was conceived initially as a bridge between Number Theory and Automorphic Representations, and has now expanded into such areas as Geometry and Quantum Field Theory, weaving together seemingly unrelated disciplines into a web of tantalizing conjectures. This book provides a new chapter in this grand project. It develops the geometric Langlands Correspondence for Loop Groups, a new approach, from a unique perspective offered by affine Kac–Moody algebras. The theory offers fresh insights into the world of Langlands dualities, with many applications to Representation Theory of Infinite-dimensional Algebras, and Quantum Field Theory. This introductory text builds the theory from scratch, with all necessary concepts defined and the essential results proved along the way. Based on courses taught by the author at Berkeley, the book provides many open problems which could form the basis for future research, and is accessible to advanced undergraduate students and beginning graduate students.

CAMBRIDGE STUDIES IN ADVANCED MATHEMATICS

All the titles listed below can be obtained from good booksellers or from Cambridge University Press. For a complete series listing visit: http://www.cambridge.org/series/sSeries.asp?code=CSAM

Already published

60 M. P. Brodmann & R.Y. Sharp *Local cohomology*
61 J. D. Dixon *et al. Analytic pro-p groups*
62 R. Stanley *Enumerative combinatorics II*
63 R. M. Dudley *Uniform central limit theorems*
64 J. Jost & X. Li-Jost *Calculus of variations*
65 A. J. Berrick & M. E. Keating *An introduction to rings and modules*
66 S. Morosawa *Holomorphic dynamics*
67 A. J. Berrick & M. E. Keating *Categories and modules with K-theory in view*
68 K. Sato *Levy processes and infinitely divisible distributions*
69 H. Hida *Modular forms and Galois cohomology*
70 R. Iorio & V. Iorio *Fourier analysis and partial differential equations*
71 R. Blei *Analysis in integer and fractional dimensions*
72 F. Borceaux & G. Janelidze *Galois theories*
73 B. Bollobás *Random graphs*
74 R. M. Dudley *Real analysis and probability*
75 T. Sheil-Small *Complex polynomials*
76 C. Voisin *Hodge theory and complex algebraic geometry, I*
77 C. Voisin *Hodge theory and complex algebraic geometry, II*
78 V. Paulsen *Completely bounded maps and operator algebras*
79 F. Gesztesy & H. Holden *Soliton Equations and Their Algebro-Geometric Solutions, I*
81 S. Mukai *An Introduction to Invariants and Moduli*
82 G. Tourlakis *Lectures in Logic and Set Theory, I*
83 G. Tourlakis *Lectures in Logic and Set Theory, II*
84 R. A. Bailey *Association Schemes*
85 J. Carlson, S. Müller-Stach & C. Peters *Period Mappings and Period Domains*
86 J. J. Duistermaat & J. A. C. Kolk *Multidimensional Real Analysis I*
87 J. J. Duistermaat & J. A. C. Kolk *Multidimensional Real Analysis II*
89 M. Golumbic & A. Trenk *Tolerance Graphs*
90 L. Harper *Global Methods for Combinatorial Isoperimetric Problems*
91 I. Moerdijk & J. Mrcun *Introduction to Foliations and Lie Groupoids*
92 J. Kollar, K. E. Smith & A. Corti *Rational and Nearly Rational Varieties*
93 D. Applebaum *Levy Processes and Stochastic Calculus*
94 B. Conrad *Modular Forms and the Ramanujan Conjecture*
95 M. Schechter *An Introduction to Nonlinear Analysis*
96 R. Carter *Lie Algebras of Finite and Affine Type*
97 H. L. Montgomery, R.C Vaughan & M. Schechter *Multiplicative Number Theory I*
98 I. Chavel *Riemannian Geometry*
99 D. Goldfeld *Automorphic Forms and L-Functions for the Group GL(n,R)*
100 M. Marcus & J. Rosen *Markov Processes, Gaussian Processes, and Local Times*
101 P. Gille & T. Szamuely *Central Simple Algebras and Galois Cohomology*
102 J. Bertoin *Random Fragmentation and Coagulation Processes*
103 E. Frenkel *Langlands Correspondence for Loop Groups*
104 A. Ambrosetti & A. Malchiodi *Nonlinear Analysis and Semilinear Elliptic Problems*
105 T. Tao, V. H. Vu *Additive Combinatorics*
106 E. B. Davies *Linear Operators and their Spectra*

Langlands Correspondence for Loop Groups

EDWARD FRENKEL
University of California, Berkeley

CAMBRIDGE
UNIVERSITY PRESS

University Printing House, Cambridge CB2 8BS, United Kingdom

One Liberty Plaza, 20th Floor, New York, NY 10006, USA

477 Williamstown Road, Port Melbourne, VIC 3207, Australia

314-321, 3rd Floor, Plot 3, Splendor Forum, Jasola District Centre, New Delhi - 110025, India

103 Penang Road, #05-06/07, Visioncrest Commercial, Singapore 238467

Cambridge University Press is part of the University of Cambridge.

It furthers the University's mission by disseminating knowledge in the pursuit of education, learning and research at the highest international levels of excellence.

www.cambridge.org
Information on this title: www.cambridge.org/9780521854436

First published 2007

A catalogue record for this publication is available from the British Library

ISBN 978-0-521-85443-6 Hardback

For my parents

Concentric geometries of transparency slightly
joggled sink through algebras of proud

inwardlyness to collide spirally with iron arithmethics...

– E.E. Cummings, "W [ViVa]" (1931)

Contents

Preface		*page* xi
1	**Local Langlands correspondence**	1
	1.1 The classical theory	1
	1.2 Langlands parameters over the complex field	9
	1.3 Representations of loop groups	18
2	**Vertex algebras**	31
	2.1 The center	31
	2.2 Basics of vertex algebras	38
	2.3 Associativity in vertex algebras	52
3	**Constructing central elements**	61
	3.1 Segal–Sugawara operators	62
	3.2 Lie algebras associated to vertex algebras	70
	3.3 The center of a vertex algebra	76
	3.4 Jet schemes	81
	3.5 The center in the case of $\mathfrak{sl}_2$	87
4	**Opers and the center for a general Lie algebra**	101
	4.1 Projective connections, revisited	101
	4.2 Opers for a general simple Lie algebra	107
	4.3 The center for an arbitrary affine Kac–Moody algebra	117
5	**Free field realization**	126
	5.1 Overview	126
	5.2 Finite-dimensional case	128
	5.3 The case of affine algebras	135
	5.4 Vertex algebra interpretation	140
	5.5 Computation of the two-cocycle	145
	5.6 Comparison of cohomology classes	151
6	**Wakimoto modules**	161
	6.1 Wakimoto modules of critical level	162
	6.2 Deforming to other levels	167

6.3 Semi-infinite parabolic induction 178
6.4 Appendix: Proof of the Kac–Kazhdan conjecture 189

7 Intertwining operators 192
7.1 Strategy of the proof 192
7.2 The case of $\mathfrak{sl}_2$ 197
7.3 Screening operators for an arbitrary $\mathfrak{g}$ 205

8 Identification of the center with functions on opers 214
8.1 Description of the center of the vertex algebra 215
8.2 Identification with the algebra of functions on opers 226
8.3 The center of the completed universal enveloping algebra 237

9 Structure of $\widehat{\mathfrak{g}}$-modules of critical level 243
9.1 Opers with regular singularity 244
9.2 Nilpotent opers 247
9.3 Miura opers with regular singularities 256
9.4 Categories of representations of $\widehat{\mathfrak{g}}$ at the critical level 265
9.5 Endomorphisms of the Verma modules 273
9.6 Endomorphisms of the Weyl modules 279

10 Constructing the Langlands correspondence 295
10.1 Opers vs. local systems 297
10.2 Harish–Chandra categories 301
10.3 The unramified case 305
10.4 The tamely ramified case 326
10.5 From local to global Langlands correspondence 353

Appendix 366
A.1 Lie algebras 366
A.2 Universal enveloping algebras 366
A.3 Simple Lie algebras 368
A.4 Central extensions 370
A.5 Affine Kac–Moody algebras 371

References 373

Index 377

Preface

The Langlands Program has emerged in recent years as a blueprint for a Grand Unified Theory of Mathematics. Conceived initially as a bridge between Number Theory and Automorphic Representations, it has now expanded into such areas as Geometry and Quantum Field Theory, weaving together seemingly unrelated disciplines into a web of tantalizing conjectures. The Langlands correspondence manifests itself in a variety of ways in these diverse areas of mathematics and physics, but the same salient features, such as the appearance of the Langlands dual group, are always present. This points to something deeply mysterious and elusive, and that is what makes this correspondence so fascinating.

One of the prevalent themes in the Langlands Program is the interplay between the *local* and *global* pictures. In the context of Number Theory, for example, "global" refers to a number field (a finite extension of the field of rational numbers) and its Galois group, while "local" means a local field, such as the field of p-adic numbers, together with its Galois group. On the other side of the Langlands correspondence we have, in the global case, automorphic representations, and, in the local case, representations of a reductive group, such as GL_n, over the local field.

In the geometric context the cast of characters changes: on the Galois side we now have vector *bundles with flat connection* on a complex Riemann surface X in the global case, and on the punctured disc $D^\times$ around a point of X in the local case. The definition of the objects on the other side of the geometric Langlands correspondence is more subtle. It is relatively well understood (after works of A. Beilinson, V. Drinfeld, G. Laumon and others) in the special case when the flat connection on our bundle has no singularities. Then the corresponding objects are the so-called "Hecke eigensheaves" on the moduli spaces of vector bundles on X. These are the geometric analogues of unramified automorphic functions. The unramified global geometric Langlands correspondence is then supposed to assign to a flat connection on our bundle (without singularities) a Hecke eigensheaf. (This is discussed in a recent review [F7], among other places, where we refer the reader for more details.)

However, in the more general case of *connections with ramification*, that is, with singularities, the geometric Langlands correspondence is much more mysterious, both in the local and in the global case. Actually, the impetus now shifts more to the local story. This is because the flat connections that we consider have finitely many singular points on our Riemann surface. The global ramified correspondence is largely determined by what happens on the punctured discs around those points, which is in the realm of the local correspondence. So the question really becomes: *what is the geometric analogue of the local Langlands correspondence?*

Now, the classical local Langlands correspondence relates representations of p-adic groups and Galois representations. In the geometric version we should replace a p-adic group by the (formal) *loop group* $G((t))$, the group of maps from the (formal) punctured disc $D^\times$ to a complex reductive algebraic group G. Galois representations should be replaced by vector bundles on $D^\times$ with a flat connection. These are the local geometric Langlands parameters. To each of them we should be able to attach a representation of the formal loop group.

Recently, Dennis Gaitsgory and I have made a general proposal describing these representations of loop groups. An important new element in our proposal is that, in contrast to the classical correspondence, the loop group now acts on categories rather than vector spaces. Thus, the Langlands correspondence for loop groups is categorical: we associate *categorical representations* of $G((t))$ to local Langlands parameters. We have proposed how to construct these categories using representations of the affine Kac–Moody algebra $\widehat{\mathfrak{g}}$, which is a central extension of the loop Lie algebra $\mathfrak{g}((t))$. Therefore the local geometric Langlands correspondence appears as the result of a successful marriage of the Langlands philosophy and the representation theory of affine Kac–Moody algebras.

Affine Kac–Moody algebras have a parameter, called the level. For a special value of this parameter, called the *critical level*, the completed enveloping algebra of an affine Kac–Moody algebra acquires an unusually large *center*. In 1991, Boris Feigin and I showed that this center is canonically isomorphic to the algebra of functions on the space of *opers* on $D^\times$. Opers are bundles on $D^\times$ with a flat connection and an additional datum (as defined by Drinfeld–Sokolov and Beilinson–Drinfeld). Remarkably, their structure group turns out to be not G, but the *Langlands dual group* ${}^L G$, in agreement with the general Langlands philosophy. This is the central result, which implies that the same salient features permeate both the representation theory of p-adic groups and the (categorical) representation theory of loop groups.

This result had been conjectured by V. Drinfeld, and it plays an important role in his and A. Beilinson's approach to the global geometric Langlands correspondence, via quantization of the Hitchin systems. The isomorphism between the center and functions on opers means that the category of representations of $\widehat{\mathfrak{g}}$ of critical level "lives" over the space of ^LG-opers on $D^\times$, and the loop group $G((t))$ acts "fiberwise" on this category. In a nutshell, the proposal of Gaitsgory and myself is that the "fibers" of this category are the sought-after categorical representations of $G((t))$ corresponding to the local Langlands parameters underlying the ^LG-opers. This has many non-trivial consequences, both for the local and global geometric Langlands correspondence, and for the representation theory of $\widehat{\mathfrak{g}}$. We hope that further study of these categories will give us new clues and insights into the mysteries of the Langlands correspondence.†

The goal of this book is to present a systematic and self-contained introduction to the local geometric Langlands correspondence for loop groups and the related representation theory of affine Kac–Moody algebras. It covers the research done in this area over the last twenty years and is partially based on the graduate courses that I have taught at UC Berkeley in 2002 and 2004. In the book, the entire theory is built from scratch, with all necessary concepts defined and all essential results proved along the way. We introduce such concepts as the Weil-Deligne group, Langlands dual group, affine Kac–Moody algebras, vertex algebras, jet schemes, opers, Miura opers, screening operators, etc., and illustrate them by detailed examples. In particular, many of the results are first explained in the simplest case of SL_2. Practically no background beyond standard college algebra is required from the reader (except possibly in the last chapter); we even explain some standard notions, such as universal enveloping algebras, in the Appendix.

In the opening chapter, we present a pedagogical overview of the classical Langlands correspondence and a motivated step-by-step passage to the geometric setting. This leads us to the study of affine Kac–Moody algebras and in particular the center of the completed enveloping algebra. We then review in great detail the construction of a series of representations of affine Kac–Moody algebras, called *Wakimoto modules*. They were defined by Feigin and myself in the late 1980s following the work of M. Wakimoto. These modules give us an effective tool for developing the representation theory of affine algebras. In particular, they are crucial in our proof of the isomorphism between the spectrum of the center and opers. A detailed exposition of the Wakimoto modules and the proof of this isomorphism constitute the main part of this book. These results allow us to establish a deep link between the representation theory of affine Kac–Moody algebras of critical level and the geometry of opers. In the closing chapter, we review the results and conjectures of

† We note that A. Beilinson has another proposal [Bei] for the local geometric Langlands correspondence, using representations of affine Kac–Moody algebras of integral levels less than critical. It would be interesting to understand the connection between his proposal and ours.

Gaitsgory and myself describing the representation categories associated to opers in the framework of the Langlands correspondence. I also discuss the implications of this for the global geometric Langlands correspondence. These are only the first steps of a new theory, which we hope will ultimately help us reveal the secrets of Langlands duality.

Contents

Here is a more detailed description of the contents of this book.

Chapter 1 is the introduction to the subject matter of this book. We begin by giving an overview of the local and global Langlands correspondence in the classical setting. Since the global case is discussed in great detail in my recent review [F7], I concentrate here mostly on the local case. Next, I explain what changes one needs to make in order to transport the local Langlands correspondence to the realm of geometry and the representation theory of loop groups. I give a pedagogical account of Galois groups, principal bundles with connections and central extensions, among other topics. This discussion leads us to the following question: how to attach to each local geometric Langlands parameter an abelian category equipped with an action of the formal loop group?

In Chapter 2 we take up this question in the context of the representation theory of affine Kac–Moody algebras. This motivates us to study the center of the completed enveloping algebra of $\widehat{\mathfrak{g}}$. First, we do that by elementary means, but very quickly we realize that we need a more sophisticated technique. This technique is the theory of vertex algebras. We give a crash course on vertex algebras (following [FB]), summarizing all necessary concepts and results.

Armed with these results, we begin in Chapter 3 a more in-depth study of the center of the completed enveloping algebra of $\widehat{\mathfrak{g}}$ at the critical level (we find that the center is trivial away from the critical level). We describe the center in the case of the simplest affine Kac–Moody algebra $\widehat{\mathfrak{sl}}_2$ and the quasi-classical analogue of the center for an arbitrary $\widehat{\mathfrak{g}}$.

In Chapter 4 we introduce the key geometric concept of opers, introduced originally in [DS, BD1]. We state the main result, due to Feigin and myself [FF6, F4], that the center at the critical level, corresponding to $\widehat{\mathfrak{g}}$, is isomorphic to the algebra of functions on ${}^L G$-opers.

In order to prove this result, we need to develop the theory of Wakimoto modules. This is done in Chapters 5 and 6, following [F4]. We start by explaining the analogous theory for finite-dimensional simple Lie algebras, which serves as a prototype for our construction. Then we explain the non-trivial elements of the infinite-dimensional case, such as the cohomological obstruction to realizing a loop algebra in the algebra of differential operators on a loop space. This leads to a conceptual explanation of the non-triviality of the critical level. In Chapter 6 we complete the construction of Wakimoto modules, both at the critical and non-critical levels. We prove some useful results on representations of affine Kac–Moody algebras, such as the Kac-Kazhdan conjecture.

Having built the theory of Wakimoto modules, we are ready to tackle the isomorphism between the center and the algebra of functions on opers. At the beginning of Chapter 7 we give a detailed overview of the proof of this isomorphism. In the rest of Chapter 7 we introduce an important class of intertwining operators between Wakimoto modules called the screening operators. We use these operators and some results on associated graded algebras in Chapter 8 to complete the proof of our main result and to identify the center with functions on opers (here we follow [F4]). In particular, we clarify the origins of the appearance of the Langlands dual group in this isomorphism, tracing it back to a certain duality between vertex algebras known as the $\mathcal{W}$-algebras. At the end of the chapter we discuss the vertex Poisson structure on the center and identify the action of the center on Wakimoto modules with the Miura transformation.

In Chapter 9 we undertake a more in-depth study of representations of affine Kac–Moody algebras of critical level. We first introduce (following [BD1] and [FG2]) certain subspaces of the space of opers on $D^\times$: opers with regular singularities and nilpotent opers, and explain the interrelations between them. We then discuss Miura opers with regular singularities and the action of the Miura transformation on them, following [FG2]. Finally, we describe the results of [FG2] and [FG6] on the algebras of endomorphisms of the Verma modules and the Weyl modules of critical level.

In Chapter 10 we bring together the results of this book to explain the proposal for the local geometric Langlands correspondence made by Gaitsgory and myself. We review the results and conjectures of our works [FG1]–[FG6], emphasizing the analogies between the geometric and the classical Langlands correspondence. We discuss in detail the interplay between opers and local systems. We then consider the simplest local system; namely, the trivial one. The corresponding categorical representations of $G((t))$ are the analogues of unramified representations of p-adic groups. Already in this case we will see rather non-trivial elements, which emulate the corresponding elements of the classical theory and at the same time generalize them in a non-trivial way. The next, and considerably more complicated, example is that of local systems on $D^\times$ with regular singularity and unipotent monodromy. These are the analogues of the tamely ramified representations of the Galois group of a p-adic field. The corresponding categories turn out to be closely related to categories of quasicoherent sheaves on the Springer fibers, which are algebraic subvarieties of the flag variety of the Langlands dual group. We summarize the conjectures and results of [FG2] concerning these categories and illustrate them by explicit computations in the case of $\widehat{\mathfrak{sl}}_2$. We also formulate some open problems in this direction. Finally, we discuss the implications of this approach for the global Langlands correspondence.

Acknowledgments

I thank Boris Feigin and Dennis Gaitsgory for their collaboration on our joint works reviewed in this book.

I have benefited over the years from numerous conversations with Alexander Beilinson, David Ben-Zvi, Joseph Bernstein, Vladimir Drinfeld, Victor Kac and David Kazhdan on the Langlands correspondence and representations of affine Kac–Moody algebras. I am grateful to all of them. I also thank R. Bezrukavnikov and V. Ginzburg for their useful comments on some questions related to the material of the last chapter.

I thank Alex Barnard for letting me use his notes of my graduate course at UC Berkeley in preparation of Chapters 2 and 3. I am also indebted to Arthur Greenspoon, Reimundo Heluani, Valerio Toledano Laredo and Xinwen Zhu for their valuable comments on the draft of the book. I am grateful to David Tranah of Cambridge University Press for his expert guidance and friendly advice throughout the publication process.

In the course of writing this book I received generous support from DARPA through its Program "Fundamental Advances in Theoretical Mathematics". I was also partially supported by the National Science Foundation through the grant DMS-0303529. I gratefully acknowledge this support.

1
Local Langlands correspondence

In this introductory chapter we explain in detail what we mean by "local Langlands correspondence for loop groups." We begin by giving a brief overview of the local Langlands correspondence for reductive groups over local non-archimedian fields, such as $\mathbb{F}_q((t))$, the field of Laurent power series over a finite field $\mathbb{F}_q$. We wish to understand an analogue of this correspondence when $\mathbb{F}_q$ is replaced by the field $\mathbb{C}$ of complex numbers. The role of the reductive group will then be played by the formal loop group $G(\mathbb{C}((t)))$. We discuss, following [FG2], how the setup of the Langlands correspondence should change in this case. This discussion will naturally lead us to categories of representations of the corresponding affine Kac–Moody algebra $\widehat{\mathfrak{g}}$ equipped with an action of $G(\mathbb{C}((t)))$, the subject that we will pursue in the rest of this book.

1.1 The classical theory

The local Langlands correspondence relates smooth representations of reductive algebraic groups over local fields and representations of the Galois group of this field. We start out by defining these objects and explaining the main features of this correspondence. As the material of this section serves motivational purposes, we will only mention those aspects of this story that are most relevant for us. For a more detailed treatment, we refer the reader to the informative surveys [V, Ku] and references therein.

1.1.1 Local non-archimedian fields

Let F be a local non-archimedian field. In other words, F is the field $\mathbb{Q}_p$ of p-adic numbers, or a finite extension of $\mathbb{Q}_p$, or F is the field $\mathbb{F}_q((t))$ of formal Laurent power series with coefficients in $\mathbb{F}_q$, the finite field with q elements.

We recall that for any q of the form p^n, where p is a prime, there is a unique, up to isomorphism, finite field of characteristic p with q elements. An element of $\mathbb{F}_q((t))$ is an expression of the form

$$\sum_{n \in \mathbb{Z}} a_n t^n, \qquad a_n \in \mathbb{F}_q,$$

such that $a_n = 0$ for all n less than some integer N. In other words, these are power series infinite in the positive direction and finite in the negative direction. Recall that a p-adic number may also be represented by a series

$$\sum_{n \in \mathbb{Z}} b_n p^n, \qquad b_n \in \{0, 1, \ldots, p-1\},$$

such that $b_n = 0$ for all n less than some integer N. We see that elements of $\mathbb{Q}_p$ look similar to elements of $\mathbb{F}_p((t))$. Both fields are complete with respect to the topology defined by the norm taking value α^{-N} on the above series if $a_N \neq 0$ and $a_n = 0$ for all $n < N$, where α is a fixed positive real number between 0 and 1. But the laws of addition and multiplication in the two fields are different: with "carry" to the next digit in the case of $\mathbb{Q}_p$, but without "carry" in the case of $\mathbb{F}_p((t))$. In particular, $\mathbb{Q}_p$ has characteristic 0, while $\mathbb{F}_q((t))$ has characteristic p. More generally, elements of a finite extension of $\mathbb{Q}_p$ look similar to elements of $\mathbb{F}_q((t))$ for some $q = p^n$, but, again, the rules of addition and multiplication, as well as their characteristics, are different.

1.1.2 Smooth representations of $GL_n(F)$

Now consider the group $GL_n(F)$, where F is a local non-archimedian field. A representation of $GL_n(F)$ on a complex vector space V is a homomorphism $\pi : GL_n(F) \to \operatorname{End} V$ such that $\pi(gh) = \pi(g)\pi(h)$ and $\pi(1) = \mathrm{Id}$. Define a topology on $GL_n(F)$ by stipulating that the base of open neighborhoods of $1 \in GL_n(F)$ is formed by the congruence subgroups $K_N, N \in \mathbb{Z}_+$. In the case when $F = \mathbb{F}_q((t))$, the group K_N is defined as follows:

$$K_N = \{g \in GL_n(\mathbb{F}_q[[t]]) \,|\, g \equiv 1 \mod t^N\},$$

and for $F = \mathbb{Q}_p$ it is defined in a similar way. For each $v \in V$ we obtain a map $\pi(\cdot)v : GL_n(F) \to V, g \mapsto \pi(g)v$. A representation (V, π) is called **smooth** if the map $\pi(\cdot)v$ is continuous for each v, where we give V the discrete topology. In other words, V is smooth if for any vector $v \in V$ there exists $N \in \mathbb{Z}_+$ such that

$$\pi(g)v = v, \qquad \forall g \in K_N.$$

We are interested in describing the equivalence classes of irreducible smooth representations of $GL_n(F)$. Surprisingly, those turn out to be related to objects of a different kind: n-dimensional representations of the Galois group of F.

1.1.3 The Galois group

Suppose that F is a subfield of K. Then the **Galois group** $\mathrm{Gal}(K/F)$ consists of all automorphisms σ of the field K such that $\sigma(y) = y$ for all $y \in F$.

Let F be a field. The algebraic closure of F is a field obtained by adjoining to F the roots of all polynomials with coefficients in F. In the case when $F = \mathbb{F}_q((t))$ some of the extensions of F may be non-separable. An example of such an extension is the field $\mathbb{F}_q((t^{1/p}))$. The polynomial defining this extension is $x^p - t$, but in $\mathbb{F}_q((t^{1/p}))$ it has multiple roots because

$$x^p - (t^{1/p})^p = (x - t^{1/p})^p.$$

The Galois group $\mathrm{Gal}(\mathbb{F}_q((t^{1/p})), \mathbb{F}_q((t)))$ of this extension is trivial, even though the degree of the extension is p.

This extension should be contrasted to the separable extensions $\mathbb{F}_q((t^{1/n}))$, where n is not divisible by p. This extension is defined by the polynomial $x^n - t$, which now has no multiple roots:

$$x^n - t = \prod_{i=0}^{n-1} (x - \zeta^i t^{1/n}),$$

where ζ is a primitive nth root of unity in the algebraic closure of $\mathbb{F}_q$. The corresponding Galois group is identified with the group $(\mathbb{Z}/n\mathbb{Z})^\times$, the group of invertible elements of $\mathbb{Z}/n\mathbb{Z}$.

We wish to avoid the non-separable extensions, because they do not contribute to the Galois group. (There are no non-separable extensions if F has characteristic zero, e.g., for $F = \mathbb{Q}_p$.) Let $\overline{F}$ be the maximal separable extension inside a given algebraic closure of F. It is uniquely defined up to isomorphism.

Let $\mathrm{Gal}(\overline{F}/F)$ be the **absolute Galois group** of F. Its elements are the automorphisms σ of the field $\overline{F}$ such that $\sigma(y) = y$ for all $y \in F$.

To gain some experience with Galois groups, let us look at the Galois group $\mathrm{Gal}(\overline{\mathbb{F}}_q/\mathbb{F}_q)$. Here $\overline{\mathbb{F}}_q$ is the algebraic closure of $\mathbb{F}_q$, which can be defined as the inductive limit of the fields $\mathbb{F}_{q^N}, N \in \mathbb{Z}_+$, with respect to the natural embeddings $\mathbb{F}_{q^N} \hookrightarrow \mathbb{F}_{q^M}$ for N dividing M. Therefore $\mathrm{Gal}(\overline{\mathbb{F}}_q/\mathbb{F}_q)$ is isomorphic to the inverse limit of the Galois groups $\mathrm{Gal}(\mathbb{F}_{q^N}/\mathbb{F}_q)$ with respect to the natural surjections

$$\mathrm{Gal}(\mathbb{F}_{q^M}/\mathbb{F}_q) \twoheadrightarrow \mathrm{Gal}(\mathbb{F}_{q^N}/\mathbb{F}_q), \qquad \forall N|M.$$

The group $\mathrm{Gal}(\mathbb{F}_{q^N}/\mathbb{F}_q)$ is easy to describe: it is generated by the **Frobenius automorphism** $x \mapsto x^q$ (note that it stabilizes $\mathbb{F}_q$), which has order N, so that $\mathrm{Gal}(\mathbb{F}_{q^N}/\mathbb{F}_q) \simeq \mathbb{Z}/N\mathbb{Z}$. Therefore we find that

$$\mathrm{Gal}(\overline{\mathbb{F}}_q/\mathbb{F}_q) \simeq \widehat{\mathbb{Z}} \stackrel{\mathrm{def}}{=} \varprojlim \mathbb{Z}/N\mathbb{Z},$$

where we have the surjective maps $\mathbb{Z}/M\mathbb{Z} \twoheadrightarrow \mathbb{Z}/N\mathbb{Z}$ for $N|M$. The group $\widehat{\mathbb{Z}}$ contains $\mathbb{Z}$ as a subgroup.

Let $F = \mathbb{F}_q((t))$. Observe that we have a natural map $\mathrm{Gal}(\overline{F}/F) \to \mathrm{Gal}(\overline{\mathbb{F}}_q/\mathbb{F}_q)$ obtained by applying an automorphism of $\overline{F}$ to $\overline{\mathbb{F}}_q \subset \overline{F}$. A similar map also exists when F has characteristic 0. Let W_F be the preimage of the subgroup $\mathbb{Z} \subset \mathrm{Gal}(\overline{\mathbb{F}}_q/\mathbb{F}_q)$. This is the **Weil group** of F. Let ν be the corresponding homomorphism $W_F \to \mathbb{Z}$. Let $W'_F = W_F \ltimes \mathbb{C}$ be the semi-direct product of W_F and the one-dimensional complex additive group $\mathbb{C}$, where W_F acts on $\mathbb{C}$ by the formula

$$\sigma x \sigma^{-1} = q^{\nu(\sigma)} x, \qquad \sigma \in W_F, x \in \mathbb{C}. \tag{1.1.1}$$

This is the **Weil–Deligne group** of F.

An n-dimensional complex representation of W'_F is by definition a homomorphism $\rho' : W'_F \to GL_n(\mathbb{C})$, which may be described as a pair (ρ, u), where ρ is an n-dimensional representation of W_F, $u \in \mathfrak{gl}_n(\mathbb{C})$, and we have $\rho(\sigma) u \rho(\sigma)^{-1} = q^{\nu(\sigma)} u$ for all $\sigma \in W_F$. The group W_F is topological, with respect to the Krull topology (in which the open neighborhoods of the identity are the normal subgroups of finite index). The representation (ρ, u) is called **admissible** if ρ is continuous (equivalently, factors through a finite quotient of W_F) and semisimple, and u is a nilpotent element of $\mathfrak{gl}_n(\mathbb{C})$.

The group W'_F was introduced by P. Deligne [De2]. The idea is that by adjoining the nilpotent element u to W_F we obtain a group whose complex admissible representations are the same as continuous ℓ-adic representations of W_F (where $\ell \neq p$ is a prime).

1.1.4 The local Langlands correspondence for GL_n

Now we are ready to state the local Langlands correspondence for the group GL_n over a local non-archimedian field F. It is a bijection between two different sorts of data. One is the set of equivalence classes of irreducible smooth representations of $GL_n(F)$. The other is the set of equivalence classes of n-dimensional admissible representations of W'_F. We represent it schematically as follows:

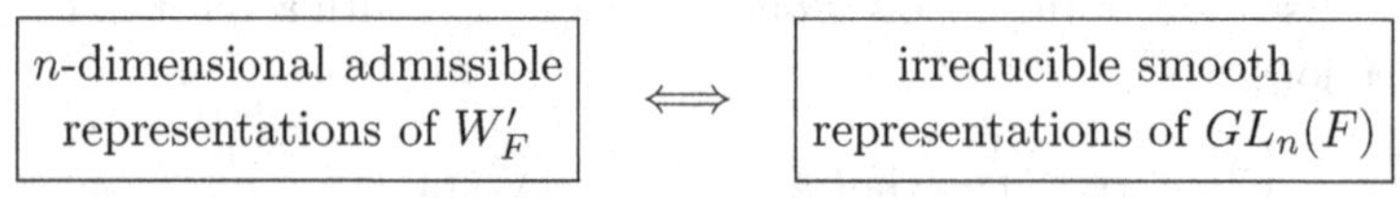

This correspondence is supposed to satisfy an overdetermined system of constraints which we will not recall here (see, e.g., [Ku]).

The local Langlands correspondence for GL_n is a theorem. In the case when $F = \mathbb{F}_q((t))$ it has been proved in [LRS], and when $F = \mathbb{Q}_p$ or its finite extension in [HT] and also in [He]. We refer the readers to these papers and to the review [C] for more details.

Despite an enormous effort made in the last two decades to understand it, the local Langlands correspondence still remains a mystery. We do know that the above bijection exists, but we cannot yet explain in a completely satisfactory way *why* it exists. We do not know the deep underlying reasons that make such a correspondence possible. One way to try to understand it is to see how general it is. In the next section we will discuss one possible generalization of this correspondence, where we replace the group GL_n by an arbitrary reductive algebraic group defined over F.

1.1.5 Generalization to other reductive groups

Let us replace the group GL_n by an arbitrary connected reductive group G over a local non-archimedian field F. The group $G(F)$ is also a topological group, and there is a notion of smooth representation of $G(F)$ on a complex vector space. It is natural to ask whether we can relate irreducible smooth representations of $G(F)$ to representations of the Weil–Deligne group W'_F. This question is addressed in the general local Langlands conjectures. It would take us too far afield to try to give here a precise formulation of these conjectures. So we will only indicate some of the objects involved, referring the reader to the articles [V, Ku], where these conjectures are described in great detail.

Recall that in the case when $G = GL_n$ the irreducible smooth representations are parametrized by admissible homomorphisms $W'_F \to GL_n(\mathbb{C})$. In the case of a general reductive group G, the representations are conjecturally parametrized by admissible homomorphisms from W'_F to the so-called Langlands dual group ${}^LG$, which is defined over $\mathbb{C}$.

In order to explain the notion of the Langlands dual group, consider first the group G over the closure $\overline{F}$ of the field F. All maximal tori T of this group are conjugate to each other and are necessarily split, i.e., we have an isomorphism $T(\overline{F}) \simeq (\overline{F}^\times)^n$. For example, in the case of GL_n, all maximal tori are conjugate to the subgroup of diagonal matrices. We associate to $T(\overline{F})$ two lattices: the weight lattice $X^*(T)$ of homomorphisms $T(\overline{F}) \to \overline{F}^\times$ and the coweight lattice $X_*(T)$ of homomorphisms $\overline{F}^\times \to T(\overline{F})$. They contain the sets of roots $\Delta \subset X^*(T)$ and coroots $\Delta^\vee \subset X_*(T)$, respectively. The quadruple $(X^*(T), X_*(T), \Delta, \Delta^\vee)$ is called the root data for G over $\overline{F}$. The root data determines G up to an isomorphism defined over $\overline{F}$. The choice of a Borel subgroup $B(\overline{F})$ containing $T(\overline{F})$ is equivalent to a choice of a basis in Δ; namely, the set of simple roots Δ_s, and the corresponding basis $\Delta_s^\vee$ in $\Delta^\vee$.

Now given $\gamma \in \mathrm{Gal}(\overline{F}/F)$, there is $g \in G(\overline{F})$ such that $g(\gamma(T(\overline{F}))g^{-1} = T(\overline{F})$ and $g(\gamma(B(\overline{F}))g^{-1} = B(\overline{F})$. Then g gives rise to an automorphism of the based root data $(X^*(T), X_*(T), \Delta_s, \Delta_s^\vee)$. Thus, we obtain an action of $\mathrm{Gal}(\overline{F}/F)$ on the based root data.

Let us now exchange the lattices of weights and coweights and the sets of simple roots and coroots. Then we obtain the based root data

$$(X_*(T), X^*(T), \Delta_s^\vee, \Delta_s)$$

of a reductive algebraic group over $\mathbb{C}$, which is denoted by ${}^L G^\circ$. For instance, the group GL_n is self-dual, the dual of SO_{2n+1} is Sp_{2n}, the dual of Sp_{2n} is SO_{2n+1}, and SO_{2n} is self-dual.

The action of $\mathrm{Gal}(\overline{F}/F)$ on the based root data gives rise to its action on ${}^L G^\circ$. The semi-direct product ${}^L G = \mathrm{Gal}(\overline{F}/F) \ltimes {}^L G^\circ$ is called the **Langlands dual group** of G.

The local Langlands correspondence for the group $G(F)$ relates the equivalence classes of irreducible smooth representations of $G(F)$ to the equivalence classes of admissible homomorphisms $W'_F \to {}^L G$. However, in general this correspondence is much more subtle than in the case of GL_n. In particular, we need to consider simultaneously representations of all inner forms of G, and a homomorphism $W'_F \to {}^L G$ corresponds in general not to a single irreducible representation of $G(F)$, but to a finite set of representations called an ***L*-packet**. To distinguish between them, we need additional data (see [V] for more details; some examples are presented in Section 10.4.1 below). But in the first approximation we can say that the essence of the local Langlands correspondence is that

irreducible smooth representations of $G(F)$ are parameterized in terms of admissible homomorphisms $W'_F \to {}^L G$.

1.1.6 On the global Langlands correspondence

We close this section with a brief discussion of the global Langlands correspondence and its connection to the local one. We will return to this subject in Section 10.5.

Let X be a smooth projective curve over $\mathbb{F}_q$. Denote by F the field $\mathbb{F}_q(X)$ of rational functions on X. For any closed point x of X we denote by F_x the completion of F at x and by $\mathcal{O}_x$ its ring of integers. If we choose a local coordinate t_x at x (i.e., a rational function on X which vanishes at x to order one), then we obtain isomorphisms $F_x \simeq \mathbb{F}_{q_x}((t_x))$ and $\mathcal{O}_x \simeq \mathbb{F}_{q_x}[[t_x]]$, where $\mathbb{F}_{q_x}$ is the residue field of x; in general, it is a finite extension of $\mathbb{F}_q$ containing $q_x = q^{\deg(x)}$ elements.

Thus, we now have a local field attached to each point of X. The ring $\mathbb{A} = \mathbb{A}_F$ of **adèles** of F is by definition the *restricted* product of the fields F_x, where x runs over the set $|X|$ of all closed points of X. The word "restricted" means that we consider only the collections $(f_x)_{x \in |X|}$ of elements of F_x in which $f_x \in \mathcal{O}_x$ for all but finitely many x. The ring $\mathbb{A}$ contains the field F, which is embedded into $\mathbb{A}$ diagonally, by taking the expansions of rational functions on X at all points.

While in the local Langlands correspondence we considered irreducible smooth representations of the group GL_n over a local field, in the global Langlands correspondence we consider irreducible **automorphic representations** of the group $GL_n(\mathbb{A})$. The word "automorphic" means, roughly, that the representation may be realized in a reasonable space of functions on the quotient $GL_n(F)\backslash GL_n(\mathbb{A})$ (on which the group $GL_n(\mathbb{A})$ acts from the right).

On the other side of the correspondence we consider n-dimensional representations of the Galois group $\text{Gal}(\overline{F}/F)$, or, more precisely, the Weil group W_F, which is a subgroup of $\text{Gal}(\overline{F}/F)$ defined in the same way as in the local case.

Roughly speaking, the global Langlands correspondence is a bijection between the set of equivalence classes of n-dimensional representations of W_F and the set of equivalence classes of irreducible automorphic representations of $GL_n(\mathbb{A})$:

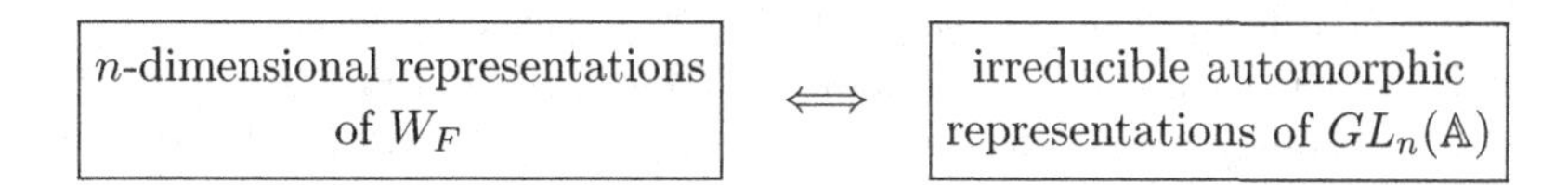

The precise statement is more subtle. For example, we should consider the so-called ℓ-adic representations of the Weil group (while in the local case we considered the admissible complex representations of the Weil–Deligne group; the reason is that in the local case those are equivalent to the ℓ-adic representations). Moreover, under this correspondence important invariants attached to the objects appearing on both sides (Frobenius eigenvalues on the Galois side and the Hecke eigenvalues on the other side) are supposed to match. We refer the reader to Part I of the review [F7] for more details.

The global Langlands correspondence has been proved for GL_2 in the 1980's by V. Drinfeld [Dr1]–[Dr4] and more recently by L. Lafforgue [Laf] for GL_n with an arbitrary n.

The global and local correspondences are compatible in the following sense. We can embed the Weil group W_{F_x} of each of the local fields F_x into the global Weil group W_F. Such an embedding is not unique, but it is well-defined up to conjugation in W_F. Therefore an equivalence class σ of n-dimensional representations of W_F gives rise to a well-defined equivalence class σ_x of n-dimensional representations of W_{F_x} for each $x \in X$. By the local Langlands correspondence, to σ_x we can attach an equivalence class of irreducible smooth representations of $GL_n(F_x)$. Choose a representation π_x in this equivalence class. Then the automorphic representation of $GL_n(\mathbb{A})$ corresponding to σ

is isomorphic to the restricted tensor product $\bigotimes'_{x \in X} \pi_x$. This is a very non-trivial statement, because a priori it is not clear why this tensor product may be realized in the space of functions on the quotient $GL_n(F)\backslash GL_n(\mathbb{A})$.

As in the local story, we may also wish to replace the group GL_n by an arbitrary reductive algebraic group defined over F. The general global Langlands conjecture predicts, roughly speaking, that irreducible automorphic representations of $G(\mathbb{A})$ are related to homomorphisms $W_F \to {}^L G$. But, as in the local case, the precise formulation of the conjecture for a general reductive group is much more intricate (see [Art]).

Finally, the global Langlands conjectures can also be stated over number fields (where they in fact originated). Then we take as the field F a finite extension of the field $\mathbb{Q}$ of rational numbers. Consider for example the case of $\mathbb{Q}$ itself. It is known that the completions of $\mathbb{Q}$ are (up to isomorphism) the fields of p-adic numbers $\mathbb{Q}_p$ for all primes p (non-archimedian) and the field $\mathcal{R}$ of real numbers (archimedian). So the primes play the role of points of an algebraic curve over a finite field (and the archimedian completion corresponds to an infinite point, in some sense). The ring of adèles $\mathbb{A}_{\mathbb{Q}}$ is defined in the same way as in the function field case, and so we can define the notion of an automorphic representation of $GL_n(\mathbb{A}_{\mathbb{Q}})$ or a more general reductive group. Conjecturally, to each equivalence class of n-dimensional representations of the Galois group $\mathrm{Gal}(\overline{\mathbb{Q}}/\mathbb{Q})$ we can attach an equivalence class of irreducible automorphic representations of $GL_n(\mathbb{A}_{\mathbb{Q}})$, but this correspondence is not expected to be a bijection because in the number field case it is known that some of the automorphic representations do not correspond to any Galois representations.

The Langlands conjectures in the number field case lead to very important and unexpected results. Indeed, many interesting representations of Galois groups can be found in "nature". For example, the group $\mathrm{Gal}(\overline{\mathbb{Q}}/\mathbb{Q})$ will act on the geometric invariants (such as the étale cohomologies) of an algebraic variety defined over $\mathbb{Q}$. Thus, if we take an elliptic curve E over $\mathbb{Q}$, then we will obtain a two-dimensional Galois representation on its first étale cohomology. This representation contains a lot of important information about the curve E, such as the number of points of E over $\mathbb{Z}/p\mathbb{Z}$ for various primes p. The Langlands correspondence is supposed to relate these Galois representations to automorphic representations of $GL_2(\mathbb{A}_F)$ in such a way that the data on the Galois side, like the number of points of $E(\mathbb{Z}/p\mathbb{Z})$, are translated into something more tractable on the automorphic side, such as the coefficients in the q-expansion of the modular forms that encapsulate automorphic representations of $GL_2(\mathbb{A}_{\mathbb{Q}})$. This leads to some startling consequences, such as the Taniyama-Shimura conjecture. For more on this, see [F7] and references therein.

The Langlands correspondence has proved to be easier to study in the function field case. The main reason is that in the function field case we can use the geometry of the underlying curve and various moduli spaces associated to this curve. A curve can also be considered over the field of complex numbers. Some recent results show that a version of the global Langlands correspondence also exists for such curves. The local counterpart of this correspondence is the subject of this book.

1.2 Langlands parameters over the complex field

We now wish to find a generalization of the local Langlands conjectures in which we replace the field $F = \mathbb{F}_q((t))$ by the field $\mathbb{C}((t))$. We would like to see how the ideas and patterns of the Langlands correspondence play out in this new context, with the hope of better understanding the deep underlying structures behind this correspondence.

So from now on G will be a connected reductive group over $\mathbb{C}$, and $G(F)$ the group $G((t)) = G(\mathbb{C}((t)))$, also known as the **loop group**; more precisely, the formal loop group, with the word "formal" referring to the algebra of formal Laurent power series $\mathbb{C}((t))$ (as opposed to the group $G(\mathbb{C}[t,t^{-1}])$, where $\mathbb{C}[t,t^{-1}]$ is the algebra of Laurent polynomials, which may be viewed as the group of maps from the unit circle $|t| = 1$ to G, or "loops" in G).

Thus, we wish to study smooth representations of the loop group $G((t))$ and try to relate them to some "Langlands parameters," which we expect, by analogy with the case of local non-archimedean fields described above, to be related to the Galois group of $\mathbb{C}((t))$ and the Langlands dual group ${}^LG$.

The local Langlands correspondence for loop groups that we discuss in this book may be viewed as the first step in carrying the ideas of the Langlands Program to the realm of complex algebraic geometry. In particular, it has far-reaching consequences for the global geometric Langlands correspondence (see Section 10.5 and [F7] for more details). This was in fact one of the motivations for this project.

1.2.1 The Galois group and the fundamental group

We start by describing the Galois group $\mathrm{Gal}(\overline{F}/F)$ for $F = \mathbb{C}((t))$. Observe that the algebraic closure $\overline{F}$ of F is isomorphic to the inductive limit of the fields $\mathbb{C}((t^{1/n})), n \geq 0$, with respect to the natural inclusions $\mathbb{C}((t^{1/n})) \hookrightarrow \mathbb{C}((t^{1/m}))$ for n dividing m. Hence $\mathrm{Gal}(\overline{F}/F)$ is the inverse limit of the Galois groups

$$\mathrm{Gal}(\mathbb{C}((t^{1/n}))/C((t))) \simeq \mathbb{Z}/n\mathbb{Z},$$

where $\overline{k} \in \mathbb{Z}/n\mathbb{Z}$ corresponds to the automorphism of $\mathbb{C}((t^{1/n}))$ sending $t^{1/n}$ to $e^{2\pi ik/n}t^{1/n}$. The result is that

$$\mathrm{Gal}(\overline{F}/F) \simeq \widehat{\mathbb{Z}},$$

where $\widehat{\mathbb{Z}}$ is the profinite completion of $\mathbb{Z}$ that we have encountered before in Section 1.1.3.

Note, however, that in our study of the Galois group of $\mathbb{F}_q((t))$ the group $\widehat{\mathbb{Z}}$ appeared as its quotient corresponding to the Galois group of the field of coefficients $\mathbb{F}_q$. Now the field of coefficients is $\mathbb{C}$, hence algebraically closed, and $\widehat{\mathbb{Z}}$ is the *entire* Galois group of $\mathbb{C}((t))$.

The naive analogue of the Langlands parameter would be an equivalence class of homomorphisms $\mathrm{Gal}(\overline{F}/F) \to {}^L G$, i.e., a homomorphism $\widehat{\mathbb{Z}} \to {}^L G$. Since G is defined over $\mathbb{C}$ and hence all of its maximal tori are split, the group $G((t))$ also contains a split torus $T((t))$, where T is a maximal torus of G (but it also contains non-split maximal tori, as the field $\mathbb{C}((t))$ is not algebraically closed). Therefore the Langlands dual group ${}^L G$ is the direct product of the Galois group and the group ${}^L G^\circ$. Because it is a direct product, we may, and will, restrict our attention to ${}^L G^\circ$. In order to simplify our notation, from now on we will denote ${}^L G^\circ$ simply by ${}^L G$.

A homomorphism $\widehat{\mathbb{Z}} \to {}^L G$ necessarily factors through a finite quotient $\widehat{\mathbb{Z}} \to \mathbb{Z}/n\mathbb{Z}$. Therefore the equivalence classes of homomorphisms $\widehat{\mathbb{Z}} \to {}^L G$ are the same as the conjugacy classes of ${}^L G$ of finite order. There are too few of these to have a meaningful generalization of the Langlands correspondence. Therefore we look for a more sensible alternative.

Let us recall the connection between Galois groups and fundamental groups. Let X be an algebraic variety over $\mathbb{C}$. If $Y \to X$ is a covering of X, then the field $\mathbb{C}(Y)$ of rational functions on Y is an extension of the field $F = \mathbb{C}(X)$ of rational functions on X. The deck transformations of the cover, i.e., automorphisms of Y which induce the identity on X, give rise to automorphisms of the field $\mathbb{C}(Y)$ preserving $\mathbb{C}(X) \subset \mathbb{C}(Y)$. Hence we identify the Galois group $\mathrm{Gal}(\mathbb{C}(Y)/\mathbb{C}(X))$ with the group of deck transformations. If our cover is unramified, then this group may be identified with a quotient of the fundamental group of X. Otherwise, this group is isomorphic to a quotient of the fundamental group of X with the ramification divisor thrown out.

In particular, we obtain that the Galois group of the maximal unramified extension of $\mathbb{C}(X)$ (which we can view as the field of functions of the "maximal unramified cover" of X) is the profinite completion of the fundamental group $\pi_1(X)$ of X. Likewise, for any divisor $D \subset X$ the Galois group of the maximal extension of $\mathbb{C}(X)$ unramified away from D is the profinite completion of $\pi_1(X \backslash D)$. We denote it by $\pi_1^{\mathrm{alg}}(X \backslash D)$. The algebraic closure of $\mathbb{C}(X)$ is the inductive limit of the fields of functions on the maximal covers of X ramified at various divisors $D \subset X$ with respect to natural inclusions corresponding to the inclusions of the divisors. Hence the Galois group $\mathrm{Gal}(\overline{\mathbb{C}(X)}/\mathbb{C}(X))$ is the inverse limit of the groups $\pi_1^{\mathrm{alg}}(X \backslash D)$ with respect to the maps $\pi_1^{\mathrm{alg}}(X \backslash D') \to \pi_1^{\mathrm{alg}}(X \backslash D)$ for $D \subset D'$.

Strictly speaking, in order to define the fundamental group of X we need to pick a reference point $x \in X$. In the above discussion we have tacitly picked as the reference point the generic point of X, i.e., $\operatorname{Spec} \mathbb{C}(X) \hookrightarrow X$. But for different choices of x the corresponding fundamental groups will be isomorphic anyway, and so the equivalences classes of their representations (which is what we are after) will be the same. Therefore we will ignore the choice of a reference point.

Going back to our setting, we see that $\operatorname{Gal}(\overline{F}/F)$ for $F = \mathbb{C}((t))$ is indeed $\widehat{\mathbb{Z}}$, the profinite completion of the "topological" fundamental group of the punctured disc $D^\times$. Thus, the naive Langlands parameters correspond to homomorphisms $\pi_1^{\text{alg}}(D^\times) \to {}^L G$. But, in the complex setting, homomorphisms $\pi_1(X) \to {}^L G$ can be obtained from more geometrically refined data; namely, from bundles with flat connections, which we will discuss presently.

1.2.2 Flat bundles

Suppose that X is a real manifold and E is a complex vector bundle over X. Thus, we are given a map $p : E \to X$, satisfying the following condition: each point $x \in X$ has a neighborhood U such that the preimage of U in E under p is isomorphic to $U \times \mathbb{C}^n$. However, the choice of such a trivialization of E over U is not unique. Two different trivializations differ by a smooth map $U \to GL_n(\mathbb{C})$. We can describe E more concretely by choosing a cover of X by open subsets $U_\alpha, \alpha \in A$, and picking trivializations of E on each U_α. Then the difference between the two induced trivializations on the overlap $U_\alpha \cap U_\beta$ will be accounted for by a function $g_{\alpha\beta} : U_\alpha \cap U_\beta \to GL_n(\mathbb{C})$, which is called the **transition function**. The transition functions satisfy the important transitivity property: on triple overlaps $U_\alpha \cap U_\beta \cap U_\gamma$ we have $g_{\alpha\gamma} = g_{\alpha\beta} g_{\beta\gamma}$.

This describes the data of an ordinary vector bundle over X. The data of a **flat vector bundle** E on X are a transitive system of preferred trivializations of E over a sufficiently small open subset of any point of X, with the difference between two preferred trivializations given by a *constant* $GL_n(\mathbb{C})$-valued function. Thus, concretely, a flat bundle may be described by choosing a sufficiently fine cover of X by open subsets $U_\alpha, \alpha \in A$, and picking trivializations of E on each U_α that belong to our preferred system of trivializations. Then the transition functions $g_{\alpha\beta} : U_\alpha \cap U_\beta \to GL_n(\mathbb{C})$ are *constant functions*, which satisfy the transitivity property. This should be contrasted to the case of ordinary bundles, for which the transition functions may be arbitrary smooth $GL_n(\mathbb{C})$-valued functions.

From the perspective of differential geometry, the data of a preferred system of trivializations are neatly expressed by the data of a **flat connection**. These are precisely the data necessary to differentiate the sections of E.

More precisely, let $\mathcal{O}_X$ be the sheaf of smooth functions on X and $\mathcal{T}$ the sheaf of smooth vector fields on X. Denote by $\operatorname{End}(E)$ the bundle of endomorphisms of E. By a slight abuse of notation we will denote by the same symbol $\operatorname{End}(E)$ the sheaf of its smooth sections. A **connection** on E is a map

$$\nabla : \mathcal{T} \longrightarrow \mathrm{End}(E),$$

which takes a vector field $\xi \in \mathcal{T}(U)$ defined on an open subset $U \subset X$ to an endomorphism ∇_ξ of E over U. It should have the following properties:

- It is $\mathcal{O}_X$ linear: $\nabla_{\xi+\eta} = \nabla_\xi + \nabla_\eta$ and $\nabla_{f\xi} = f\nabla_\xi$ for $f \in \mathcal{O}_X$.
- It satisfies the Leibniz rule: $\nabla_\xi(f\phi) = f\nabla_\xi(\phi) + \phi \cdot \xi(f)$ for $f \in \mathcal{O}_X$ and $\phi \in \mathrm{End}(E)$.

A connection is called **flat** if it has the additional property that ∇ is a Lie algebra homomorphism

$$[\nabla_\xi, \nabla_\chi] = \nabla_{[\xi,\chi]}, \tag{1.2.1}$$

where we consider the natural Lie algebra structures on $\mathcal{T}$ and $\mathrm{End}(E)$.

Let us discuss a more concrete realization of connections. On a small enough open subset U we can pick coordinates $x_i, i = 1, \ldots, N$, and trivialize the bundle E. Then by $\mathcal{O}_X$-linearity, to define the restriction of a connection ∇ to U it is sufficient to write down the operators $\nabla_{\partial_{x_i}}$, where we use the notation $\partial_{x_i} = \dfrac{\partial}{\partial x_i}$. The Leibniz rule implies that these operators must have the form

$$\nabla_{\partial_{x_i}} = \partial_{x_i} + A_i(\mathbf{x}),$$

where A_i is an $n \times n$ matrix-valued smooth function on U.

The flatness condition then takes the form

$$[\nabla_{\partial_{x_i}}, \nabla_{\partial_{x_j}}] = \partial_{x_i} A_j - \partial_{x_j} A_i + [A_i, A_j] = 0. \tag{1.2.2}$$

So, for a very concrete description of a flat connection in the real case we can simply specify matrices A_i locally, satisfying the above equations. But we should also make sure that they transform in the correct way under changes of coordinates and changes of trivialization of the bundle E. If we introduce new coordinates such that $x_i = x_i(\mathbf{y})$, then we find that

$$\partial_{y_i} = \sum_{j=1}^{n} \frac{\partial x_j}{\partial y_i} \partial_{x_j},$$

and so

$$\nabla_{\partial_{y_j}} = \sum_{j=1}^{n} \frac{\partial x_j}{\partial y_i} \nabla_{\partial_{x_j}}.$$

A given section of E that appears as a $\mathbb{C}^n$-valued function $f(\mathbf{x})$ on U with respect to the old trivialization will appear as the function $g(\mathbf{x}) \cdot f(\mathbf{x})$ with respect to a new trivialization, where $g(\mathbf{x})$ is the $GL_n(\mathbb{C})$-valued transition function. Therefore the connection operators will become

$$g\nabla_{\partial_{x_i}} g^{-1} = \nabla_{\partial_{x_i}} + gA_i g^{-1} - (\partial_{x_i} g) g^{-1}.$$

Using the connection operators, we construct a transitive system of preferred local trivializations of our bundle as follows: locally, on a small open subset U, consider the system of the first order differential equations

$$\nabla_\xi \Phi(\mathbf{x}) = 0, \qquad \xi \in \mathcal{T}(U). \tag{1.2.3}$$

For this system to make sense the flatness condition (1.2.2) should be satisfied. (Otherwise, the right hand side of (1.2.2), which is a matrix-valued function, would have to annihilate $\Phi(\mathbf{x})$. Therefore we would have to restrict $\Phi(\mathbf{x})$ to the kernel of this matrix-valued function.) In this case, the standard theorems of existence and uniqueness of solutions of linear differential equations tell us that for each point $\mathbf{x}_0 \in U$ and each vector v in the fiber $E_{\mathbf{x}_0}$ of E at $\mathbf{x}_0$ it has a unique local solution Φ_v satisfying the initial condition $\Phi(\mathbf{x}_0) = v$. The solutions Φ_v for different $v \in E_{\mathbf{x}_0}$ are sections of E over U, which are called **horizontal**. They give us a transitive system of preferred identifications of the fibers of E over the points of U: $E_{\mathbf{x}} \simeq E_{\mathbf{y}}, \mathbf{x}, \mathbf{y} \in U$. If we identify one of these fibers, say the one at $\mathbf{x}_0$, with $\mathbb{C}^n$, then we obtain a trivialization of E over U. Changing the identification $E_{\mathbf{x}_0} \simeq \mathbb{C}^n$ would change this trivialization only by a constant function $U \to GL_n$. Thus, we obtain the desired preferred system of trivializations of E.

Put differently, the data of a flat connection give us a preferred system of identifications of nearby fibers of E, which are locally transitive. But if we use them to identify the fibers of E lying over a given path in X starting and ending at a point $\mathbf{x}_0$, then we may obtain a non-trivial automorphism of the fiber $E_{\mathbf{x}_0}$, the **monodromy** of the flat connection along this path. (More concretely, this monodromy may be found by considering solutions of the system (1.2.3) defined in a small tubular neighborhood of our path.) This automorphism will depend only on the homotopy class of our path, and so any flat connection on E gives rise to a homomorphism $\pi_1(X, \mathbf{x}_0) \to \operatorname{Aut} E_{\mathbf{x}_0}$, where $\operatorname{Aut} E_{\mathbf{x}_0}$ is the group of automorphisms of $E_{\mathbf{x}_0}$. If we identify $E_{\mathbf{x}_0}$ with $\mathbb{C}^n$, we obtain a homomorphism $\pi_1(X, \mathbf{x}_0) \to GL_n(\mathbb{C})$.

Thus, we assign to a flat bundle on X of rank n an equivalence class of homomorphisms $\pi_1(X, \mathbf{x}_0) \to GL_n(\mathbb{C})$.

In the case when X is the punctured disc $D^\times$, or a smooth projective complex curve, equivalence classes of homomorphisms $\pi_1(X, \mathbf{x}_0) \to GL_n(\mathbb{C})$ are precisely what we wish to consider as candidates for the Langlands parameters in the complex setting. That is why we are interested in their reformulations as bundles with flat connections.

1.2.3 Flat bundles in the holomorphic setting

Suppose now that X is a complex algebraic variety and E is a **holomorphic vector bundle** over X. Recall that giving a vector bundle E over X the structure of a holomorphic vector bundle is equivalent to specifying which sections of E are considered to be holomorphic. On an open subset U of X with holomorphic coordinates $z_i, i = 1, \ldots, N$, and anti-holomorphic coordinates $\overline{z}_i, i = 1, \ldots, N$, specifying the holomorphic sections is equivalent to defining the $\overline{\partial}$-operators $\nabla_{\partial_{\overline{z}_i}}, i = 1, \ldots, N$, on the sections of E. The holomorphic sections of E are then precisely those annihilated by these operators. Thus, we obtain an action of the anti-holomorphic vector fields on all smooth sections of E. This action of anti-holomorphic vector fields gives us "half" of the data of a flat connection. So, to specify a flat connection on a holomorphic vector bundle E we only need to define an action of the (local) holomorphic vector fields on X on the holomorphic sections of E.

On a sufficiently small open subset U of X we may now choose a trivialization compatible with the holomorphic structure, so that we have $\nabla_{\partial_{\overline{z}_i}} = \partial_{\overline{z}_i}$. Then extending these data to the data of a flat connection on E amounts to constructing operators

$$\nabla_{\partial_{z_i}} = \partial_{z_i} + A_i(\mathbf{z}),$$

where the flatness condition demands that the A_i's be holomorphic matrix valued functions of $z_j, j = 1, \ldots, N$, such that

$$[\nabla_{\partial_{z_i}}, \nabla_{\partial_{z_j}}] = \partial_{z_i} A_j - \partial_{z_j} A_i + [A_i, A_j] = 0.$$

In particular, if X is a complex curve, i.e., $N = 1$, any holomorphic connection on a holomorphic vector bundle gives rise to a flat connection, and hence an equivalence class of homomorphisms $\pi_1(X) \to GL_n(\mathbb{C})$.†

This completes the story of flat connections on vector bundles, or equivalently, principal GL_n-bundles. Next, we define flat connections on principal G-bundles, where G is a complex algebraic group.

1.2.4 Flat G-bundles

Recall that a **principal G-bundle** over a manifold X is a manifold $\mathcal{P}$ which is fibered over X such that there is a natural fiberwise right action of G on $\mathcal{P}$, which is simply-transitive along each fiber. In addition, it is locally trivial: each point has a sufficient small open neighborhood $U \subset X$ over which the bundle $\mathcal{P}$ may be trivialized, i.e., there is an isomorphism $t_U : \mathcal{P}|_U \xrightarrow{\sim} U \times G$ commuting with the right actions of G. More concretely, we may describe $\mathcal{P}$ by choosing a cover of X by open subsets $U_\alpha, \alpha \in A$, and picking trivializations of $\mathcal{P}$ on each U_α. Then the difference between the two induced trivializations on the overlap $U_\alpha \cap U_\beta$ will be described by the transition functions $g_{\alpha\beta} : U_\alpha \cap U_\beta \to G$, which satisfy the transitivity condition, as in the case of GL_n.

† Note, however, that some holomorphic vector bundles on curves do not carry any holomorphic connections; for instance, a line bundle carries a holomorphic connection if and only if it has degree 0.

A flat structure on such a bundle $\mathcal{P}$ is a preferred system of locally transitive identifications of the fibers of $\mathcal{P}$, or equivalently, a preferred system of transitive local trivializations (this means that the transition functions $g_{\alpha\beta}$ are constant G-valued functions). In order to give a differential geometric realization of flat connections, we use the Tannakian formalism. It says, roughly, that G may be reconstructed from the category of its finite-dimensional representations equipped with the structure of the tensor product, satisfying various compatibilities, and the "fiber functor" to the category of vector spaces.

Given a principal G-bundle $\mathcal{P}$ on X and a finite-dimensional representation V of G we can form the associated vector bundle

$$V_{\mathcal{P}} = \mathcal{P} \underset{G}{\times} V.$$

Thus we obtain a functor from the category of finite-dimensional representations of G to the category of vector bundles on X. Both of these categories are tensor categories and it is clear that the above functor is actually a tensor functor. The Tannakian formalism allows us to reconstruct the principal G-bundle $\mathcal{P}$ from this functor. In other words, the data of a principal G-bundle on X are encoded by the data of a collection of vector bundles on X labeled by finite-dimensional representations of G together with the isomorphisms $(V \otimes W)_{\mathcal{P}} \simeq V_{\mathcal{R}} \otimes W_{\mathcal{P}}$ for each pair of representations V, W and various compatibilities stemming from the tensor properties of this functor.

In order to define a flat connection on a holomorphic principal G-bundle $\mathcal{P}$ we now need to define a flat connection on each of the associated vector bundles $V_{\mathcal{P}}$ in a compatible way. What this means is that on each sufficiently small open subset $U \subset X$, after choosing a system of local holomorphic coordinates $z_i, i = 1, \dots, N$, on U and a holomorphic trivialization of $\mathcal{P}$ over U, the data of a flat connection on $\mathcal{P}$ are given by differential operators

$$\nabla_{\partial_{z_i}} = \partial_{z_i} + A_i(\mathbf{z}),$$

where now the A_i's are holomorphic functions on U with values in the Lie algebra $\mathfrak{g}$ of G.

The transformation properties of the operators $\nabla_{\partial_{z_i}}$ under changes of coordinates and trivializations are given by the same formulas as above. In particular, under a change of trivialization given by a holomorphic G-valued function g on U we have

$$\partial_{z_i} + A_i \mapsto \partial_{z_i} + gA_ig^{-1} - (\partial_{z_i}g)g^{-1}. \tag{1.2.4}$$

Such transformations are called the **gauge transformations**.

The meaning of this formula is as follows: the expression $gA_ig^{-1} - (\partial_{z_i}g)g^{-1}$ appearing in the right hand side is a well-defined element of $\operatorname{End} V$ for any finite-dimensional representation V of G. The Tannakian formalism discussed above then implies that there is a "universal" element of $\mathfrak{g}$ whose action on V is given by this formula.

1.2.5 Regular vs. irregular singularities

In the same way as in the case of GL_n, we obtain that a holomorphic principal G-bundle on a complex variety X with a holomorphic flat connection gives rise to an equivalence class of homomorphisms from the fundamental group of X to G. But does this set up a bijection of the corresponding equivalence classes? If X is compact, this is indeed the case, but, if X is not compact, then there are more flat bundles than there are representations of $\pi_1(X)$. In order to obtain a bijection, we need to impose an additional condition on the connection; we need to require that it has **regular singularities** at infinity.

If X is a curve obtained from a projective (hence compact) curve $\overline{X}$ by removing finitely many points, this condition means that the connection operator $\partial_z + A(z)$ has a pole of order at most one at each of the removed points (this condition is generalized to a higher-dimensional variety X in a straightforward way by restricting the connection to all curves lying in X).

For example, consider the case when $X = \mathbf{A}^\times = \operatorname{Spec} \mathbb{C}[t, t^{-1}]$ and assume that the rank of E is equal to one, so it is a line bundle. We can embed X into the projective curve $\mathbb{P}^1$, so that there are two points at infinity: $t = 0$ and $t = \infty$. Any holomorphic line bundle on $\mathbf{A}^\times$ can be trivialized. A general connection operator on $\mathbf{A}^\times$ then reads

$$\nabla_{\partial_t} = \partial_t + A(t), \qquad A(t) = \sum_{i=N}^{M} A_i t^i. \tag{1.2.5}$$

The group of invertible functions on $\mathbf{A}^\times$ consists of the elements $\alpha t^n, \alpha \in \mathbb{C}^\times, n \in \mathbb{Z}$. It acts on such operators by gauge transformations (1.2.4), and αt^n acts as follows:

$$A(t) \mapsto A(t) - \frac{n}{t}.$$

The set of equivalence classes of line bundles with a flat connection on $\mathbf{A}^\times$ is in bijection with the quotient of the set of operators (1.2.5) by the group of gauge transformations, or, equivalently, the quotient of $\mathbb{C}[t, t^{-1}]$ by the additive action of the group $\mathbb{Z} \cdot \frac{1}{t}$. Thus, we see that this set is huge.

Let us now look at connections with regular singularities only. The condition that the connection (1.2.5) has a regular singularity at $t = 0$ means that $N \geq -1$. To see what the condition at $t = \infty$ is, we perform a change of variables $u = t^{-1}$. Then, since $\partial_t = -u^2 \partial_u$, we find that

$$\nabla_{\partial_u} = \partial_u - u^{-2} A(u^{-1}).$$

Hence the condition of regular singularity at ∞ is that $M \leq -1$, and so to satisfy both conditions, $A(t)$ must have the form $A(t) = \frac{a}{t}$.

Taking into account the gauge transformations, we conclude that the set of equivalence classes of line bundles with a flat connection with regular singularity on $\mathbf{A}^\times$ is isomorphic to $\mathbb{C}/\mathbb{Z}$. Given a connection (1.2.5) with $A(t) = \frac{a}{t}$, the solutions of the equation

$$\left(\partial_t + \frac{a}{t}\right)\Phi(t) = 0$$

are $\Phi_C(t) = C\exp(-at)$, where $C \in \mathbb{C}$. These are the horizontal sections of our line bundle. Hence the monodromy of this connection along the loop going counterclockwise around the origin in $\mathbf{A}^\times$ is equal to $\exp(-2\pi ia)$. We obtain a one-dimensional representation of $\pi_1(\mathbf{A}^\times) \simeq \mathbb{Z}$ sending $1 \in \mathbb{Z}$ to $\exp(-2\pi ia)$. Thus, the exponential map sets up a bijection between the set of equivalence classes of connections with regular singularities on $\mathbf{A}^\times$ and representations of the fundamental group of $\mathbf{A}^\times$.

But if we allow irregular singularities, we obtain many more connections. For instance, consider the connection $\partial_t + \frac{1}{t^2}$. It has an irregular singularity at $t = 0$. The corresponding horizontal sections have the form $C\exp(-1/t), C \in \mathbb{C}$, and hence the monodromy representation of $\pi_1(\mathbf{A}^\times)$ corresponding to this connection is trivial, and so it is the same as for the connection $\nabla_{\partial_t} = \partial_t$. But this connection is not equivalent to the connection ∂_t under the action of the *algebraic* gauge transformations. We can of course obtain one from the other by the gauge action with the function $\exp(-1/t)$, but this function is not algebraic. Thus, if we allow irregular singularities, there are many more algebraic gauge equivalence classes of flat ${}^LG$-connections than there are homomorphisms from the fundamental group to ${}^LG$.

1.2.6 Connections as the Langlands parameters

Our goal is to find a geometric enhancement of the set of equivalence classes of homomorphisms $\pi_1(D^\times) \to {}^LG$, of which there are too few. The above discussion suggests a possible way to do it. We have seen that such homomorphisms are the same as ${}^LG$-bundles on $D^\times$ with a connection with regular singularity at the origin. Here by **punctured disc** we mean the scheme $D^\times = \operatorname{Spec}\mathbb{C}((t))$, so that the algebra of functions on $D^\times$ is, by definition, $\mathbb{C}((t))$. Any ${}^LG$-bundle on $D^\times$ can be trivialized, and so defining a connection on it amounts to giving a first order operator

$$\partial_t + A(t), \qquad A(t) \in {}^L\mathfrak{g}((t)), \tag{1.2.6}$$

where ${}^L\mathfrak{g}$ is the Lie algebra of the Langlands dual group ${}^LG$.

Changing trivialization amounts to a gauge transformation (1.2.4) with $g \in {}^LG((t))$. The set of equivalence classes of ${}^LG$-bundles with a connection on $D^\times$ is in bijection with the set of gauge equivalence classes of operators (1.2.6). We denote this set by $\operatorname{Loc}_{{}^LG}(D^\times)$. Thus, we have

$$\operatorname{Loc}_{{}^LG}(D^\times) = \{\partial_t + A(t),\ A(t) \in {}^L\mathfrak{g}((t))\}/{}^LG((t)). \tag{1.2.7}$$

A connection (1.2.6) has regular singularity if and only if $A(t)$ has a pole of order at most one at $t = 0$. If A_{-1} is the residue of $A(t)$ at $t = 0$, then the corresponding monodromy of the connection around the origin is an element of ${}^LG$ equal to $\exp(2\pi i A_{-1})$. Two connections with regular singularity are (algebraically) gauge equivalent to each other if and only if their monodromies are conjugate to each other. Hence the set of equivalence classes of connections with regular singularities is just the set of conjugacy classes of ${}^LG$, or, equivalently, homomorphisms $\mathbb{Z} \to {}^LG$. These are the naive Langlands parameters that we saw before (here we ignore the difference between the topological and algebraic fundamental groups). But now we generalize this by allowing connections with *arbitrary*, that is, regular and irregular, singularities at the origin. Then we obtain many more gauge equivalence classes.

We will often refer to points of $\mathrm{Loc}_{{}^LG}(D^\times)$ as **local systems** on $D^\times$, meaning the de Rham version of the notion of a local system; namely, a principal bundle with a connection that has a pole of an arbitrary order at the origin.

Thus, we come to the following proposal: the local Langlands parameters in the complex setting should be the points of $\mathrm{Loc}_{{}^LG}(D^\times)$: the equivalence classes of flat ${}^LG$-bundles on $D^\times$ or, more concretely, the gauge equivalence classes (1.2.7) of first order differential operators.

We note that the Galois group of the local field $\mathbb{F}_q((t))$ has a very intricate structure: apart from the part coming from $\mathrm{Gal}(\overline{\mathbb{F}}_q/\mathbb{F}_q)$ and the tame inertia, which is analogous to the monodromy that we observe in the complex case, there is also the wild inertia subgroup whose representations are very complicated. It is an old idea that in the complex setting flat connections with irregular singularities in some sense play the role of representations of the Galois group whose restriction to the wild inertia subgroup is non-trivial. Our proposal simply exploits this idea in the context of the Langlands correspondence.

1.3 Representations of loop groups

Having settled the issue of the Langlands parameters, we have to decide what it is that we will be parameterizing. Recall that in the classical setting the homomorphism $W'_F \to {}^LG$ parameterized irreducible smooth representations of the group $G(F)$, $F = \mathbb{F}_q((t))$. We start by translating this notion to the representation theory of loop groups.

1.3.1 Smooth representations

The loop group $G((t))$ contains the congruence subgroups

$$K_N = \{g \in G[[t]] \,|\, g \equiv 1 \bmod t^N\}, \qquad N \in \mathbb{Z}_+. \tag{1.3.1}$$

It is natural to call a representation of $G((t))$ on a complex vector space V **smooth** if for any vector $v \in V$ there exists $N \in \mathbb{Z}_+$ such that $K_N \cdot v = v$. This condition may be interpreted as the continuity condition, if we define a topology on $G((t))$ by taking as the base of open neighborhoods of the identity the subgroups $K_N, N \in \mathbb{Z}_+$, as before.

But our group G is now a complex Lie group (not a finite group), and so $G((t))$ is an infinite-dimensional Lie group. More precisely, we view $G((t))$ as an ind-group, i.e., as a group object in the category of ind-schemes. At first glance, it is natural to consider the algebraic representations of $G((t))$. We observe that $G((t))$ is generated by the "parahoric" algebraic groups P_i, corresponding to the affine simple roots. For these subgroups the notion of algebraic representation makes perfect sense. A representation of $G((t))$ is then said to be algebraic if its restriction to each of the P_i's is algebraic.

However, this naive approach leads us to the following discouraging fact: an irreducible smooth representation of $G((t))$, which is algebraic, is necessarily one-dimensional. To see that, we observe that an algebraic representation of $G((t))$ gives rise to a representation of its Lie algebra $\mathfrak{g}((t)) = \mathfrak{g} \otimes \mathbb{C}((t))$, where $\mathfrak{g}$ is the Lie algebra of G. Representations of $\mathfrak{g}((t))$ obtained from algebraic representations of $G((t))$ are called **integrable**. A smooth representation of $G((t))$ gives rise to a smooth representation of $\mathfrak{g}((t))$, i.e., for any vector v there exists $N \in \mathbb{Z}_+$ such that

$$\mathfrak{g} \otimes t^N \mathbb{C}[[t]] \cdot v = 0. \tag{1.3.2}$$

We can decompose $\mathfrak{g} = \mathfrak{g}_{ss} \oplus \mathfrak{r}$, where $\mathfrak{r}$ is the center and $\mathfrak{g}_{ss}$ is the semi-simple part. An irreducible representation of $\mathfrak{g}((t))$ is therefore the tensor product of an irreducible representation of $\mathfrak{g}_{ss}$ and an irreducible representation of $\mathfrak{r}((t))$, which is necessarily one-dimensional since $\mathfrak{r}((t))$ is abelian. Now we have the following

Lemma 1.3.1 *A smooth integrable representation of the Lie algebra $\mathfrak{g}((t))$, where $\mathfrak{g}$ is semi-simple, is trivial.*

Proof. We follow the argument of [BD1], 3.7.11(ii). Let V be such a representation. Then there exists $N \in \mathbb{Z}_+$ such that (1.3.2) holds. Let us pick a Cartan decomposition $\mathfrak{g} = \mathfrak{n}_- \oplus \mathfrak{h} \oplus \mathfrak{n}_+$ (see Appendix A.3 below) of the constant subalgebra $\mathfrak{g} \subset \mathfrak{g}((t))$. Let H be the Cartan subgroup of G corresponding to $\mathfrak{h} \subset \mathfrak{g}$. By our assumption, V is an algebraic representation of H, and hence it decomposes into a direct sum of weight spaces $V = \bigoplus V_\chi$ over the integral weights χ of H. Thus, any vector $v \in V$ may be decomposed as the sum $v = \sum_{\chi \in \mathfrak{h}^*} v_\chi$, where $v_\chi \in V_\chi$.

Now observe that there exists an element $h \in H((t))$ such that

$$\mathfrak{n}_+ \subset h(\mathfrak{g} \otimes t^N \mathbb{C}[[t]])h^{-1}.$$

Then we find that the vector $h \cdot v$ is invariant under $\mathfrak{n}_+$. Since the action of $\mathfrak{n}_+$ is compatible with the grading by the integral weights of H, we obtain that each vector $h \cdot v_\chi$ is also invariant under $\mathfrak{n}_+$. Consider the $\mathfrak{g}$-submodule of V generated by the vector $h \cdot v_\chi$. By our assumption, the action of $\mathfrak{g}$ on V integrates to an algebraic representation of G on V. Hence V is a direct sum of finite-dimensional irreducible representations of G with dominant integral weights. Therefore $h \cdot v_\chi = 0$, and hence $v_\chi = 0$, unless χ is a dominant integral weight. Next, observe that $h^{-1}(\mathfrak{g} \otimes t^N \mathbb{C}[[t]])h$ contains $\mathfrak{n}_-$. Applying the same argument to $h^{-1} \cdot v$, we find that $v_\chi = 0$ unless χ is an anti-dominant integral weight.

Therefore we obtain that $v = v_0$, and hence $h \cdot v$ is invariant under $\mathfrak{h} \oplus \mathfrak{n}_+$. But then the $\mathfrak{g}$-submodule generated by $h \cdot v$ is trivial. Hence $h \cdot v$ is $\mathfrak{g}$-invariant, and so v is $h^{-1}\mathfrak{g}h$-invariant. But the Lie algebras $\mathfrak{g} \otimes t^N \mathbb{C}[[t]]$ and $h^{-1}\mathfrak{g}h$, where h runs over those elements of $H((t))$ for which $h(\mathfrak{g} \otimes t^N \mathbb{C}[[t]])h^{-1}$ contains $\mathfrak{n}_+$, generate the entire Lie algebra $\mathfrak{g}((t))$. Therefore v is $\mathfrak{g}((t))$-invariant. We conclude that V is a trivial representation of $\mathfrak{g}((t))$. $\square$

Thus, we find that the class of algebraic representations of loop groups turns out to be too restrictive. We could relax this condition and consider differentiable representations, i.e., the representations of $G((t))$ considered as a Lie group. But it is easy to see that the result would be the same. Replacing $G((t))$ by its central extension $\widehat{G}$ would not help us much either: irreducible integrable representations of $\widehat{G}$ are parameterized by dominant integral weights, and there are no extensions between them [K2]. These representations are again too sparse to be parameterized by the geometric data considered above. Therefore we should look for other types of representations.

Going back to the original setup of the local Langlands correspondence, we recall that there we considered representations of $G(\mathbb{F}_q((t)))$ on $\mathbb{C}$-vector spaces, so we could not possibly use the algebraic structure of $G(\mathbb{F}_q((t)))$ as an ind-group over $\mathbb{F}_q$. Therefore we cannot expect the class of algebraic (or differentiable) representations of the complex loop group $G((t))$ to be meaningful from the point of view of the Langlands correspondence. We should view the loop group $G((t))$ as an abstract topological group, with the topology defined by means of the congruence subgroups; in other words, consider its smooth representations as an *abstract* group.

To give an example of such a representation, consider the vector space of finite linear combinations $\sum \delta_x$, where x runs over the set of points of the quotient $G((t))/K_N$. The group $G((t))$ naturally acts on this space: $g\left(\sum \delta_x\right) = \sum \delta_{g \cdot x}$. It is clear that for any $g \in G((t))$ the subgroup gK_Ng^{-1} contains K_M for large enough M, and so this is indeed a smooth representation of $G((t))$. But it is certainly not an algebraic representation (nor is it differentiable). In fact, it has absolutely nothing to do with the algebraic structure of $G((t))$, which is a serious drawback. After all, the whole point of trying to generalize the Langlands correspondence from the setting of finite fields to that of the complex field was to be able to use the powerful tools of complex algebraic geometry. If our representations are completely unrelated to geometry, there is not much that we can learn from them.

So we need to search for some geometric objects that encapsulate representations of our groups and make sense both over a finite field and over the complex field.

1.3.2 From functions to sheaves

We start by revisiting smooth representations of the group $G(F)$, where $F = \mathbb{F}_q((t))$. We realize such representations more concretely by considering their matrix coefficients. Let (V, π) be an irreducible smooth representation of $G(F)$. We define the **contragredient** representation $V^\vee$ as the linear span of all smooth vectors in the dual representation V^*. This span is stable under the action of $G(F)$ and so it admits a smooth representation $(V^\vee, \pi^\vee)$ of $G(F)$. Now let ϕ be a K_N-invariant vector in $V^\vee$. Then we define a linear map

$$V \to C(G(F)/K_N), \qquad v \mapsto f_v,$$

where

$$f_v(g) = \langle \pi^\vee(g)\phi, v\rangle.$$

Here $C(G(F)/K_N)$ denotes the vector space of $\mathbb{C}$-valued locally constant functions on $G(F)/K_N$. The group $G(F)$ naturally acts on this space by the formula $(g \cdot f)(h) = f(g^{-1}h)$, and the above map is a morphism of representations, which is non-zero, and hence injective, if (V, π) is irreducible.

Thus, we realize our representation in the space of functions on the quotient $G(F)/K_N$. More generally, we may realize representations in spaces of functions on the quotient $G((t))/K$ with values in a finite-dimensional vector space, by considering a finite-dimensional subrepresentation of K inside V rather than the trivial one.

An important observation here is that $G(F)/K$, where $F = \mathbb{F}_q((t))$ and K is a compact subgroup of $G(F)$, is not only a set, but it is the set of points of an algebraic variety (more precisely, an ind-scheme) defined over the field $\mathbb{F}_q$. For example, for $K_0 = G(\mathbb{F}_q[[t]])$, which is the maximal compact subgroup, the quotient $G(F)/K_0$ is the set of $\mathbb{F}_q$-points of the ind-scheme called the **affine Grassmannian**.

Next, we recall an important idea going back to Grothendieck that functions on the set of $\mathbb{F}_q$-points on an algebraic variety X defined over $\mathbb{F}_q$ can often be viewed as the "shadows" of the so-called ℓ-adic sheaves on X.

Let us discuss them briefly. Let ℓ be a prime that does not divide q. The definition of the category of ℓ-adic sheaves on X involves several steps (see, e.g., [Mi, FK]). First we consider locally constant $\mathbb{Z}/\ell^m\mathbb{Z}$-sheaves on X in the étale topology (in which the role of open subsets is played by étale morphisms $U \to X$). A $\mathbb{Z}_\ell$-sheaf on X is by definition a system $(\mathcal{F}_m)$ of locally constant $\mathbb{Z}/\ell^m\mathbb{Z}$-sheaves satisfying natural compatibilities. Then we define the category of $\mathbb{Q}_\ell$-sheaves by killing the torsion sheaves in the category of $\mathbb{Z}_\ell$-sheaves. In a similar fashion we define the category of E-sheaves on X, where E is a finite extension of $\mathbb{Q}_\ell$. Finally, we take the direct limit of the categories of E-sheaves on X, and the objects of this category are called the locally constant ℓ-adic sheaves on X. Such a sheaf of rank n is the same as an n-dimensional representation of the Galois group of the field of functions on X that is everywhere unramified. Thus, locally constant ℓ-adic sheaves are the analogues of the usual local systems (with respect to the analytic topology) in the case of complex algebraic varieties.

We may generalize the above definition of an ℓ-adic local system on X by allowing the $\mathbb{Z}/\ell^n\mathbb{Z}$-sheaves $\mathcal{F}_n$ to be constructible, i.e., for which there exists a stratification of X by locally closed subvarieties X_i such that the sheaves $\mathcal{F}|_{X_i}$ are locally constant. As a result, we obtain the notion of a constructible ℓ-adic sheaf on X, or an ℓ-adic sheaf, for brevity.

The key step in the geometric reformulation of this notion is the Grothendieck **fonctions-faisceaux** dictionary (see, e.g., [La]). Let $\mathcal{F}$ be an ℓ-adic sheaf and x be an $\mathbb{F}_{q_1}$-point of X, where $q_1 = q^m$. Then we have the Frobenius conjugacy class Fr_x acting on the stalk $\mathcal{F}_x$ of $\mathcal{F}$ at x. Hence we can define a function $\mathbf{f}_{q_1}(\mathcal{F})$ on the set of $\mathbb{F}_{q_1}$-points of X, whose value at x is $\mathrm{Tr}(\mathrm{Fr}_x, \mathcal{F}_x)$. This function takes values in the algebraic closure $\overline{\mathbb{Q}}_\ell$ of $\mathbb{Q}_\ell$. But there is not much of a difference between $\overline{\mathbb{Q}}_\ell$-valued functions and $\mathbb{C}$-valued functions: since they have the same cardinality, $\overline{\mathbb{Q}}_\ell$ and $\mathbb{C}$ may be identified as abstract fields. Besides, in most interesting cases, the values actually belong to $\overline{\mathbb{Q}}$, which is inside both $\overline{\mathbb{Q}}_\ell$ and $\mathbb{C}$.

More generally, if $\mathcal{F}$ is a complex of ℓ-adic sheaves, we define a function $\mathbf{f}_{q_1}(\mathcal{F})$ on $X(\mathbb{F}_{q_1})$ by taking the alternating sums of the traces of Fr_x on the stalk cohomologies of $\mathcal{F}$ at x. The map $\mathcal{F} \to \mathbf{f}_{q_1}(\mathcal{F})$ intertwines the natural operations on sheaves with natural operations on functions (see [La], Section 1.2). For example, pull-back of a sheaf corresponds to the pull-back of a function, and push-forward of a sheaf *with compact support* corresponds to the fiberwise integration of a function (this follows from the Grothendieck–Lefschetz trace formula).

Let $K_0(\mathcal{S}h_X)$ be the complexified Grothendieck group of the category of ℓ-adic sheaves on X. Then the above construction gives us a map

$$K_0(\mathcal{S}h_X) \to \prod_{m \geq 1} X(\mathbb{F}_{q^m}),$$

and it is known that this map is injective (see [La]).

Therefore we may hope that the functions on the quotients $G(F)/K_N$, which realize our representations, arise, via this construction, from ℓ-adic sheaves, or, more generally, from complexes of ℓ-adic sheaves, on X.

Now, the notion of a constructible sheaf (unlike the notion of a function) has a transparent and meaningful analogue for a complex algebraic variety X; namely, those sheaves of $\mathbb{C}$-vector spaces whose restrictions to the strata of a stratification of the variety X are locally constant. The affine Grassmannian and more general ind-schemes underlying the quotients $G(F)/K_N$ may be defined both over $\mathbb{F}_q$ and $\mathbb{C}$. Thus, it is natural to consider the categories of such sheaves (or, more precisely, their derived categories) on these ind-schemes over $\mathbb{C}$ as the replacements for the vector spaces of functions on their points realizing smooth representations of the group $G(F)$.

We therefore naturally come to the idea, advanced in [FG2], that the representations of the loop group $G((t))$ that we need to consider are not realized on vector spaces, but on **categories**, such as the derived category of coherent sheaves on the affine Grassmannian. Of course, such a category has a Grothendieck group, and the group $G((t))$ will act on the Grothendieck group as well, giving us a representation of $G((t))$ on a vector space. But we obtain much more structure by looking at the categorical representation. The objects of the category, as well as the action, will have a geometric meaning, and thus we will be using the geometry as much as possible.

Let us summarize: to each local Langlands parameter $\chi \in \mathrm{Loc}_{{}^L G}(D^\times)$ we wish to attach a category $\mathcal{C}_\chi$ equipped with an action of the loop group $G((t))$. But what kind of categories should these $\mathcal{C}_\chi$ be and what properties do we expect them to satisfy?

To get closer to answering these questions, we wish to discuss two more steps that we can make in the above discussion to get to the types of categories with an action of the loop group that we will consider in this book.

1.3.3 A toy model

At this point it is instructive to detour slightly and consider a toy model of our construction. Let G be a split reductive group over $\mathbb{Z}$, and B its Borel subgroup. A natural representation of $G(\mathbb{F}_q)$ is realized in the space of complex- (or $\overline{\mathbb{Q}}_\ell$-) valued functions on the quotient $G(\mathbb{F}_q)/B(\mathbb{F}_q)$. It is natural to ask what is the "correct" analogue of this representation if we replace the field $\mathbb{F}_q$ by the complex field and the group $G(\mathbb{F}_q)$ by $G(\mathbb{C})$. This may be viewed as a simplified version of our quandary, since instead of considering $G(\mathbb{F}_q((t)))$ we now look at $G(\mathbb{F}_q)$.

The quotient $G(\mathbb{F}_q)/B(\mathbb{F}_q)$ is the set of $\mathbb{F}_q$-points of the algebraic variety defined over $\mathbb{Z}$ called the flag variety of G and denoted by Fl. Our discussion in the previous section suggests that we first need to replace the notion of a function on $\mathrm{Fl}(\mathbb{F}_q)$ by the notion of an ℓ-adic sheaf on the variety $\mathrm{Fl}_{\mathbb{F}_q} = \mathrm{Fl} \underset{\mathbb{Z}}{\otimes} \mathbb{F}_q$.

Next, we replace the notion of an ℓ-adic sheaf on Fl, considered as an algebraic variety over $\mathbb{F}_q$, by the notion of a constructible sheaf on $\mathrm{Fl}_{\mathbb{C}} = \mathrm{Fl} \underset{\mathbb{Z}}{\otimes} \mathbb{C}$, which is an algebraic variety over $\mathbb{C}$. The complex algebraic group $G_{\mathbb{C}}$ naturally acts on $\mathrm{Fl}_{\mathbb{C}}$ and hence on this category. Now we make two more reformulations of this category.

First of all, for a smooth complex algebraic variety X we have a **Riemann–Hilbert correspondence**, which is an equivalence between the derived category of constructible sheaves on X and the derived category of $\mathcal{D}$-modules on X that are holonomic and have regular singularities.

Here we consider the sheaf of algebraic differential operators on X and sheaves of modules over it, which we simply refer to as $\mathcal{D}$-modules. The simplest example of a $\mathcal{D}$-module is the sheaf of sections of a vector bundle on X equipped with a flat connection. The flat connection enables us to multiply any section by a function and we can use the flat connection to act on sections by vector fields. The two actions generate an action of the sheaf of differential operators on the sections of our bundle. The sheaf of horizontal sections of this bundle is then a locally constant sheaf on X. We have seen above that there is a bijection between the set of isomorphism classes of rank n bundles on X with connection having regular singularities and the set of isomorphism classes of locally constant sheaves on X of rank n, or, equivalently, n-dimensional representations of $\pi_1(X)$. This bijection may be elevated to an equivalence of the corresponding categories, and the general Riemann–Hilbert correspondence is a generalization of this equivalence of categories that encompasses more general $\mathcal{D}$-modules.

The Riemann–Hilbert correspondence allows us to associate to any holonomic $\mathcal{D}$-module on X a complex of constructible sheaves on X, and this gives us a functor between the corresponding derived categories, which turns out to be an equivalence if we restrict ourselves to the holonomic $\mathcal{D}$-modules with regular singularities (see [Bor2, GM] for more details).

Thus, over $\mathbb{C}$ we may pass from constructible sheaves to $\mathcal{D}$-modules. In our case, we consider the category of (regular holonomic) $\mathcal{D}$-modules on the flag variety $\mathrm{Fl}_{\mathbb{C}}$. This category carries a natural action of $G_{\mathbb{C}}$.

Finally, let us observe that the Lie algebra $\mathfrak{g}$ of $G_{\mathbb{C}}$ acts on the flag variety infinitesimally by vector fields. Therefore, given a $\mathcal{D}$-module $\mathcal{F}$ on $\mathrm{Fl}_{\mathbb{C}}$, the space of its global sections $\Gamma(\mathrm{Fl}_{\mathbb{C}}, \mathcal{F})$ has the structure of $\mathfrak{g}$-module. We obtain a functor Γ from the category of $\mathcal{D}$-modules on $\mathrm{Fl}_{\mathbb{C}}$ to the category of $\mathfrak{g}$-modules. A. Beilinson and J. Bernstein have proved that this functor is an equivalence between the category of all $\mathcal{D}$-modules on $\mathrm{Fl}_{\mathbb{C}}$ (not necessarily regular holonomic) and the category $\mathcal{C}_0$ of $\mathfrak{g}$-modules on which the center of the universal enveloping algebra $U(\mathfrak{g})$ acts through the augmentation character.

Thus, we can now answer our question as to what is a meaningful geometric analogue of the representation of the finite group $G(\mathbb{F}_q)$ on the space of functions on the quotient $G(\mathbb{F}_q)/B(\mathbb{F}_q)$. The answer is the following: it is a **category** equipped with an action of the algebraic group $G_{\mathbb{C}}$. This category has two incarnations: one is the category of $\mathcal{D}$-modules on the flag variety $\mathrm{Fl}_{\mathbb{C}}$, and the other is the category $\mathcal{C}_0$ of modules over the Lie algebra $\mathfrak{g}$ with trivial central character. Both categories are equipped with natural actions of the group $G_{\mathbb{C}}$.

Let us pause for a moment and spell out what exactly we mean when we say that the group $G_{\mathbb{C}}$ acts on the category $\mathcal{C}_0$. For simplicity, we will describe the action of the corresponding group $G(\mathbb{C})$ of $\mathbb{C}$-points of $G_{\mathbb{C}}$.† This means the following: each element $g \in G$ gives rise to a functor F_g on $\mathcal{C}_0$ such that F_1 is the identity functor, and the functor $F_{g^{-1}}$ is quasi-inverse to F_g. Moreover, for any pair $g, h \in G$ we have a fixed isomorphism of functors $i_{g,h} : F_{gh} \to F_g \circ F_h$ so that for any triple $g, h, k \in G$ we have the equality $i_{h,k} i_{g,hk} = i_{g,h} i_{gh,k}$ of isomorphisms $F_{ghk} \to F_g \circ F_h \circ F_k$. (We remark that the last condition could be relaxed: we could ask only that $i_{h,k} i_{g,hk} = \gamma_{g,h,k} i_{g,h} i_{gh,k}$, where $\gamma_{g,h,k}$ is a non-zero complex number for each triple $g, h, k \in G$; these numbers then must satisfy a three-cocycle condition. However, we will only consider the situation where $\gamma_{g,h,k} \equiv 1$.)

The functors F_g are defined as follows. Given a representation (V, π) of $\mathfrak{g}$ and an element $g \in G(\mathbb{C})$, we define a new representation $F_g((V, \pi)) = (V, \pi_g)$, where by definition $\pi_g(x) = \pi(\mathrm{Ad}_g(x))$. Suppose that (V, π) is irreducible. Then it is easy to see that $(V, \pi_g) \simeq (V, \pi)$ if and only if (V, π) is integrable, i.e., is obtained from an algebraic representation of G. This is equivalent to this representation being finite-dimensional. But a general representation (V, π) is infinite-dimensional, and so it will not be isomorphic to (V, π_g), at least for some $g \in G$.

Now we consider morphisms in $\mathcal{C}_0$, which are just $\mathfrak{g}$-homomorphisms. Given a $\mathfrak{g}$-homomorphism between representations (V, π) and (V', π'), i.e., a linear map $T : V \to V'$ such that $T\pi(x) = \pi'(x)T$ for all $x \in \mathfrak{g}$, we set $F_g(T) = T$. The isomorphisms $i_{g,h}$ are all equal to the identity in this case.

The simplest examples of objects of the category $\mathcal{C}_0$ are the Verma modules induced from one-dimensional representations of a Borel subalgebra $\mathfrak{b} \subset \mathfrak{g}$. The corresponding $\mathcal{D}$-module is the $\mathcal{D}$-module of "delta-functions" supported at the point of $\mathrm{Fl}_{\mathbb{C}}$ stabilized by $\mathfrak{b}$. In the Grothendieck group of the category $\mathcal{C}_0$ the classes of these objects span a subrepresentation of $G(\mathbb{C})$ (considered now as a discrete group!), which looks exactly like the representation we defined at the beginning of Section 1.3.2. What we have achieved is that we have

† More generally, for any $\mathbb{C}$-algebra R, we have an action of $G(R)$ on the corresponding base-changed category over R. Thus, we are naturally led to the notion of an algebraic group (or, more generally, a group scheme) acting on an abelian category, which is spelled out in [FG2], Section 20.

replaced this representation by something that makes sense from the point of view of the representation theory of the complex *algebraic* group G (rather than the corresponding discrete group); namely, the category $\mathcal{C}_0$.

1.3.4 Back to loop groups

In our quest for a complex analogue of the local Langlands correspondence we need to decide what will replace the notion of a smooth representation of the group $G(F)$, where $F = \mathbb{F}_q((t))$. As the previous discussion demonstrates, we should consider representations of the complex loop group $G((t))$ on various categories of $\mathcal{D}$-modules on the ind-schemes $G((t))/K$, where K is a "compact" subgroup of $G((t))$, such as $G[[t]]$ or the Iwahori subgroup (the preimage of a Borel subgroup $B \subset G$ under the homomorphism $G[[t]] \to G$), or the categories of representations of the Lie algebra $\mathfrak{g}((t))$. Both scenarios are viable, and they lead to interesting results and conjectures, which we will discuss in detail in Chapter 10, following [FG2]. In this book we will concentrate on the second scenario and consider categories of (projective) modules ove the loop algebra $\mathfrak{g}((t))$.

The group $G((t))$ acts on the category of representations of $\mathfrak{g}((t))$ in the way that we described in the previous section. In order to make the corresponding representation of $G((t))$ smooth, we need to restrict ourselves to those representations on which the action of the Lie subalgebra $\mathfrak{g} \otimes t^N \mathbb{C}[[t]]$ is integrable for some $N > 0$. Indeed, on such representations the action of $\mathfrak{g} \otimes t^N \mathbb{C}[[t]]$ may be exponentiated to an action of its Lie group, which is the congruence subgroup K_N. If (V, π) is such a representation, then for each $g \in K_N$ the operator of the action of g on V will provide an isomorphism between (V, π) and $F_g((V, \pi))$. Therefore we may say that (V, π) is "stable" under F_g.

As the following lemma shows, this condition is essentially equivalent to the condition of a $\mathfrak{g}((t))$-module being smooth, i.e., such that any vector is annihilated by the Lie algebra $\mathfrak{g} \otimes t^M \mathbb{C}[[t]]$ for sufficiently large M.

Lemma 1.3.1 *Suppose that (V, π) is a smooth finitely generated module over the Lie algebra $\mathfrak{g}((t))$. Then there exists $N > 0$ such that the action of the Lie subalgebra $\mathfrak{g} \otimes t^N \mathbb{C}[[t]]$ on V is integrable.*

Proof. Let $v_1, \ldots, v_k$ be a generating set of vectors in V. Since (V, π) is smooth, each vector v_i is annihilated by a Lie subalgebra $\mathfrak{g} \otimes t^{N_i} \mathbb{C}[[t]]$ for some N_i. Let M_i be the induced representation

$$M_i = \operatorname{Ind}_{\mathfrak{g} \otimes t^{N_i} \mathbb{C}[[t]]}^{\mathfrak{g}((t))} \mathbb{C} = U(\mathfrak{g}((t))) \underset{\mathfrak{g} \otimes t^{N_i} \mathbb{C}[[t]]}{\otimes} \mathbb{C}.$$

Then we have a surjective homomorphism $\bigoplus_{i=1}^k M_i \to V$ sending the generating vector of M_i to v_i for each $i = 1, \ldots, k$. Let N be the largest number among $N_1, \ldots, N_k$. Then the action of $\mathfrak{g} \otimes t^N \mathbb{C}[[t]]$ on each M_i is integrable, and hence the same is true for V. □

1.3.5 From the loop algebra to its central extension

Thus, we can take as the categorical analogue of a smooth representation of a reductive GOP over a local non-archimedian field the category of smooth finitely generated modules over the Lie algebra $\mathfrak{g}((t))$ equipped with a natural action of the loop group $G((t))$. (In what follows we will drop the condition of being finitely generated.)

Let us observe however that we could choose instead the category of smooth representations of a central extension of $\mathfrak{g}((t))$. The group $G((t))$ still acts on such a central extension via the adjoint action. Since the action of the group $G((t))$ on the category comes through its adjoint action, no harm will be done if we extend $\mathfrak{g}((t))$ by a central subalgebra.

The notion of a central extension of a Lie algebra is described in detail in Appendix A.4. In particular, it is explained there that the equivalence classes of central extensions of $\mathfrak{g}((t))$ are described by the second cohomology $H^2(\mathfrak{g}((t)), \mathbb{C})$.†

We will use the decomposition $\mathfrak{g} = \mathfrak{g}_{ss} \oplus \mathfrak{r}$, where $\mathfrak{r}$ is the center and $\mathfrak{g}_{ss}$ is the semi-simple part. Then it is possible to show that

$$H^2(\mathfrak{g}((t)), \mathbb{C}) \simeq H^2(\mathfrak{g}_{ss}((t)), \mathbb{C}) \oplus H^2(\mathfrak{r}((t)), \mathbb{C}).$$

In other words, the central extension of $\mathfrak{g}((t))$ is determined by its restriction to $\mathfrak{g}_{ss}((t))$ and to $\mathfrak{r}((t))$. The central extensions that we will consider will be trivial on the abelian part $\mathfrak{r}((t))$, and so our central extension will be a direct sum $\widehat{\mathfrak{g}}_{ss} \oplus \mathfrak{r}((t))$. An irreducible representation of this Lie algebra is isomorphic to the tensor product of an irreducible representation of $\widehat{\mathfrak{g}}_{ss}$ and that of $\mathfrak{r}((t))$. But the Lie algebra $\mathfrak{r}((t))$ is abelian, and so its irreducible representations are one-dimensional. It is not hard to deal with these representations separately, and so from now on we will focus on representations of central extensions of $\mathfrak{g}((t))$, where $\mathfrak{g}$ is a semi-simple Lie algebra.

If $\mathfrak{g}$ is semi-simple, it can be decomposed into a direct sum of simple Lie algebras $\mathfrak{g}_i, i = 1, \ldots, m$, and we have

$$H^2(\mathfrak{g}((t)), \mathbb{C}) \simeq \bigoplus_{i=1}^{m} H^2(\mathfrak{g}_i((t)), \mathbb{C}).$$

Again, without loss of generality we can treat each simple factor $\mathfrak{g}_i$ separately.

† Here we need to take into account our topology on the loop algebra $\mathfrak{g}((t))$ in which the base of open neighborhoods of zero is given by the Lie subalgebras $\mathfrak{g} \otimes t^N \mathbb{C}[[t]]$, $N \in \mathbb{Z}_+$. Therefore we should restrict ourselves to the corresponding continuous cohomology. We will use the same notation $H^2(\mathfrak{g}((t)), \mathbb{C})$ for it.

Hence from now on we will assume that $\mathfrak{g}$ is a simple Lie algebra. In this case we have the following description of $H^2(\mathfrak{g}((t)), \mathbb{C})$.

Recall that a inner product κ on a Lie algebra $\mathfrak{g}$ is called **invariant** if

$$\kappa([x, y], z) + \kappa(y, [x, z]) = 0, \qquad \forall x, y, z \in \mathfrak{g}.$$

This formula is the infinitesimal version of the formula $\kappa(\mathrm{Ad}\, g \cdot y, \mathrm{Ad}\, g \cdot z) = \kappa(y, z)$ expressing the invariance of the form with respect to the Lie group action (with g being an element of the corresponding Lie group). The vector space of non-degenerate invariant inner products on a finite-dimensional simple Lie algebra $\mathfrak{g}$ is one-dimensional. One can produce such a form starting with any non-trivial finite-dimensional representation $\rho_V : \mathfrak{g} \to \mathrm{End}\, V$ of $\mathfrak{g}$ by the formula

$$\kappa_V(x, y) = \mathrm{Tr}_V(\rho_V(x)\rho_V(y)).$$

The standard choice is the adjoint representation, which gives rise to the **Killing form** $\kappa_{\mathfrak{g}}$. The following result is well-known.

Lemma 1.3.2 *For a simple Lie algebra $\mathfrak{g}$ the space $H^2(\mathfrak{g}((t)), \mathbb{C})$ is one-dimensional and is identified with the space of invariant bilinear forms on $\mathfrak{g}$. Given such a form κ, the corresponding central extension can be constructed from the cocycle*

$$c(A \otimes f(t), B \otimes g(t)) = -\kappa(A, B)\, \mathrm{Res}_{t=0}\, f dg. \qquad (1.3.3)$$

Here for a formal Laurent power series $a(t) = \sum_{n \in \mathbb{Z}} a_n t^n$ we set

$$\mathrm{Res}_{t=0}\, a(t)dt = a_{-1}.$$

1.3.6 Affine Kac–Moody algebras and their representations

The central extension

$$0 \to \mathbb{C}\mathbf{1} \to \widehat{\mathfrak{g}}_\kappa \to \mathfrak{g}((t)) \to 0$$

corresponding to a non-zero cocycle κ is called the **affine Kac–Moody algebra** . We denote it by $\widehat{\mathfrak{g}}_\kappa$. It is customary to refer to κ as the **level.** As a vector space, it is equal to the direct sum $\mathfrak{g}((t)) \oplus \mathbb{C}\mathbf{1}$, and the commutation relations read

$$[A \otimes f(t), B \otimes g(t)] = [A, B] \otimes f(t)g(t) - (\kappa(A, B)\, \mathrm{Res}\, f dg)\mathbf{1}, \qquad (1.3.4)$$

where $\mathbf{1}$ is a central element, which commutes with everything else. Note that the Lie algebra $\widehat{\mathfrak{g}}_\kappa$ and $\widehat{\mathfrak{g}}_{\kappa'}$ are isomorphic for non-zero inner products κ, κ'. Indeed, in this case we have $\kappa = \lambda\kappa'$ for some $\lambda \in \mathbb{C}^\times$, and the map $\widehat{\mathfrak{g}}_\kappa \to \widehat{\mathfrak{g}}_{\kappa'}$, which is equal to the identity on $\mathfrak{g}((t))$ and sends $\mathbf{1}$ to $\lambda\mathbf{1}$, is an isomorphism. By Lemma 1.3.2, the Lie algebra $\widehat{\mathfrak{g}}_\kappa$ with non-zero κ is in fact a universal central extension of $\mathfrak{g}((t))$. (The Lie algebra $\widehat{\mathfrak{g}}_0$ is by definition the split extension $\mathfrak{g}((t)) \oplus \mathbb{C}\mathbf{1}$.)

Note that the restriction of the cocycle (1.3.3) to the Lie subalgebra $\mathfrak{g} \otimes t^N \mathbb{C}[[t]], N \in \mathbb{Z}_+$ is equal to 0, and so it remains a Lie subalgebra of $\widehat{\mathfrak{g}}_\kappa$. A **smooth** representation of $\widehat{\mathfrak{g}}_\kappa$ is a representation such that every vector is annihilated by this Lie subalgebra for sufficiently large N. Note that the statement of Lemma 1.3.1 remains valid if we replace $\mathfrak{g}((t))$ by $\widehat{\mathfrak{g}}_\kappa$.

Thus, we define the category $\widehat{\mathfrak{g}}_\kappa$-mod whose objects are smooth $\widehat{\mathfrak{g}}_\kappa$-modules on which the central element $\mathbf{1}$ acts as the identity. The morphisms are homomorphisms of representations of $\widehat{\mathfrak{g}}_\kappa$. Throughout this book, unless specified otherwise, by a "$\widehat{\mathfrak{g}}_\kappa$-module" we will always mean a module on which the central element $\mathbf{1}$ acts as the identity.†

The group $G((t))$ acts on the Lie algebra $\widehat{\mathfrak{g}}_\kappa$ for any κ. Indeed, the adjoint action of the central extension of $G((t))$ factors through the action of $G((t))$. It is easy to compute this action and to find that

$$g \cdot (A(t) + c\mathbf{1}) = \left(gA(t)g^{-1} + \mathrm{Res}_{t=0}\, \kappa((\partial_t g)g^{-1}, A(t))\mathbf{1}\right).$$

It is interesting to observe that the dual space to $\widehat{\mathfrak{g}}_\kappa$ (more precisely, a hyperplane in the dual space) may be identified with the space of connections on the trivial G-bundle on $D^\times$ so that the coadjoint action gets identified with the gauge action of $G((t))$ on the space of such connections (for more on this, see [FB], Section 16.4).

We use the action of $G((t))$ on $\widehat{\mathfrak{g}}_\kappa$ to construct an action of $G((t))$ on the category $\widehat{\mathfrak{g}}_\kappa$-mod, in the same way as in Section 1.3.3. Namely, suppose we are given an object (M, ρ) of $\widehat{\mathfrak{g}}_\kappa$-mod, where M is a vector space and $\rho : \widehat{\mathfrak{g}}_\kappa \to \mathrm{End}\, M$ is a Lie algebra homomorphism making M into a smooth $\widehat{\mathfrak{g}}_\kappa$-module. Then, for each $g \in G((t))$, we define a new object (M, ρ_g) of $\widehat{\mathfrak{g}}_\kappa$-mod, where by definition $\rho_g(x) = \pi(\mathrm{Ad}_g(x))$. If we have a morphism $(M, \rho) \to (M', \rho')$ in $\widehat{\mathfrak{g}}_\kappa$-mod, then we obtain an obvious morphism $(M, \rho_g) \to (M', \rho'_g)$. Thus, we have defined a functor $F_g : \widehat{\mathfrak{g}}_\kappa\text{-mod} \to \widehat{\mathfrak{g}}_\kappa\text{-mod}$. These functors satisfy the conditions of Section 1.3.3. Thus, we obtain an action of $G((t))$ on the category $\widehat{\mathfrak{g}}_\kappa$-mod.

Recall the space $\mathrm{Loc}_{{}^L G}(D^\times)$ of the Langlands parameters that we defined in Section 1.2.6. Elements of $\mathrm{Loc}_{{}^L G}(D^\times)$ have a concrete description as gauge equivalence classes of first order operators $\partial_t + A(t), A(t) \in {}^L\mathfrak{g}((t))$, modulo the action of ${}^L G((t))$ (see formula (1.2.7)).

We can now formulate the local Langlands correspondence over $\mathbb{C}$ as the following problem:

† Note that we could have $\mathbf{1}$ act instead as λ times the identity for $\lambda \in \mathbb{C}^\times$; but the corresponding category would just be equivalent to the category $\widehat{\mathfrak{g}}_{\lambda\kappa}$-mod.

> *To each local Langlands parameter* $\chi \in \mathrm{Loc}_{^L G}(D^\times)$ *associate a subcategory* $\widehat{\mathfrak{g}}_\kappa\text{-mod}_\chi$ *of* $\widehat{\mathfrak{g}}_\kappa\text{-mod}$ *which is stable under the action of the loop group* $G((t))$.

We wish to think of the category $\widehat{\mathfrak{g}}_\kappa\text{-mod}$ as "fibering" over the space of local Langlands parameters $\mathrm{Loc}_{^L G}(D^\times)$, with the categories $\widehat{\mathfrak{g}}_\kappa\text{-mod}_\chi$ being the "fibers" and the group $G((t))$ acting along these fibers. From this point of view the categories $\widehat{\mathfrak{g}}_\kappa\text{-mod}_\chi$ should give us a "spectral decomposition" of the category $\widehat{\mathfrak{g}}_\kappa\text{-mod}$ over $\mathrm{Loc}_{^L G}(D^\times)$.

In Chapter 10 of this book we will present a concrete proposal made in [FG2] describing these categories in the special case when $\kappa = \kappa_c$ is the *critical level*, which is minus one half of $\kappa_{\mathfrak{g}}$, the Killing form defined above. This proposal is based on the fact that at the critical level the center of the category $\widehat{\mathfrak{g}}_{\kappa_c}\text{-mod}$, which is the same as the center of the completed universal enveloping algebra of $\widehat{\mathfrak{g}}_{\kappa_c}$, is very large. It turns out that its spectrum is closely related to the space $\mathrm{Loc}_{^L G}(D^\times)$ of Langlands parameters, and this will enable us to define $\widehat{\mathfrak{g}}_{\kappa_c}\text{-mod}_\chi$ as, roughly speaking, the category of smooth $\widehat{\mathfrak{g}}_{\kappa_c}$-modules with a fixed central character.

In order to explain more precisely how this works, we need to develop the representation theory of affine Kac–Moody algebras and in particular describe the structure of the center of the completed enveloping algebra of $\widehat{\mathfrak{g}}_{\kappa_c}$. In the next chapter we will start on a long journey towards this goal. This will occupy the main part of this book. Then in Chapter 10 we will show, following the papers [FG1]–[FG6], how to use these results in order to construct the local geometric Langlands correspondence for loop groups.

2
Vertex algebras

Let $\mathfrak{g}$ be a simple finite-dimensional Lie algebra and $\widehat{\mathfrak{g}}_\kappa$ the corresponding affine Kac–Moody algebra (the central extension of $\mathfrak{g}((t))$), introduced in Section 1.3.6. We have the category $\widehat{\mathfrak{g}}_\kappa$-mod, whose objects are smooth $\widehat{\mathfrak{g}}_\kappa$-modules on which the central element $\mathbf{1}$ acts as the identity. As explained at the end of the previous chapter, we wish to show that this category "fibers" over the space of Langlands parameters, which are gauge equivalence classes of ${}^LG$-connections on the punctured disc $D^\times$ (or perhaps, something similar). Moreover, the loop group $G((t))$ should act on this category "along the fibers."

Any abelian category may be thought of as "fibering" over the spectrum of its center. Hence the first idea that comes to mind is to describe the center of the category $\widehat{\mathfrak{g}}_\kappa$-mod in the hope that its spectrum is related to the Langlands parameters. As we will see, this is indeed the case for a particular value of κ.

In order to show that, however, we first need to develop a technique for dealing with the completed universal enveloping algebra of $\widehat{\mathfrak{g}}_\kappa$. This is the formalism of vertex algebras. In this chapter we will first motivate the necessity of vertex algebras and then introduce the basics of the theory of vertex algebras.

2.1 The center

2.1.1 The case of simple Lie algebras

Let us first recall what is the center of an abelian category. Let $\mathcal{C}$ be an abelian category over $\mathbb{C}$. The center $Z(\mathcal{C})$ is by definition the set of endomorphisms of the identity functor on $\mathcal{C}$. Let us recall that such an endomorphism is a system of endomorphisms $e_M \in \operatorname{Hom}_{\mathcal{C}}(M, M)$, for each object M of $\mathcal{C}$, which is compatible with the morphisms in $\mathcal{C}$: for any morphism $f : M \to N$ in $\mathcal{C}$ we have $f \circ e_M = e_N \circ f$. It is clear that $Z(\mathcal{C})$ has a natural structure of a commutative algebra over $\mathbb{C}$.

Let $S = \operatorname{Spec} Z(\mathcal{C})$. This is an affine algebraic variety such that $Z(\mathcal{C})$ is the algebra of functions on S. Each point $s \in S$ defines an algebra homomorphism (equivalently, a character) $\rho_s : Z(\mathcal{C}) \to \mathbb{C}$ (evaluation of a function at the point s). We define the full subcategory $\mathcal{C}_s$ of $\mathcal{C}$ whose objects are the objects of $\mathcal{C}$ on which $Z(\mathcal{C})$ acts according to the character ρ_s. It is instructive to think of the category $\mathcal{C}$ as "fibering" over S, with the fibers being the categories $\mathcal{C}_s$.

Now suppose that $\mathcal{C} = A$-mod is the category of left modules over an associative $\mathbb{C}$-algebra A. Then A itself, considered as a left A-module, is an object of $\mathcal{C}$, and so we obtain a homomorphism

$$Z(\mathcal{C}) \to Z(\mathrm{End}_A A) = Z(A^{\mathrm{opp}}) = Z(A),$$

where $Z(A)$ is the center of A. On the other hand, each element of $Z(A)$ defines an endomorphism of each object of A-mod, and so we obtain a homomorphism $Z(A) \to Z(\mathcal{C})$. It is easy to see that these maps define mutually inverse isomorphisms between $Z(\mathcal{C})$ and $Z(A)$.

If $\mathfrak{g}$ is a Lie algebra, then the category $\mathfrak{g}$-mod of $\mathfrak{g}$-modules coincides with the category $U(\mathfrak{g})$-mod of $U(\mathfrak{g})$-modules, where $U(\mathfrak{g})$ is the universal enveloping algebra of $\mathfrak{g}$ (see Appendix A.2). Therefore the center of the category $\mathfrak{g}$-mod is equal to the center of $U(\mathfrak{g})$, which by abuse of notation we denote by $Z(\mathfrak{g})$.

Let us recall the description of $Z(\mathfrak{g})$ in the case when $\mathfrak{g}$ is a finite-dimensional simple Lie algebra over $\mathbb{C}$ of rank ℓ (see Section A.3). It can be proved by a combination of results of Harish-Chandra and Chevalley (see [Di]).

Theorem 2.1.1 *The center $Z(\mathfrak{g})$ is a polynomial algebra $\mathbb{C}[P_i]_{i=1,\ldots,\ell}$ generated by elements $P_i, i = 1, \ldots, \ell$, of orders $d_i + 1$, where d_i are the exponents of $\mathfrak{g}$.*

An element P of $U(\mathfrak{g})$ is said to have order i if it belongs to the ith term $U(\mathfrak{g})_{\leq i}$ of the Poincaré–Birkhoff–Witt filtration on $U(\mathfrak{g})$ described in Appendix A.2, but does not belong to $U(\mathfrak{g})_{\leq (i-1)}$.

The exponents form a set of positive integers attached to each simple Lie algebra. For example, for $\mathfrak{g} = \mathfrak{sl}_n$ this set is $\{1, \ldots, n-1\}$. There are several equivalent definitions, and the above theorem may be taken as one of them (we will encounter another definition in Section 4.2.4 below).

The first exponent of any simple Lie algebra $\mathfrak{g}$ is always 1, so there is always a quadratic element in the center of $U(\mathfrak{g})$. This element is called the **Casimir element** and can be constructed as follows. Let $\{J^a\}$ be a basis for $\mathfrak{g}$ as a vector space. Fix any non-zero invariant inner product κ_0 on $\mathfrak{g}$ and let $\{J_a\}$ be the dual basis to $\{J^a\}$ with respect to κ_0. Then the Casimir element is given by the formula

$$P = \frac{1}{2} \sum_{a=1}^{\dim \mathfrak{g}} J^a J_a.$$

Note that it does not depend on the choice of the basis $\{J^a\}$. Moreover, changing κ_0 to κ_0' would simply multiply P by a scalar λ such that $\kappa = \lambda\kappa_0'$.

It is a good exercise to compute this element in the case of $\mathfrak{g} = \mathfrak{sl}_2$. We have the standard generators of $\mathfrak{sl}_2$

$$e = \begin{pmatrix} 0 & 1 \\ 0 & 0 \end{pmatrix}, \qquad h = \begin{pmatrix} 1 & 0 \\ 0 & -1 \end{pmatrix}, \qquad f = \begin{pmatrix} 0 & 0 \\ 1 & 0 \end{pmatrix}$$

and the inner product

$$\kappa_0(a,b) = \operatorname{Tr} ab.$$

Then the Casimir element is

$$P = \frac{1}{2}\left(ef + fe + \frac{1}{2}h^2\right).$$

2.1.2 The case of affine Lie algebras

We wish to obtain a result similar to Theorem 2.1.1 describing the universal enveloping algebra of an affine Kac–Moody algebra.

The first step is to define an appropriate enveloping algebra whose category of modules coincides with the category $\widehat{\mathfrak{g}}_\kappa$-mod. Let us recall from Section 1.3.6 that objects of $\widehat{\mathfrak{g}}_\kappa$-mod are $\widehat{\mathfrak{g}}_\kappa$-modules M on which the central element $\mathbf{1}$ acts as the identity and which are *smooth*, that is, for any vector $v \in M$ we have

$$(\mathfrak{g} \otimes t^N \mathbb{C}[[t]]) \cdot v = 0 \tag{2.1.1}$$

for sufficiently large N.

As a brief aside, let us remark that we have a polynomial version $\widehat{\mathfrak{g}}_\kappa^{\mathrm{pol}}$ of the affine Lie algebra, which is the central extension of the polynomial loop algebra $\mathfrak{g}[t,t^{-1}]$. Let $\widehat{\mathfrak{g}}_\kappa^{\mathrm{pol}}$-mod be the category of smooth $\widehat{\mathfrak{g}}_\kappa^{\mathrm{pol}}$-modules, defined in the same way as above, with the Lie algebra $\mathfrak{g} \otimes t^N \mathbb{C}[t]$ replacing $\mathfrak{g} \otimes t^N \mathbb{C}[[t]]$. The smoothness condition allows us to extend the action of $\widehat{\mathfrak{g}}_\kappa^{\mathrm{pol}}$ on any object of this category to an action of $\widehat{\mathfrak{g}}_\kappa$. Therefore the categories $\widehat{\mathfrak{g}}_\kappa^{\mathrm{pol}}$ and $\widehat{\mathfrak{g}}_\kappa$-mod coincide.

Here it is useful to explain why we prefer to work with the Lie algebra $\widehat{\mathfrak{g}}_\kappa$ as opposed to $\widehat{\mathfrak{g}}_\kappa^{\mathrm{pol}}$. This is because $\widehat{\mathfrak{g}}_\kappa$ is naturally attached to the (formal) punctured disc $D^\times = \operatorname{Spec} \mathbb{C}((t))$, whereas as $\widehat{\mathfrak{g}}_\kappa^{\mathrm{pol}}$ is attached to $\mathbb{C}^\times = \operatorname{Spec} \mathbb{C}[t,t^{-1}]$ (or the unit circle). This means in particular that, unlike $\widehat{\mathfrak{g}}_\kappa^{\mathrm{pol}}$, the Lie algebra $\widehat{\mathfrak{g}}_\kappa$ is not tied to a particular coordinate t, but we may replace t by any other formal coordinate on the punctured disc. This means, as we will do in Section 3.5.2, that $\widehat{\mathfrak{g}}_\kappa$ may be attached to the formal neighborhood of any point on a smooth algebraic curve, a property that is very important in applications that we have in mind, such as passing from the local to global

Langlands correspondence. In contrast, the polynomial version $\widehat{\mathfrak{g}}_\kappa^{\mathrm{pol}}$ is forever tied to $\mathbb{C}^\times$.

Going back to the category $\widehat{\mathfrak{g}}_\kappa$-mod, we see that there are two properties that its objects satisfy. Therefore it does not coincide with the category of all modules over the universal enveloping algebra $U(\widehat{\mathfrak{g}}_\kappa)$ (which is the category of all $\widehat{\mathfrak{g}}_\kappa$-modules). We need to modify this algebra.

First of all, since $\mathbf{1}$ acts as the identity, the action of $U(\widehat{\mathfrak{g}}_\kappa)$ factors through the quotient

$$U_\kappa(\widehat{\mathfrak{g}}) \stackrel{\mathrm{def}}{=} U(\widehat{\mathfrak{g}}_\kappa)/(\mathbf{1}-1).$$

Second, the smoothness condition (2.1.1) implies that the action of $U_\kappa(\widehat{\mathfrak{g}})$ extends to an action of its completion defined as follows.

Define a linear topology on $U_\kappa(\widehat{\mathfrak{g}})$ by using as the basis of neighborhoods for 0 the following left ideals:

$$I_N = U_\kappa(\widehat{\mathfrak{g}})(\mathfrak{g} \otimes t^N \mathbb{C}[[t]]), \qquad N \geqslant 0.$$

Let $\widetilde{U}_\kappa(\widehat{\mathfrak{g}})$ be the completion of $U_\kappa(\widehat{\mathfrak{g}})$ with respect to this topology. We call it that **completed universal enveloping algebra** of $\widehat{\mathfrak{g}}_\kappa$. Note that, equivalently, we can write

$$\widetilde{U}_\kappa(\widehat{\mathfrak{g}}) = \varprojlim U_\kappa(\widehat{\mathfrak{g}})/I_N.$$

Even though the I_N's are only left ideals (and not two-sided ideals), one checks that the associative product structure on $U_\kappa(\widehat{\mathfrak{g}})$ extends by continuity to an associative product structure on $\widetilde{U}_\kappa(\widehat{\mathfrak{g}})$ (this follows from the fact that the Lie bracket on $U_\kappa(\widehat{\mathfrak{g}})$ is continuous in the above topology). Thus, $\widetilde{U}_\kappa(\widehat{\mathfrak{g}})$ is a complete topological algebra. It follows from the definition that the category $\widehat{\mathfrak{g}}_\kappa$-mod coincides with the category of discrete modules over $\widetilde{U}_\kappa(\widehat{\mathfrak{g}})$ on which the action of $\widetilde{U}_\kappa(\widehat{\mathfrak{g}})$ is pointwise continuous (this is precisely equivalent to the condition (2.1.1)).

It is now easy to see that the center of our category $\widehat{\mathfrak{g}}_\kappa$-mod is equal to the center of the algebra $\widetilde{U}_\kappa(\widehat{\mathfrak{g}})$, which we will denote by $Z_\kappa(\widehat{\mathfrak{g}})$. The argument is similar to the one we used above: though $\widetilde{U}_\kappa(\widehat{\mathfrak{g}})$ itself is not an object of $\widehat{\mathfrak{g}}_\kappa$-mod, we have a collection of objects $\widetilde{U}_\kappa(\widehat{\mathfrak{g}})/I_N$. Using this collection, we obtain an isomorphism between the center of the category $\widehat{\mathfrak{g}}_\kappa$-mod and the inverse limit of the algebras $Z(\mathrm{End}_{\widehat{\mathfrak{g}}_\kappa} \widetilde{U}_\kappa(\widehat{\mathfrak{g}})/I_N)$, which, by definition, coincides with $Z_\kappa(\widehat{\mathfrak{g}})$.

Now we can formulate our first question:

describe the center $Z_\kappa(\widehat{\mathfrak{g}})$ for all levels κ.

We will see that the center $Z_\kappa(\widehat{\mathfrak{g}})$ is trivial (i.e., equal to the scalars) unless $\kappa = \kappa_c$, the critical level. However, at the critical level $Z_{\kappa_c}(\widehat{\mathfrak{g}})$ is large, and its structure is reminiscent to that of $Z(\mathfrak{g})$ described in Theorem 2.1.1. For the precise statement, see Theorem 4.3.6.

2.1.3 The affine Casimir element

Before attempting to describe the entire center, let us try to construct some central elements "by hand." In the finite-dimensional case the simplest generator of $Z(\mathfrak{g})$ was particularly easy to construct. So we start by attempting to define a similar operator in the affine case.

Let A be any element of $\mathfrak{g}$ and n an integer. Then $A \otimes t^n$ is an element of $\mathfrak{g} \otimes \mathbb{C}((t))$ and hence of $\widehat{\mathfrak{g}}_\kappa$. We denote this element by A_n. We collect all of the elements associated to $A \in \mathfrak{g}$ into a single formal power series in an auxiliary variable z:

$$A(z) = \sum_{n \in \mathbb{Z}} A_n z^{-n-1}.$$

The shift by one in the exponent for z may seem a little strange at first but it is convenient to have. For example, we have the following formula

$$A_n = \mathrm{Res}_{z=0}\, A(z) z^n dz.$$

Note that this is just a formal notation. None of the power series we use actually has to converge anywhere.

Now, an obvious guess for an equivalent to the Casimir operator is the formal power series

$$\frac{1}{2} \sum_{a=1}^{\dim \mathfrak{g}} J^a(z) J_a(z). \tag{2.1.2}$$

There are, however, many problems with this expression. If we extract the coefficients of this sum, we see that they are two-way infinite sums. The infinity, by itself, is not a problem, because our completed enveloping algebra $\widetilde{U}_\kappa(\widehat{\mathfrak{g}})$ contains infinite sums. But, unfortunately, here we encounter a "wrong infinity," which needs to be corrected.

To be more concrete, let us look in detail at the case of $\mathfrak{g} = \mathfrak{sl}_2$. Then our potential Casimir element is

$$P(z) = \frac{1}{2}\left(e(z)f(z) + f(z)e(z) + \frac{1}{2} h(z)h(z) \right). \tag{2.1.3}$$

It is easy to write down the coefficients in front of particular powers of z in this series. If we write

$$P(z) = \sum_{n \in \mathbb{Z}} P_N z^{-N-2},$$

then

$$P_N = \sum_{m+n=N} (e_m f_n + f_m e_n + \frac{1}{2} h_m h_n). \tag{2.1.4}$$

None of these expressions belongs to $\widetilde{U}_\kappa(\widehat{\mathfrak{sl}}_2)$.

To see this, let us observe that, by definition, an element of $\widetilde{U}_\kappa(\widehat{\mathfrak{sl}}_2)$ may be written in the form

$$K + \sum_{n \geq 0} (Q_n e_n + R_n f_n + S_n h_n),$$

where K, Q_n, R_n, S_n are *finite* linear combinations of monomials in the generators $e_m, f_m, h_m, m \in \mathbb{Z}$.

Let us examine the first term in (2.1.4). It may be written as the sum of two terms

$$\sum_{n+m=N; n \geq 0} e_m f_n + \sum_{n+m=N; n<0} e_m f_n. \tag{2.1.5}$$

The first of them belongs to $\widetilde{U}_\kappa(\widehat{\mathfrak{sl}}_2)$, but the second one does not: the order of the two factors is wrong! This means that this element does not give rise to a well-defined operator on a module from the category $\widehat{\mathfrak{sl}}_{2,\kappa}$-mod. Indeed, we can write

$$e_m f_n = f_n e_m + [e_m, f_n] = f_n e_m + h_{m+n}.$$

Thus, the price to pay for switching the order is the commutator between the two factors, which is non-zero. Therefore, while the sum

$$\sum_{n+m=N; n<0} f_n e_m$$

belongs to $\widetilde{U}_\kappa(\widehat{\mathfrak{sl}}_2)$ and its action is well-defined on any module from $\widehat{\mathfrak{sl}}_{2,\kappa}$-mod, the sum

$$\sum_{n+m=N; n<0} e_m f_n$$

that we are given differs from it by h_{m+n} added up infinitely many times, which is meaningless.

To resolve this problem, we need to redefine our operators P_N so as to make them fit into the completion $\widetilde{U}_\kappa(\widehat{\mathfrak{sl}}_2)$. There is an obvious way to do this: we just switch by hand the order in the second summation in (2.1.5) to comply with the requirements:

$$\sum_{n+m=N; n \geq 0} e_m f_n + \sum_{n+m=N; n<0} f_n e_m. \tag{2.1.6}$$

Note, however, that this is not the only way to modify the definition. Another way to do it is

$$\sum_{n+m=N; m<0} e_m f_n + \sum_{n+m=N; m \geq 0} f_n e_m. \tag{2.1.7}$$

Since n and m are constrained by the equation $n + m = N$, it is easy to see that this expression also belongs to $\widetilde{U}_\kappa(\widehat{\mathfrak{sl}}_2)$. But it is different from (2.1.6) because the order of *finitely many* terms is switched. For example, if $N \geq 0$, the terms $e_m f_n, 0 \leq m \leq N$, in (2.1.6) are replaced by $f_n e_m$ in (2.1.7). So the difference between the two expressions is $(N+1)h_{n+m}$, which is of course a well-defined element of the enveloping algebra.

Thus, the upshot is that there are several inequivalent ways to "regularize" the meaningless expression (2.1.5), which differ at finitely many places. In what follows we will use the second scenario. To write it down in a more convenient way, we introduce the notion of **normal ordering**. For any $A, B \in \mathfrak{sl}_2$ (or an arbitrary simple Lie algebra $\mathfrak{g}$) we will set

$$:A_m B_n: \stackrel{\text{def}}{=} \begin{cases} A_m B_n, & m < 0, \\ B_n A_m, & m \geq 0. \end{cases}$$

Then (2.1.7) may be rewritten as

$$\sum_{n+m=N} :e_m f_n:,$$

which is the z^{-N-2} coefficient of $:e(z)f(z):$, where we apply the normal ordering by linearity.

Now we apply the normal ordering to the formal power series $P(z)$ given by formula (2.1.3). The result is another formal power series

$$S(z) = \sum_{N \in \mathbb{Z}} S_N z^{-N-2} = \frac{1}{2}\left(:e(z)f(z): + :f(z)e(z): + \frac{1}{2} :h(z)h(z): \right). \quad (2.1.8)$$

The corresponding coefficients S_N are now well-defined elements of the completion $\widetilde{U}_\kappa(\widehat{\mathfrak{sl}}_2)$.

For a general simple Lie algebra, we write

$$S(z) = \sum_{N \in \mathbb{Z}} S_N z^{-N-2} = \frac{1}{2} \sum_{a=1}^{\dim \mathfrak{g}} :J^a(z) J_a(z):. \quad (2.1.9)$$

Note that $S(z)$ is independent of the choice of the basis $\{J^a\}$. The coefficients $S_N \in \widetilde{U}_\kappa(\widehat{\mathfrak{g}})$ of $S(z)$ are called the **Segal–Sugawara operators**. Are they central elements of $\widetilde{U}_\kappa(\widehat{\mathfrak{g}})$? To answer this question, we need to compute the commutators

$$[S_n, A_m] = S_n A_m - A_m S_n$$

in $\widetilde{U}_\kappa(\widehat{\mathfrak{g}})$ for all $A \in \mathfrak{g}$. The elements S_n are central if and only if these commutators vanish.

This computation is not an easy task. Let us first give the answer. Let κ_c be the **critical** invariant inner product on $\mathfrak{g}$ defined by the formula

$$\kappa_c(A, B) = -\frac{1}{2} \operatorname{Tr}_{\mathfrak{g}} \operatorname{ad} A \operatorname{ad} B. \quad (2.1.10)$$

Then we have

$$[S_n, A_m] = -\frac{\kappa - \kappa_c}{\kappa_0} n A_{n+m}. \tag{2.1.11}$$

Since the invariant inner products on a simple Lie algebra $\mathfrak{g}$ form a one-dimensional vector space, the ratio appearing in this formula is well-defined (recall that $\kappa_0 \neq 0$ by our assumption).

Formula (2.1.11) comes as a surprise. It shows that the Segal–Sugawara operators are indeed central for one specific value of κ, but this value is not $\kappa = 0$, as one might naively expect, but the *critical* one, $\kappa = \kappa_c$! This may be thought of as a "quantum correction" due to our regularization scheme (the normal ordering). In fact, if we just formally compute the commutators between the coefficients of the original (unregularized) series (2.1.2) and A_m, we will find that they are central elements at $\kappa = 0$, as naively expected. But these coefficients are not elements of our completion $\widetilde{U}_\kappa(\widehat{\mathfrak{g}})$, so they cannot possibly define central elements in $\widetilde{U}_\kappa(\widehat{\mathfrak{g}})$. (They belong to a different completion of $U_\kappa(\widehat{\mathfrak{g}})$, one that does not act on smooth $\widehat{\mathfrak{g}}_\kappa$-modules and hence is irrelevant for our purposes.) The regularized elements S_n become central only after we shift the level by κ_c. This is the first indication of the special role that the critical level κ_c plays in representation theory of affine Kac–Moody algebras.

How does one prove formula (2.1.11)? A direct calculation is tedious and not very enlightening. Even if we do make it, the next question will be to compute the commutation relations between the S_n's (this is related to the Poisson structure on the center, as we will see below), which is a still harder calculation if we approach it with "bare hands." This suggests that we need to develop some more serious tools in order to perform calculations of this sort. After all, we are now only discussing the quadratic Casimir element. But what about higher-order central elements?

The necessary tools are provided in the theory of **vertex algebras**, which in particular gives us nice and compact formulas for computing the commutation relations such as (2.1.11). The idea, roughly, is that the basic objects are not the elements J_n^a of $\widehat{\mathfrak{g}}_\kappa$ and the topological algebra $\widetilde{U}_\kappa(\widehat{\mathfrak{g}})$ that they generate, but rather the generating series $J^a(z)$ and the vertex algebra that they generate. We will take up this theory in the next section.

2.2 Basics of vertex algebras

In this section we give a crash course on the theory of vertex algebras, following [FB], where we refer the reader for more details.

Vertex algebras were originally defined by R. Borcherds [Bo], and the foundations of the theory were laid down in [FLM, FHL]. The formalism that we will use in this book is close to that of [FB, K3], and all results on vertex algebras presented below are borrowed from these two books. We also note that vertex algebras have geometric counterparts: chiral algebras and factorization algebras introduced in [BD2]. The connection between them and vertex algebras is explained in [FB].

2.2.1 Fields

Let R be an algebra over $\mathbb{C}$; a **formal power series** over R in the variables $z_1, \ldots, z_n$ is a sum of the form

$$\sum_{i_1,\ldots,i_n \in \mathbb{Z}} A_{i_1 \cdots i_n} z_1^{i_1} \cdots z_n^{i_n}.$$

The set of all such formal power series is denoted by $R[[z_1^{\pm 1}, \ldots, z_n^{\pm 1}]]$.

Note carefully the difference between formal power series $R[[z^{\pm 1}]]$ (which can have arbitrarily large and small powers of z), Taylor power series $R[[z]]$ (which have no negative powers of z) and Laurent power series $R((z))$ (which have negative powers of z bounded from below).

What operations can be performed on formal power series? We can certainly add them, differentiate them, multiply them by polynomials. However, we cannot multiply them by other formal power series. The reason for this is that the coefficients of the product will consist of infinite sums, e.g.,

$$\left(\sum_n A_n z^n\right)\left(\sum_m B_m z^m\right) = \sum_n z^n \left(\sum_{i+j=n} A_i B_j\right).$$

Nevertheless, we can multiply two formal power series if the variables they are in are disjoint, so for example $f(z)g(w)$ makes sense as a formal power series in the two variables z and w.

A particularly important example of a field is the **formal delta-function**. This is denoted by $\delta(z-w)$† and is defined by the formula

$$\delta(z-w) = \sum_{n \in \mathbb{Z}} z^n w^{-n-1}.$$

It has the following easy to check properties:

(i) $A(z)\delta(z-w) = A(w)\delta(z-w)$,
(ii) $(z-w)\delta(z-w) = 0$,
(iii) $(z-w)^{n+1}\partial_w^n \delta(z-w) = 0$.

The first of these properties tells us that

$$\mathrm{Res}_{z=0}\left(A(z)\delta(z-w)dz\right) = A(w),$$

which is a property we would expect the delta-function to have.

† It is not a function of $z-w$. This is just notation indicating that the properties of this formal delta-function correspond closely with properties of the usual delta-function.

We can make the analogy between $\mathbb{C}[[z^{\pm 1}]]$ and distributions more precise in the following way. Given a formal power series $A(z)$ in z, define a linear functional ("distribution") ϕ_A on the space of polynomials $\mathbb{C}[z, z^{-1}]$ by the formula

$$\phi_A(f(z)) = \operatorname{Res} A(z)f(z)dz.$$

Recall that the product $A(z)f(z)$ is well-defined, since $f(z)$ is a polynomial and so only finite sums turn up in the product. Conversely, given a distribution ϕ on $\mathbb{C}[z, z^{-1}]$, define a formal power series A_ϕ by

$$A_\phi(z) = \sum_{n \in \mathbb{Z}} \phi(z^n) z^{-n-1}.$$

It is easy to see that these two operations are inverse to each other. Hence $\mathbb{C}[[z^{\pm 1}]]$ is exactly the space of all distributions on the space of polynomials $\mathbb{C}[z, z^{-1}]$.

We can now think of $\delta(z - w)$ as being a formal power series in the variable z with $w \in \mathbb{C}^\times$ any non-zero complex number. Under the above identification it is easy to see that we get a delta-function corresponding to w in the usual sense.

Next we define **fields** as special types of formal power series. Let V be a vector space over $\mathbb{C}$, so $\operatorname{End} V$ is an algebra over $\mathbb{C}$. A field is a formal power series in $\operatorname{End} V[[z^{\pm 1}]]$. We write the field as follows

$$A(z) = \sum_{n \in \mathbb{Z}} A_n z^{-n-1}.$$

The power of z chosen is one that will later make much of the notation simpler. Fields must satisfy the following additional property: *For each $v \in V$ there is an integer $N \geqslant 0$ such that $A_n \cdot v = 0$ for all $n \geqslant N$.* We may rephrase this condition as saying that $A(z) \cdot v$ is a Laurent polynomial for any $v \in V$.

If the vector space V is $\mathbb{Z}$-graded, i.e., $V = \bigoplus_{n \in \mathbb{Z}} V_n$, then we have the usual concept of homogeneous elements of V as well as homogeneous endomorphisms: $\phi \in \operatorname{End} V$ is homogeneous of degree m if $\phi(V_n) \subset V_{n+m}$ for all n. In the case of vertex algebras it is common to call the homogeneity degree of a vector in V the **conformal dimension**.

2.2.2 Definition

Now we are ready to give the definition of a vertex algebra.

A **vertex algebra** consists of the following data:

(i) A vector space V (the **space of states**);
(ii) A vector in V denoted by $|0\rangle$ (the **vacuum vector**);
(iii) An endomorphism $T : V \to V$ (the **translation operator**);

(iv) A linear map $Y(\cdot, z) : V \to \operatorname{End} V[[z^{\pm 1}]]$ sending vectors in V to fields on V (also called **vertex operators**)

$$A \in V \mapsto Y(A, z) = \sum_{n \in \mathbb{Z}} A_{(n)} z^{-n-1}.$$

(the **state-field correspondence**).

These satisfy the following axioms:

(i) $Y(|0\rangle, z) = \mathrm{id}_V$;
(ii) $Y(A, z)|0\rangle = A + z(\ldots) \in V[[z]]$;
(iii) $[T, Y(A, z)] = \partial_z Y(A, z)$;
(iv) $T|0\rangle = 0$;
(v) (**locality**) For any two vectors $A, B \in \mathbb{Z}$ there is a non-negative integer N such that

$$(z - w)^N [Y(A, z), Y(B, w)] = 0.$$

It follows from the axioms for a vertex algebra that the action of T may be defined by the formula

$$T(A) = A_{(-2)} |0\rangle,$$

so T is not an independent datum. However, we have included it in the set of data, because this makes axioms more transparent and easy to formulate.

A vertex algebra is called $\mathbb{Z}$- (or $\mathbb{Z}_+$-) graded if V is a $\mathbb{Z}$- (resp., $\mathbb{Z}_+$-) graded vector space, $|0\rangle$ is a vector of degree 0, T is a linear operator of degree 1, and for $A \in V_m$ the field $Y(A, z)$ has conformal dimension m, i.e.,

$$\deg A_{(n)} = -n + m - 1.$$

A particularly simple example of a vertex algebra can be constructed from a commutative associative unital algebra with a derivation.

Let V be a commutative associative unital algebra with a derivation T. We define the vertex algebra structure as follows:

$$Y(|0\rangle, z) = \mathrm{Id}_V,$$

$$Y(A, z) = \sum_{n \geqslant 0} \frac{z^n}{n!} \mathrm{mult}(T^n A) = \mathrm{mult}(e^{Tz} A), \tag{2.2.1}$$

$$T = T. \tag{2.2.2}$$

Here the operators $\mathrm{mult}(A)$ in the power series are left multiplication by A.

It is an easy exercise to check that this is a vertex algebra structure. The vertex algebra structure is particularly simple because the axiom of locality has become a form of commutativity

$$[Y(A, z), Y(B, w)] = 0.$$

This is a very special property. Any vertex algebra with this property is called

commutative. Another property that this vertex algebra structure has that is unusual is that the formal power series that occur have only non-negative powers of z. It turns out that these two properties are equivalent.

Lemma 2.2.1 *A vertex algebra is commutative if and only if* $Y(A,z) \in \operatorname{End} V[[z]]$ *for all* $A \in V$.

Proof. If V is commutative then

$$Y(A,z)Y(B,w)\left|0\right\rangle = Y(B,w)Y(A,z)\left|0\right\rangle.$$

Expanding these in powers of w and taking the constant coefficient in w using axiom (ii), we see that $Y(A,z)B \in V[[z]]$ for any A and B. This shows that $Y(A,z) \in \operatorname{End} V[[z]]$.

Conversely, if $Y(A,z) \in \operatorname{End} V[[z]]$ for all $A \in V$, then $Y(A,z)Y(B,w) \in \operatorname{End} V[[z,w]]$. Locality then says that

$$(z-w)^N Y(A,z)Y(B,w) = (z-w)^N Y(B,w)Y(A,z).$$

As $(z-w)^N$ has no divisors of zero in $\operatorname{End} V[[z,w]]$, it follows that V is commutative. □

So we have seen that commutative associative unital algebra with a derivation gives rise to a commutative vertex algebra. It is easy to see that the above construction can be run in the other direction and so these two categories are equivalent. In particular, $\mathbb{Z}$-graded commutative vertex algebras correspond to $\mathbb{Z}$-graded commutative associative algebras with a derivation of degree 1.

2.2.3 More on locality

We have just seen that the property of commutativity in a vertex algebra is very restrictive. The point of the theory of vertex algebras is that we replace it by a more general axiom; namely, locality. In this sense, one may think of the notion of vertex algebra as generalizing the familiar notion of commutative algebra. In this section we will look closely at the locality axiom and try to gain some insights into its meaning.

Let $v \in V$ be a vector in V and $\phi : V \to \mathbb{C}$ a linear functional on V. Given $A, B \in V$ we can form two formal power series

$$\langle \phi, Y(A,z)Y(B,w)v\rangle \qquad \text{and} \qquad \langle \phi, Y(B,w)Y(A,z)v\rangle$$

These two formal power series, which are *a priori* elements of $\mathbb{C}[[z^{\pm 1}, w^{\pm 1}]]$, actually belong to the subspaces $\mathbb{C}((z))((w))$ and $\mathbb{C}((w))((z))$, respectively. These two subspaces are different: the first consists of bounded below powers of w, but powers of z are not uniformly bounded, whereas the second consists of bounded below powers of z but not uniformly bounded powers of w.

The intersection of the two spaces consists of those series in z and w which have bounded below powers in both z and w. In other words, we have

$$\mathbb{C}((z))((w)) \cap \mathbb{C}((w))((z)) = \mathbb{C}[[z,w]][z^{-1},w^{-1}]. \tag{2.2.3}$$

Note that $\mathbb{C}((z))((w))$ and $\mathbb{C}((w))((z))$ are closed under multiplication and are actually fields (here we use the terminology "field" in the usual sense!). Their intersection is a subalgebra $\mathbb{C}[[z,w]][z^{-1},w^{-1}]$. Therefore within each of the two fields we have the fraction field of $\mathbb{C}[[z,w]][z^{-1},w^{-1}]$. This fraction field is denoted by $\mathbb{C}((z,w))$ and consists of ratios $f(z,w)/g(z,w)$, where f, g are in $\mathbb{C}[[z,w]]$.

However, the embeddings $\mathbb{C}((z,w))$ into $\mathbb{C}((z))((w))$ and $\mathbb{C}((w))((z))$ are different. These embeddings are easy to describe; we simply take Laurent power series expansions assuming one of the variables is "small." We will illustrate how this works using the element $\frac{1}{z-w} \in \mathbb{C}((z,w))$.

Assume that w is the "small" variable, and so $|w| < |z|$. Then we can expand

$$\frac{1}{z-w} = \frac{1}{z(1-\frac{w}{z})} = z^{-1}\sum_{n\geqslant 0}\left(\frac{w}{z}\right)^n$$

in positive powers of w/z. Note that the result will have bounded below powers of w and so will lie in $\mathbb{C}((z))((w))$.

Assume now that z is the "small" variable, and so $|z| < |w|$. We can then expand

$$\frac{1}{z-w} = -\frac{1}{w(1-\frac{z}{w})} = -z^{-1}\sum_{n<0}\left(\frac{w}{z}\right)^n$$

in negative powers of w/z because $|z| < |w|$. Note that the result will have bounded below powers of z and so will lie in $\mathbb{C}((w))((z))$.

It is instructive to think of the two rings $\mathbb{C}((w))((z))$ and $\mathbb{C}((z))((w))$ as representing functions in two variables, which have one of their variables much smaller than the other. The "domains of definition" of these functions are $|w| \gg |z|$ and $|z| \gg |w|$, respectively.

So we have now seen that elements in $\mathbb{C}((w))((z))$ and $\mathbb{C}((z))((w))$, although they can look very different, may in fact be representing the same element of $\mathbb{C}((z,w))$. This is very similar to the idea of *analytic continuation* from complex analysis. When these two different elements come from the same rational function in z and w we could think of them as "representing the same function" (we could even think of the rational function as being the fundamental object rather than the individual representations).

What locality is telling us precisely that the formal power series

$$\langle \phi, Y(A,z)Y(B,w)v\rangle \qquad \text{and} \qquad \langle \phi, Y(B,w)Y(A,z)v\rangle \tag{2.2.4}$$

represent the same rational function in z and w in $\mathbb{C}((w))((z))$ and $\mathbb{C}((z))((w))$, respectively.

Indeed, the locality axiom states that

$$(z-w)^N \langle \phi, Y(A,z)Y(B,w)v\rangle \qquad \text{and} \qquad (z-w)^N \langle \phi, Y(B,w)Y(A,z)v\rangle$$

are equal to each other, as elements of $\mathbb{C}[[z^{\pm 1}, w^{\pm 1}]]$. Due to the equality (2.2.3), we find that both of them actually belong to $\mathbb{C}[[z,w]][z^{-1}, w^{-1}]$. So, the formal power series (2.2.4) are representations of the same element of $\mathbb{C}[[z,w]][z^{-1}, w^{-1}, (z-w)^{-1}]$ in $\mathbb{C}((w))((z))$ and $\mathbb{C}((z))((w))$, respectively. If we ask in addition that as v and ϕ vary there is a universal bound on the power of $(z-w)$ that can occur in the denominator, then we obtain an equivalent form of the locality axiom. This is a reformulation that will be useful in what follows.†

2.2.4 Vertex algebra associated to $\widehat{\mathfrak{g}}_\kappa$

We now present our main example of a non-commutative vertex algebra, based on the affine Kac–Moody algebra $\widehat{\mathfrak{g}}_\kappa$. Many of the interesting properties of vertex algebras will be visible in this example.

Let $\mathfrak{g}$ be a finite-dimensional complex simple Lie algebra with an ordered basis $\{J^a\}$, where $a = 1, \ldots, \dim \mathfrak{g}$ (see Section A.3). Recall that the affine Kac–Moody algebra $\widehat{\mathfrak{g}}$ has a basis consisting of the elements J^a_n, $a = 1, \ldots, \dim \mathfrak{g}$, $n \in \mathbb{Z}$, and $\mathbf{1}$.

Previously, we grouped the elements associated to J^a into a formal power series

$$J^a(z) = \sum_{n \in \mathbb{Z}} J^a_n z^{-n-1}.$$

This gives us a hint about what some of the fields in this vertex algebra should be.

We should first describe the vector space V on which the vertex algebra is built. We know that we will need a special vector $|0\rangle$ in V to be the vacuum vector. We also know that if $J^a(z)$ are indeed vertex operators, then the J^a_n's should be linear operators on V and, by axiom (i), the elements J^a_n for $n \geqslant 0$ should annihilate the vacuum vector $|0\rangle$.

Notice that the set of Lie algebra elements which are supposed to annihilate $|0\rangle$ form the Lie subalgebra $\mathfrak{g}[[t]]$ of $\widehat{\mathfrak{g}}_\kappa$. Thus, $\mathbb{C}|0\rangle$ is the trivial one-dimensional representation of $\mathfrak{g}[[t]]$. We also define an action of the central element $\mathbf{1}$ on $|0\rangle$ as follows: $\mathbf{1}|0\rangle = 1$. Let us denote the resulting representation of $\mathfrak{g}[[t]] \oplus \mathbb{C}\mathbf{1}$ by $\mathbb{C}_\kappa$. We can now define a $\widehat{\mathfrak{g}}_\kappa$-module by using the

† It may seem slightly strange that we only allow three types of singularities: at $z = 0, w = 0$, and $z = w$. But these are the only equations that do not depend on the choice of coordinate, which is a valuable property for us as we will need a coordinate-free description of vertex operators and algebras.

induction functor:

$$V_\kappa(\mathfrak{g}) = \mathrm{Ind}_{\mathfrak{g}[[t]]\oplus\mathbb{C}\mathbf{1}}^{\widehat{\mathfrak{g}}_\kappa} \mathbb{C}_\kappa = U(\widehat{\mathfrak{g}}_\kappa) \underset{U(\mathfrak{g}[[t]]\oplus\mathbb{C}\mathbf{1})}{\otimes} \mathbb{C}_\kappa.$$

It is called **the vacuum Verma module** of level κ.

Recall that κ is unique up to a scalar. Therefore it is often convenient to fix a particular invariant inner product κ_0 and write an arbitrary one as $\kappa = k\kappa_0, k \in \mathbb{C}$. This is the point of view taken, for example, in [FB], where as κ_0 we take the inner product with respect to which the squared length of the maximal root is equal to 2, and denote the corresponding vacuum module by $V_k(\mathfrak{g})$.

The structure of the module $V_\kappa(\mathfrak{g})$ is easy to describe. By the Poincaré–Birkhoff–Witt theorem, $V_\kappa(\mathfrak{g})$ is isomorphic to $U(\mathfrak{g}\otimes t^{-1}\mathbb{C}[t^{-1}])|0\rangle$. Therefore it has a basis of lexicographically ordered monomials of the form

$$J^{a_1}_{n_1} \dots J^{a_m}_{n_m}|0\rangle, \tag{2.2.5}$$

where $n_1 \leq n_2 \leq \dots \leq n_m < 0$, and if $n_i = n_{i+1}$, then $a_i \leq a_{i+1}$. We define a $\mathbb{Z}$-grading on $\widehat{\mathfrak{g}}_\kappa$ and on $V_\kappa(\mathfrak{g})$ by the formula $\deg J^a_n = -n, \deg|0\rangle = 0$. The homogeneous graded components of $V_\kappa(\mathfrak{g})$ are finite-dimensional and they are non-zero only in non-negative degrees. Here is the picture of the first few homogeneous components of $V_\kappa(\mathfrak{g})$:

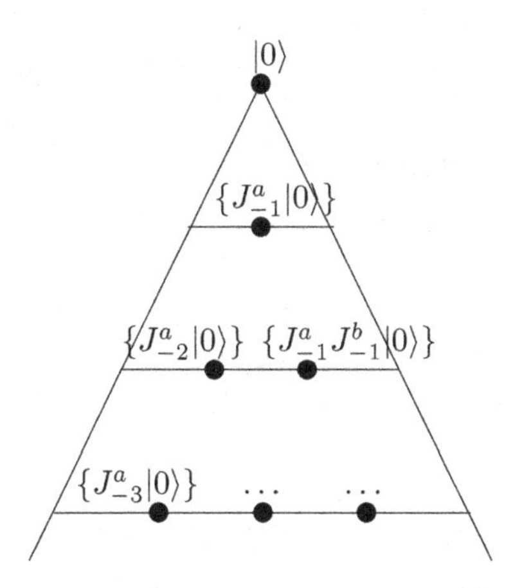

The action of $\widehat{\mathfrak{g}}_\kappa$ is described as follows. The action of J^a_n with $n < 0$ is just the obvious action on $U(\mathfrak{g}\otimes t^{-1}\mathbb{C}[t^{-1}]) \simeq V_\kappa(\mathfrak{g})$. To apply J^a_n with $n \geqslant 0$, we use the commutation relations in the Lie algebra $\widehat{\mathfrak{g}}_\kappa$ to move this term through to the vacuum vector, which the J^a_n's with $n \geqslant 0$ annihilate.

Let us illustrate the structure of $V_\kappa(\mathfrak{g})$ in the case of the affine Lie algebra

$\widehat{\mathfrak{sl}}_2$. The elements J^a are now denoted by e, f, h and the commutation relations between them are

$$[h, e] = 2e, \qquad [h, f] = -2f, \qquad [e, f] = h.$$

For example, consider the vector

$$e_{-1}f_{-2}\,|0\rangle \in V_\kappa(\mathfrak{sl}_2).$$

If we apply h_1 to it, we obtain

$$\begin{aligned} h_1 e_{-1} f_{-2}\,|0\rangle &= ([h_1, e_{-1}] + e_{-1}h_1)\, f_{-2}\,|0\rangle \\ &= (2e_0 + e_{-1}h_1)\, f_{-2}\,|0\rangle \\ &= 2\,([e_0, f_{-2}] + f_{-2}e_0)\,|0\rangle + e_{-1}\,([h_1, f_{-2}] + f_{-2}h_1)\,|0\rangle \\ &= 2h_{-2}\,|0\rangle + 0 - 2e_{-1}f_{-1}\,|0\rangle + 0 \\ &= 2\,(h_{-2} - e_{-1}f_{-1})\,|0\rangle\,. \end{aligned}$$

So, at each step we are simply moving the annihilation operators closer and closer to the vacuum until we have made them all disappear.

We now have the vector space $V = V_\kappa(\mathfrak{g})$ and vacuum vector $|0\rangle$ for our vertex algebra. We still need to define the translation operator T and the state-field correspondence $Y(\cdot, z)$.

The translation operator T is defined by interpreting it as the vector field $-\partial_t$ (the reason for this will become more clear later on). This vector field naturally acts on the Lie algebra $\mathfrak{g}((t))$ and preserves the Lie subalgebra $\mathfrak{g}[[t]]$. Therefore it acts on $V_\kappa(\mathfrak{g})$. Concretely, this means that we have the commutation relations

$$[T, J^a_n] = -nJ^a_{n-1},$$

and the vacuum vector is invariant under T,

$$T\,|0\rangle = 0,$$

as required by axiom (iv) of vertex algebras. These two conditions uniquely specify the action of T on the vector space $V_\kappa(\mathfrak{g})$.

2.2.5 Defining vertex operators

Finally, we need to define the state-field correspondence. For the vacuum vector this is determined by axiom (i):

$$Y(|0\rangle, z) = \mathrm{Id}\,.$$

The elements of the next degree, namely 1, are of the form $J^a_{-1}\,|0\rangle$. To guess the form of the vertex operators $Y(J^a_{-1}\,|0\rangle, z)$ corresponding to them, we first look at the associated graded space of $V_\kappa(\mathfrak{g})$, which is a commutative vertex algebra.

To describe this associated graded space, we observe that the Poincaré–Bikhoff–Witt filtration on $U(\mathfrak{g}_\kappa)$ induces one on $V_\kappa(\mathfrak{g})$. The ith term of this filtration, which we denote by $V_\kappa(\mathfrak{g})_{\leq i}$, is the span of all monomials (2.2.5) with $m \leq i$. The associated graded algebra $\mathrm{gr}\, V_\kappa(\mathfrak{g})$ with respect to this filtration is the symmetric algebra with generators corresponding to J^a_n with $n < 0$. To distinguish them from the actual J^a_n's, we will denote these generators by $\overline{J}^a_n$.

Recall that for any Lie algebra $\mathfrak{g}$, the associated graded $\mathrm{gr}\, U(\mathfrak{g})$ is isomorphic to $\mathrm{Sym}\, \mathfrak{g}$. This implies that for any κ we have

$$\mathrm{gr}\, V_\kappa(\mathfrak{g}) = \mathrm{Sym}(\mathfrak{g}((t))/\mathfrak{g}[[t]]) \simeq \mathrm{Sym}(t^{-1}\mathfrak{g}[[t^{-1}]]).$$

Thus, $\mathrm{gr}\, V_\kappa(\mathfrak{g})$ is a commutative unital algebra with a derivation T corresponding to the vector field $-\partial_t$. It is uniquely determined by the formula $T \cdot \overline{J}^a_n = -n\overline{J}^a_{n-1}$. Therefore, according to the discussion of Section 2.2.2, $\mathrm{gr}\, V_\kappa(\mathfrak{g})$ has a natural structure of a commutative vertex algebra. By definition (see formula (2.2.1)), in this vertex algebra we have

$$Y(\overline{J}^a_{-1}, z) = \sum_{n \geqslant 0} \frac{z^n}{n!} \mathrm{mult}(T^n \cdot \overline{J}^a_{-1}) = \sum_{n \geqslant 0} \mathrm{mult}(\overline{J}^a_{-n-1}) z^n.$$

By abusing notation, we will write this as

$$Y(\overline{J}^a_{-1}, z) = \sum_{m<0} \overline{J}^a_m z^{-m-1} \tag{2.2.6}$$

with the understanding that on the right hand side $\overline{J}^a_m$ stands for the corresponding operator of multiplication acting on $\mathrm{gr}\, V_\kappa(\mathfrak{g})$.

In formula (2.2.6) only "half" of the generators of $\widehat{\mathfrak{g}}_\kappa$ is involved; namely, those with $m < 0$. This ensures that the resulting sum has no negative powers of z, as expected in a commutative vertex algebra. Now we generalize this formula to the case of a *non-commutative* vertex algebra $V_\kappa(\mathfrak{g})$, in which we are allowed to have negative powers of z appearing in the vertex operators. This leads us to the following proposal for the vertex operator corresponding to $J^a_{-1} |0\rangle \in V_\kappa(\mathfrak{g})$:†

$$Y(J^a_{-1} |0\rangle, z) = \sum_{n \in \mathbb{Z}} J^a_n z^{-n-1} = J^a(z).$$

It is easy to see that these vertex operators satisfy the relations

$$Y(J^a_{-1} |0\rangle, z) |0\rangle = J^a_{-1} |0\rangle + z(\ldots), \qquad \left[T, Y(J^a_{-1} |0\rangle, z)\right] = \partial_z Y(J^a_{-1} |0\rangle, z)$$

required by the axioms of vertex algebras.

We should also check that these vertex operators satisfy the locality axiom. To do this we use the commutation relations

$$[J^a_n, J^b_m] = [J^a, J^b]_{n+m} + n\kappa(J^a, J^b)\delta_{n,-m}\mathbf{1} \tag{2.2.7}$$

† note that J^a_n refers here to the operator of action of J^a_n on the representation $V_\kappa(\mathfrak{g})$

in $\widehat{\mathfrak{g}}_\kappa$ to evaluate the commutator of $J^a(z)$ and $J^b(w)$:

$$\left[J^a(z), J^b(w)\right] = \left[J^a, J^b\right](w)\delta(z-w) + \kappa(J^a, J^b)\partial_w \delta(z-w).$$

Now, recalling from (2.2.1) that $(z-w)^2$ annihilates both $\delta(z-w)$ and its derivative, we see that the locality axiom is satisfied.

We now need to define the vertex operators corresponding to the more general elements of $V_\kappa(\mathfrak{g})$. In fact, we will see in Theorem 2.2.5 below that the data that we have already defined: $|0\rangle$, T and $Y(J^a_{-1}|0\rangle, z)$ uniquely determine the entire vertex algebra structure on $V_\kappa(\mathfrak{g})$ (provided that it exists!). The reason is that the vectors $J^a_{-1}|0\rangle$ *generate* $V_\kappa(\mathfrak{g})$ in the following sense: $V_\kappa(\mathfrak{g})$ is spanned by the vectors obtained by successively applying the coefficients of the vertex operators $Y(J^a_{-1}|0\rangle, z)$ to the vacuum vector $|0\rangle$.

Here we will motivate the remaining structure from that on the associated graded algebra $\operatorname{gr} V_\kappa(\mathfrak{g})$. First of all, we find by an explicit calculation that in $\operatorname{gr} V_\kappa(\mathfrak{g})$ we have

$$Y(\overline{J}^a_n, z) = \frac{1}{(-n-1)!}\partial_z^{-n-1} \sum_{m<0} \overline{J}^a_m z^{-m-1}.$$

This motivates the formula

$$Y(J^a_n|0\rangle, z) = \frac{1}{(-n-1)!}\partial_z^{-n-1} J^a(z)$$

in $V_\kappa(\mathfrak{g})$.

Next, observe that in any commutative vertex algebra V we have the following simple identity

$$Y(AB, z) = Y(A, z)Y(B, z),$$

which follows from formula (2.2.1) and the Leibniz rule for the derivation T (here on the left hand side AB stands for the ordinary product with respect to the ordinary commutative algebra structure).

Therefore it is tempting to set, for example,

$$Y(J^a_{-1}J^b_{-a}|0\rangle, z) = Y(J^a_{-1}, z)Y(J^b_{-1}, z) = J^a(z)J^b(z).$$

However, we already know that the coefficients of the product on the right hand side are not well-defined as endomorphisms of $V_\kappa(\mathfrak{g})$. The problem is that the annihilation operators do not appear to the right of the creation operators. We had also suggested a cure: switching the order of some of the terms.

We will now define this procedure, called normal ordering, in a more systematic way. Let us define, for a formal power series

$$f(z) = \sum_{n\in\mathbb{Z}} f_n z^n \in R[[z^{\pm 1}]],$$

where R is any $\mathbb{C}$-algebra, the series $f(z)_+$ to be the part with non-negative powers of z and $f(z)_-$ to be the part with strictly negative powers of z:

$$f_+(z) = \sum_{n \geq 0} f_n z^n, \qquad f_-(z) = \sum_{n<0} f_n z^n.$$

The **normally ordered product** of two fields $A(z)$ and $B(z)$ is then defined to be

$$:A(z)B(z): \stackrel{\text{def}}{=} A(z)_+B(z) + B(z)A(z)_-.$$

It is an easy exercise (left for the reader) to check that $:A(z)B(z):$ is again a field (although the coefficients of z may be infinite sums, when applied to a vector they become finite sums). When we have more than two fields, the normally ordered product is defined from right to left, e.g.,

$$:A(z)B(z)C(z): = :A(z)\,(:B(z)C(z):)::.$$

This formula looks somewhat *ad hoc*, but we will see in Theorem 2.2.5 below that it is uniquely determined by the axioms of vertex algebra.

Another way to write the definition for normally ordered product is to use residues. For a function in two variables $F(z,w)$ define $F_{|z|>|w|}$ to be the expansion assuming z is the "large" variable (and similarly for $F_{|w|>|z|}$). For example

$$\left(\frac{1}{z-w}\right)_{|z|>|w|} = z^{-1}\sum_{n=0}^{\infty}\left(\frac{w}{z}\right)^n, \quad \left(\frac{1}{z-w}\right)_{|w|>|z|} = -w^{-1}\sum_{n=0}^{\infty}\left(\frac{z}{w}\right)^n.$$

Then we can represent the normally ordered product as

$$:A(z)B(z): = \\ \mathrm{Res}_{w=0}\left(\left(\frac{1}{w-z}\right)_{|w|>|z|} A(w)B(z) - \left(\frac{1}{w-z}\right)_{|z|>|w|} B(z)A(w)\right) dw.$$

This identity is easy to see from the following formulas

$$\begin{aligned} \mathrm{Res}_{z=0}\left(A(z)\left(\frac{1}{z-w}\right)_{|z|>|w|}\right) dz &= A(w)_+, \\ \mathrm{Res}_{z=0}\left(A(z)\left(\frac{1}{z-w}\right)_{|w|>|z|}\right) dz &= -A(w)_-, \end{aligned}$$

and these may be proved by simple calculation.

Using the normal ordering by induction, we arrive at the following guess for a general vertex operator in $V_\kappa(\mathfrak{g})$:

$$Y(J^{a_1}_{n_1} \dots J^{a_m}_{n_m} |0\rangle, z) = \\ \frac{1}{(-n_1-1)!} \cdots \frac{1}{(-n_m-1)!} :\partial_z^{-n_1-1} J^{a_1}(z) \dots \partial_z^{-n_m-1} J^{a_m}(z):. \quad (2.2.8)$$

Theorem 2.2.2 *The above formulas define the structure of a $\mathbb{Z}_+$-graded vertex algebra on $V_\kappa(\mathfrak{g})$.*

2.2.6 Proof of the Theorem

We have $Y(|0\rangle, z) = \mathrm{Id}$ by definition. Next, we need to check

$$Y(A,z)\,|0\rangle = A + z(\ldots)$$

to see that the vacuum axiom holds. This is clearly true when $A = |0\rangle$. We then prove it in general by induction: assuming it to hold for $Y(B,z)$, where $B \in V_\kappa(\mathfrak{g})_{\leq i}$, we find for any $A \in \mathfrak{g}$ that

$$Y(A_{-n}B\,|0\rangle, z)\,|0\rangle = \frac{1}{(-n-1)!}{:}\partial_z^{-n-1}A(z)\cdot Y(B,z){:}\,|0\rangle = A_{-n}B + z(\ldots).$$

The translation axiom (iii) boils down to proving the identity

$$[T, {:}A(z)B(z){:}] = \partial_z{:}A(z)B(z){:},$$

assuming that $[T, A(z)] = \partial_z A(z)$ and $[T, B(z)] = \partial_z B(z)$. We leave this to the reader, as well as the fact that the $\mathbb{Z}_+$-grading is compatible with the vertex operators.

The locality follows from Dong's lemma presented below, which shows that normally ordered products of local fields remain local. It is also clear that derivatives of local fields are still local. So, all we need to do is check that the generating fields $J^a(z)$ are mutually local. From the commutation relations we obtain that

$$\left[J^a(z), J^b(w)\right] = \left[J^a, J^b\right](w)\delta(z-w) - \kappa(J^a, J^b)\partial_w\delta(z-w).$$

From this it is clear that

$$(z-w)^2\left[J^a(z), J^b(w)\right] = 0,$$

and so we have the required locality.

This completes the proof modulo the following lemma.

Lemma 2.2.3 (Dong) *If $A(z)$, $B(z)$, $C(z)$ are mutually local fields, then $:A(z)B(z):$ and $C(z)$ are mutually local as well.*

Proof. The result will follow (by taking $\mathrm{Res}_{x=0}$) if we can show that the following two expressions are equal after multiplying by a suitable power of $(y-z)$:

$$F = \left(\frac{1}{x-y}\right)_{|x|>|y|} A(x)B(y)C(z) - \left(\frac{1}{x-y}\right)_{|y|>|x|} B(y)A(x)C(z),$$

$$G = \left(\frac{1}{x-y}\right)_{|x|>|y|} C(z)A(x)B(y) - \left(\frac{1}{x-y}\right)_{|y|>|x|} C(z)B(y)A(x).$$

As A, B and C are mutually local, we know that there is an integer N such that

$$\begin{aligned}(x-y)^N A(x)B(y) &= (x-y)^N B(y)A(x)\\ (y-z)^N B(y)C(z) &= (y-z)^N C(z)B(y)\\ (x-z)^N A(x)C(z) &= (x-z)^N C(z)A(x).\end{aligned}$$

We will now show that

$$(y-z)^{3N}F = (y-z)^{3N}G.$$

The binomial identity gives

$$(y-z)^{3N} = \sum_{n=0}^{2N}\binom{2N}{n}(y-x)^{2N-n}(x-z)^n(y-z)^N.$$

Now, if $0 \leqslant n < N$ the power of $(y-x)$ is large enough that we can swap $A(x)$ and $B(y)$; the two terms in F (and G) then cancel, so these do not contribute. For $n \geq N$ the powers of $(x-z)$ and $(y-z)$ are large enough that we can swap $A(x)$, $C(z)$ as well as $B(y)$, $C(z)$. This allows us to make terms in F the same as those in G. Hence

$$(y-z)^{3N}F = (y-z)^{3N}G,$$

and we are done. □

The proof of Theorem 2.2.2 works in more generality than simply the case of an affine Kac–Moody algebra. In this form it is called the (weak) **reconstruction theorem.**

Theorem 2.2.4 (Weak Reconstruction) *Let V be a vector space, $|0\rangle$ a vector of V, and T an endomorphism of V. Let*

$$a^\alpha(z) = \sum_{n\in\mathbb{Z}} a^\alpha_{(n)} z^{-n-1},$$

where α runs over an ordered set I, be a collection of fields on V such that

(i) $[T, a^\alpha(z)] = \partial_z a^\alpha(z)$;
(ii) $T\,|0\rangle = 0$, $a^\alpha(z)\,|0\rangle = a^\alpha + z(\ldots)$;
(iii) *$a^\alpha(z)$ and $a^\beta(z)$ are mutually local;*
(iv) *the lexicographically ordered monomials $a^{\alpha_1}_{(n_1)} \cdots a^{\alpha_m}_{(n_m)}\,|0\rangle$ with $n_i < 0$ form a basis of V.*

Then the formula

$$Y(a^{\alpha_1}_{(n_1)} \cdots a^{\alpha_m}_{(n_m)}\,|0\rangle\,, z) = \\ \frac{1}{(-n_1-1)!}\cdots\frac{1}{(-n_m-1)!}{:}\partial^{-n_1-1}a^{\alpha_1}(z)\cdots\partial^{-n_m-1}a^{\alpha_n}(z){:}\,, \quad (2.2.9)$$

where $n_i < 0$, defines a vertex algebra structure on V such that $|0\rangle$ is the vacuum vector, T is the translation operator and $Y(a^\alpha, z) = a^\alpha(z)$ for all $\alpha \in I$.

To prove this we simply repeat the proof used for the case of $V = V_\kappa(g)$.

The reconstruction theorem is actually true in a much less restrictive case: we do not have to assume that the vectors in (iv) form a basis, merely that they span V. And even in this case it is possible to derive that the resulting vertex algebra structure is unique.

Theorem 2.2.5 (Strong Reconstruction)

Let V be a vector space, equipped with the structures of Theorem 2.2.4 satisfying conditions (i)–(iii) and the condition

(iv') The vectors $a^{\alpha_1}_{(-j_1-1)} \cdots a^{\alpha_n}_{(-j_n-1)} |0\rangle$ with $j_s \geqslant 0$ span V. Then these structures together with formula (2.2.9) define a vertex algebra structure on V.

Moreover, this is the unique vertex algebra structure on V satisfying conditions (i)–(iii) and (iv') and such that $Y(a^\alpha, z) = a^\alpha(z)$ for all $\alpha \in I$.

We will not prove this result here, referring the reader to [FB], Theorem 4.4.1. In particular, we see that there there was no arbitrariness in our assignment of vertex operators in Section 2.2.5.

2.3 Associativity in vertex algebras

We have now made it about half way towards our goal of finding a proper formalism for the Segal–Sugawara operators S_n defined by formula (2.1.9). We have introduced the notion of a vertex algebra and have seen that vertex algebras neatly encode the structures related to the formal power series such as the generating series $S(z)$ of the Segal–Sugawara operators. In fact, the series $S(z)$ itself is one of the vertex operators in the vertex algebra $V_\kappa(\mathfrak{g})$:

$$S(z) = \frac{1}{2} \sum_{a=1} Y(J^a_{-1} J_{a,-1}|0\rangle, z).$$

What we need to do now is to learn how to relate properties of the vertex operators such as $S(z)$, emerging from the vertex algebra structure, and properties of their Fourier coefficients S_n, which are relevant to the description of the center of the completed enveloping algebra of $\widehat{\mathfrak{g}}_\kappa$ that we are interested in. In order to do that, we need to develop the theory of vertex algebras a bit further, and in particular understand the meaning of associativity in this context.

2.3.1 Three domains

We have already seen that locality axiom for a vertex algebra is telling us that the two formal power series

$$Y(A,z)Y(B,w)C \qquad \text{and} \qquad Y(B,w)Y(A,z)C$$

are expansions of the same element from $V[[z,w]][z^{-1},w^{-1},(z-w)^{-1}]$ in two different domains. One of these expansions, $V((z))((w))$, corresponds to w being "small;" the other, $V((w))((z))$, corresponds to z being "small." If we think of the points z and w as being complex numbers (in other words, points on the Riemann sphere), we can think of these two expansions as being done in the domains "w is very close to 0" and "w is very close to ∞." There is now an obvious third choice: "w is very close to z," which we have not yet discussed.

Algebraically, the space corresponding to the domain "w is very close to z" is $V((w))((z-w))$ (or alternatively, $V((z))((z-w))$; these are actually identical). The expression, in terms of vertex algebras, that we expect to live in this space is

$$Y(Y(A,z-w)B,w)C.$$

To see this, we look at the case of commutative vertex algebras. In a commutative vertex algebra, locality is equivalent to commutativity,

$$Y(A,z)Y(B,w)C = Y(B,w)Y(A,z)C,$$

but we also have the identity

$$Y(A,z)Y(B,w)C = Y(Y(A,z-w)B,w)C$$

which expresses the associativity property of the underlying commutative algebra. In a general vertex algebra, locality holds only in the sense of "analytic continuation" from different domains. Likewise, we should expect that the associativity property also holds in a similar sense. More precisely, we expect that $Y(Y(A,z-w)B,w)C$ is the expansion of the same element of $V[[z,w]][z^{-1},w^{-1},(z-w)^{-1}]$, as for the other two expressions, but now in the space $V((w))((z-w))$ (i.e., assuming $z-w$ to be "small"). We will now show that this is indeed the case.

2.3.2 Some useful facts

We start with a couple of basic but useful results about vertex algebras.

Lemma 2.3.1 *Suppose that U is a vector space, $f(z) \in U[[z]]$ is a power series, and $R \in \operatorname{End} U$ is an endomorphism. If $f(z)$ satisfies*

$$\partial_z f(z) = Rf(z),$$

then it is uniquely determined by the value of $f(0)$.

Proof. A simple induction shows that $f(z)$ must be of the form

$$f(z) = K + R(K)z + \frac{R^2}{2!}(K)z^2 + \frac{R^3}{3!}(K)z^3 + \ldots,$$

so it is determined by $K = f(0)$. □

Corollary 2.3.1 *For any vector A in a vertex algebra V we have*

$$Y(A,z)|0\rangle = e^{zT}A.$$

Proof. Both sides belong to $V[[z]]$. Therefore, in view of Lemma 2.3.1, it suffices to show that they satisfy the same differential equation, as their constant terms are both equal to A. It follows from the vertex algebra axioms that

$$\partial_z Y(A,z)|0\rangle = [T, Y(A,z)]|0\rangle = TY(A,z)|0\rangle.$$

On the other hand, we obviously have $\partial_z e^{zT}A = Te^{zT}A$. □

Lemma 2.3.2 *In any vertex algebra we have*

$$e^{wT}Y(A,z)e^{-wT} = Y(A, z+w),$$

where negative powers of $(z+w)$ are expanded as power series assuming that w is "small" (i.e., in positive powers of w/z).

Proof. In any Lie algebra we have the following identity:

$$e^{wT}Ge^{-wT} = \sum_{n\geqslant 0} \frac{w^n}{n!}(\mathrm{ad}\, T)^n G$$

So, in our case, using the formula $[T, Y(A,z)] = \partial_z Y(A,z)$ and the fact that $[T, \partial_z] = 0$, we find that

$$\begin{aligned} e^{wT}Y(A,z)e^{-wT} &= \sum_{n\geqslant 0} \frac{w^n}{n!}(\mathrm{ad}\, T)^n Y(A,z) \\ &= \sum_{n\geqslant 0} \frac{w^n}{n!}\partial_z^n Y(A,z) \\ &= e^{w\partial_z}Y(A,z). \end{aligned}$$

To complete the proof, we use the identity $e^{w\partial_z}f(z) = f(z+w)$ in $R[[z^{\pm 1}, w]]$, which holds for any $f(z) \in R[[z^{\pm 1}]]$ and any $\mathbb{C}$-algebra R. □

This lemma tells us that exponentiating the infinitesimal translation operator T really does give us a translation operator $z \mapsto z + w$.

Proposition 2.3.2 (Skew Symmetry) *In any vertex algebra we have the identity*

$$Y(A,z)B = e^{zT}Y(B,-z)A$$

in $V((z))$.

Proof. By locality, we know that there is a large integer N such that

$$(z-w)^N Y(A,z)Y(B,w)\,|0\rangle = (z-w)^N Y(B,w)Y(A,z)\,|0\rangle\,.$$

This is actually an equality in $V[[z,w]]$ (note that there are no negative powers of w on the left and no negative powers of z on the right). Now, by the above results we compute

$$\begin{array}{rrcl} & (z-w)^N Y(A,z)Y(B,w)\,|0\rangle & = & (z-w)^N Y(B,w)Y(A,z)\,|0\rangle \\ \Rightarrow & (z-w)^N Y(A,z)e^{wT}B & = & (z-w)^N Y(B,w)e^{zT}A \\ \Rightarrow & (z-w)^N Y(A,z)e^{wT}B & = & (z-w)^N e^{zT}Y(B,w-z)A \\ \Rightarrow & (z)^N Y(A,z)B & = & (z)^N e^{zT}Y(B,-z)A \\ \Rightarrow & Y(A,z)B & = & e^{zT}Y(B,-z)A \end{array}$$

In the fourth line we have set $w=0$, which is allowed as there are no negative powers of w in the above expressions. □

In terms of the Fourier coefficients, we can write the skew symmetry property as

$$A_{(n)}B = (-1)^{n+1}\left(B_{(n)}A - T(B_{(n-1)}A) + \frac{1}{2}T^2(B_{(n-2)}A) - \dots\right).$$

2.3.3 *Proof of associativity*

We now have enough to prove the associativity property.

Theorem 2.3.3 *In any vertex algebra the expressions*

$$Y(A,z)Y(B,w)C, \qquad Y(B,w)Y(A,z)C, \qquad \text{and} \qquad Y(Y(A,z-w)B,w)C$$

are the expansions, in

$$V((z))((w)), \qquad V((w))((z)), \qquad \text{and} \qquad V((w))((z-w)),$$

respectively, of one and the same element of $V[[z,w]][z^{-1},w^{-1},(z-w)^{-1}]$.

Proof. We already know this from locality for the first two expressions, so we only need to show it for the first and the last one. Using Proposition 2.3.2 and Lemma 2.3.2, we compute

$$\begin{aligned} Y(A,z)Y(B,w)C &= Y(A,z)e^{wT}Y(C,-w)B \\ &= e^{wT}Y(A,z-w)Y(C,-w)B \end{aligned}$$

(note that it is okay to multiply on the left by the power series e^{wT} as it has no negative powers of w). In this final expression we expand negative powers of $(z-w)$ assuming that w is "small." This defines a map $V((z-w))((w)) \to V((z))((w))$, which is easily seen to be an isomorphism that intertwines the embeddings of $V[[z,w]][z^{-1},w^{-1},(z-w)^{-1}]$ into the two spaces.

On the other hand, we compute, again using Proposition 2.3.2,

$$\begin{aligned} Y(Y(A,z-w)B,w)C &= Y\left(\sum_n A_{(n)}B(z-w)^{-n-1},w\right)C \\ &= \sum_n Y(A_{(n)}B,w)C(z-w)^{-n-1} \\ &= \sum_n e^{wT}Y(C,-w)A_{(n)}B(z-w)^{-n-1} \\ &= e^{wT}Y(C,-w)Y(A,z-w)B. \end{aligned}$$

This calculation holds in $V((w))((z-w))$.

By locality we know that

$$Y(C,-w)Y(A,z-w)B \qquad \text{and} \qquad Y(A,z-w)Y(C,-w)B$$

are expansions of the same element of $V[[z,w]][z^{-1},w^{-1},(z-w)^{-1}]$. This implies that $Y(Y(A,z-w)B,w)C$ and $Y(A,z)Y(B,w)C$ are also expansions of the same element. □

We have now established the associativity property of vertex algebras. It is instructive to think of this property as saying that

$$Y(A,z)Y(B,w) = Y(Y(A,z-w)B,w) = \sum_{n\in\mathbb{Z}} Y(A_{(n)}B,w)(z-w)^{-n-1}. \quad (2.3.1)$$

This is very useful as it gives a way to represent the product of two vertex operators as a linear combination of vertex operators. However, we have to be careful in interpreting this formula. The two sides are formal power series in two different vector spaces, $V((z))$ and $V((w))((z-w))$. They do "converge" when we apply the terms to a fixed vector $C \in V$, but even then the expressions are not equal as they are expansions of a common "function" in two different domains.

Written in the above form, and with the above understanding, the equality (2.3.1) is known as the **operator product expansion** (or **OPE**). Formulas of this type originally turned up in the physics literature on conformal field theory. One of the motivations for developing the theory of vertex algebras was to find a mathematically rigorous interpretation of these formulas.

2.3.4 Corollaries of associativity

We now look at consequences of the associativity law. In particular, we will see that our previous definitions of normally ordered product and the formula for the vertex operators in the case of $V_\kappa(\mathfrak{g})$ are basically unique.

Lemma 2.3.4 *Suppose that $\phi(z)$ and $\psi(w)$ are two fields. Then the following are equivalent:*

(i) $$[\phi(z), \psi(w)] = \sum_{i=0}^{N-1} \frac{1}{i!}\gamma_i(w)\partial_w^i \delta(z-w);$$

(ii) $$\phi(z)\psi(w) = \sum_{i=0}^{N-1} \gamma_i(w) \left(\frac{1}{(z-w)^{i+1}}\right)_{|z|>|w|} + {:}\phi(z)\psi(w){:}$$

and $$\psi(w)\phi(z) = \sum_{i=0}^{N-1} \gamma_i(w) \left(\frac{1}{(z-w)^{i+1}}\right)_{|w|>|z|} + {:}\phi(z)\psi(w){:} \quad ,$$

where the $\gamma_i(w)$ are fields, N is a non-negative integer and

$$ {:}\phi(z)\psi(w){:} = \phi_+(z)\psi(w) + \psi(w)\phi_-(z).$$

Proof. Assuming (ii) we see the commutator $[\phi(z), \psi(w)]$ is the difference of expansions of $(z-w)^{-i-1}$ in the domains $|z| > |w|$ and $|w| > |z|$. It is clear that

$$\left(\frac{1}{z-w}\right)_{|z|>|w|} - \left(\frac{1}{z-w}\right)_{|w|>|z|} = \delta(z-w).$$

Differentiating i times with respect to w, we obtain

$$\left(\frac{1}{(z-w)^{i+1}}\right)_{|z|>|w|} - \left(\frac{1}{(z-w)^{i+1}}\right)_{|w|>|z|} = \frac{1}{i!}\partial_w^i \delta(z-w).$$

This gives us (i).

Conversely, assume (i) and write

$$\begin{aligned}
\phi(z)\psi(w) &= (\phi(z)_+ + \phi(z)_-)\,\psi(w) \\
&= (\phi(z)_+\psi(w) + \psi(w)\phi(z)_-) + (\phi(z)_-\psi(w) - \psi(w)\phi(z)_-) \\
&= {:}\phi(z)\psi(w){:} + [\phi(z)_-, \psi(w)]\,.
\end{aligned}$$

Taking the terms with powers negative in z in the right hand side of (i), we obtain the first formula in (ii). The same calculation works for the product $\psi(w)\phi(z)$. □

Now suppose that the fields $\phi(z)$ and $\psi(w)$ are vertex operators $Y(A, z)$ and $Y(B, w)$ in a vertex algebra V. Then locality tells us that the commutator $[Y(A, z), Y(B, w)]$ is annihilated by $(z-w)^N$ for some N. It is easy to see that the kernel of the operator of multiplication by $(z-w)^N$ in $\operatorname{End} V[[z^{\pm 1}, w^{\pm 1}]]$ is linearly generated by the series of the form $\gamma(w)\partial_w^i \delta(z-w), i = 0, \ldots, N-1$. Thus, we can write

$$[Y(A,z), Y(B,w)] = \sum_{i=0}^{N-1} \frac{1}{i!} \gamma_i(w) \partial_w^i \delta(z-w),$$

where $\gamma_i(w), i = 0, \ldots, N-1$, are some fields on V (we do not know yet that they are also vertex operators). Therefore we obtain that $Y(A,z)$ and $Y(B,w)$ satisfy condition (i) of Lemma 2.3.4. Hence they also satisfy condition (ii):

$$Y(A,z)Y(B,w) = \sum_{i=0}^{N-1} \frac{\gamma_i(w)}{(z-w)^{i+1}} + :Y(A,z)Y(B,w): , \tag{2.3.2}$$

where by $(z-w)^{-1}$ we understand its expansion in positive powers of w/z.

We obtain that for any $C \in V$ the series $Y(A,z)Y(B,w)C \in V(\!(z)\!)(\!(w)\!)$ is an expansion of

$$\left(\sum_{i=0}^{N-1} \frac{\gamma_i(w)}{(z-w)^{i+1}} + :Y(A,z)Y(B,w): \right) C \in V[[z,w]][z^{-1}, w^{-1}, (z-w)^{-1}].$$

Using the Taylor formula, we obtain that the expansion of this element in $V(\!(w)\!)(\!(z-w)\!)$ is equal to

$$\left(\sum_{i=0}^{N-1} \frac{\gamma_i(w)}{(z-w)^{i+1}} + \sum_{m \geq 0} \frac{(z-w)^m}{m!} :\partial_w^m Y(A,w) \cdot Y(B,w): \right) C. \tag{2.3.3}$$

The coefficient in front of $(z-w)^k, k \in \mathbb{Z}$, in the right hand side of (2.3.3) must be equal to the corresponding term in the right hand side of formula (2.3.1). Let us first look at the terms with $k \geq 0$. Comparison of the two formulas gives

$$Y(A_{(n)}B, z) = \frac{1}{(-n-1)!} :\partial_z^{-n-1} Y(A,z) \cdot Y(B,z): , \qquad n < 0. \tag{2.3.4}$$

In particular, setting $B = |0\rangle$, $n = -2$ and recalling that $A_{(-2)}|0\rangle = TA$, we find that

$$Y(TA, z) = \partial_z Y(A,z). \tag{2.3.5}$$

Using formula (2.3.4), we obtain the following corollary by induction on m (recall that our convention is that the normal ordering is nested from right to left):

Corollary 2.3.3 *For any $A^1, \ldots, A^m \in V$, and $n_1, \ldots, n_m < 0$, we have*

$$Y(A^1_{(n_1)} \ldots A^k_{(n_m)}|0\rangle, z)$$
$$= \frac{1}{(-n_1-1)!} \cdots \frac{1}{(-n_m-1)!} :\partial_z^{-n_1-1} Y(A^1,z) \cdot \ldots \cdot \partial_z^{-n_m-1} Y(A^m,z): .$$

This justifies formulas (2.2.8) and (2.2.9).

Now we compare the coefficients in front of $(z-w)^k, k<0$, in formulas (2.3.1) and (2.3.3). We find that

$$\gamma_i(w) = Y(A_{(i)}B, w), \qquad i \geq 0,$$

and so formula (2.3.2) can be rewritten as

$$Y(A,z)Y(B,w) = \sum_{n\geq 0} \frac{Y(A_{(n)}B,w)}{(z-w)^{n+1}} + :Y(A,z)Y(B,w):. \tag{2.3.6}$$

Note that unlike formula (2.3.1), this identity makes sense in End $V[[z^{-1},w^{-1}]]$ if we expand $(z-w)^{-1}$ in positive powers of w/z.

Now Lemma 2.3.4 implies the following commutation relations:

$$[Y(A,z),Y(B,w)] = \sum_{n\geq 0} \frac{1}{n!} Y(A_{(n)}B,w)\partial_w^n \delta(z-w). \tag{2.3.7}$$

Expanding both sides of (2.3.7) as formal power series and using the equality

$$\frac{1}{n!}\partial_w^n \delta(z-w) = \sum_{m\in\mathbb{Z}} \binom{m}{n} z^{-m-1} w^{m-n},$$

we obtain the following identity for the commutators of Fourier coefficients of arbitrary vertex operators:

$$[A_{(m)}, B_{(k)}] = \sum_{n\geq 0} \binom{m}{n} (A_{(n)}B)_{(m+k-n)}. \tag{2.3.8}$$

Here, by definition, for any $m \in \mathbb{Z}$,

$$\binom{m}{n} = \frac{m(m-1)\dots(m-n+1)}{n!}, \quad n \in \mathbb{Z}_{>0}; \qquad \binom{m}{0} = 1.$$

Several important remarks should be made about this formula. It shows that the collection of all Fourier coefficients of all vertex operators form a Lie algebra; and the commutators of Fourier coefficients depend only on the singular terms in the OPE.

We will usually simplify the notation in the above lemma, writing

$$\phi(z)\psi(w) = \sum_{i=0}^{N-1} \gamma_i(w) \left(\frac{1}{(z-w)^{i+1}} \right)_{|z|>|w|} + :\phi(z)\psi(w):$$

as

$$\phi(z)\psi(w) \sim \sum_{i=0}^{N-1} \frac{\gamma_i(w)}{(z-w)^{i+1}}.$$

In the physics literature this is what is usually written down for an OPE. What we have basically done is to remove all the terms which are non-singular at

$z = w$. We can do this because the structure of all the commutation relations depends *only* on the singular terms.

3

Constructing central elements

In the previous chapter we started investigating the center of the completed enveloping algebra $\widetilde{U}_\kappa(\widehat{\mathfrak{g}})$ of $\widehat{\mathfrak{g}}_\kappa$. After writing some explicit formulas we realized that we needed to develop some techniques in order to perform the calculations. This led us to the concept of vertex algebras and vertex operators. We will now use this formalism in order to describe the center.

In Section 3.1 we will use the general commutation relations (2.3.8) between the Fourier coefficients of vertex operators in order to prove formula (2.1.11) for the commutation relations between the Segal–Sugawara operators and the generators of $\widehat{\mathfrak{g}}_\kappa$. However, we quickly realize that this proof gives us the commutation relations between these operators not in $\widetilde{U}_\kappa(\widehat{\mathfrak{g}})$, as we wanted, but in the algebra of endomorphisms of $V_\kappa(\mathfrak{g})$. In order to prove that the same formula holds in $\widetilde{U}_\kappa(\widehat{\mathfrak{g}})$, we need to work a little harder. To this end we associate in Section 3.2 to an arbitrary vertex algebra V a Lie algebra $U(V)$. We then discuss the relationship between $U(V_\kappa(\mathfrak{g}))$ and $\widetilde{U}_\kappa(\widehat{\mathfrak{g}})$. The conclusion is that formula (2.1.11) does hold in the completed enveloping algebra $\widetilde{U}_\kappa(\widehat{\mathfrak{g}})$. In particular, we obtain that the Segal–Sugawara operators are central elements of $\widetilde{U}_\kappa(\widehat{\mathfrak{g}})$ when $\kappa = \kappa_c$, the critical level.

Next, in Section 3.3 we introduce the notion of center of a vertex algebra. We show that the center of the vertex algebra $V_\kappa(\mathfrak{g})$ is a commutative algebra that is naturally realized as a quotient of the center of $\widetilde{U}_\kappa(\widehat{\mathfrak{g}})$. We show that the center of $V_\kappa(\mathfrak{g})$ is trivial if $\kappa \neq \kappa_c$. Our goal is therefore to find the center $\mathfrak{z}(\widehat{\mathfrak{g}})$ of $V_{\kappa_c}(\mathfrak{g})$. We will see below that this will enable us to describe the center of $\widetilde{U}_\kappa(\widehat{\mathfrak{g}})$ as well. As the first step, we discuss the associated graded analogue of $\mathfrak{z}(\widehat{\mathfrak{g}})$. In order to describe it we need to make a detour to the theory of jet schemes in Section 3.4.

The results of Section 3.4 are already sufficient to find $\mathfrak{z}(\widehat{\mathfrak{g}})$ for $\mathfrak{g} = \mathfrak{sl}_2$: we show in Section 3.5 that $\mathfrak{z}(\widehat{\mathfrak{sl}}_2)$ is generated by the Segal–Sugawara operators. However, we are not satisfied with this result as it does not give us a coordinate-independent description of the center. Therefore we need to find how the group of changes of coordinates on the disc acts on $\mathfrak{z}(\widehat{\mathfrak{sl}}_2)$. As the result of this computation, performed in Section 3.5, we find that $\mathfrak{z}(\widehat{\mathfrak{sl}}_2)$ is canonically isomorphic to the algebra of functions on the space of projective connections on the disc. This is the prototype for the description of the center $\mathfrak{z}(\widehat{\mathfrak{g}})$ obtained in the next chapter, which is one of the central results of this book.

3.1 Segal–Sugawara operators

Armed with the technique developed in the previous chapter, we are now ready to prove formula (2.1.11) for the commutation relations between the Segal–Sugawara operators S_n and the generators J^a_n of $\widehat{\mathfrak{g}}_\kappa$.

3.1.1 Commutation relations with $\widehat{\mathfrak{g}}_\kappa$

We have already found that the commutation relations (2.2.7) between the J^a_n's may be recorded nicely as the commutation relations for the fields $J^a(z)$:

$$\left[J^a(z), J^b(w)\right] = \left[J^a, J^b\right](w)\delta(z-w) + \kappa(J^a, J^b)\partial_w\delta(z-w).$$

Using the formalism of the previous section, we rewrite this in the equivalent form of the OPE

$$J^a(z)J^b(w) \sim \frac{\left[J^a, J^b\right](w)}{(z-w)} + \frac{\kappa(J^a, J^b)}{(z-w)^2}.$$

Now recall that

$$S(z) = \frac{1}{2}\sum_a Y(J^a_{-1}J_{a,-1}|0\rangle, z) = \frac{1}{2}\sum_a {:}J^a(z)J_a(z){:}$$

where $\{J_a\}$ is the dual basis to $\{J^a\}$ under some fixed inner product κ_0. We know from our previous discussions about normally ordered products that this is a well-defined field, so we would now like to work out the commutation relations of the Fourier coefficients of $S(z)$ and the $J^a_{(n)}$'s, which are the Fourier coefficients of $J^a(z)$.

Let

$$S = \frac{1}{2}\sum_a J^a_{(-1)}J_{a,(-1)}\left|0\right\rangle, \tag{3.1.1}$$

so that S is the element of $V_\kappa(\mathfrak{g})$ generating the field $S(z)$:

$$S(z) = Y(S, z) = \sum_{n\in\mathbb{Z}} S_{(n)}z^{-n-1}.$$

Note that since we defined the operators S_n by the formula

$$S(z) = \sum_{n\in\mathbb{Z}} S_n z^{-n-2},$$

we have

$$S_n = S_{(n+1)}.$$

We need to compute the OPE

$$Y(J^b, z)Y(S, w) \sim \sum_{n \geqslant 0} \frac{Y(J^b_{(n)} S, w)}{(z-w)^{n+1}}.$$

This means we need to compute the elements

$$\frac{1}{2} J^b_{(n)} \sum_a J^a_{(-1)} J_{a,(-1)} |0\rangle \in V_\kappa(\mathfrak{g}), \qquad n \geq 0.$$

As we will see, this is much easier than trying to compute the commutators of the infinite sum representing S_n with J^b_m.

The vertex algebra $V_\kappa(\mathfrak{g})$ is $\mathbb{Z}_+$-graded, and the degree of S is equal to 2. Therefore we must have $n \leqslant 2$, for otherwise the resulting element is of negative degree, and there are no such non-zero elements. Thus, we have three cases to consider:

n=2. In this case, repeatedly using the commutation relations and the invariance of the inner product κ quickly shows that the result is

$$\frac{1}{2} \sum_a \kappa\left(J^b, [J^a, J_a]\right) |0\rangle .$$

If we now choose $\{J^a\}$ to be an orthonormal basis, which we may do without loss of generality, so that $J^a = J_a$, we see that this is 0.

n=1. We obtain

$$\sum_a \kappa(J^b, J_a) J^a_{-1} |0\rangle + \frac{1}{2} \sum_a [[J^b, J^a], J_a]_{-1} |0\rangle .$$

The first of these terms looks like the decomposition of J^b_{-1} using the basis J^a_{-1}. However, we have to be careful that κ is not necessarily the same inner product as κ_0; they are multiples of each other. So, the first term is just $\frac{\kappa}{\kappa_0} J^b_{-1} |0\rangle$.

The second term is the action of the Casimir element of $U(\mathfrak{g})$ on the adjoint representation of $\mathfrak{g}$, realized as a subspace $\mathfrak{g} \otimes t^{-1}|0\rangle \subset V_\kappa(\mathfrak{g})$. This is a central element and so it acts as a scalar, λ, say. Any finite-dimensional representation V of $\mathfrak{g}$ gives rise to a natural inner product by

$$\kappa_V(x, y) = \operatorname{Tr} \rho_V(x) \rho_V(y).$$

In particular $\kappa_{\mathfrak{g}}$ is the inner product from the adjoint representation known as the **Killing form** which we have seen before in Section 1.3.5. If we pick $\kappa_0 = \kappa_{\mathfrak{g}}$, it is obvious that the scalar λ should be $1/2$ (due to the overall factor $1/2$ in the formula). So, for a general κ_0 the second term is $\dfrac{1}{2} \dfrac{\kappa_{\mathfrak{g}}}{\kappa_0} J^b_{-1} |0\rangle$.

n=0. We get

$$\frac{1}{2}\sum_a \left(J^a_{-1}[J^b, J_a]_{-1} + [J^b, J^a]_{-1}J_{a,-1}\right).$$

This corresponds to taking the commutator with the Casimir element and hence is 0 (as the Casimir is central).

Hence the OPE is

$$J^b(z)S(w) \sim \frac{\kappa + \frac{1}{2}\kappa_{\mathfrak{g}}}{\kappa_0} \frac{J^b(w)}{(z-w)^2}.$$

So, we find that there is a **critical value** for the inner product κ, for which these two fields, $S(z)$ and $J^b(z)$ (and hence their Fourier coefficients), commute with each other. The critical value is $\kappa_c = -\frac{1}{2}\kappa_{\mathfrak{g}}$.

Using formula (2.3.8) and the first form of the OPE, we immediately obtain the commutation relations

$$[J^a_n, S_{(m)}] = \frac{\kappa - \kappa_c}{\kappa_0} n J^a_{n+m-1}.$$

Recall that we define S_n to be $S_{(n+1)}$. In other words, we shift the grading by 1, so that the degree of S_n becomes $-n$ (but note that we have $J^a_n = J^a_{(n)}$). Then we obtain the following formula, which coincides with (2.1.11):

$$[S_n, J^a_m] = -\frac{\kappa - \kappa_c}{\kappa_0} n J^a_{n+m}.$$

Thus, we find that when $\kappa = \kappa_c$ the operators S_n commute with the action of the Lie algebra $\widehat{\mathfrak{g}}$, which is what we wanted to show.

We remark that the ratio $\frac{1}{2}\frac{\kappa_{\mathfrak{g}}}{\kappa_0}$ is related to the **dual Coxeter number** of $\mathfrak{g}$. More precisely, if κ_0 is chosen so that the maximal root has squared length 2, which is the standard normalization [K2], then this ratio is equal to $h^\vee$, the dual Coxeter number of $\mathfrak{g}$ (for example, it is equal to n for $\mathfrak{g} = \mathfrak{sl}_n$). For this choice of κ_0 the above formula becomes

$$[S_n, J^a_m] = -(k + h^\vee) n J^a_{n+m}.$$

Even away from the critical value the commutation relations are quite nice. To simplify our formulas, for $\kappa \neq \kappa_c$ we rescale S and $S(w)$ by setting

$$\widetilde{S} = \frac{\kappa_0}{\kappa + \frac{1}{2}\kappa_{\mathfrak{g}}} S, \qquad \widetilde{S}(z) = \frac{\kappa_0}{\kappa + \frac{1}{2}\kappa_{\mathfrak{g}}} S(z).$$

Then we have the OPE

$$J^a(z)\widetilde{S}(w) \sim \frac{J^a(w)}{(z-w)^2}.$$

Note that often this is written in the opposite order, which can be achieved by using locality, swapping z and w and using Taylor's formula:

$$\widetilde{S}(z)J^a(w) \sim \frac{J^a(w)}{(z-w)^2} + \frac{\partial_w J^a(w)}{(z-w)}. \tag{3.1.2}$$

Using formula (2.3.8) and the first form of the OPE, we find that

$$[J_n^a, \widetilde{S}_{(m)}] = nJ_{n+m-1}^a.$$

Let us set $\widetilde{S}_n = \widetilde{S}_{(n+1)}$. Then the commutation relations may be rewritten as follows:

$$[\widetilde{S}_n, J_m^a] = -mJ_{n+m}^a.$$

These commutation relations are very suggestive. Recall that J_n^a stands for the element $J^a \otimes t^n$ in $\widehat{\mathfrak{g}}_\kappa$. Thus, the operations given by $\operatorname{ad}(\widetilde{S}_n)$ look very much like

$$\operatorname{ad}(\widetilde{S}_n) = -t^{n+1}\partial_t.$$

One can now ask if the operators $\operatorname{ad}(\widetilde{S}_n)$ actually generate the Lie algebra of vector fields of this form. It will turn out that this is almost true ("almost" because we will actually obtain a central extension of this Lie algebra).

To see that, we now compute the OPE for $\widetilde{S}(z)\widetilde{S}(w)$.

3.1.2 Relations between Segal–Sugawara operators

To perform the calculation we need to compute the element of $V_\kappa(\mathfrak{g})$ given by the formula

$$\widetilde{S}_{(n)}\frac{1}{2}\frac{\kappa_0}{\kappa - \kappa_c}\sum_a J_{(-1)}^a J_{a,(-1)} |0\rangle$$

for $n \geq 0$. From the grading considerations we find that we must have $n \leqslant 3$ (otherwise the resulting element is of negative degree). So there are four cases to consider:

n=3 In this case, repeatedly using the above commutation relations quickly shows that the result is

$$\frac{1}{2}\frac{\kappa_0}{\kappa - \kappa_c}\sum_a \kappa\left(J^a, J_a\right)|0\rangle.$$

Hence we obtain $\frac{1}{2}\frac{\kappa}{\kappa-\kappa_c}\dim\mathfrak{g}\,|0\rangle$.

n=2 We obtain

$$\frac{1}{2}\frac{\kappa_0}{\kappa - \kappa_c}\sum_a [J^a, J_a]_{(-1)}|0\rangle,$$

which is 0, as we can see by picking $J^a = J_a$.

n=1 We obtain

$$\frac{1}{2}\frac{\kappa_0}{\kappa - \kappa_c}\sum_a 2J_{(-1)}^a J_{a,(-1)}|0\rangle,$$

which is just $2\widetilde{S}$.

n=0 We obtain

$$\frac{1}{2}\frac{\kappa_0}{\kappa-\kappa_c}\sum_a \left(J^a_{(-1)}J_{a,(-2)} + J^a_{(-2)}J_{a,(-1)}\right)|0\rangle\,,$$

which is just $T(\widetilde{S})$.

Thus, we see that the OPE is

$$\widetilde{S}(z)\widetilde{S}(w) \sim \frac{\frac{\kappa}{\kappa-\kappa_c}\dim\mathfrak{g}/2}{(z-w)^4} + \frac{2\widetilde{S}(w)}{(z-w)^2} + \frac{\partial_w\widetilde{S}(w)}{(z-w)}.$$

We denote the constant occurring in the first term as $c_\kappa/2$ (later we will see that c_κ is the so-called central charge). We then use this OPE and formula (2.3.8) to compute the commutation relations among the coefficients $\widetilde{S}_n = \widetilde{S}_{(n+1)}$:

$$[\widetilde{S}_n, \widetilde{S}_m] = (n-m)\widetilde{S}_{n+m} + \frac{n^3-n}{12}c_\kappa\delta_{n,-m}. \tag{3.1.3}$$

These are the defining relations of the Virasoro algebra, which we discuss in the next section.

3.1.3 The Virasoro algebra

Let $\mathcal{K} = \mathbb{C}((t))$ and consider the Lie algebra

$$\operatorname{Der}\mathcal{K} = \mathbb{C}((t))\partial_t$$

of (continuous) derivations of $\mathcal{K}$ with the usual Lie bracket

$$[f(t)\partial_t, g(t)\partial_t] = (f(t)g'(t) - f'(t)g(t))\partial_t.$$

The **Virasoro algebra** is the central extension of $\operatorname{Der}\mathcal{K}$. The central element is denoted by C and so we have the short exact sequence

$$0 \longrightarrow \mathbb{C}C \longrightarrow Vir \longrightarrow \operatorname{Der}\mathcal{K} \longrightarrow 0.$$

As a vector space, $Vir = \operatorname{Der}\mathcal{K} \oplus \mathbb{C}C$, so Vir has a topological basis given by

$$L_n = -t^{n+1}\partial_t, \qquad n \in \mathbb{Z},$$

and C (we include the minus sign in order to follow standard convention). The Lie bracket in the Virasoro algebra is given by the formula

$$[L_n, L_m] = (n-m)L_{n+m} + \frac{n^3-n}{12}\delta_{n+m}C. \tag{3.1.4}$$

In fact, it is known that Vir is the universal central extension of $\operatorname{Der}\mathcal{K}$, just like $\widehat{\mathfrak{g}}_\kappa$ is the universal central extension of $\mathfrak{g}((t))$.

The relations (3.1.4) become exactly the relations between the operators $\widetilde{S}_n$ found in formula (3.1.3) if we identify C with c_κ. This means that the $\widetilde{S}_n$'s generate an action of the Virasoro algebra on $V_\kappa(\mathfrak{g})$ for $\kappa \neq \kappa_c$.

We now construct a vertex algebra associated to the Virasoro algebra in a similar way to the affine Kac–Moody algebra case.

Let $\mathcal{O} = \mathbb{C}[[t]]$ and consider the Lie algebra $\operatorname{Der}\mathcal{O} = \mathbb{C}[[t]]\partial_t \subset \operatorname{Der}\mathcal{K}$. It is topologically generated by $\{L_{-1}, L_0, L_1, \dots\}$. The Lie algebra $\operatorname{Der}\mathcal{O} \oplus \mathbb{C}C$ has a one-dimensional representation $\mathbb{C}_c$ where $\operatorname{Der}\mathcal{O}$ acts by 0 and the central element C acts as the scalar $c \in \mathbb{C}$. Inducing this representation to the Virasoro algebra gives us a Vir-module, which we call the vacuum module of central charge c,

$$\operatorname{Vir}_c = \operatorname{Ind}_{\operatorname{Der}\mathcal{O}\oplus\mathbb{C}C}^{Vir} \mathbb{C}_c = U(Vir) \underset{U(\operatorname{Der}\mathcal{O}\oplus\mathbb{C}C)}{\otimes} \mathbb{C}_c.$$

This will be the vector space underlying our vertex algebra. By the Poincaré–Birkhoff–Witt theorem, a basis for this module is given by the monomials

$$L_{n_1} L_{n_2} \cdots L_{n_m} |0\rangle \, ,$$

where $n_1 \leqslant n_2 \leqslant \cdots \leqslant n_m < -1$ and $|0\rangle$ denotes a generating vector for the representation $\mathbb{C}_c$. This will be the vacuum vector of our vertex algebra.

We define a $\mathbb{Z}$-grading on Vir by setting $\deg L_n = -n$. Setting $\deg|0\rangle = 0$, we obtain a $\mathbb{Z}_+$-grading on Vir_c. Note that the subspace of degree 1 is zero, and the subspace of degree 2 is one-dimensional, spanned by $L_{-2}|0\rangle$.

Recall that the operator L_{-1} realizes $-\partial_t$, and so we choose it to be the translation operator T. We then need to define the state-field correspondence. By analogy with the vertex algebra $V_\kappa(\mathfrak{g})$, we define

$$Y(L_{-2}|0\rangle, z) = T(z) \overset{\text{def}}{=} \sum_{n\in\mathbb{Z}} L_n z^{-n-2}.$$

Note that the power of z in front of L_n has to be $-n-2$ rather than $-n-1$, because we want Vir_c to be a $\mathbb{Z}_+$-graded vertex algebra (in the conventions of Section 2.2.2). The commutation relations (3.1.4) imply the following formula:

$$[T(z), T(w)] = \frac{C}{12}\partial_w^3 \delta(z-w) + 2T(w)\partial_w \delta_w(z-w) + \partial_w T(w) \cdot \delta(z-w).$$

Therefore

$$(z-w)^4 [T(z), T(w)] = 0.$$

Thus, we obtain that the vertex operator $Y(L_{-2}|0\rangle, z)$ is local with respect to itself. Hence we can apply the reconstruction theorem to give Vir_c a vertex algebra structure.

Theorem 3.1.1 *The above structures endow* Vir_c *with a vertex algebra structure. The defining OPE is*

$$T(z)T(w) \sim \frac{c/2}{(z-w)^4} + \frac{T(w)}{(z-w)^2} + \frac{\partial_w T(w)}{(z-w)}.$$

The number c is known as the **central charge**.

3.1.4 Conformal vertex algebras

The Virasoro algebra plays a prominent role in the theory of vertex algebras and in conformal field theory. Its Lie subalgebra $\operatorname{Der} \mathcal{O} = \mathbb{C}[[t]]\partial_t$ may be viewed as the Lie algebra of infinitesimal changes of variables on the disc $D = \operatorname{Spec} \mathbb{C}[[t]]$, which is an important symmetry that we will extensively use below. Moreover, the full Virasoro algebra "uniformizes" the moduli spaces of pointed curves (see Chapter 17 of [FB]). Often this symmetry is realized as an "internal symmetry" of a vertex algebra V, meaning that there is an action of the Virasoro algebra on V that is generated by Fourier coefficients of a vertex operator $Y(\omega, z)$ for some $\omega \in V$. This prompts the following definition.

A $\mathbb{Z}$-graded vertex algebra V is called **conformal**, of **central charge** c, if we are given a non-zero **conformal vector** $\omega \in V_2$ such that the Fourier coefficients L_n^V of the corresponding vertex operator

$$Y(\omega, z) = \sum_{n \in \mathbb{Z}} L_n^V z^{-n-2} \tag{3.1.5}$$

satisfy the defining relations of the Virasoro algebra with central charge c, and in addition we have $L_{-1}^V = T$, $L_0^V|_{V_n} = n \, \mathrm{Id}$. Note that we have $L_n^V = \omega_{(n+1)}$.

An obvious example of a conformal vertex algebra is Vir_c itself, with $\omega = L_{-2}|0\rangle$. Another example is $V_\kappa(\mathfrak{g})$, with $\kappa \neq \kappa_c$. In this case, as we have seen above, $\omega = \widetilde{S}$ gives it the structure of a conformal vertex algebra with central charge c_κ.

It is useful to note that a conformal vertex algebra V is automatically equipped with a homomorphism $\phi : \mathrm{Vir}_c \to V$, which is uniquely determined by the property that $\phi(L_{-2}|0\rangle) = \omega$.

Here by a **vertex algebra homomorphism** we understand a linear map $\phi : V \to V'$, where V and V' are vertex algebras (homogeneous of degree 0 if V and V' are $\mathbb{Z}$-graded), such that

(i) ϕ sends $|0\rangle$ to $|0\rangle'$;
(ii) ϕ intertwines T and T', i.e., $\phi \circ T = T' \circ \phi$;
(iii) ϕ intertwines Y and Y', i.e.,

$$\phi \circ Y(A, z)B = Y'(\phi(A), z)\phi(B).$$

Lemma 3.1.2 *A $\mathbb{Z}_+$-graded vertex algebra V is conformal, of central charge c, if and only if it contains a non-zero vector $\omega \in V_2$ such that the Fourier coefficients L_n^V of the corresponding vertex operator*

$$Y(\omega, z) = \sum_{n \in \mathbb{Z}} L_n^V z^{-n-2}$$

satisfy the following conditions: $L_{-1}^V = T$, L_0^V is the grading operator and $L_2^V \cdot \omega = \frac{c}{2}|0\rangle$.

Moreover, in that case there is a unique vertex algebra homomorphism $\mathrm{Vir}_c \to V$ *such that* $|0\rangle \mapsto |0\rangle$ *and* $L_{-2}v_c \mapsto \omega$.

The proof is left to the reader (see [FB], Lemma 3.4.5, where one needs to correct the statement by replacing "$\mathbb{Z}$-graded" by "$\mathbb{Z}_+$-graded").

3.1.5 Digression: Why do central extensions keep appearing?

It seems like an interesting coincidence that the Lie algebras associated to the punctured disc $D^\times = \operatorname{Spec} \mathbb{C}((t))$, such as the formal loop algebra $\mathfrak{g}((t))$ and the Lie algebra $\operatorname{Der}\mathcal{K}$ of vector fields on $D^\times$ come with a canonical central extension. It is therefore natural to ask whether there is a common reason behind this phenomenon.

It turns out that the answer is "yes." This is best understood in terms of a certain "master Lie algebra" that encompasses all of the above examples. This Lie algebra is $\widetilde{\mathfrak{gl}}_\infty$, whose elements are infinite matrices $(a_{ij})_{i,j\in\mathbb{Z}}$ with finitely many non-zero diagonals. The commutator is given by the usual formula $[A,B] = AB - BA$. This is bigger than the naïve definition of $\mathfrak{gl}_\infty$ as the direct limit of the Lie algebras $\mathfrak{gl}_n$ (whose elements are infinite matrices with finitely many non-zero *entries*).

If $\mathfrak{g}$ is a finite-dimensional Lie algebra, then $\mathfrak{g}\otimes\mathbb{C}[t,t^{-1}]$ acts on itself via the adjoint representation. Choosing a basis in $\mathfrak{g}$, we obtain a basis in $\mathfrak{g}\otimes\mathbb{C}[t,t^{-1}]$ labeled by the integers. Therefore to each element of $\mathfrak{g}\otimes\mathbb{C}[t,t^{-1}]$ we attach an infinite matrix, and it is clear that this matrix has finitely many non-zero diagonals (but infinitely many non-zero entries!). Thus, we obtain a Lie algebra homomorphism $\mathfrak{g}\otimes\mathbb{C}[t,t^{-1}] \to \widetilde{\mathfrak{gl}}_\infty$, which is actually injective.

Likewise, the adjoint action of the Lie algebra $\operatorname{Der}\mathbb{C}[t,t^{-1}] = \mathbb{C}[t,t^{-1}]\partial_t$ on itself also gives rise to an injective Lie algebra homomorphism $\operatorname{Der}\mathbb{C}[t,t^{-1}] \to \widetilde{\mathfrak{gl}}_\infty$.

Now the point is that $\widetilde{\mathfrak{gl}}_\infty$ has a one-dimensional universal central extension, and restricting this central extension to the Lie subalgebras $\mathfrak{g}\otimes\mathbb{C}[t,t^{-1}]$ and $\operatorname{Der}\mathbb{C}[t,t^{-1}]$, we obtain the central extensions corresponding to the affine Kac–Moody and Virasoro algebras, respectively.†

The two-cocycle defining this central extension may be defined as follows. Let us write the matrix A as

$$\left(\begin{array}{c|c} A_{-+} & A_{++} \\ \hline A_{--} & A_{+-} \end{array}\right).$$

Then the two-cocycle for the extensions is given by

$$\gamma(A,B) = \operatorname{Tr}(A_{++}B_{--} - B_{++}A_{--}).$$

† More precisely, of their Laurent polynomial versions, but if we pass to a completion of $\widetilde{\mathfrak{gl}}_\infty$, we obtain in the same way the central extensions of the corresponding complete topological Lie algebras.

It is easy to check that the expression in brackets has finitely many non-zero entries, and so the trace is well-defined. Since we can regard our Lie algebras as embedded inside $\widetilde{\mathfrak{gl}}_\infty$, we obtain a central extension for each of them by using the central extension of $\widetilde{\mathfrak{gl}}_\infty$. If we work out what these extensions are in our cases, they turn out to be exactly the central extensions we have been using. This gives us an explanation of the similarity between the structures of the Virasoro and Kac–Moody central extensions.

3.2 Lie algebras associated to vertex algebras

Let us summarize where we stand now in terms of reaching our goal of describing the center of the completed universal enveloping algebra $\widetilde{U}_\kappa(\widehat{\mathfrak{g}})$.

We have constructed explicitly quadratic elements S_n of $\widetilde{U}_\kappa(\widehat{\mathfrak{g}})$, the Segal–Sugawara operators, which were our first candidates for central elements. Moreover, we have computed in Section 3.1 the commutation relations between the S_n's and $\widehat{\mathfrak{g}}_\kappa$ and between the S_n's themselves. In particular, we saw that these operators commute with $\widehat{\mathfrak{g}}_\kappa$ when $\kappa = \kappa_c$.

However, there is an important subtlety which we have not yet discussed. Our computation in Section 3.1 gave the commutation relations between Fourier coefficients of vertex operators coming from the vertex algebra $V_\kappa(\mathfrak{g})$. By definition, those Fourier coefficients are defined as endomorphisms of $V_\kappa(\mathfrak{g})$, not as elements of $\widetilde{U}_\kappa(\widehat{\mathfrak{g}})$. Thus, *a priori* our computation is only valid in the algebra $\operatorname{End} V_\kappa(\mathfrak{g})$, not in $\widetilde{U}_\kappa(\widehat{\mathfrak{g}})$. Nevertheless, we will see in this section that the same result *is valid* in $\widetilde{U}_\kappa(\widehat{\mathfrak{g}})$. In particular, this will imply that the Segal–Sugawara operators are indeed central elements of $\widetilde{U}_\kappa(\widehat{\mathfrak{g}})$.

This is not obvious. What is fairly obvious is that there is a natural Lie algebra homomorphism $\widetilde{U}_\kappa(\widehat{\mathfrak{g}}) \to \operatorname{End} V_\kappa(\mathfrak{g})$. If this homomorphism were injective, then knowing the commutation relations in $\operatorname{End} V_\kappa(\mathfrak{g})$ would suffice. But precisely at the critical level $\kappa = \kappa_c$ this homomorphism is *not* injective. Indeed, the Segal–Sugawara operators $S_n, n \geq -1$, annihilate the vacuum vector $|0\rangle \in V_{\kappa_c}(\mathfrak{g})$, and, because they are central, they are mapped to 0 in $\operatorname{End} V_{\kappa_c}(\mathfrak{g})$. Therefore we need to find a way to interpret our computations of the relations between Fourier coefficients of vertex operators in such a way that it makes sense inside $\widetilde{U}_\kappa(\widehat{\mathfrak{g}})$ rather than $\operatorname{End} V_{\kappa_c}(\mathfrak{g})$.

In this section we follow closely the material of [FB], Sects. 4.1–4.3.

3.2.1 Lie algebra of Fourier coefficients

Let us define, following [Bo], for each vertex algebra V, a Lie algebra $U(V)$ that is spanned by formal symbols $A_{[n]}, A \in V, n \in \mathbb{Z}$, and whose Lie bracket emulates formula (2.3.8).

Here is a more precise definition. Let $U'(V)$ be the vector space

$$V \otimes \mathbb{C}[t,t^{-1}]/\operatorname{Im}\partial, \quad \text{where} \quad \partial = T \otimes 1 + 1 \otimes \partial_t.$$

We denote by $A_{[n]}$ the projection of $A \otimes t^n \in V \otimes \mathbb{C}[t,t^{-1}]$ onto $U'(V)$. Then $U'(V)$ is spanned by these elements modulo the relations

$$(TA)_{[n]} = -nA_{[n-1]}. \tag{3.2.1}$$

We define a bilinear bracket on $U'(V)$ by the formula

$$[A_{[m]}, B_{[k]}] = \sum_{n \geq 0} \binom{m}{n} (A_{(n)}B)_{[m+k-n]} \tag{3.2.2}$$

(compare with formula (2.3.8)). Using the identity

$$[T, A_{(n)}] = -nA_{(n-1)}$$

in $\operatorname{End} V$, which follows from the translation axiom, one checks easily that this map is well-defined.

If V is $\mathbb{Z}$–graded, then we introduce a $\mathbb{Z}$–gradation on $U'(V)$, by setting, for each homogeneous element A of V, $\deg A_{[n]} = -n + \deg A - 1$. The map (3.2.2) preserves this gradation.

We also consider a completion $U(V)$ of $U'(V)$ with respect to the natural topology on $\mathbb{C}[t,t^{-1}]$:

$$U(V) = (V \otimes \mathbb{C}((t)))/\operatorname{Im}\partial.$$

It is spanned by linear combinations $\sum_{n \geq N} f_n A_{[n]}$, $f_n \in \mathbb{C}, A \in V, N \in \mathbb{Z}$, modulo the relations which follow from the identity (3.2.1). The bracket given by formula (3.2.2) is clearly continuous. Hence it gives rise to a bracket on $U(V)$.

We have a linear map $U(V) \to \operatorname{End} V$,

$$\sum_{n \geq N} f_n A_{[n]} \mapsto \sum_{n \geq N} f_n A_{(n)} = \operatorname{Res}_{z=0} Y(A,z) f(z) dz,$$

where $f(z) = \sum_{n \geq N} f_n z^n \in \mathbb{C}((z))$.

Proposition 3.2.1 *The bracket* (3.2.2) *defines Lie algebra structures on* $U'(V)$ *and on* $U(V)$. *Furthermore, the natural maps* $U'(V) \to \operatorname{End} V$ *and* $U(V) \to \operatorname{End} V$ *are Lie algebra homomorphisms.*

Proof. Let us define $U'(V)_0 \subset U(V)$ as the quotient

$$U'(V)_0 = V/\operatorname{Im} T = (V \otimes 1)/\operatorname{Im}\partial,$$

so it is the subspace of $U'(V)$ spanned by the elements $A_{[0]}$ only.

The bracket (3.2.2) restricts to a bracket on $U'(V)_0$:

$$[A_{[0]}, B_{[0]}] = (A_{(0)}B)_{[0]}. \tag{3.2.3}$$

We first prove that it endows $U'(V)_0$ with the structure of a Lie algebra under the defined bracket. We need to show that it is antisymmetric and satisfies the Jacobi identity. Recall the identity

$$Y(A,z)B = e^{zT}Y(B,-z)A$$

from Proposition 2.3.2. Looking at the z^{-1} coefficient gives

$$A_{(0)}B = -B_{(0)}A + T(\cdots).$$

Taking the [0] part of this and recalling that we mod out by the image of T, we see that

$$(A_{(0)}B)_{[0]} = -(B_{(0)}A)_{[0]}.$$

In other words, the bracket is antisymmetric.

The Jacobi identity is equivalent to

$$[C_{[0]},[A_{[0]},B_{[0]}]] = [[C_{[0]},A_{[0]}],B_{[0]}] + [A_{[0]},[C_{[0]},B_{[0]}]].$$

The right hand side of this gives

$$((C_{(0)}A)_{(0)}B)_{[0]} + (A_{(0)}(C_{(0)}B))_{[0]}.$$

But from the commutation relations (2.3.8) for the Fourier coefficients of vertex operators we know that

$$[C_{(0)},A_{(0)}] = (C_{(0)}A)_{(0)}.$$

This shows the Jacobi identity.

We will now derive from this that the bracket (3.2.2) on $U'(V)$ satisfies the axioms of a Lie algebra by constructing a bigger vertex algebra W such that $U'(V)$ is $U'(W)_0$ (which we already know to be a Lie algebra).

As $\mathbb{C}[t,t^{-1}]$ is a commutative, associative, unital algebra with the derivation $T = -\partial_t$, we can make it into a commutative vertex algebra, as explained in Section 2.2.2. Let $W = V \otimes \mathbb{C}[t,t^{-1}]$ with the vertex algebra structure defined in the obvious way. Then $U'(V) = U'(W)_0$. We now show that formula (3.2.2) in $U'(V)$ coincides with formula (3.2.3) in $U'(W)_0$. The latter reads

$$[(A\otimes t^n)_{[0]},(B\otimes t^m)_{[0]}] = ((A\otimes t^n)_{(0)}(B\otimes t^m))_{[0]}.$$

Suppose that $f(t)$ is an element of $\mathbb{C}[t,t^{-1}]$. Then the vertex operator associated to $f(t)$ is

$$Y(f(t),z) = \mathrm{mult}(e^{zT}f(t)) = \mathrm{mult}(f(t+z)),$$

where necessary expansions are performed in the domain $|z|<|t|$. In our case this means that we need the z^{-1} coefficient of

$$Y(A,z)\otimes(t+z)^m,$$

which is

$$\sum_{n\geq 0}\binom{m}{n}A_{(n)}\otimes t^{m-n}.$$

Putting this together gives

$$[(A\otimes t^m)_{[0]},(B\otimes t^k)_{[0]}]=\sum_{n\geq 0}\binom{m}{n}(A_{(n)}B)\otimes t^{m+k-n},$$

which coincides with formula (3.2.2) we want for $U'(V)$.

The fact that the map $U'(V)\to \operatorname{End} V$ is a Lie algebra homomorphism follows from the comparison of formulas (2.3.8) and (3.2.2). The fact that the Lie bracket on $U'(V)$ extends to a Lie bracket on $U(V)$ and the corresponding map $U(V)\to \operatorname{End} V$ is also a Lie algebra homomorphism follow by continuity. This completes the proof. □

Given a vertex algebra homomorphism $\phi: V\to V'$, we define $\widetilde{\phi}$ to be the map $U(V)\to U'(V)$ sending $A_{[n]}\mapsto \phi(A)_{[n]}$ (and extended by continuity). This is a Lie algebra homomorphism. To see that, we need to check that

$$\widetilde{\phi}([A_{[n]},B_{[m]}])=[\widetilde{\phi}(A_{[n]}),\widetilde{\phi}(B_{[m]})].$$

Expanding the left hand side, using the definition of the bracket, and applying the definition of $\widetilde{\phi}$ to each side reduces it to

$$\phi(A_{(n)}B)=\phi(A)_{(n)}\phi(B).$$

But this follows from the fact that ϕ is a vertex algebra homomorphism.

Thus, we obtain that the assignments $V\mapsto U'(V)$ and $V\mapsto U(V)$ define functors from the category of vertex algebras to the categories of Lie algebras and complete topological Lie algebras, respectively.

Finally, we note that if V is a $\mathbb{Z}$-graded vertex algebra, then the Lie algebra $U'(V)$ is also $\mathbb{Z}$-graded with respect to the degree assignment $\deg A_{[n]} = -n+\deg A-1$.

3.2.2 From vertex operators to the enveloping algebra

Let us now look more closely at the Lie algebra $U(V_\kappa(\mathfrak{g}))$. By Theorem 2.2.2, the vertex operators corresponding to elements of $V_\kappa(\mathfrak{g})$ are finite linear combinations of the formal power series

$$:\partial_z^{n_1}J^{a_1}(z)\cdots\partial_z^{n_m}J^{a_m}(z):.$$

This means that we may think of elements of $U(V_\kappa(\mathfrak{g}))$ as finite linear combinations of Fourier coefficients of these series. Each of these coefficients is, in general, an infinite series in terms of the basis elements J^a_n of $\widehat{\mathfrak{g}}$. However, one checks by induction that each of these series becomes finite modulo the left ideal $\widetilde{U}_\kappa(\mathfrak{g})\cdot\mathfrak{g}\otimes t^N\mathbb{C}[[t]]$, for any $N\geq 0$. Therefore each of them gives rise to a well-defined element of $\widetilde{U}_\kappa(\widehat{\mathfrak{g}})$. We have already seen that in the case of the Segal–Sugawara elements (2.1.9): they are bona fide elements of $\widetilde{U}_\kappa(\widehat{\mathfrak{g}})$.

Thus, we have a linear map $V \otimes \mathbb{C}((t)) \to \widetilde{U}_\kappa(\widehat{\mathfrak{g}})$, and it is clear that it vanishes on the image of ∂. Therefore we obtain a linear map $U(V_\kappa(\mathfrak{g})) \to \widetilde{U}_\kappa(\widehat{\mathfrak{g}})$. It is not difficult to show that it is actually injective, but we will not use this fact here. The key result which will allow us to transform our previous computations from $\operatorname{End} V$ to $\widetilde{U}_\kappa(\widehat{\mathfrak{g}})$ is the following:

Proposition 3.2.1 *The map $U(V_\kappa(\mathfrak{g})) \to \widetilde{U}_\kappa(\widehat{\mathfrak{g}})$ is a Lie algebra homomorphism.*

The proof involves some rather tedious calculations, so we will omit it, referring the reader instead to [FB], Proposition 4.2.2.

Now we see that all of our computations in Section 3.1 are in fact valid in $\widetilde{U}_\kappa(\widehat{\mathfrak{g}})$.

Corollary 3.2.2 *For $\kappa = \kappa_c$ the Segal–Sugawara operators S_n are central in $\widetilde{U}_\kappa(\widehat{\mathfrak{g}})$. For $\kappa \neq \kappa_c$ they generate the Virasoro algebra inside $\widetilde{U}_\kappa(\widehat{\mathfrak{g}})$.*

3.2.3 Enveloping algebra associated to a vertex algebra

Here we construct an analogue of the topological associative algebra $\widetilde{U}_\kappa(\widehat{\mathfrak{g}})$ for an arbitrary $\mathbb{Z}$-graded vertex algebra V.

Note that the image of $U(V_\kappa(\mathfrak{g}))$ in $\widetilde{U}_\kappa(\widehat{\mathfrak{g}})$ is not closed under multiplication. For instance, $U(V_\kappa(\mathfrak{g}))$ contains $\widehat{\mathfrak{g}}$ as a Lie subalgebra. It is spanned by $A_n = (A_{-1}|0\rangle)_{[n]}, A \in \mathfrak{g}, n \in \mathbb{Z}$, and $(|0\rangle)_{[-1]}$. But $U(V_\kappa(\mathfrak{g}))$ does not contain products of elements of $\widehat{\mathfrak{g}}$, such as $A_n B_k$. In fact, all quadratic elements in $U(V_\kappa(\mathfrak{g}))$ are infinite series.

However, we can obtain $\widetilde{U}_\kappa(\widehat{\mathfrak{g}})$ as a completion of the universal enveloping algebra of $U(V_\kappa(\mathfrak{g}))$ modulo certain natural relations. An analogous construction may in fact be performed for any $\mathbb{Z}$-graded vertex algebra V in the following way. Denote by $U(U(V))$ the universal enveloping algebra of the Lie algebra $U(V)$. Define its completion

$$\widetilde{U}(U(V)) = \varprojlim U(U(V))/I_N,$$

where I_N is the left ideal generated by $A_{[n]}, A \in V_m, n \geq N + m$.† Let $\widetilde{U}(V)$ be the quotient of $\widetilde{U}(U(V))$ by the two-sided ideal generated by the Fourier coefficients of the series

$$Y[A_{(-1)}B, z] - {:}Y[A, z]Y[B, z]{:}, \qquad A, B \in V,$$

† Here we correct an error in [FB], Section 4.3.1, where this condition was written as $n > N$; the point is that $\deg A_{[n]} = -n + m - 1$ (see the end of Section 3.2.1).

where we set

$$Y[A,z] = \sum_{n\in\mathbb{Z}} A_{[n]} z^{-n-1},$$

and the normal ordering is defined in the same way as for the vertex operators $Y(A,z)$.

Clearly, $\widetilde{U}(V)$ is a complete topological associative algebra (with the topology for which a basis of open neighborhoods of 0 is given by the completions of the left ideals $I_N, N > 0$). The assignment $V \mapsto \widetilde{U}(V)$ gives rise to a functor from the category of vertex algebras to the category of complete topological associative algebras.

The associative algebra $\widetilde{U}(V)$ may be viewed as a generalization of the completed enveloping algebra $\widetilde{U}_\kappa(\widehat{\mathfrak{g}})$, because, as the following lemma shows, for $V = V_\kappa(\mathfrak{g})$ we have $\widetilde{U}(V_\kappa(\mathfrak{g})) \simeq \widetilde{U}_\kappa(\widehat{\mathfrak{g}})$.

Lemma 3.2.2 *There is a natural isomorphism $\widetilde{U}(V_\kappa(\mathfrak{g})) \simeq \widetilde{U}_\kappa(\widehat{\mathfrak{g}})$ for all $\kappa \in \mathbb{C}$.*

Proof. We need to define mutually inverse algebra homomorphisms between $\widetilde{U}(V_\kappa(\mathfrak{g}))$ and $\widetilde{U}_\kappa(\widehat{\mathfrak{g}})$. The Lie algebra homomorphism $U(V_\kappa(\mathfrak{g})) \to \widetilde{U}_\kappa(\widehat{\mathfrak{g}})$ of Proposition 3.2.1 gives rise to a homomorphism of the universal enveloping algebra of the Lie algebra $U(V_\kappa(\mathfrak{g}))$ to $\widetilde{U}_\kappa(\widehat{\mathfrak{g}})$. It is clear that under this homomorphism any element of the ideal I_N is mapped to the left ideal of $U(V_\kappa(\mathfrak{g}))$ generated by $A_n, A \in \mathfrak{g}, n > M$, for sufficiently large M. Therefore this homomorphism extends to a homomorphism from the completion of the universal enveloping algebra of $U(V_\kappa(\mathfrak{g}))$ to $\widetilde{U}_\kappa(\widehat{\mathfrak{g}})$. But according to the definitions, this map sends the series $Y[A_{(-1)}B, z]$ precisely to the series $:Y[A,z]Y[B,z]:$ for all $A, B \in V_\kappa(\mathfrak{g})$. Hence we obtain a homomorphism $\widetilde{U}(V_\kappa(\mathfrak{g})) \to \widetilde{U}_\kappa(\widehat{\mathfrak{g}})$.

To construct the inverse homomorphism, recall that $\widehat{\mathfrak{g}}$ is naturally a Lie subalgebra of $U(V_\kappa(\mathfrak{g}))$. Hence we have a homomorphism from $U_\kappa(\widehat{\mathfrak{g}})$ to the universal enveloping algebra of the Lie algebra $U(V_\kappa(\mathfrak{g}))$. This homomorphism extends by continuity to a homomorphism from $\widetilde{U}_\kappa(\widehat{\mathfrak{g}})$ to the completion of the universal enveloping algebra of $U(V_\kappa(\mathfrak{g}))$, and hence to $\widetilde{U}(V_\kappa(\mathfrak{g}))$. Now observe that the resulting map $\widetilde{U}_\kappa(\widehat{\mathfrak{g}}) \to \widetilde{U}(V_\kappa(\mathfrak{g}))$ is surjective, because each series $Y[A,z]$ is a linear combination of normally ordered products of the generating series $J^a(z) = \sum_{n\in\mathbb{Z}} J_a^n z^{-n-1}$ of the elements J_n^a of $\widehat{\mathfrak{g}}$. Furthermore, it is clear from the construction that the composition $\widetilde{U}_\kappa(\widehat{\mathfrak{g}}) \to \widetilde{U}(V_\kappa(\mathfrak{g})) \to \widetilde{U}_\kappa(\widehat{\mathfrak{g}})$ is the identity. This implies that the map $\widetilde{U}_\kappa(\widehat{\mathfrak{g}}) \to \widetilde{U}(V_\kappa(\mathfrak{g}))$ is also injective. Therefore the two maps indeed define mutually inverse isomorphisms. □

3.3 The center of a vertex algebra

According to Corollary 3.2.2, the Segal–Sugawara elements are indeed central elements of the completed enveloping algebra $\widetilde{U}_\kappa(\widehat{\mathfrak{g}})$ for a special level $\kappa = \kappa_c$. Now we would like to see if there are any other central elements, and, if so, how to construct them.

3.3.1 Definition of the center

In order to do this we should think about why the elements we constructed were central in the first place.

Lemma 3.3.1 *The elements $A_{[n]}, A \in V_\kappa(\mathfrak{g}), n \in \mathbb{Z}$, of $\widetilde{U}_\kappa(\widehat{\mathfrak{g}})$ are central if A is annihilated by all operators J_n^a with $n \geqslant 0$.*

Proof. This follows immediately from the commutation relations formula (3.2.2). □

The elements $J_n^a, n \geq 0$, span the Lie algebra $\mathfrak{g}[[t]]$. This leads us to define the center $\mathfrak{z}(V_\kappa(\mathfrak{g}))$ of the vertex algebra $V_\kappa(\mathfrak{g})$ as its subspace of $\mathfrak{g}[[t]]$-invariants:

$$\mathfrak{z}(V_\kappa(\mathfrak{g})) = V_\kappa(\mathfrak{g})^{\mathfrak{g}[[t]]}. \tag{3.3.1}$$

In order to generalize this definition to other vertex algebras, we note that any element $B \in \mathfrak{z}(V_\kappa(\mathfrak{g}))$ satisfies $A_{(n)}B = 0$ for all $A \in V_\kappa(\mathfrak{g})$ and $n \geq 0$. This follows from formula (2.2.8) and the definition of normal ordering by induction on m.

Thus, we define the **center** of an arbitrary vertex algebra V to be

$$\mathcal{Z}(V) = \{B \in V \mid A_{(n)}B = 0 \text{ for all } A \in V, n \geq 0\}.$$

By the OPE formula (2.3.7), an equivalent definition is

$$\mathcal{Z}(V) = \{B \in V \mid [Y(A,z), Y(B,w)] = 0 \text{ for all } A \in V\}.$$

Formula (3.2.2) implies that if $B \in \mathcal{Z}(V)$, then $B_{[n]}$ is a central element in $U(V)$ and in $\widetilde{U}(V)$ for all $n \in \mathbb{Z}$.

The space $\mathcal{Z}(V)$ is clearly non-zero, because the vacuum vector $|0\rangle$ is contained in $\mathcal{Z}(V)$.

Lemma 3.3.2 *The center $\mathcal{Z}(V)$ of a vertex algebra V is a commutative vertex algebra.*

Proof. We already know that $|0\rangle$ is in the center. A simple calculation shows that T maps $\mathcal{Z}(V)$ into $\mathcal{Z}(V)$:

$$A_{(n)}(TB) = T(A_{(n)}B) - [T, A_{(n)}]B = nA_{(n-1)}B = 0.$$

Let $B, C \in \mathcal{Z}(V)$ be central. Then

$$A_{(n)}B_{(m)}C = B_{(m)}A_{(n)}C + [A_{(n)}, B_{(m)}]C = 0.$$

Therefore $\mathcal{Z}(V)$ is closed under the state-field correspondence. The commutativity is obvious. □

3.3.2 The center of the affine Kac–Moody vertex algebra

So our plan is now to study the center of the vertex algebra $V_\kappa(\mathfrak{g})$ and then to show that we obtain all central elements for $\widetilde{U}_\kappa(\widehat{\mathfrak{g}})$ from the elements in $\mathcal{Z}(V_\kappa(\mathfrak{g}))$ by taking their Fourier coefficients (viewed as elements of $\widetilde{U}_\kappa(\widehat{\mathfrak{g}})$).

Since $V_\kappa(\mathfrak{g})$ is much smaller than $\widetilde{U}_\kappa(\widehat{\mathfrak{g}})$, this looks like a good strategy. We start with the case when $\kappa \neq \kappa_c$.

Proposition 3.3.3 *If $\kappa \neq \kappa_c$, then the center of $V_\kappa(\mathfrak{g})$ is trivial, i.e.,*

$$\mathcal{Z}(V_\kappa(\mathfrak{g})) = \mathbb{C} \cdot |0\rangle \,.$$

Proof. Suppose that A is a homogeneous element in the center $\mathcal{Z}(V_\kappa(\mathfrak{g}))$. Let $\widetilde{S}$ be the vector introduced above generating the Virasoro field

$$Y(\widetilde{S}, z) = \sum_{n \in \mathbb{Z}} L_n z^{-n-2}.$$

Recall that we need $\kappa \neq \kappa_c$ for this element to be well-defined as we needed to divide by this factor. Now, if $B \in \mathcal{Z}(V_\kappa(\mathfrak{g}))$, then $A_{(n)}B = 0$ for all $A \in V_\kappa(\mathfrak{g})$ and $n \geq 0$, as we saw above. In particular, taking $A = \widetilde{S}$ and $n = 1$ we find that $L_0 B = 0$ (recall that $L_0 = \widetilde{S}_{(1)}$). But L_0 is the grading operator, so we see that B must have degree 0, hence it must be a multiple of the vacuum vector $|0\rangle$. □

Note that *a priori* it does not follow that the center of $\widetilde{U}_\kappa(\widehat{\mathfrak{g}})$ for $\kappa \neq \kappa_c$ is trivial. It only follows that we cannot obtain non-trivial central elements of $\widetilde{U}_\kappa(\widehat{\mathfrak{g}})$ from the vertex algebra $V_\kappa(\mathfrak{g})$ by the above construction. However, we will prove later, in Proposition 4.3.9, that the center of $\widetilde{U}_\kappa(\widehat{\mathfrak{g}})$ is in fact trivial.

From now on we will focus our attention on the critical level $\kappa = \kappa_c$. We will see that the center of $\widetilde{U}_{\kappa_c}(\widehat{\mathfrak{g}})$ may be generated from the center of the vertex algebra $V_{\kappa_c}(\mathfrak{g})$. From now on, in order to simplify the notation we will denote the latter by $\mathfrak{z}(\widehat{\mathfrak{g}})$, so that

$$\mathfrak{z}(\widehat{\mathfrak{g}}) = \mathcal{Z}(V_{\kappa_c}(\mathfrak{g})).$$

We now discuss various interpretations of $\mathfrak{z}(\widehat{\mathfrak{g}})$.

The commutative vertex algebra structure on $\mathfrak{z}(\widehat{\mathfrak{g}})$ gives rise to an ordinary commutative algebra structure on it. Note that it is $\mathbb{Z}_+$-graded. It is natural to ask what the meaning of this algebra structure is from the point of view of $V_{\kappa_c}(\mathfrak{g})$.

According to (3.3.1), we have

$$\mathfrak{z}(\widehat{\mathfrak{g}}) = V_{\kappa_c}(\mathfrak{g})^{\mathfrak{g}[[t]]}. \tag{3.3.2}$$

This enables us to identify $\mathfrak{z}(\widehat{\mathfrak{g}})$ with the algebra of $\widehat{\mathfrak{g}}_{\kappa_c}$-endomorphisms of $V_{\kappa_c}(\mathfrak{g})$. Indeed, a $\mathfrak{g}[[t]]$-invariant vector $A \in V_{\kappa_c}(\mathfrak{g})$ gives rise to a non-trivial endomorphism of V, which maps the highest weight vector v_{κ_c} to A. On the other hand, given an endomorphism e of $V_{\kappa_c}(\mathfrak{g})$, we obtain a $\mathfrak{g}[[t]]$-invariant vector $e(|0\rangle)$. It is clear that the two maps are inverse to each other, and so we obtain an isomorphism

$$V_{\kappa_c}(\mathfrak{g})^{\mathfrak{g}[[t]]} \simeq \operatorname{End}_{\widehat{\mathfrak{g}}_{\kappa_c}} V_{\kappa_c}(\mathfrak{g}).$$

Now let A_1 and A_2 be two $\mathfrak{g}[[t]]$-invariant vectors in $V_{\kappa_c}(\mathfrak{g})$. Let e_1 and e_2 be the corresponding endomorphisms of $V_{\kappa_c}(\mathfrak{g})$. Then the image of $|0\rangle$ under the composition $e_1 e_2$ equals

$$e_1(A_2) = e_1 \circ (A_2)_{(-1)}|0\rangle = (A_2)_{(-1)} \circ e_1(|0\rangle) = (A_2)_{(-1)} A_1.$$

The last expression is nothing but the product $A_2 A_1$ coming from the commutative algebra structure on $\mathfrak{z}(\widehat{\mathfrak{g}})$. Therefore we find that as an algebra, $\mathfrak{z}(\widehat{\mathfrak{g}})$ is isomorphic to the algebra $\operatorname{End}_{\widehat{\mathfrak{g}}_{\kappa_c}} V_{\kappa_c}(\mathfrak{g})$ with the opposite multiplication. But since $\mathfrak{z}(\widehat{\mathfrak{g}})$ is commutative, we obtain an isomorphism of algebras

$$\mathfrak{z}(\widehat{\mathfrak{g}}) \simeq \operatorname{End}_{\widehat{\mathfrak{g}}_{\kappa_c}} V_{\kappa_c}(\mathfrak{g}). \tag{3.3.3}$$

In particular, we find that the algebra $\operatorname{End}_{\widehat{\mathfrak{g}}_{\kappa_c}} V_{\kappa_c}(\mathfrak{g})$ is commutative, which is *a priori* not obvious. (The same argument also works for other levels κ, but in those cases we find that $\mathfrak{z}(V_\kappa(\mathfrak{g})) \simeq \operatorname{End}_{\widehat{\mathfrak{g}}_\kappa} V_\kappa(\mathfrak{g}) = \mathbb{C}$, according to Proposition 3.3.3.)

Finally, we remark that $V_{\kappa_c}(\mathfrak{g})$ is isomorphic to the universal enveloping algebra $U(\mathfrak{g} \otimes t^{-1}\mathbb{C}[t^{-1}])$ (note however that this isomorphism is not canonical: it depends on the choice of the coordinate t). Under this isomorphism, we obtain an injective map

$$\mathfrak{z}(\widehat{\mathfrak{g}}) \hookrightarrow U(\mathfrak{g} \otimes t^{-1}\mathbb{C}[t^{-1}]).$$

It is clear from the above calculation that this is a homomorphism of algebras. Thus, $\mathfrak{z}(\widehat{\mathfrak{g}})$ may be viewed as a commutative subalgebra of $U(\mathfrak{g} \otimes t^{-1}\mathbb{C}[t^{-1}])$.

3.3.3 The associated graded algebra

Our goal is to describe the center $\mathfrak{z}(\widehat{\mathfrak{g}})$ of $V_{\kappa_c}(\mathfrak{g})$. It is instructive to describe first its graded analogue. Recall that in Section 2.2.5 we defined a filtration on $V_\kappa(\mathfrak{g})$ that is preserved by the action of $\widehat{\mathfrak{g}}_\kappa$ and such that the associated graded space is

$$\operatorname{gr} V_\kappa(\mathfrak{g}) = \operatorname{Sym}(\mathfrak{g}((t))/\mathfrak{g}[[t]]).$$

The topological dual space to $\mathfrak{g}((t))/\mathfrak{g}[[t]]$ is naturally identified with the space $\mathfrak{g}^*[[t]]dt$ with respect to the pairing

$$\langle \phi(t)dt, A(t)\rangle = \operatorname{Res}_{t=0}\langle \phi(t), A(t)\rangle dt.$$

This pairing is canonical in the sense that it does not depend on the choice of coordinate t.

However, to simplify our formulas below, we will use a coordinate t to identify $\mathfrak{g}^*[[t]]dt$ with $\mathfrak{g}^*[[t]]$. This is the inverse limit of finite-dimensional vector spaces

$$\mathfrak{g}^*[[t]] = \varprojlim \mathfrak{g}^* \otimes \mathbb{C}[[t]]/(t^N),$$

and so the algebra of regular functions on it is the direct limit of free polynomial algebras

$$\operatorname{Fun} \mathfrak{g}^*[[t]] = \varinjlim \operatorname{Fun}(\mathfrak{g}^* \otimes \mathbb{C}[[t]]/(t^N)).$$

Since $\mathfrak{g}^*[[t]]$ is isomorphic to the topological dual vector space to $\mathfrak{g}((t))/\mathfrak{g}[[t]]$, we have a natural isomorphism

$$\operatorname{Sym}(\mathfrak{g}((t))/\mathfrak{g}[[t]]) \simeq \operatorname{Fun} \mathfrak{g}^*[[t]],$$

and so we have

$$\operatorname{gr} V_\kappa(\mathfrak{g}) \simeq \operatorname{Fun} \mathfrak{g}^*[[t]].$$

Now the filtration on $V_{\kappa_c}(\mathfrak{g})$ induces a filtration on $\mathfrak{z}(\widehat{\mathfrak{g}})$. Let

$$\operatorname{gr} \mathfrak{z}(\widehat{\mathfrak{g}}) \subset \operatorname{Fun} \mathfrak{g}^*[[t]]$$

be the associated graded space. The commutative algebra structure on $\mathfrak{z}(\widehat{\mathfrak{g}})$ gives rise to such a structure on $\operatorname{gr} \mathfrak{z}(\widehat{\mathfrak{g}})$. Note also that the Lie algebra $\mathfrak{g}[[t]]$ acts on itself via the adjoint representation, and this induces its action on $\operatorname{Fun} \mathfrak{g}^*[[t]]$.

Lemma 3.3.1 *We have an injective homomorphism of commutative algebras*

$$\operatorname{gr} \mathfrak{z}(\widehat{\mathfrak{g}}) \hookrightarrow (\operatorname{Fun} \mathfrak{g}^*[[t]])^{\mathfrak{g}[[t]]}.$$

Proof. The fact that the embedding $\operatorname{gr} \mathfrak{z}(\widehat{\mathfrak{g}}) \hookrightarrow \operatorname{Fun} \mathfrak{g}^*[[t]]$ is a homomorphism of algebras follows from the isomorphism (3.3.3).

Suppose now that $A \in V_{\kappa_c}(\mathfrak{g})_{\leqslant i}$, and denote the projection of A onto $\operatorname{gr}_i(V_{\kappa_c}(\mathfrak{g})) = V_{\kappa_c}(\mathfrak{g})_{\leqslant i}/V_{\kappa_c}(\mathfrak{g})_{\leqslant (i-1)}$ by $\operatorname{Symb}_i(A)$. It is easy to see that

$$\operatorname{Symb}_i(x \cdot A) = x \cdot \operatorname{Symb}_i(A) \text{ for } x \in \mathfrak{g}[[t]].$$

If $A \in \mathfrak{z}(\widehat{\mathfrak{g}})$, then $x \cdot A = 0$ and so $x \cdot \operatorname{Symb}_i(A) = 0$. □

Thus, the symbols of elements of $\mathfrak{z}(\widehat{\mathfrak{g}})$ are realized as $\mathfrak{g}[[t]]$-invariant functions on $\mathfrak{g}^*[[t]]$. We now begin to study the algebra of these functions, which we denote by

$$\operatorname{Inv} \mathfrak{g}^*[[t]] = (\operatorname{Fun} \mathfrak{g}^*[[t]])^{\mathfrak{g}[[t]]}.$$

Note that the action of the Lie algebra $\mathfrak{g}[[t]]$ on $\operatorname{Fun} \mathfrak{g}^*[[t]]$ comes from the coadjoint action of $\mathfrak{g}$ on $\mathfrak{g}^*$. This action may be exponentiated to the adjoint action of the group $G[[t]]$, where G is the connected simply-connected Lie group with Lie algebra $\mathfrak{g}$. Therefore $G[[t]]$ acts on $\operatorname{Fun} \mathfrak{g}^*[[t]]$. Since $G[[t]]$ is connected, it follows that

$$\operatorname{Inv} \mathfrak{g}^*[[t]] = (\operatorname{Fun} \mathfrak{g}^*[[t]])^{G[[t]]}.$$

If we can work out the size of this algebra of invariant functions, then we will be able to place an upper bound on the size of $\mathfrak{z}(\widehat{\mathfrak{g}})$ and this will turn out to be small enough to allow us to conclude that the elements we subsequently construct give us the entire center of $V_{\kappa_c}(\mathfrak{g})$ (and ultimately of $\widetilde{U}_{\kappa_c}(\widehat{\mathfrak{g}})$).

3.3.4 Symbols of central elements

It is instructive to look at what happens in the case of a finite-dimensional simple Lie algebra $\mathfrak{g}$. In this case the counterpart of $\mathfrak{z}(\widehat{\mathfrak{g}})$ is the center $Z(\mathfrak{g})$ of the universal enveloping algebra $U(\mathfrak{g})$ of $\mathfrak{g}$. Let $\operatorname{gr} U(\mathfrak{g})$ be the associated graded algebra with respect to the Poincaré–Birkhoff–Witt filtration. Then

$$\operatorname{gr} U(\mathfrak{g}) = \operatorname{Sym} \mathfrak{g} = \operatorname{Fun} \mathfrak{g}^*.$$

Let $\operatorname{gr} Z(\mathfrak{g})$ be the associated graded algebra of $Z(\mathfrak{g})$ with respect to the induced filtration. In the same way as in Lemma 3.3.1 we obtain an injective homomorphism

$$\operatorname{gr} Z(\mathfrak{g}) \hookrightarrow \operatorname{Inv} \mathfrak{g}^* = (\operatorname{Fun} \mathfrak{g}^*)^{\mathfrak{g}} = (\operatorname{Fun} \mathfrak{g}^*)^G.$$

It is actually an isomorphism, because $U(\mathfrak{g})$ and $\operatorname{Sym} \mathfrak{g}$ are isomorphic as $\mathfrak{g}$-modules (with respect to the adjoint action). This follows from the fact that both are direct limits of finite-dimensional representations, and there are no non-trivial extensions between such representations.

Recall that by Theorem 2.1.1 we have

$$Z(\mathfrak{g}) = \mathbb{C}[P_i]_{i=1,\ldots,\ell}.$$

Let $\overline{P}_i$ be the symbol of P_i in $\operatorname{Fun} \mathfrak{g}^*$. Then

$$\operatorname{Inv} \mathfrak{g}^* = \mathbb{C}[\overline{P}_i]_{i=1,\ldots,\ell}.$$

According to Theorem 2.1.1, $\deg \overline{P}_i = d_i + 1$.

We will now use the elements $\overline{P}_i$ to create a large number of elements in $\operatorname{Inv} \mathfrak{g}^*[[t]]$. We will use the generators $\overline{J}^a_n, n < 0$, of $\operatorname{Sym}(\mathfrak{g}((t))/\mathfrak{g}[[t]]) =$

$\operatorname{Fun} \mathfrak{g}^*[[t]]$, which are the symbols of $J^a_n|0\rangle \in V_{\kappa_c}(\mathfrak{g})$. These are linear functions on $\mathfrak{g}^*[[t]]$ defined by the formula

$$\overline{J}^a_n(\phi(t)) = \operatorname{Res}_{t=0}\langle \phi(t), J^a\rangle t^n dt. \tag{3.3.4}$$

We will also write

$$\overline{J}^a(z) = \sum_{n<0} \overline{J}^a_n z^{-n-1}.$$

Suppose that we write $\overline{P}_i$ as a polynomial in the linear elements, $\overline{P}_i(\overline{J}^a)$. Define a set of elements $\overline{P}_{i,n} \in \operatorname{Fun} \mathfrak{g}^*[[t]]$ by the formula

$$\overline{P}_i(\overline{J}^a(z)) = \sum_{n<0} \overline{P}_{i,n} z^{-n-1}.$$

Note that each of the elements $\overline{P}_{i,n}$ is a finite polynomial in the $\overline{J}^a_n$'s.

Lemma 3.3.4 *The polynomials $\overline{P}_{i,n}$ are in* $\operatorname{Inv} \mathfrak{g}^*[[t]]$.

The proof is straightforward and is left to the reader.

Thus, we have a homomorphism

$$\mathbb{C}[\overline{P}_{i,n}]_{i=1,\ldots,\ell;n<0} \longrightarrow \operatorname{Inv} \mathfrak{g}^*[[t]].$$

We wish to show that this is actually an isomorphism, but in order to do this we need to develop some more technology.

3.4 Jet schemes

Let X be a complex algebraic variety. A jet scheme of X is an algebraic object that represent the "space of formal paths on X", i.e., morphisms from the disc $D = \operatorname{Spec} \mathbb{C}[[t]]$ to X. We wish to treat it as a scheme, i.e., something glued from affine algebraic varieties defined by polynomial equations in an affine space. In this section we give a precise definition of these objects. We are interested in these objects because of the simple fact that $\mathfrak{g}^*[[t]]$ is the jet scheme of $\mathfrak{g}^*$, considered as an affine algebraic variety. The technique of jet schemes will allow us to describe the algebra of invariant functions on $\mathfrak{g}^*[[t]]$.

3.4.1 Generalities on schemes

In algebraic geometry it is often convenient to describe the functor of points of an algebraic variety (or a scheme). For example, an affine algebraic variety X over a field k is completely determined by the ring of regular functions on it. Let us denote it by $\operatorname{Fun} X$, so we can write $X = \operatorname{Spec}(\operatorname{Fun} X)$. Now, given a k-algebra R, we define the set of R-points of X, denoted by $X(R)$, as the set of morphisms $\operatorname{Spec} R \to X$. This is the same as the set of homomorphisms $\operatorname{Fun} X \to R$ of k-algebras.

If $R = k$, we recover the usual notion of k-points of X, because each homomorphism $\operatorname{Fun} X \to k$ corresponds to taking the value of a function at a point $x \in X(k)$. However, the set of k-points by itself does not capture the entire structure of X in general. For example, if k is not algebraically closed, the set $X(k)$ may well be empty, but this does not mean that X is empty. Consider the case of $X = \operatorname{Spec} \mathbb{R}[x]/(x^2+1)$. Then $X(\mathbb{R})$ is empty, but $X(\mathbb{C})$ consists of one element. Therefore we see that we may well lose information about X if we look at its k-points. If X is an affine algebraic variety, then it is the algebra $\operatorname{Fun} X$ which captures the structure of X, not the set $X(k)$. For a general algebraic variety X, it is the structure of a scheme on X, which essentially amounts to a covering of X by affine algebraic varieties, each described by its algebra of regular functions.

Now, fixing X, but letting the k-algebra R vary, we obtain a functor from the category of commutative k-algebras to the category of sets which satisfies some natural conditions (it is a "sheaf" on the category of k-algebras, in the appropriate sense). These conditions ensure that it extends uniquely to a functor from the category of schemes over k (as those are the objects "glued" from the affine algebraic varieties) to the category of sets. This is what is referred to as the "functor of points."

One defines in a similar way the functor of points of an arbitrary algebraic variety X (not necessarily affine) over k. An important result is that this functor determines X uniquely (up to isomorphism). Such functors from the category of k-algebras (or from the category of schemes over k) to the category of sets are called *representable*, and in that case we say that the variety X *represents* the corresponding functor. Thus, if we want to define a "would be" algebraic variety X, we do not lose any information if we start by defining its "would be" functor of points. *A priori*, it may be unclear whether the functor is a functor of points (i.e., that it may be realized by the above construction). So the question we need to answer is whether this functor is representable. If it is, then the object that represents it is the sought-after algebraic variety.

3.4.2 Definition of jet schemes

Let us see now how this works in the case of jet schemes. In this case the "would be" functor of points is easy to define. Indeed, let X is an algebraic variety, which for definiteness we will assume to be defined over $\mathbb{C}$. Then an R-point of X is a morphism $\operatorname{Spec} R \to X$, or equivalently a homomorphism of algebras $\operatorname{Fun} X \to R$. An R-point of the "would be" jet scheme JX should be viewed as a morphism from the disc over $\operatorname{Spec}(R)$, which is $\operatorname{Spec}(R[[t]])$, to X. Thus, we define the functor of points of JX as the functor sending a $\mathbb{C}$-algebra R to the set of morphisms $\operatorname{Spec}(R[[t]]) \to X$, and homomorphisms $R \to R'$ to obvious maps between these sets. It is not difficult to see that this functor is a "sheaf," and so in particular extends in a unique way to the category of all schemes over $\mathbb{C}$. In what follows we will just look at this functor on the category of $\mathbb{C}$-algebras.

If X is an affine variety, then a morphism $\operatorname{Spec} R[[t]] \to X$ is the same as a homomorphism $\operatorname{Fun} X \to R[[t]]$, so we obtain a very concrete realization of the "would be" functor of points of the set scheme JX.

For example, suppose that X is the affine space $\mathbb{A}^N$. Choosing coordinate functions $x_1, \ldots, x_N$ on $\mathbb{A}^N$, we obtain that $\operatorname{Fun} \mathbb{A}^N = \mathbb{C}[x_1, \ldots, x_N]$. In other words, $\mathbb{A}^N = \operatorname{Spec} \mathbb{C}[x_1, \ldots, x_N]$. Thus, $\mathbb{C}$-points of $\mathbb{A}^N$ are homomorphisms $\phi : \mathbb{C}[x_1, \ldots, x_N] \to \mathbb{C}$. Those are given by n-tuples of complex numbers $(a_1, \ldots, a_N)$, such that $\phi(x_i) = a_i$, as expected. Likewise, its R-points are n-tuples of elements of R.

By definition, the set of R-points of the jet scheme $J\mathbb{A}^N$ should now be given by homomorphisms

$$\mathbb{C}[x_1, \ldots, x_N] \to R[[t]].$$

Such a homomorphism is uniquely determined by the images of the generators x_i, which are now formal power series

$$a_i(t) = \sum_{n<0} a_{i,n} t^{-n-1}, \qquad a_{i,n} \in R, i = 1, \ldots, N \tag{3.4.1}$$

(the indices are chosen so as to agree with the convention of our previous formulas). Note that there are no conditions of convergence on the formal power series. In other words, it is really "formal" paths that we are describing.

We can equivalently think of the collection of N formal power series as a collection of all of their Fourier coefficients. Hence the set of R-points of $J\mathbb{A}^N$ is

$$\{a_{i,n} \in R \,|\, i = 1, \ldots, N; n < 0\}.$$

The corresponding functor is representable by

$$J\mathbb{A}^N \stackrel{\text{def}}{=} \operatorname{Spec} \mathbb{C}[x_{i,n}]_{i=1,\ldots,N; n<0}.$$

Thus we obtain the definition of the jet scheme $J\mathbb{A}^N$ of $\mathbb{A}^N$. As expected, $J\mathbb{A}^N$ is an infinite-dimensional affine space (inverse limit of finite-dimensional ones).

Suppose now that we have a finite-dimensional affine algebraic variety X. What does its jet scheme look like? We may write

$$X = \operatorname{Spec}\ (\mathbb{C}[x_1, \ldots, x_N] / \langle F_1, \ldots, F_M \rangle),$$

where the F_i's are polynomials in the variables $x_1, \ldots, x_N$. This means that the complex points of X are n-tuples of complex numbers $(a_1, \ldots, a_N)$ satisfying the equations

$$F_i(a_1, \ldots, a_N) = 0 \quad \text{for all } i.$$

By definition, the set of R-points of the jet scheme JX of X is the set of homomorphisms

$$\mathbb{C}[x_1, \ldots, x_N] / \langle F_1, \ldots, F_M \rangle \to R[[t]].$$

Such a homomorphism is again uniquely determined by the images of the x_i's which are formal power series $a_i(t)$ as in formula (3.4.1), but now they have to satisfy the equations

$$F_i(a_1(t), \ldots, a_N(t)) \equiv 0, \qquad i = 1, \ldots, N. \tag{3.4.2}$$

We obtain the equations on the Fourier coefficients $a_{i,n}$ by reading off the coefficients of t^m.

A compact way to organize these equations is as follows. Define a derivation T of the algebra $\mathbb{C}[x_{j,n}]$ by the formula

$$T(x_{j,n}) = -n x_{j,n-1}.$$

Let us replace x_j by $x_{j,-1}$ in the defining polynomials F_i and call these F_i again. Then the equations that come from (3.4.2) are precisely

$$T^m F_i = 0, \qquad i = 1, \ldots, N; m \geq 0.$$

Therefore we find that this functor of points is representable by the scheme

$$JX = \operatorname{Spec}\left(\mathbb{C}[x_{j,n}]_{j=1,\ldots,N; n<0} / \langle T^m F_i \rangle_{i=1,\ldots,M; m\geqslant 0} \right).$$

This is the **jet scheme** of X.

In particular, we see that the algebra of functions on JX carries a derivation T, inherited from $\mathbb{C}[x_{j,n}]_{j=1,\ldots,N;n<0}$. Therefore it is a commutative vertex algebra.

Now we have the definition of the jet scheme of an arbitrary affine scheme. What about more general schemes? An arbitrary scheme has a covering by affine schemes. To define the jet scheme of X we simply need to work out how to glue together the jet schemes for the affine pieces. For example $\mathbb{P}^1$ can be decomposed into two affine schemes $\mathbb{A}^1 \cup \mathbb{A}^1$, where these are the complements of ∞ and 0 respectively. Therefore they are given by $\operatorname{Spec}(\mathbb{C}[x])$ and $\operatorname{Spec}(\mathbb{C}[y])$ with identification given by $x \leftrightarrow y^{-1}$. We already know how to get the jet schemes of these affine components. The obvious way to generalize the gluing conditions is to substitute formal power series into them, i.e.,

$$x(t) \longleftrightarrow y(t)^{-1}.$$

In order to write $y(t)^{-1}$ as a formal power series in t it is only necessary to invert y_{-1}, which is invertible on the overlap of the two affine lines in $\mathbb{P}^1$. Therefore this formula is well-defined on the overlap, and it glues together the jet schemes of the two affine lines into the jet scheme of $\mathbb{P}^1$.

A similar statement is true in general and makes the gluing possible (the only difficulties come from the inverted variables, but these are compatible as they were compatible for the original scheme). Therefore we have constructed the scheme JX associated to any scheme X (of finite type).

Let

$$D_n = \mathrm{Spec}(\mathbb{C}[[t]]/(t^{n+1}))$$

be the nth order neighborhood of a point. We can construct **nth order jet schemes** J_nX which represent the maps from D_n to X. The natural maps $D_{n-1} \to D_n$ give rise to maps $J_nX \to J_{n-1}X$. We may then represent the (infinite order) jet scheme JX as the inverse limit of the finite jet schemes:

$$JX = \varprojlim J_nX.$$

This gives a realization of the jet scheme as an inverse limit of schemes of finite type.

The following lemma follows immediately from the criterion of smoothness: a morphism $\phi : X \to Y$ of schemes of finite type over $\mathbb{C}$ is smooth if and only if for any $\mathbb{C}$-algebra R and its nilpotent enlargement R_{nilp} (i.e., a $\mathbb{C}$-algebra whose quotient by the ideal generated by nilpotent elements is R) and a morphism $p : \mathrm{Spec}(R) \to Y$, which lifts to X via ϕ, any extension of p to a morphism $\mathrm{Spec}(R_{\mathrm{nilp}}) \to Y$ also lifts to X.

We will say that a morphism of $\mathbb{C}$-schemes $X \to Y$ is surjective if the corresponding map of the sets of $\mathbb{C}$-points $X(\mathbb{C}) \to Y(\mathbb{C})$ is surjective.

Lemma 3.4.1 *If a morphism $\phi : X \to Y$ is smooth and surjective, then the corresponding morphisms of jet schemes $J_n\phi : J_nX \to J_nY$ and $J\phi : JX \to JY$ are (formally) smooth and surjective.*

3.4.3 Description of invariant functions on $J\mathfrak{g}$

We now use the notion of jet schemes to prove the following theorem about the invariant functions on $\mathfrak{g}^*[[t]]$, which is due to A. Beilinson and V. Drinfeld [BD1] (see also Proposition A.1 of [EF]). Recall that $\mathfrak{g}$ is a simple Lie algebra of rank ℓ.

Theorem 3.4.2 *The algebra* $\mathrm{Inv}\,\mathfrak{g}^*[[t]]$ *is equal to* $\mathbb{C}[\overline{P}_{i,n}]_{i=1,\ldots,\ell;n<0}$.

Proof. Recall that $\mathrm{Inv}\,\mathfrak{g}^*[[t]]$ is the algebra of $G[[t]]$-invariant functions on $\mathfrak{g}^*[[t]]$. In terms of jet schemes, we can write $\mathfrak{g}^*[[t]] = J\mathfrak{g}^*, G[[t]] = JG$, and so the space $\mathrm{Inv}\,\mathfrak{g}^*[[t]] = (\mathrm{Fun}\,J\mathfrak{g}^*)^{JG}$ should be thought of as the algebra of functions on the quotient $J\mathfrak{g}^*/JG$. Unfortunately, this is not a variety, as the structure of the orbits of JG in $J\mathfrak{g}^*$ is rather complicated.

More precisely, define $\mathcal{P}$ to be $\mathrm{Spec}\,\mathrm{Inv}\,\mathfrak{g}^* = \mathrm{Spec}\mathbb{C}[\overline{P}_i]_{i=1,\ldots,\ell}$. Thus, $\mathcal{P}$ is the affine space with coordinates $\overline{P}_i, i = 1, \ldots, \ell$. The inclusion $\mathrm{Inv}\,\mathfrak{g}^* \hookrightarrow \mathrm{Fun}\,\mathfrak{g}^*$ gives rise to a surjective map $p : \mathfrak{g}^* \to \mathcal{P}$. The group G acts along the fibers, but the fibers do not always consist of single G-orbits.

For example, the inverse image of 0 is the **nilpotent cone**. Let us identify $\mathfrak{g}$ with $\mathfrak{g}^*$ by means of a non-degenerate invariant inner product. Then in the case of $\mathfrak{g} = \mathfrak{sl}_n$ the nilpotent cone consists of all nilpotent $n \times n$ matrices in $\mathfrak{sl}_n$. The orbits of nilpotent matrices are parameterized by the Jordan forms, or, equivalently, partitions of n. There is one dense orbit consisting of matrices with one Jordan block and several smaller orbits corresponding to other Jordan block forms. It is these smaller orbits that are causing the problems. In particular, the morphism $\mathfrak{g}^* \to \mathcal{P}$ is not smooth. So we throw these orbits away by defining the open dense subset of regular elements

$$\mathfrak{g}^*_{\text{reg}} = \{x \in \mathfrak{g}^* \mid \dim \mathfrak{g}_x = \ell\}$$

where $\mathfrak{g}_x$ is the centralizer of x in $\mathfrak{g}$.†

The map obtained by restriction

$$p_{\text{reg}} : \mathfrak{g}^*_{\text{reg}} \longrightarrow \mathcal{P}$$

is already smooth. Furthermore, it is surjective and each fiber is a single G-orbit [Ko].

Applying the functor of jet schemes, we obtain a morphism

$$Jp_{\text{reg}} : J\mathfrak{g}^*_{\text{reg}} \longrightarrow J\mathcal{P},$$

which is (formally) smooth and surjective, by Lemma 3.4.1. Here

$$J\mathcal{P} = \operatorname{Spec} \mathbb{C}[\overline{P}_{i,n}]_{i=1,\dots,\ell;n<0}$$

is the infinite affine space with coordinates $\overline{P}_{i,n}, i = 1, \dots, \ell; n < 0$.

We want to show that the fibers of Jp_{reg} also consist of single JG-orbits. To prove that, consider the map

$$G \times \mathfrak{g}^*_{\text{reg}} \longrightarrow \mathfrak{g}^*_{\text{reg}} \underset{\mathcal{P}}{\times} \mathfrak{g}^*_{\text{reg}}.$$

This is again a smooth morphism. But note now that its surjectivity is equivalent to the fact that the fibers of $\mathfrak{g}^*_{\text{reg}} \longrightarrow \mathcal{P}$ consist of single orbits. Taking jet schemes, we obtain a morphism

$$JG \times J\mathfrak{g}^*_{\text{reg}} \longrightarrow J\mathfrak{g}^*_{\text{reg}} \underset{J\mathcal{P}}{\times} J\mathfrak{g}^*_{\text{reg}}.$$

It is (formally) smooth and surjective by Lemma 3.4.1. This shows that each fiber of Jp_{reg} is a single JG-orbit.

This implies that the algebra of JG-invariant functions on $J\mathfrak{g}^*_{\text{reg}}$ is equal to $\text{Fun}(J\mathcal{P}) = \mathbb{C}[\overline{P}_{i,n}]_{i=1,\dots,\ell;n<0}$.

To be precise, we prove first the corresponding statement for finite jet schemes. For each $N \geq 0$ we have a morphism $J_N\mathfrak{g}^*_{\text{reg}} \to J_N\mathcal{P}$, equipped with a fiberwise action of J_NG. We obtain in the same way as above that each fiber of this morphism is a single J_NG-orbit. Proposition 0.2 of [Mu]

† This is the minimal possible dimension the centralizer can have.

then implies that the algebra of $J_N G$-invariant functions on $J_N \mathfrak{g}^*_{\text{reg}}$ is equal to $\text{Fun}(J_N \mathcal{P}) = \mathbb{C}[\overline{P}_{i,n}]_{i=1,\ldots,\ell;-N-1\leq n<0}$.

Now, any JG-invariant function on $J\mathfrak{g}^*_{\text{reg}}$ comes by pull-back from a $J_N G$-invariant function on $J_N \mathfrak{g}^*_{\text{reg}}$. Hence we obtain that the algebra of JG-invariant functions on $J\mathfrak{g}^*_{\text{reg}}$ is equal to $\text{Fun}(J\mathcal{P}) = \mathbb{C}[\overline{P}_{i,n}]_{i=1,\ldots,\ell;n<0}$.

But regular elements are open and dense in $\mathfrak{g}^*$, and hence $J_N \mathfrak{g}^*_{\text{reg}}$ is open and dense in $J_N \mathfrak{g}^*$, for both finite and infinite N. Therefore all functions on $J_N \mathfrak{g}^*$ are uniquely determined by their restrictions to $J_N \mathfrak{g}^*_{\text{reg}}$, for both finite and infinite N. Therefore the algebra of $J_N G$-invariant functions on $J_N \mathfrak{g}^*$ is equal to $\text{Fun}(J_N \mathcal{P}) = \mathbb{C}[\overline{P}_{i,n}]_{i=1,\ldots,\ell;-N-1\leq n<0}$. □

3.5 The center in the case of $\mathfrak{sl}_2$

Thus, we have now been able to describe the graded analogue $\text{gr}\,\mathfrak{z}(\widehat{\mathfrak{g}})$ of the center $\mathfrak{z}(\widehat{\mathfrak{g}})$.

It is now time to describe the center $\mathfrak{z}(\widehat{\mathfrak{g}})$ itself. We start with the simplest case, when $\mathfrak{g} = \mathfrak{sl}_2$.

3.5.1 *First description*

We know from Lemma 3.3.1 that there is an injection from the associated graded $\text{gr}\,\mathfrak{z}(\widehat{\mathfrak{g}})$ of the center $\mathfrak{z}(\widehat{\mathfrak{g}})$ of $V_{\kappa_c}(\mathfrak{g})$ into the algebra $\text{Inv}\,\mathfrak{g}^*[[t]]$ of invariant functions on $\mathfrak{g}^*[[t]]$. We have just shown that the algebra $\text{Inv}\,\mathfrak{g}^*[[t]]$ is equal to the polynomial ring $\mathbb{C}[\overline{P}_{i,n}]_{i=1,\ldots,\ell;n<0}$. In the case of $\mathfrak{g} = \mathfrak{sl}_2(\mathbb{C})$ this is already sufficient to determine the structure of the center $\mathfrak{z}(\widehat{\mathfrak{sl}}_2)$ of the vertex algebra $V_{\kappa_c}(\mathfrak{sl}_2)$. Namely, we prove the following:

Theorem 3.5.1 *The center $\mathfrak{z}(\widehat{\mathfrak{sl}}_2)$ of the vertex algebra $V_{\kappa_c}(\mathfrak{sl}_2)$ is equal to $\mathbb{C}[S_n]_{n\leq -2}|0\rangle$. Here, the S_n's are the Segal–Sugawara operators (note that the operators $S_n, n \geq -1$, annihilate the vacuum vector $|0\rangle$).*

Proof. We already know that the Segal–Sugawara operators $S_n, n \in \mathbb{Z}$, are central elements of the completed enveloping algebra of critical level. Therefore we have a map

$$\mathbb{C}[S_n]_{n\leq -2}|0\rangle \to V_{\kappa_c}(\mathfrak{sl}_2)^{\mathfrak{sl}_2[[t]]} = \mathfrak{z}(\widehat{\mathfrak{sl}}_2). \tag{3.5.1}$$

We need to prove that this map is an equality, i.e., the operators $S_n, n \leq -2$, are algebraically independent and there are no other elements in $\mathfrak{z}(\widehat{\mathfrak{sl}}_2)$. For that it is sufficient to show that the map of associated graded algebras is an isomorphism.

Recall that

$$S_n = \frac{1}{2} \sum_a \sum_{j+k=n} :J^a_j J_{a,k}:. \tag{3.5.2}$$

For $n \leq -2$ some of these terms will have both j and k negative and the rest will have at least one of j and k positive; the former will act

$$V(\mathfrak{g})_{\leqslant i} \longrightarrow V(\mathfrak{g})_{\leqslant (i+2)}$$

(because they are creation operators) and the rest act

$$V(\mathfrak{g})_{\leqslant i} \longrightarrow V(\mathfrak{g})_{\leqslant (i+1)}$$

(because they have at least one annihilation operator). Hence the latter terms will be removed when looking at the symbol, and we find that the symbol of the monomial $S_{n_1} \ldots S_{n_m}$ with $n_i \leq -2$ is equal to $\overline{P}_{1,n_1+1} \ldots \overline{P}_{1,n_m+1}$. Here

$$\overline{P}_1(z) = \sum_{n<0} P_{1,n} z^{-n-1} = \frac{1}{2} \sum_a \overline{J}^a(z) \overline{J}_a(z)$$

is the first (and the only, for $\mathfrak{g} = \mathfrak{sl}_2$) of the generating series $\overline{P}_i(z)$ considered above.

In particular, we see that the map (3.5.1) is compatible with the natural filtrations, and the corresponding map of associated graded spaces factors as follows

$$\mathrm{gr}\, \mathbb{C}[S_n]_{n \leq -2} |0\rangle \simeq \mathbb{C}[\overline{P}_n]_{n \leq -2} \rightarrow (\mathrm{gr}\, V_{\kappa_c}(\mathfrak{sl}_2))^{\mathfrak{sl}_2[[t]]} \simeq \mathrm{Inv}(\mathfrak{sl}_2)^*[[t]],$$

where the first map sends $S_n \mapsto \overline{P}_{1,n}$. We obtain from Theorem 3.4.2 that the second map is an isomorphism. Therefore we have

$$\mathrm{gr}\, \mathbb{C}[S_n]_{n \leq -2} |0\rangle \simeq \mathrm{Inv}(\mathfrak{sl}_2)^*[[t]].$$

But we know that

$$\mathrm{gr}\, \mathbb{C}[S_n]_{n \leq -2} |0\rangle \subset \mathrm{gr}\, \mathfrak{z}(\widehat{\mathfrak{sl}}_2) \subset \mathrm{Inv}\, (\mathfrak{sl}_2)^*[[t]]$$

(see Lemma 3.3.1). Therefore both of the above inclusions are equalities, and so $\mathrm{gr}\, \mathbb{C}[S_n]_{n \leq -2} |0\rangle = \mathrm{gr}\, \mathfrak{z}(\widehat{\mathfrak{sl}}_2)$. Since we know that $\mathbb{C}[S_n]_{n \leq -2} |0\rangle \subset \mathfrak{z}(\widehat{\mathfrak{sl}}_2)$, this implies that $\mathbb{C}[S_n]_{n \leq -2} |0\rangle = \mathfrak{z}(\widehat{\mathfrak{sl}}_2)$. □

A similar result may be obtained for the center of the completed universal enveloping algebra $\widetilde{U}_{\kappa_c}(\widehat{\mathfrak{sl}}_2)$: the center is topologically generated by $S_n, n \in \mathbb{Z}$ (so this time all of the S_n's are used). The argument is essentially the same, but we will postpone this proof till Section 4.3.2.

We will see below that both of these results remain true in the case of a more general Lie algebra $\mathfrak{g}$. In other words, we will show that $\mathfrak{z}(\widehat{\mathfrak{g}})$ is "as large as possible," that is, $\mathrm{gr}\, \mathfrak{z}(\widehat{\mathfrak{g}}) = \mathrm{Inv}\, \mathfrak{g}^*[[t]]$. This is equivalent to the existence of central elements $S_i \in \mathfrak{z}(\widehat{\mathfrak{g}}) \subset V_{\kappa_c}(\mathfrak{g})$, whose symbols are equal to $\overline{P}_{i,-1} \in \mathrm{Inv}\, \mathfrak{g}^*[[t]]$. It will then follow in the same way as for $\mathfrak{g} = \mathfrak{sl}_2$ that

$$\mathfrak{z}(\widehat{\mathfrak{g}}) = \mathbb{C}[S_{i,[n]}]_{i=1,\ldots,\ell; n<0},$$

where the $S_{i,[n]}$'s are the Fourier coefficients of the formal power series $Y[S_i, z]$. Moreover, it will follow that the center of $\widetilde{U}_{\kappa_c}(\widehat{\mathfrak{g}})$ is topologically generated by $S_{i,[n]}, i = 1, \ldots, \ell; n \in \mathbb{Z}$.

We have already constructed the first of these elements; namely, the quadratic element $\frac{1}{2} J^a_{-1} J_{a,-1}|0\rangle \in \mathfrak{z}(\widehat{\mathfrak{g}})$ whose symbol is $\frac{1}{2} \overline{J}^a_{-1} \overline{J}_{a,-1}$, corresponding to the quadratic Casimir element $\overline{P}_1$. Unfortunately, the formulas for the central elements corresponding to the higher order Casimir elements $\overline{P}_{i,-1}, i > 1$, are unknown in general (the leading order terms are obvious, but it is finding the lower order terms that causes lots of problems). So we cannot prove this result directly, as in the case of $\mathfrak{g} = \mathfrak{sl}_2$. We will have to use a different argument instead.

3.5.2 Coordinate-independence

Our current approach suffers from another kind of deficiency: everything we have done so far has been coordinate-dependent. We started with a Lie algebra $\mathfrak{g}$ and constructed the Lie algebra $\widehat{\mathfrak{g}}$ as the central extension of $\mathfrak{g} \otimes \mathbb{C}((t))$. However, there are many local fields which are isomorphic to the field $\mathbb{C}((t))$, but not naturally.

For example, let X be a smooth algebraic curve over $\mathbb{C}$, and x any point on X. The choice of x gives rise to a valuation $|\cdot|_x$ on the field of rational functions $\mathbb{C}(X)$. The valuation is defined to be the order of vanishing of f at x and is coordinate-independent. The completion of the algebra of functions under this valuation is denoted by $\mathcal{K}_x$. Let us pick a local coordinate at x, that is, a rational function vanishing to order 1 at x (more generally, we may choose a "formal coordinate," which is an arbitrary element of $\mathcal{K}_x$ vanishing at x to order 1). If we denote this function by t, then we obtain a map $\mathbb{C}((t)) \to \mathcal{K}_x$ which is an isomorphism. Thus, we identify $\mathcal{K}_x$ with $\mathbb{C}((t))$ for each choice of a coordinate at x. But usually there is no preferred choice, and therefore there is no canonical isomorphism $\mathcal{K}_x \simeq \mathbb{C}((t))$.

However, all of the objects that we have discussed so far may be attached canonically to $\mathcal{K}_x$ without any reference to a particular coordinate. Indeed, we define the affine Kac–Moody algebra $\widehat{\mathfrak{g}}_{\kappa,x}$ attached to $\mathcal{K}_x$ as the central extension

$$0 \longrightarrow \mathbb{C}\mathbf{1} \longrightarrow \widehat{\mathfrak{g}}_{\kappa,x} \longrightarrow \mathfrak{g} \otimes \mathcal{K}_x \longrightarrow 0$$

given by the two-cocycle

$$c(A \otimes f, B \otimes g) = -\kappa(A, B) \operatorname{Res}_x(f\, dg).$$

This formula is coordinate-independent as $f\, dg$ is a one-form, and so its residue is canonically defined. We then define a coordinate-independent vacuum $\widehat{\mathfrak{g}}_{\kappa,x}$-module associated to the point x as

$$V_\kappa(\mathfrak{g})_x = \operatorname{Ind}^{\widehat{\mathfrak{g}}_{\kappa,x}}_{\mathfrak{g}\otimes\mathcal{O}_x \oplus \mathbb{C}\mathbf{1}} \mathbb{C}, \tag{3.5.3}$$

where $\mathcal{O}_x$ is the ring of integers of $\mathcal{K}_x$ (comprising those elements of $\mathcal{K}_x$ which have no pole at x).

Now our task is to describe the algebra

$$\mathfrak{z}(\widehat{\mathfrak{g}})_x = \operatorname{End}_{\widehat{\mathfrak{g}}_x} V_{\kappa_c}(\mathfrak{g})_x = (V_{\kappa_c}(\mathfrak{g})_x)^{\mathfrak{g}\otimes\mathcal{O}_x}, \tag{3.5.4}$$

and we should try to do this in a coordinate-independent way. Unfortunately, everything we have done so far has been pegged to the field $\mathbb{C}((t))$, and so to interpret our results for an arbitrary local field $\mathcal{K}_x$ we need to pick a coordinate t at x. The problem is that we do not know yet what will happen if we choose another coordinate t'. Therefore at the moment we cannot describe $\mathfrak{z}(\widehat{\mathfrak{g}})_x$ using Theorem 3.5.1, for example, because it does not describe the center in intrinsic geometric terms. For that we need to know how the Segal–Sugawara operators generating the center transform under changes of coordinates. We will address these questions in the next section.

3.5.3 The group of coordinate changes

We start off by doing something which looks coordinate-dependent, but it will eventually allow us to do things in a coordinate-independent way.

Let $\mathcal{K}$ denote the complete topological algebra $\mathbb{C}((t))$, and $\mathcal{O} = \mathbb{C}[[t]]$ with the induced topology. We study the group of continuous automorphisms of $\mathcal{O}$, which we denote by $\operatorname{Aut}\mathcal{O}$. It can be thought of as the group of automorphisms of the **disc** $D = \operatorname{Spec}\mathcal{O}$.† As t is a topological generator of the algebra $\mathcal{O}$, a continuous automorphism of $\mathcal{O}$ is determined by its action on t. So we write

$$\rho(t) = a_0 + a_1 t + a_2 t^2 + \dots .$$

In order for this to be an automorphism it is necessary and sufficient that $a_0 = 0$ and $a_1 \neq 0$. Note the composition formula in $\operatorname{Aut}\mathcal{O}$:

$$\rho(t) \circ \phi(t) = \phi(\rho(t)).$$

Let x be a smooth point of a complex curve X. We have the local field $\mathcal{K}_x$ and its ring of integers $\mathcal{O}_x$ defined in the previous section. The ring $\mathcal{O}_x$ is a complete local ring, with the maximal ideal $\mathfrak{m}_x$ consisting of those elements that vanish at x. By definition, a **formal coordinate** at x is a topological generator of $\mathfrak{m}_x$. We define $\mathcal{A}ut_x$ to be the set of all formal coordinates at x. The group $\operatorname{Aut}\mathcal{O}$ naturally acts on $\mathcal{A}ut_x$ by the formula $t \mapsto \rho(t)$. Note that according to the above composition formula this is a right action.

Recall that if G is a group, then a **G-torsor** is a non-empty set S with a right action of G, which is simply transitive. By choosing a point s_0 in S,

† Sometimes D is referred to as a "formal disc," but we prefer the term "disc," reserving "formal disc" for the formal scheme $\operatorname{Spf}\mathcal{O}$ (this is explained in detail in Appendix A.1.1. of [FB]).

we obtain an obvious isomorphism between G and $S : g \mapsto g(s_0)$. However, *a priori* there is no natural choice of such a point, and hence no natural isomorphism between the two (just the G-action).

Given a G-torsor S and a representation M of G we can **twist** M **by** S by defining

$$\mathcal{M} = S \underset{G}{\times} M.$$

It is instructive to think of a G-torsor S as a principal G-bundle over a point. Then $\mathcal{M}$ is the vector bundle over the point associated to the representation M.

Now observe that the above action of $\operatorname{Aut} \mathcal{O}$ makes $\mathcal{A}ut_x$ into an $\operatorname{Aut} \mathcal{O}$-torsor. Hence we can twist representations of $\operatorname{Aut} \mathcal{O}$. The most obvious representations are $\mathbb{C}[[t]]$ and $\mathbb{C}((t))$. Not surprisingly, we find that

$$\mathcal{O}_x = \mathcal{A}ut_x \underset{\operatorname{Aut} \mathcal{O}}{\times} \mathbb{C}[[t]], \qquad \mathcal{K}_x = \mathcal{A}ut_x \underset{\operatorname{Aut} \mathcal{O}}{\times} \mathbb{C}((t)). \tag{3.5.5}$$

In other words, we can recover the local field $\mathcal{K}_x$ and its ring of integers $\mathcal{O}_x$ from $\mathbb{C}((t))$ and $\mathbb{C}[[t]]$ using the action of the group $\operatorname{Aut} \mathcal{O}$. This gives us a hint as to how to deal with the issue of coordinate-independence that we discussed in the previous section: if we formulate all our results for the field $\mathbb{C}((t))$ *keeping track* of the action of its group of symmetries $\operatorname{Aut} \mathcal{O}$, then we can seamlessly pass from $\mathbb{C}((t))$ to an arbitrary local field $\mathcal{K}_x$ by using the "twisting by the torsor" construction. In order to explain this more precisely, we discuss the types of representations of $\operatorname{Aut} \mathcal{O}$ that we should allow.

3.5.4 Action of coordinate changes

Let $\operatorname{Aut}_+ \mathcal{O}$ be the subgroup of $\operatorname{Aut} \mathcal{O}$ consisting of those $\rho(t)$ as above that have $a_1 = 1$. The group $\operatorname{Aut} \mathcal{O}$ is the semi-direct product

$$\operatorname{Aut} \mathcal{O} = \mathbb{C}^\times \ltimes \operatorname{Aut}_+ \mathcal{O},$$

where $\mathbb{C}^\times$ is realized as the group of rescalings $t \mapsto at$.

Denote the Lie algebra of $\operatorname{Aut} \mathcal{O}$ (resp., $\operatorname{Aut}_+ \mathcal{O}$) by $\operatorname{Der}_0 \mathcal{O}$ (resp., $\operatorname{Der}_+ \mathcal{O}$). It is easy to see that

$$\operatorname{Der}_0 \mathcal{O} = t\mathbb{C}[[t]]\partial_t, \qquad \operatorname{Der}_+ \mathcal{O} = t^2\mathbb{C}[[t]]\partial_t.$$

In particular, this shows that $\operatorname{Der}_+ \mathcal{O}$ is a pro-nilpotent Lie algebra and hence $\operatorname{Aut}_+ \mathcal{O}$ is a pro-unipotent group.†

We wish to describe the category of continuous representations of $\operatorname{Aut} \mathcal{O}$ on vector spaces with discrete topology. As $\operatorname{Aut}_+ \mathcal{O}$ is a pro-unipotent group we

† It may appear strange that the Lie algebra of $\operatorname{Aut} \mathcal{O}$ is $\operatorname{Der}_0 \mathcal{O}$ and not $\operatorname{Der} \mathcal{O} = \mathbb{C}[[t]]\partial_t$. The reason is that the part of $\operatorname{Der} \mathcal{O}$ spanned by the translation vector field ∂_t cannot be exponentiated within the realm of algebraic groups, but only ind-groups, see [FB], Section 6.2.3, for more details.

know that the exponential map from $\mathrm{Der}_+ \mathcal{O}$ to $\mathrm{Aut}_+ \mathcal{O}$ is an isomorphism. Hence a discrete representation of $\mathrm{Aut}_+ \mathcal{O}$ is the same as a discrete representation of $\mathrm{Der}_+ \mathcal{O}$, i.e., such that for any vector v we have $t^n \partial_t \cdot v = 0$ for n greater than a positive integer N.

The Lie algebra of the group $\mathbb{C}^\times = \{t \mapsto at\}$ of rescalings is $\mathbb{C}t\partial_t \simeq \mathbb{C}$. Here we have a discrepancy as the Lie algebra has irreducible representations parameterized by complex numbers, whereas $\mathbb{C}^\times$ only has representations parameterized by integers: on the representation corresponding to $n \in \mathbb{Z}$ the element $a \in \mathbb{C}^\times$ acts as a^n. Furthermore, any linear transformation T on a vector space V defines a representation of $\mathbb{C}$, $\lambda \mapsto T$ (in particular, non-trivial extensions are possible because of non-trivial Jordan forms), but any representation of $\mathbb{C}^\times$ is the direct sum of one-dimensional representations labeled by integers. So the only representations of the Lie algebra $\mathbb{C}$ that lift to representations of $\mathbb{C}^\times$ are those on which the element $1 \in \mathbb{C}$ (which corresponds to the vector field $t\partial_t$ in our case) acts semi-simply and with integer eigenvalues.

To summarize, a discrete representation of $\mathrm{Aut}\,\mathcal{O}$ is the same as a discrete representation of the Lie algebra $\mathrm{Der}_0 \mathcal{O}$ such that the generator $t\partial_t$ acts semi-simply with integer eigenvalues. In most cases we consider, the action of the Lie algebra $\mathrm{Der}_0 \mathcal{O}$ may be augmented to an action of $\mathrm{Der}\,\mathcal{O} = \mathbb{C}[[t]]\partial_t$. (The action of the additional generator ∂_t is important, as it plays the role of a connection that allows us to identify the objects associated to nearby points x, x' of a curve X.)

In Section 3.1.3 we introduced a topological basis of the Virasoro algebra. The elements $L_n = -t^{n+1}\frac{\partial}{\partial t}$ of this basis with $n \geq 0$ (resp., $n > 0$) generate $\mathrm{Der}_0 \mathcal{O}$ (resp., $\mathrm{Der}_+ \mathcal{O}$). The Lie algebra $\mathbb{C}$ is generated by $L_0 = -t\partial_t$. Suppose that V is a representation of $\mathrm{Der}_0 \mathcal{O}$, on which L_0 acts semi-simply with integer eigenvalues. Thus, we obtain a $\mathbb{Z}$-grading on V. The commutation relations in $\mathrm{Der}_0 \mathcal{O}$ imply that L_n shifts the grading by $-n$. Therefore, if the $\mathbb{Z}$-grading is bounded from below, then any vector in V is annihilated by L_n for sufficiently large n, and so V is a discrete representation of $\mathrm{Der}_0 \mathcal{O}$, which then may be exponentiated to $\mathrm{Aut}\,\mathcal{O}$.

An example of such a representation is given by the vacuum module $V_\kappa(\mathfrak{g})$. The Lie algebra $\mathrm{Der}_0 \mathcal{O}$ (and even the larger Lie algebra $\mathrm{Der}\,\mathcal{O}$) naturally acts on it because it acts on $\mathfrak{g}((t))$ and preserves the Lie subalgebra $\mathfrak{g}[[t]] \oplus \mathbb{C}\mathbf{1}$ from which $V_\kappa(\mathfrak{g})$ is induced. In particular, L_0 becomes the vertex algebra grading operator, and since this grading takes only non-negative values, we find that the action of $\mathrm{Der}_0 \mathcal{O}$ may be exponentiated to $\mathrm{Aut}\,\mathcal{O}$. The center $\mathfrak{z}(\widehat{\mathfrak{g}})$ is a subspace of $V_{\kappa_c}(\mathfrak{g})$ preserved by $\mathrm{Aut}\,\mathcal{O}$, and hence it is a subrepresentation of $V_{\kappa_c}(\mathfrak{g})$.

Thus, we can apply the twisting construction to $V_\kappa(\mathfrak{g})$ and $\mathfrak{z}(\widehat{\mathfrak{g}})$. Using formulas (3.5.5), we find that

$$\mathcal{A}ut_x \underset{\mathrm{Aut}\,\mathcal{O}}{\times} V_\kappa(\mathfrak{g}) = V_\kappa(\mathfrak{g})_x, \qquad \mathcal{A}ut_x \underset{\mathrm{Aut}\,\mathcal{O}}{\times} \mathfrak{z}(\widehat{\mathfrak{g}}) = \mathfrak{z}(\widehat{\mathfrak{g}})_x,$$

where $V_\kappa(\mathfrak{g})_x$ and $\mathfrak{z}(\widehat{\mathfrak{g}})_x$ are defined by formulas (3.5.3) and (3.5.4), respectively.

Thus, we have found a way to describe all spaces $\mathfrak{z}(\widehat{\mathfrak{g}})_x$, for different local fields $\mathcal{K}_x$, simultaneously: we simply need to describe the action of $\operatorname{Der}\mathcal{O}$ on $\mathfrak{z}(\widehat{\mathfrak{g}})$. We will now do this for $\mathfrak{g} = \mathfrak{sl}_2$.

3.5.5 Warm-up: Kac–Moody fields as one-forms

Let us start with a simpler example: consider the algebra $\operatorname{gr} V_\kappa(\mathfrak{g})$. As we saw previously, it is naturally isomorphic to

$$\operatorname{Sym}(\mathfrak{g}((t))/\mathfrak{g}[[t]]) = \operatorname{Fun}\mathfrak{g}^*[[t]]dt.$$

In Section 3.3.3 we had identified $\mathfrak{g}^*[[t]]dt$ with $\mathfrak{g}^*[[t]]$, but now that we wish to formulate our results in a coordinate-independent way, we are not going to do that.

Given a smooth point x of a curve X, we have the disc $D_x = \operatorname{Spec}\mathcal{O}_x$ and the punctured disc $D_x^\times = \operatorname{Spec}\mathcal{K}_x$ at x. Let $\Omega_{\mathcal{O}_x}$ be the $\mathcal{O}_x$-module of differentials. We may think of elements of $\Omega_{\mathcal{O}_x}$ as one-forms on the disc D_x. Consider the $\widehat{\mathfrak{g}}_{\kappa,x}$-module $V_\kappa(\mathfrak{g})_x$ attached to x and its associated graded $\operatorname{gr} V_\kappa(\mathfrak{g})_x$. Now observe that

$$\mathcal{A}ut_x \underset{\operatorname{Aut}\mathcal{O}}{\times} (\mathfrak{g}^*[[t]]dt) = \mathfrak{g}^* \otimes \Omega_{\mathcal{O}_x}.$$

Thus, $\operatorname{gr} V_\kappa(\mathfrak{g})_x$ is *canonically* identified with the algebra of functions on the space of $\mathfrak{g}^*$-valued one-forms on D_x.

Now we wish to obtain this result from an explicit computation. This computation will serve as a model for the computation performed in the next section, by which we will identify $\mathfrak{z}(\widehat{\mathfrak{sl}}_2)_x$ with the algebra of functions on the space of projective connections on D_x.

Let us represent an element $\phi \in \mathfrak{g}^*[[t]]dt$ by the formal power series

$$\phi(J^a) = \overline{J}^a(t) = \sum_{n<0} \overline{J}^a_n t^{-n-1}, \qquad a = 1, \ldots, \dim\mathfrak{g}.$$

Then the $\overline{J}^a_n$'s are coordinates on $\mathfrak{g}^*[[t]]dt$, and we have

$$\operatorname{Fun}\mathfrak{g}^*[[t]]dt \simeq \mathbb{C}[\overline{J}^a_n]_{a=1,\ldots,\dim\mathfrak{g};n<0}.$$

The Lie algebra $\operatorname{Der}\mathcal{O}$ acts on $\operatorname{Fun}\mathfrak{g}^*[[t]]dt$ by derivations. Therefore it is sufficient to describe its action on the generators $\overline{J}^a_m$:

$$L_n \cdot \overline{J}^a_m = \begin{cases} -m\overline{J}^a_{n+m} & \text{if } n+m \leqslant -1 \\ 0 & \text{otherwise} \end{cases} \tag{3.5.6}$$

We already know from the previous discussion that

$$\overline{J}^a(t)dt = \sum_{m\leqslant -1} \overline{J}^a_m t^{-m-1} dt$$

is a canonical one-form on the disc $D = \operatorname{Spec} \mathcal{O}$. Now we rederive this result using formulas (3.5.6). What we need to show is that $\overline{J}^a(t)dt$ is invariant under the action of the group $\operatorname{Aut} \mathcal{O}$ of changes of coordinates on D, or, equivalently (since $\operatorname{Aut} \mathcal{O}$ is connected), under the action of the Lie algebra $\operatorname{Der}_0 \mathcal{O}$. This Lie algebra acts in two different ways: it acts on the elements $\overline{J}^a_n$ as in (3.5.6) and it acts on $t^{-n-1}dt$ in the usual way. What we claim is that these two actions cancel each other.

Indeed, the action of L_n on $t^{-m-1}dt$ is given by

$$L_n \circ (t^{-m-1}dt) = (m-n)t^{n-m-1}dt.$$

Therefore

$$\begin{aligned} L_n \cdot \overline{J}^a(t)dt &= \sum_{m \leqslant -1} (L_n \cdot \overline{J}^a_m) t^{-m-1}dt + \sum_{m \leqslant -1} \overline{J}^a_m (L_n \cdot t^{-m-1}dt) \\ &= - \sum_{m \leqslant -1; n+m \leqslant -1} m \overline{J}^a_{m+n} t^{-m-1}dt + \sum_{m \leqslant -1} \overline{J}^a_m (m-n) t^{n-m-1}dt \\ &= \sum_{k \leqslant -1} (n-k) \overline{J}^a_k t^{n-k-1}dt + \sum_{m \leqslant -1} (m-n) \overline{J}^a_m t^{n-m-1}dt \\ &= 0, \end{aligned}$$

as expected. This implies that for *any* local field $\mathcal{K}_x$ we have a canonical isomorphism $\operatorname{gr} V_\kappa(\mathfrak{g})_x \simeq \operatorname{Fun} \mathfrak{g}^* \otimes \Omega_{\mathcal{O}_x}$, but now we have derived it using nothing but the transformation formula (3.5.6).

Of course, in retrospect our calculation seems tautological. All we said was that under the action of $\operatorname{Aut} \mathcal{O}$ the generators $\overline{J}^a_n$ of $\operatorname{gr} V_\kappa(\mathfrak{g})$ (for fixed a) transform in the same way as $t^n \in \mathcal{K}/\mathcal{O}$. We have the residue pairing between $\mathcal{K}/\mathcal{O} = \mathbb{C}((t))/\mathbb{C}[[t]]$ and $\Omega_{\mathcal{O}} = \mathbb{C}[[t]]dt$ with respect to which $\{t^n\}_{n<0}$ and $\{t^{-n-1}dt\}_{n<0}$ are dual (topological) bases. The expression $\overline{J}^a(t)dt$ pairs these two bases. Since the residue pairing is $\operatorname{Aut} \mathcal{O}$-invariant, it is clear that $\overline{J}^a(t)$ is also $\operatorname{Aut} \mathcal{O}$-invariant. Thus, the fact that $\overline{J}^a(t)$ is a canonical one-form follows immediately once we realize that the generators $\{\overline{J}^a_n\}$ of $\operatorname{gr} V_\kappa(\mathfrak{g})$ transform in the same way as the functions $\{t^n\}$ under the action of $\operatorname{Aut} \mathcal{O}$. In the next section we will obtain an analogue of this statement for the Segal–Sugawara operators S_n, the generators of the center $\mathfrak{z}(\widehat{\mathfrak{sl}}_2)$. This will enable us to give a coordinate-independent description of $\mathfrak{z}(\widehat{\mathfrak{sl}}_2)$.

3.5.6 The transformation formula for the central elements

From the analysis of the previous section it is clear what we need to do to obtain a coordinate-independent description of $\mathfrak{z}(\widehat{\mathfrak{sl}}_2)$: we need to find out how the generators S_n of $\mathfrak{z}(\widehat{\mathfrak{sl}}_2)$ transform under the action of $\operatorname{Der} \mathcal{O}$. A possible way to do it is to use formula (3.5.2) and the formula

$$L_n \cdot J^a_m = -m J^a_{n+m} \qquad (3.5.7)$$

for the action of $\operatorname{Der}\mathcal{O}$ on the Kac–Moody generators J^a_n.† This is a rather cumbersome calculation, but, fortunately, there is an easier way to do it.

Let us suppose that $\kappa \neq \kappa_c$. Then we define the renormalized Segal–Sugawara operators $\widetilde{S}_n = \frac{\kappa_0}{\kappa-\kappa_c} S_n, n \geq -1$, as in Section 3.1.1. These are elements of the completed enveloping algebra $\widetilde{U}_\kappa(\widehat{\mathfrak{g}})$, which commute with the elements $\widehat{\mathfrak{g}}$ in the same way as the operators $L_n = -t^{n+1}\partial_t, n \geq -1$; see formula (3.5.7). Therefore the action of L_n on any other element X of $\widetilde{U}_\kappa(\widehat{\mathfrak{g}})$ is equal to $[\widetilde{S}_n, X]$.

We are interested in the case when $X = S_m$. The corresponding commutators have already been computed in formula (3.1.3). This formula implies the following:

$$L_n \cdot S_m = [\widetilde{S}_n, S_m] = (n-m)S_{n+m} + \frac{n^3-n}{12}\dim(\mathfrak{g})\frac{\kappa}{\kappa_0}\delta_{n,-m}.$$

Though this formula was derived for $\kappa \neq \kappa_c$, it has a well-defined limit when $\kappa = \kappa_c$, and therefore gives us the sought-after transformation formula

$$L_n \cdot S_m = (n-m)S_{n+m} + \frac{n^3-n}{12}\dim(\mathfrak{g})\frac{\kappa_c}{\kappa_0}\delta_{n,-m} \tag{3.5.8}$$

(it depends on κ_0 because κ_0 enters explicitly the definition of S_m).

Let us specialize now to the case of $\mathfrak{sl}_2$. We will use the inner product κ_0 defined by the formula $\kappa_0(A,B) = \operatorname{Tr}_{\mathbb{C}^2}(AB)$. Then $\kappa_c = -2\kappa_0$, and formula (3.5.8) becomes

$$L_n \cdot S_m = (n-m)S_{n+m} - \frac{1}{2}(n^3-n)\delta_{n,-m}.$$

This is the transformation formula for the elements S_m of $\widetilde{U}_{\kappa_c}(\widehat{\mathfrak{sl}}_2)$. In particular, it determines the action of the group $\operatorname{Aut}\mathcal{O}$ on the center of $\widetilde{U}_{\kappa_c}(\widehat{\mathfrak{sl}}_2)$ which is topologically generated by the S_m's. We would like however to understand first the action of $\operatorname{Aut}\mathcal{O}$ on the center $\mathfrak{z}(\widehat{\mathfrak{sl}}_2)$ of the vertex algebra $V_{\kappa_c}(\mathfrak{sl}_2)$.

According to Theorem 3.5.1, we have a homomorphism

$$\mathbb{C}[S_m]_{m\in\mathbb{Z}} \to \mathfrak{z}(\widehat{\mathfrak{sl}}_2), \qquad X \mapsto X|0\rangle,$$

and it induces an isomorphism between the quotient of $\mathbb{C}[S_m]_{m\in\mathbb{Z}}$ by the ideal generated by $S_m, m \geq -1$ and $\mathfrak{z}(\widehat{\mathfrak{sl}}_2)$. This ideal is preserved by the action of $\operatorname{Der}\mathcal{O}$. Thus, to find the action of $\operatorname{Der}\mathcal{O}$ on $\mathfrak{z}(\widehat{\mathfrak{sl}}_2) \simeq \mathbb{C}[S_m]_{m\leq -2}$ we simply need to set $S_m, m \geq -1$, equal to 0.

† Note the difference between formulas (3.5.6) and (3.5.7). The former describes the action of $\operatorname{Der}\mathcal{O}$ on the associated graded algebra $\operatorname{gr} V_\kappa(\mathfrak{g})$, whereas the latter describes the action on $V_\kappa(\mathfrak{g})$ or on $\widetilde{U}_\kappa(\widehat{\mathfrak{g}})$. The two actions are different: for instance, $\overline{J}^a_n, n \geq 0$, acts by 0 on $\operatorname{gr} V_\kappa(\mathfrak{g})$, but $J^a_n, n \geq 0$, act non-trivially on $V_\kappa(\mathfrak{g})$.

This results in the following transformation law:

$$L_n \cdot S_m = \begin{cases} (n-m)S_{n+m} & \text{if } n+m \leqslant -2 \\ -\frac{1}{2}(n^3-n) & \text{if } n+m=0 \\ 0 & \text{otherwise.} \end{cases} \tag{3.5.9}$$

We now want to work out what sort of geometrical object gives rise to transformation laws like this. It turns out that these objects are projective connections on the disc $D = \operatorname{Spec}\mathbb{C}[[t]]$.

3.5.7 Projective connections

A **projective connection** on $D = \operatorname{Spec}\mathbb{C}[[t]]$ is a second order differential operator

$$\rho : \Omega_{\mathcal{O}}^{-1/2} \longrightarrow \Omega_{\mathcal{O}}^{3/2} \tag{3.5.10}$$

such that the principal symbol is 1 and the subprincipal symbol is 0.

Let us explain what this means. As before, $\Omega_{\mathcal{O}}$ is the $\mathcal{O}$-module of differentials, i.e., one-forms on D. These look like this: $f(t)dt$. Now we define the $\mathcal{O}$-module $\Omega_{\mathcal{O}}^{\lambda}$ as the set of "λ-forms", that is, things which look like this: $f(t)(dt)^{\lambda}$. It is a free $\mathcal{O}$-module with one generator, but the action of vector fields $\xi(t)\partial_t \operatorname{Der}\mathcal{O}$ on it depends on λ. To find the transformation formula we apply the transformation $t \mapsto t + \epsilon\xi(t)$, considered as an element of the group $\operatorname{Aut}\mathcal{O}$ over the ring of dual numbers $\mathbb{C}[\epsilon]/(\epsilon^2)$, to $f(t)(dt)^{\lambda}$. We find that

$$f(t)(dt)^{\lambda} \mapsto f(t+\epsilon\xi(t))d((t+\epsilon\xi(t)))^{\lambda}.$$

Now we take the ϵ-linear term. The result is

$$\xi(t)\partial_t \cdot f(t)(dt)^{\lambda} = (\xi(t)f'(t) + \lambda f(t)\xi'(t))(dt)^{\lambda}.$$

Note, however, that the action of $t\partial_t$ is semi-simple, but its eigenvalues are of the form $\lambda + n, n \in \mathbb{Z}_+$. Thus, if $\lambda \notin \mathbb{Z}$, the action of $\operatorname{Der}\mathcal{O}$ cannot be exponentiated to an action of $\operatorname{Aut}\mathcal{O}$ and we cannot use our twisting construction to assign to $\Omega_{\mathcal{O}}^{\lambda}$ a line bundle on a disc $D_x = \operatorname{Spec}\mathcal{O}_x$ (or on an arbitrary smooth algebraic curve for that matter). If λ is a half-integer, for instance, $\lambda = 1/2$, we may construct $\Omega^{1/2}$ by extracting a square root from Ω. On a general curve this is ambiguous: the line bundle we wish to construct is well-defined only up to tensoring with a line bundle $\mathcal{L}$, such that $\mathcal{L}^{\otimes 2}$ is canonically trivialized. But on a disc this does not create a problem, as such $\mathcal{L}$ is then trivialized almost canonically; there are two trivializations which differ by a sign, so this is a very mild ambiguity.

In any case, we will not be interested in $\Omega_{\mathcal{O}}^{1/2}$ itself, but in linear operators (3.5.10). Once we choose one of the two possible modules $\Omega_{\mathcal{O}}^{1/2}$, we take its inverse as our $\Omega_{\mathcal{O}}^{-1/2}$ and its third tensor power (over $\mathcal{O}$) as our $\Omega_{\mathcal{O}}^{3/2}$. The space of differential operators between them already does not depend on any choices. The same construction works for an arbitrary curve, so for example if X is compact, there are 2^{2g} different choices for $\Omega^{1/2}$ (they are called theta characteristics), but the corresponding spaces of projective connections are canonically identified for all of these choices.

Now, a second order differential operator (3.5.10) is something that can be written in the form

$$\rho = v_2(t)\partial_t^2 + v_1(t)\partial_t + v_0(t).$$

Note here that each $v_i(t) \in \mathbb{C}[[t]]$, but its transformation formula under the group $\operatorname{Aut}\mathcal{O}$ is *a priori* not that of a ordinary function, as we will see below.

The **principal symbol** of this differential operator is the coefficient $v_2(t)$. So the first condition on the projective connection means that ρ is of the form

$$\rho = \partial_t^2 + v_1(t)\partial_t + v_0(t).$$

The **subprincipal symbol** of this operator is the coefficient $v_1(t)$. The vanishing of $v_1(t)$ means that the operator is of the form

$$\rho(t) = \frac{\partial}{\partial t^2} - v(t),$$

where we have set $v(t) = -v_0(t)$ for notational convenience.

How do projective connections transform under the action of a vector field $\xi(t)\partial_t \in \operatorname{Der}\mathcal{O}$? As we already know the action of $\xi(t)\partial_t$ on $\Omega^{-1/2}$ and $\Omega^{3/2}$, it is easy to find out how it acts on $\rho : \Omega^{-1/2} \to \Omega^{3/2}$. Namely, we find that

$$\xi(t)\partial_t \cdot \Big((\partial_t^2 - v(t)) \cdot f(t)(dt)^{-1/2}\Big)$$

$$\begin{aligned}
&= \xi(t)\partial_t \cdot \Big((f''(t) - v(t)f(t))(dt)^{3/2}\Big) \\
&= \Big(\xi(t)(f'''(t) - v'(t)f(t) - v(t)f'(t)) \\
&\quad + \frac{3}{2}(f''(t) - v(t)f(t))\xi'(t)\Big)\,(dt)^{3/2}
\end{aligned}$$

and

$$(\partial_t^2 - v(t)) \cdot \Big(\xi(t)\partial_t \cdot f(t)(dt)^{-1/2}\Big)$$

$$\begin{aligned}
&= (\partial_t^2 - v(t))\left(\xi(t)f'(t) - \frac{1}{2}f(t)\xi'(t)\right)(dt)^{-1/2} \\
&= \Big(\frac{3}{2}\xi'(t)f''(t) + \xi(t)f'''(t) - v(t)\xi(t)f'(t) \\
&\quad - \frac{1}{2}f(t)\xi'''(t) + \frac{1}{2}v(t)f(t)\xi'(t)\Big)\,(dt)^{3/2}.
\end{aligned}$$

The action on the projective connection will come from the difference of these two expressions:

$$-\xi(t)v'(t) - 2v(t)\xi'(t) + \frac{1}{2}\xi'''(t).$$

Thus, under the action of the infinitesimal change of coordinates $t \mapsto t + \epsilon\xi(t)$ we have

$$v(t) \mapsto v(t) + \epsilon(\xi(t)v'(t) + 2v(t)\xi'(t) - \frac{1}{2}\xi'''(t)). \tag{3.5.11}$$

It is possible to exponentiate the action of $\operatorname{Der}\mathcal{O}$ to give the action of $\operatorname{Aut}\mathcal{O}$ on projective connections. This may done by an explicit, albeit tedious, computation, but one can follow a faster route, as explained in [FB], Section 9.2. Suppose that we have two formal coordinates t and s related by the formula $t = \varphi(s)$. Suppose that with respect to the coordinate t the projective connection ρ has the form $\partial_t^2 - v(t)$. Then with respect to the coordinate s it has the form $\partial_s^2 - \widetilde{v}(s)$, where

$$\widetilde{v}(s) = v(\varphi(s))\varphi'(s)^2 - \frac{1}{2}\{\varphi, s\},$$

and

$$\{\varphi, s\} = \frac{\varphi'''}{\varphi'} - \frac{3}{2}\left(\frac{\varphi''}{\varphi'}\right)^2 \tag{3.5.12}$$

is the so-called **Schwarzian derivative** of φ.

3.5.8 Back to the center

Now we can say precisely what kinds of geometric objects the Segal–Sugawara operators S_m are: the operator S_m behaves as the function on the space of projective connections $\partial_t^2 - v(t)$ picking the t^{-m-2}-coefficient of $v(t)$.

Lemma 3.5.1 *The expression*

$$\rho_S = \partial_t^2 - \sum_{m \leqslant -2} S_m t^{-m-2}$$

defines a canonical projective connection on the disc $D = \operatorname{Spec}\mathbb{C}[[t]]$, i.e., it is independent of the choice of the coordinate t.

Proof. The proof is similar to the argument used in Section 3.5.5. We need to check that the combined action of $L_n, n \geq -1$, on ρ_S, coming from the action on the S_n's (given by formula (3.5.9)) and on the projective connection (given by formula (3.5.11)), is equal to 0.

We find that the former is

$$- \sum_{m \leqslant -2; n+m \leqslant -2} (n-m) S_{n+m} t^{-m-2} + \sum_{m \leqslant -2; n+m=0} \frac{1}{2}(n^3 - n) t^{-m-2}$$

$$= - \sum_{k \leqslant -2} (2n-k) S_k t^{n-k-2} + \frac{1}{2}(n^3-n) t^{n-2},$$

and the latter is

$$t^{n+1} v'(t) + 2(n+1) t^n v(t) - \frac{1}{2}(n^3-n) t^{n-2}$$

$$= \sum_{m \leqslant -2} \left(-(m+2) S_m t^{n-m-2} + 2(n+1) S_m t^{n-m-2} \right) - \frac{1}{2}(n^3-n) t^{n-2}$$

$$= \sum_{m \leqslant -2} (2n-m) S_m t^{n-m-2} - \frac{1}{2}(n^3-n) t^{n-2}.$$

Summing them gives 0. □

Thus, the geometric meaning of the center $\mathfrak{z}(\widehat{\mathfrak{sl}}_2)$ of the $\widehat{\mathfrak{sl}}_2$ vertex algebra at the critical level is finally revealed:

The center $\mathfrak{z}(\widehat{\mathfrak{sl}}_2)$ is isomorphic to the algebra of functions on the space $\mathcal{P}roj(D)$ of projective connections on D.

Applying the twisting by the torsor Aut_x to Theorem 3.5.1, we obtain the following result.

Theorem 3.5.2 *The center $\mathfrak{z}(\widehat{\mathfrak{sl}}_2)_x$ attached to a point x on a curve is canonically isomorphic to* Fun $\mathcal{P}roj(D_x)$, *where $\mathcal{P}roj(D_x)$ is the space of projective connections on the disc $D_x = \operatorname{Spec} \mathcal{O}_x$ around the point x.*

Likewise, we obtain a coordinate-independent description of the center $Z(\widehat{\mathfrak{sl}}_{2,x})$ of the completed universal enveloping algebra $\widetilde{U}_\kappa(\widehat{\mathfrak{sl}}_{2,x})$ (this will be proved in Section 4.3.2).

Corollary 3.5.2 *The center $Z(\widehat{\mathfrak{sl}}_2)_x$ attached to a point x on a curve X is canonically isomorphic to the algebra* Fun $\mathcal{P}roj(D_x^\times)$ *on the space of projective connections on the punctured disc $D_x^\times = \operatorname{Spec} \mathcal{K}_x$ around the point x.*

We can use Theorem 3.5.2 to construct a family of $\widehat{\mathfrak{sl}}_{2,\kappa_c,x}$-modules parameterized by projective connections on D_x. Namely, given $\rho \in \mathcal{P}roj(D_x)$ we let $\widetilde{\rho} : \mathfrak{z}(\widehat{\mathfrak{sl}}_2)_x \to \mathbb{C}$ be the corresponding character and set

$$V_\rho = V_{\kappa_c}(\mathfrak{sl}_2) / \operatorname{Im}(\operatorname{Ker} \widetilde{\rho}).$$

We will see below that all of these modules are irreducible, and these are in fact all possible irreducible unramified $\widehat{\mathfrak{sl}}_{2,\kappa_c,x}$-modules. Thus, we have been able to link representations of an affine Kac–Moody algebra to projective connections, which are, as we will see below, special kinds of connections for the group PGL_2, the Langlands dual group of SL_2. This is the first step in our quest for the local Langlands correspondence for loop groups.

But what are the analogous structures for an arbitrary affine Kac–Moody algebra of critical level? In the next chapter we introduce the notion of *opers*, which are the analogues of projective connections for general simple Lie groups. It will turn out that the center $\mathfrak{z}(\widehat{\mathfrak{g}})$ of $V_{\kappa_c}(\mathfrak{g})$ is isomorphic to the algebra of functions on the space of ${}^L G$-opers on the disc, where ${}^L G$ is the Langlands dual group to G.

4

Opers and the center for a general Lie algebra

We now wish to generalize Theorem 3.5.2 describing the center $\mathfrak{z}(\widehat{\mathfrak{g}})$ of the vertex algebra $V_{\kappa_c}(\mathfrak{g})$ for $\mathfrak{g} = \mathfrak{sl}_2$ to the case of an arbitrary simple Lie algebra $\mathfrak{g}$. As the first step, we need to generalize the notion of projective connections, which, as we have seen, are responsible for the center in the case of $\mathfrak{g} = \mathfrak{sl}_2$, to the case of an arbitrary $\mathfrak{g}$. In Section 4.1 we revisit the notion of projective connection and recast it in terms of flat PGL_2-bundles with some additional structures. This will enable us to generalize projective connections to the case of an arbitrary $\mathfrak{g}$, in the form of *opers*. In Section 4.2 we will discuss in detail various definitions and realizations of opers and the action of changes of coordinates on them.

Then we will formulate in Section 4.3 our main result: a canonical isomorphism between the center $\mathfrak{z}(\widehat{\mathfrak{g}})$ and the algebra of functions on the space of opers on the disc associated to the Langlands dual group ${}^LG$. The proof of this result occupies the main part of this book. We will use it to describe the center $Z(\widehat{\mathfrak{g}})$ of the completed enveloping algebra $\widetilde{U}_{\kappa_c}(\widehat{\mathfrak{g}})$ at the critical level in Theorem 4.3.6. Finally, we will show in Proposition 4.3.9 that the center of $\widetilde{U}_{\kappa}(\widehat{\mathfrak{g}})$ is trivial away from the critical level.

4.1 Projective connections, revisited

In this section we discuss equivalent realizations of projective connections: as projective structures and as PGL_2-opers. The last realization is most important to us, as it suggests how to generalize the notion of projective connection (which naturally arose in our description of the center $\mathfrak{z}(\widehat{\mathfrak{g}})$ for $g = \mathfrak{sl}_2$) to the case of an arbitrary simple Lie algebra $\mathfrak{g}$.

4.1.1 Projective structures

The notion of projective connections introduced in Section 3.5.7 makes perfect sense not only on the disc, but also on an arbitrary smooth algebraic curve X over $\mathbb{C}$. By definition, a projective connection on X is a second order differential operator acting between the sheaves of sections of the following line bundles:

$$\rho : \Omega_X^{-1/2} \longrightarrow \Omega_X^{3/2}, \tag{4.1.1}$$

such that the principal symbol is 1 and the subprincipal symbol is 0. Here we need to choose the square root $\Omega_X^{1/2}$ of the canonical line bundle Ω_X on X. However, as explained in Section 3.5.7, the spaces of projective connections corresponding to different choices of $\Omega_X^{1/2}$ are canonically identified.

We will denote the space of projective connections on X by $\mathcal{P}roj(X)$.

It is useful to observe that $\mathcal{P}roj(X)$ is an affine space modeled on the vector space $H^0(X, \Omega_X^2)$ of quadratic differentials. Indeed, given a projective connection, i.e., a second order operator ρ as in (4.1.1), and a quadratic differential $\varpi \in H^0(X, \Omega_X^2)$, the sum $\rho + \varpi$ is a new projective connection. Moreover, for any pair of projective connections ρ, ρ' the difference $\rho - \rho'$ is a zeroth order differential operator $\Omega_X^{-1/2} \longrightarrow \Omega_X^{3/2}$, which is the same as a section of Ω^2. One can show that for any smooth curve (either compact or not) the space $\mathcal{P}roj(X)$ is non-empty (see [FB], Section 8.2.12). Therefore it is an $H^0(X, \Omega_X^2)$-torsor.

Locally, on an open analytic subset $U_\alpha \subset X$, we may choose a coordinate z_α and trivialize the line bundle $\Omega^{1/2}$. Then, in the same way as in Section 3.5.7, we see that $\rho_\alpha = \rho|_{U_\alpha}$ may be written with respect to this trivialization as an operator of the form $\partial_{z_\alpha}^2 - v_\alpha(z_\alpha)$. In the same way as in Section 3.5.7 it follows that on the overlap $U_\alpha \cap U_\beta$, with $z_\alpha = \varphi_{\alpha\beta}(z_\beta)$, we then have the transformation formula

$$v_\beta(z_\beta) = v_\alpha(\varphi_{\alpha\beta}(z_\beta)) \left(\frac{\partial \varphi_{\alpha\beta}}{\partial z_\beta} \right)^2 - \frac{1}{2} \{\varphi_{\alpha\beta}, z_\beta\}. \tag{4.1.2}$$

We will now identify projective connections on X with a different kind of structures, which we will now define.

A projective chart on X is by definition a covering of X by open subsets $U_\alpha, \alpha \in A$, with local coordinates z_α such that the transition functions $z_\beta = f_{\alpha\beta}(z_\alpha)$ on the overlaps $U_\alpha \cap U_\beta$ are Möbius transformations

$$f(z) = \frac{az + b}{cz + d}, \qquad \begin{pmatrix} a & b \\ c & d \end{pmatrix} \in PGL_2(\mathbb{C}).$$

Two projective charts are called equivalent if their union is also a projective chart. The equivalence classes of projective charts are called **projective structures**.

Proposition 4.1.1 *There is a bijection between the set of projective structures on X and the set of projective connections on X.*

Proof. It is easy to check that for a function $\varphi(z)$ the Schwarzian derivative $\{\varphi, z\}$ is 0 if and only if $\varphi(z)$ is a Möbius transformation. Hence, given a projective structure, we can define a projective connection by assigning the second order operator $\partial_{z_\alpha}^2$ on each chart U_α. According to formula (4.1.2), these transform correctly.

Conversely, given a projective connection, consider the space of solutions of the differential equation

$$\left(\partial_{z_\alpha}^2 - v_\alpha\right)\phi(z_\alpha) = 0$$

on U_α. This has a two-dimensional space of solutions, spanned by $\phi_{1,\alpha}$ and $\phi_{2,\alpha}$. Choosing the cover to be fine enough, we may assume that ϕ_2 is never 0 and the Wronskian of the two solutions is never 0. Define then

$$\mu_\alpha = \frac{\phi_{1,\alpha}}{\phi_{2,\alpha}} : U_\alpha \longrightarrow \mathbb{C}.$$

This is well defined and has a non-zero derivative (since the Wronskian is non-zero). Hence, near the origin it gives a complex coordinate on U_α. In a different basis it is clear that the μ's will be related by a Möbius transformation. □

4.1.2 PGL_2-opers

We now rephrase slightly the definition of projective structures. This will give us another way to think about projective connections, which we will be able to generalize to a more general situation.

We can think of a Möbius transformation as giving us an element of the group $PGL_2(\mathbb{C})$. Hence, if we have a projective structure on X, then to two charts U_α and U_β which overlap we have associated a *constant* map

$$f_{\alpha\beta} : U_\alpha \cap U_\beta \longrightarrow PGL_2(\mathbb{C}).$$

As explained in Section 1.2.4, this gives us the structure of a *flat* $PGL_2(\mathbb{C})$-bundle. Equivalently, as explained in Section 1.2.3, it may be represented as a holomorphic PGL_2-bundle $\mathcal{F}$ on X with a holomorphic connection ∇ (which is automatically flat as X is a curve).

However, we have not yet used the fact that we also had coordinates z_α around.

The group $PGL_2(\mathbb{C})$ acts naturally on the projective line $\mathbb{P}^1$. Let us form the associated $\mathbb{P}^1$-bundle

$$\mathbb{P}^1_{\mathcal{F}} = \mathcal{F} \underset{PGL_2(\mathbb{C})}{\times} \mathbb{P}^1$$

on X. The flat structure on $\mathcal{F}$ induces one on $\mathbb{P}^1_{\mathcal{F}}$. Thus, we have a preferred system of identifications of nearby fibers of $\mathbb{P}^1_{\mathcal{F}}$. Now, by definition of $\mathcal{F}$, on each U_α our bundle becomes trivial, and we can use the coordinate z_α to define a local section of $\mathbb{P}^1_{\mathcal{F}}|_{U_\alpha} = U_\alpha \times \mathbb{P}^1$. We simply let the section take value $z_\alpha(x) \in \mathbb{C} \subset \mathbb{P}^1$ at the point x.

This is actually a global section of $\mathbb{P}^1_{\mathcal{F}}$, because the coordinates transform between each other by exactly the same element of $PGL_2(\mathbb{C})$ which we used as the transition function of our bundle on the overlapping open subsets. As z_α is a coordinate, it has non-vanishing derivative at all points. Therefore the coordinates are giving us a global section of the $\mathbb{P}^1$-bundle $\mathbb{P}^1_{\mathcal{F}}$ which has a non-vanishing derivative at all points (with respect to the flat connection on $\mathbb{P}^1_{\mathcal{F}}$).

One may think of this structure as that of a compatible system of local identifications of our Riemann surface with the projective line. Hence the name "projective structure."

We now define a PGL_2**-oper** on X to be a flat $PGL_2(\mathbb{C})$-bundle $\mathcal{F}$ over X together with a globally defined section of the associated $\mathbb{P}^1$-bundle $\mathbb{P}^1_{\mathcal{F}}$ which has a nowhere vanishing derivative with respect to the connection.

So, we have seen that a projective structure, or, equivalently, a projective connection on X, gives rise to a $PGL_2(\mathbb{C})$-oper on X. It is clear that this identification is reversible: namely, we use the section to define the local coordinates; then the flat PGL_2-bundle transition functions define the Möbius transformations of the local coordinates on the overlaps. Hence we see that projective structures (and hence projective connections) on X are the same thing as PGL_2-opers on X.

Now our goal is to generalize the notion of PGL_2-oper to the case of an arbitrary simple Lie group. This will enable us to give a coordinate-independent description of the center of the completed enveloping algebra associated to a general affine Kac–Moody algebra $\widehat{\mathfrak{g}}_{\kappa_c}$.

In order to generalize the concept of oper we need to work out what $\mathbb{P}^1$ has to do with the group $PGL_2(\mathbb{C})$ and what the non-vanishing condition on the derivative of the section is telling us.

The group $PGL_2(\mathbb{C})$ acts transitively on $\mathbb{P}^1$, realized as the variety of lines in $\mathbb{C}^2 = \operatorname{span}(e_1, e_2)$, and the stabilizer of the line $\operatorname{span}(e_1)$ is the **Borel subgroup** $B \subset PGL_2(\mathbb{C})$ of upper triangular matrices. Thus, $\mathbb{P}^1$ may be represented as a homogeneous space $PGL_2(\mathbb{C})/B$. The Borel subgroup is a concept that easily generalizes to other simple Lie groups, so the corresponding homogeneous space G/B should be the correct generalization of $\mathbb{P}^1$.

Now, an element of $\mathbb{P}^1$ can be regarded as defining a right coset of B in G. Hence, our section of $\mathbb{P}^1_{\mathcal{F}}$ gives us above each point $x \in X$ a right B-coset in the fiber $\mathcal{F}_x$. This means that this section gives us a subbundle of the principal G-bundle, which is a principal B-bundle.

We recall that a B**-reduction** of a principal G-bundle $\mathcal{F}$ is a B-bundle $\mathcal{F}_B$ and an isomorphism $\mathcal{F} \simeq \mathcal{F}_B \underset{B}{\times} G$. In other words, we are given a subbundle of $\mathcal{F}$, which is a principal B-bundle such that the action of B on it agrees with the action restricted from G.

So, a PGL_2-oper gives us a flat PGL_2-bundle $(\mathcal{F}, \nabla)$ with a reduction $\mathcal{F}_B$

of $\mathcal{F}$ to the Borel subgroup B. The datum of $\mathcal{F}_B$ is equivalent to the datum of a section of $\mathbb{P}^1_{\mathcal{F}}$. We now have to interpret the non-vanishing condition on the derivative of our section of $\mathbb{P}^1_{\mathcal{F}}$ that appears in the definition of PGL_2-opers in terms of $\mathcal{F}_B$. This condition is basically telling us that the flat connection on $\mathcal{F}$ does not preserve the B-subbundle $\mathcal{F}_B$ anywhere. Let us describe this more precisely.

As explained in Section 1.2.4, if we choose a local coordinate t on an open subset $U \subset X$ and trivialize the holomorphic PGL_2-bundle $\mathcal{F}$, then the connection ∇ gives rise to a first order differential operator

$$\nabla_{\partial_t} = \partial_t + \begin{pmatrix} a(t) & b(t) \\ c(t) & d(t) \end{pmatrix}, \tag{4.1.3}$$

where the matrix is in the Lie algebra $\mathfrak{g} = \mathfrak{sl}_2(\mathbb{C})$, so $a(t) + d(t) = 0$). This operator is not canonical, but it gets changed by gauge transformations under changes of trivialization of $\mathcal{F}$ (and it also gets transformed under changes of the coordinate t).

Our condition on the PGL_2-oper may be rephrased as follows: let us choose a local trivialization of $\mathcal{F}$ on U which is horizontal with respect to the connection, so $\nabla_{\partial_t} = \partial_t$. Then our section of $\mathbb{P}^1_{\mathcal{F}}$ may be represented as a function $f : U \to \mathbb{P}^1$, and the condition is that its derivative with respect to t has to be everywhere non-zero.

4.1.3 More on the oper condition

The above description is correct, but not very useful, because in general it is difficult to find a horizontal trivialization explicitly. For that one needs to solve a system of differential equations, which may not be easy.

It is more practical to choose instead a trivialization of $\mathcal{F}$ on $U \subset X$ that is induced by some trivialization of $\mathcal{F}_B$ on U. Then f becomes a constant map whose value is the coset of B in $\mathbb{P}^1 = PGL_2/B$. But the connection now has the form (4.1.3), where the matrix elements are some non-trivial functions. These functions are determined only up to gauge transformations. However, because our trivialization of $\mathcal{F}$ is induced by that of $\mathcal{F}_B$, the ambiguity consists of gauge transformations with values in B only. Any conditions on ∇_{∂_t} that we wish to make now should be invariant under these B-valued gauge transformations (and coordinate changes of t). For example, connections that preserve our B-reduction are precisely the ones in which $c(t) = 0$; this is clearly a condition that is invariant under B-valued gauge transformations.

Our condition is, to the contrary, that the derivative of our function $f : U \to \mathbb{P}^1$ (obtained from our section of $\mathbb{P}^1_{\mathcal{F}}$ using the trivialization of $\mathcal{F}_B$) with respect to ∇_{∂_t} is everywhere non-zero. Note that the value of this derivative at a point $x \in U$ is really a tangent vector to $\mathbb{P}^1$ at $f(x)$. Since $f(x) = B \in PGL_2/B = \mathbb{P}^1$, the tangent space at $f(x)$ is naturally identified with $\mathfrak{sl}_2/\mathfrak{b}$.

Therefore our derivative is represented by the matrix element $c(t)$ appearing in formula (4.1.3). So the condition is simply that $c(t)$ is nowhere vanishing.

Connections satisfying this property may be brought to a standard form by using B-valued gauge transformations. Indeed, first we use the gauge transformation by the diagonal matrix $\begin{pmatrix} c(t)^{1/2} & 0 \\ 0 & c(t)^{-1/2} \end{pmatrix}$ (note that it is well-defined in PGL_2) to bring the connection operator to the form (4.1.3) with $c(t) = 1$. Next, we follow with the upper triangular transformation

$$\begin{pmatrix} 1 & -a \\ 0 & 1 \end{pmatrix} \left(\partial_t + \begin{pmatrix} a & b \\ 1 & -a \end{pmatrix} \right) \begin{pmatrix} 1 & a(t) \\ 0 & 1 \end{pmatrix} = \partial_t + \begin{pmatrix} 0 & b + a^2 + \partial_t a \\ 1 & 0 \end{pmatrix}.$$

The result is a connection operator of the form

$$\nabla = \partial_t + \begin{pmatrix} 0 & v(t) \\ 1 & 0 \end{pmatrix}. \tag{4.1.4}$$

It is clear from the above construction that there is a unique B-valued function, the product of the above diagonal and upper triangular matrices, that transforms a given operator (4.1.3) with nowhere vanishing $c(t)$ to this form. Thus, we have used up all of the freedom available in B-valued gauge transformations in order to bring the connection to this form.

This means that on a small open subset $U \subset X$, with respect to a coordinate t (or on a disc D_x around a point $x \in X$, with respect to a choice of a formal coordinate t), the space of PGL_2-opers is identified with the space of operators of the form (4.1.4), where $v(t)$ is a function on U (respectively, $v(t) \in \mathbb{C}[[t]]$).

As is well known from the theory of matrix differential equations, the differential equation $\nabla\Phi(t) = 0$ is equivalent to the second order differential equation $(\partial_t^2 - v(t))\Phi(t) = 0$. Thus, we may identify the space of opers with the space of operators $\partial_t^2 - v(t)$. These look like projective connections, but to identify them with projective connections we still need to check that under changes of coordinates they really transform as differential operators $\Omega_X^{-1/2} \longrightarrow \Omega_X^{3/2}$.

To see this, we realize our flat PGL_2-bundle $\mathcal{F}$ as an equivalence class of rank two flat vector bundles modulo tensoring with flat line bundles. Using the freedom of tensoring with flat line bundles, we may choose a representative in this class such that its determinant is the trivial line bundle with the trivial connection. We will denote this representative also by $\mathcal{F}$. It is not unique, but may be tensored with any flat line bundle $(\mathcal{L}, \nabla)$ whose square is trivial. The B-reduction $\mathcal{F}_B$ gives rise to a line subbundle $\mathcal{F}_1 \subset \mathcal{F}$, defined up to tensoring with a flat line bundle $(\mathcal{L}, \nabla)$ as above.

A connection ∇ is a map $\mathcal{F} \otimes \mathcal{T}_X \to \mathcal{F}$, or equivalently, $\mathcal{F} \to \mathcal{F} \otimes \Omega_X$, since Ω_X and $\mathcal{T}_X$ are dual line bundles. What is the meaning of the oper condition from this point of view? It is precisely that the composition

$$\mathcal{F}_1 \xrightarrow{\nabla} \mathcal{F} \otimes \Omega_X \to (\mathcal{F}/\mathcal{F}_1) \otimes \Omega_X$$

is an isomorphism of line bundles (it corresponds to the function $c(t)$ appearing in (4.1.3)).

Since we have already identified $\det \mathcal{F} \simeq \mathcal{F}_1 \otimes (\mathcal{F}/\mathcal{F}_1)$ with $\mathcal{O}_X$, this means that $(\mathcal{F}_1)^{\otimes 2} \simeq \Omega_X$, and so $\mathcal{F}_1 \simeq \Omega_X^{1/2}$. This also implies that $\mathcal{F}/\mathcal{F}_1 \simeq \Omega_X^{-1/2}$. Both isomorphisms do not depend on the choice of $\Omega_X^{1/2}$ as $\mathcal{F}_1$ is anyway only defined modulo tensoring with the square root of the trivial line bundle. Let us pick a particular square root $\Omega_X^{1/2}$. Then we may write

$$0 \longrightarrow \Omega_X^{1/2} \longrightarrow \mathcal{F} \longrightarrow \Omega_X^{-1/2} \longrightarrow 0.$$

Assume for a moment that X is a projective curve of genus g. Then we have

$$\mathrm{Ext}^1(\Omega_X^{-1/2}, \Omega_X^{1/2}) \cong H^1(X, \Omega_X) \cong H^0(X, \mathcal{O}_X)^* \cong \mathbb{C},$$

so that there may be at most two isomorphism classes of extensions of $\Omega_X^{-1/2}$ by $\Omega_X^{1/2}$: the split and the non-split ones. If the extension were split, then there would be an induced connection on the line bundle $\Omega_X^{1/2}$. But this bundle has degree $g-1$ and so cannot carry a flat connection if $g \neq 1$. Thus, for $g \neq 1$ we have a unique oper bundle corresponding to a non-zero class in $H^1(X, \Omega_X)$. (For $g = 1$ the two extensions are isomorphic to each other, so there is again a unique oper bundle.)

For an arbitrary smooth curve X, from the form of the connection we obtain that if we have a horizontal section of $\mathcal{F}$, then its projection onto $\Omega_X^{-1/2}$ is a solution of a second order operator, which is our $\partial_t^2 - v(t)$. It is also clear that this operator acts from $\Omega_X^{-1/2}$ to $\Omega_X^{3/2}$, and hence we find that it is indeed a projective connection. Since flat bundles, as well as projective connections, are uniquely determined by their horizontal sections, or solutions, we find that PGL_2-opers X are in one-to-one correspondence with projective connections.

We have made a full circle. We started out with projective connections, have recast them as projective structures, translated that notion into the notion of PGL_2-opers, which we have now found to correspond naturally to projective connections by using canonical representatives of the oper connection. Thus, we have found three different incarnations of one and the same object: projective connections, projective structures and PGL_2-opers. It is the last notion that we will generalize to an arbitrary simple Lie algebra in place of $\mathfrak{sl}_2$.

4.2 Opers for a general simple Lie algebra

In this section we introduce opers associated to an arbitrary simple Lie algebra, or equivalently, a simple Lie group of adjoint type. We will see in the next section that opers are to the center of the vertex algebra $V_{\kappa_c}(\mathfrak{g})$ what projective connections are to the center of $V_{\kappa_c}(\mathfrak{sl}_2)$. But there is an important twist to the story: the center is isomorphic to the algebra of functions on the space of opers on the disc, associated not to G, but to the *Langlands dual group* $^L G$ of G. This appearance of the dual group is very important from the point of view of the local Langlands correspondence.

4.2.1 Definition of opers

Let G be a simple algebraic group of adjoint type, B its Borel subgroup and $N = [B,B]$ its unipotent radical, with the corresponding Lie algebras $\mathfrak{n} \subset \mathfrak{b} \subset \mathfrak{g}$.

Thus, $\mathfrak{g}$ is a simple Lie algebra, and as such it has the Cartan decomposition

$$\mathfrak{g} = \mathfrak{n}_- \oplus \mathfrak{h} \oplus \mathfrak{n}_+.$$

We will choose generators $e_1, \ldots, e_\ell$ (resp., $f_1, \ldots, f_\ell$) of $\mathfrak{n}_+$ (resp., $\mathfrak{n}_-$). We have $\mathfrak{n}_{\alpha_i} = \mathbb{C}e_i, \mathfrak{n}_{-\alpha_i} = \mathbb{C}f_i$, see Appendix A.3. We take $\mathfrak{b} = \mathfrak{h} \oplus \mathfrak{n}_+$ as the Lie algebra of B. Then $\mathfrak{n}$ is the Lie algebra of N. In what follows we will use the notation $\mathfrak{n}$ for $\mathfrak{n}_+$.

It will be useful for us to assume that we only make a choice of $\mathfrak{n} = \mathfrak{n}_+$ and $\mathfrak{b} = \mathfrak{h} \oplus \mathfrak{n}_+$, but not of $\mathfrak{h}$ or $\mathfrak{n}_-$. We then have an *abstract* Cartan Lie algebra $\mathfrak{h} = \mathfrak{b}/\mathfrak{n}$, but no embedding of $\mathfrak{h}$ into $\mathfrak{b}$. We will sometimes choose such an embedding, and each such choice then also gives us the lower nilpotent subalgebra $\mathfrak{n}_-$ (defined as the span of negative root vectors corresponding to this embedding). Likewise, we will have subgroups $N \subset B \subset G$ and the abstract Cartan group $H = G/B$, but no splitting $H \hookrightarrow B$. Whenever we do use a splitting $\mathfrak{h} \hookrightarrow \mathfrak{b}$, or equivalently, $H \hookrightarrow B$, we will explain how this choice affects our discussion.

Let $[\mathfrak{n},\mathfrak{n}]^\perp \subset \mathfrak{g}$ be the orthogonal complement of $[\mathfrak{n},\mathfrak{n}]$ with respect to a non-degenerate invariant inner product κ_0. We have

$$[\mathfrak{n},\mathfrak{n}]^\perp/\mathfrak{b} \simeq \bigoplus_{i=1}^{\ell} \mathfrak{n}_{-\alpha_i}.$$

Clearly, the group B acts on $\mathfrak{n}^\perp/\mathfrak{b}$. Our first observation is that there is an open B-orbit $\mathbf{O} \subset \mathfrak{n}^\perp/\mathfrak{b} \subset \mathfrak{g}/\mathfrak{b}$, consisting of vectors whose projection on each subspace $\mathfrak{n}_{-\alpha_i}$ is non-zero. This orbit may also be described as the B-orbit of the sum of the projections of the generators $f_i, i = 1, \ldots, \ell$, of any possible subalgebra $\mathfrak{n}_-$, onto $\mathfrak{g}/\mathfrak{b}$. The action of B on $\mathbf{O}$ factors through an action of $H = B/N$. The latter is simply transitive and makes $\mathbf{O}$ into an H-torsor (see Section 3.5.3).

Let X be a smooth curve and x a point of X. As before, we denote by $\mathcal{O}_x$ the completed local ring and by $\mathcal{K}_x$ its field of fractions. The ring $\mathcal{O}_x$ is isomorphic, but not canonically, to $\mathbb{C}[[t]]$. Then $D_x = \operatorname{Spec} \mathcal{O}_x$ is the disc without a coordinate and $D_x^\times = \operatorname{Spec} \mathcal{K}_x$ is the corresponding punctured disc.

Suppose now that we are given a principal G-bundle $\mathcal{F}$ on a smooth curve X, or D_x, or $D_x^\times$, together with a connection ∇ (automatically flat) and a reduction $\mathcal{F}_B$ to the Borel subgroup B of G. Then we define the relative position of ∇ and $\mathcal{F}_B$ (i.e., the failure of ∇ to preserve $\mathcal{F}_B$) as follows. Locally, choose any flat connection ∇' on $\mathcal{F}$ preserving $\mathcal{F}_B$, and take the difference $\nabla - \nabla'$, which is a section of $\mathfrak{g}_{\mathcal{F}_B} \otimes \Omega_X$. We project it onto $(\mathfrak{g}/\mathfrak{b})_{\mathcal{F}_B} \otimes \Omega_X$. It is clear that the resulting local section of $(\mathfrak{g}/\mathfrak{b})_{\mathcal{F}_B} \otimes \Omega_X$ is independent of the choice ∇'. These sections patch together to define a global $(\mathfrak{g}/\mathfrak{b})_{\mathcal{F}_B}$-valued one-form on X, denoted by $\nabla/\mathcal{F}_B$.

Let X be a smooth curve, or D_x, or $D_x^\times$. Suppose we are given a principal G-bundle $\mathcal{F}$ on X, a connection ∇ on $\mathcal{F}$ and a B-reduction $\mathcal{F}_B$. We will say that $\mathcal{F}_B$ is **transversal** to ∇ if the one-form $\nabla/\mathcal{F}_B$ takes values in $\mathbf{O}_{\mathcal{F}_B} \subset (\mathfrak{g}/\mathfrak{b})_{\mathcal{F}_B}$. Note that $\mathbf{O}$ is $\mathbb{C}^\times$-invariant, so that $\mathbf{O} \otimes \Omega_X$ is a well-defined subset of $(\mathfrak{g}/\mathfrak{b})_{\mathcal{F}_B} \otimes \Omega_X$.

Now, a G-**oper** on X is by definition a triple $(\mathcal{F}, \nabla, \mathcal{F}_B)$, where $\mathcal{F}$ is a principal G-bundle $\mathcal{F}$ on X, ∇ is a connection on $\mathcal{F}$ and $\mathcal{F}_B$ is a B-reduction of $\mathcal{F}$, such that $\mathcal{F}_B$ is transversal to ∇.

This definition is due to A. Beilinson and V. Drinfeld [BD1] (in the case when X is the punctured disc opers were introduced earlier by V. Drinfeld and V. Sokolov in [DS]).

It is clear that for $G = PGL_2$ we obtain the definition of PGL_2-oper from Section 4.1.2.

4.2.2 Realization as gauge equivalence classes

Equivalently, the transversality condition may be reformulated as saying that if we choose a local trivialization of $\mathcal{F}_B$ and a local coordinate t then the connection will be of the form

$$\nabla = \partial_t + \sum_{i=1}^{\ell} \psi_i(t) f_i + \mathbf{v}(t), \tag{4.2.1}$$

where each $\psi_i(t)$ is a nowhere vanishing function, and $\mathbf{v}(t)$ is a $\mathfrak{b}$-valued function. One shows this in exactly the same way as we did in Section 4.1.3 in the case of $G = PGL_2$.

If we change the trivialization of $\mathcal{F}_B$, then this operator will get transformed by the corresponding B-valued gauge transformation (see Section 1.2.4). This observation allows us to describe opers on the disc $D_x = \operatorname{Spec} \mathcal{O}_x$ and the punctured disc $D_x^\times = \operatorname{Spec} \mathcal{K}_x$ in a more explicit way. The same reasoning will work on any sufficiently small analytic subset U of any curve, equipped with a local coordinate t, or on a Zariski open subset equipped with an étale coordinate. For the sake of definiteness, we will consider now the case of the base D_x.

Let us choose a coordinate t on D_x, i.e., an isomorphism $\mathcal{O}_x \simeq \mathbb{C}[[t]]$. Then we identify D_x with $D = \operatorname{Spec} \mathbb{C}[[t]]$. The space $\operatorname{Op}_G(D)$ of G-opers on D is the quotient of the space of all operators of the form (4.2.1), where $\psi_i(t) \in \mathbb{C}[[t]], \psi_i(0) \neq 0, i = 1, \ldots, \ell$, and $\mathbf{v}(t) \in \mathfrak{b}[[t]]$, by the action of the group $B[[t]]$ of gauge transformations:

$$g \cdot (\partial_t + A(t)) = \partial_t + gA(t)g^{-1} - g^{-1}\partial_t g.$$

Let us choose a splitting $\imath : H \to B$ of the homomorphism $B \to H$. Then B becomes the product $B = H \ltimes N$. The B-orbit $\mathbf{O}$ is an H-torsor, and so we can use H-valued gauge transformations to make all functions $\psi_i(t)$ equal to 1. In other words, there is a unique element of $H[[t]]$, namely, the element $\prod_{i=1}^{\ell} \check{\omega}_i(\psi_i(t))$, where $\check{\omega}_i : \mathbb{C}^\times \to H$ is the ith fundamental coweight of G, such that the corresponding gauge transformation brings our connection operator to the form

$$\nabla = \partial_t + \sum_{i=1}^{\ell} f_i + \mathbf{v}(t), \qquad \mathbf{v}(t) \in \mathfrak{b}[[t]]. \tag{4.2.2}$$

What remains is the group of N-valued gauge transformations. Thus, we obtain that $\operatorname{Op}_G(D)$ is equal to the quotient of the space $\widetilde{\operatorname{Op}}_G(D)$ of operators of the form (4.2.2) by the action of the group $N[[t]]$ by gauge transformations:

$$\operatorname{Op}_G(D) = \widetilde{\operatorname{Op}}_G(D)/N[[t]].$$

This gives us a very concrete realization of the space of opers on the disc as gauge equivalence classes.

4.2.3 Action of coordinate changes

In the above formulas we use a particular coordinate t on our disc D_x (and we have identified D_x with D using this coordinate). As our goal is to formulate all of our results in a coordinate-independent way, we need to figure out how the gauge equivalence classes introduced above change if we choose another coordinate.

So suppose that s is another coordinate on the disc D_x such that $t = \varphi(s)$. In terms of this new coordinate the operator (4.2.2) will become

$$\nabla_{\partial_t} = \nabla_{\varphi'(s)^{-1}\partial_s} = \varphi'(s)^{-1}\partial_s + \sum_{i=1}^{\ell} f_i + \mathbf{v}(\varphi(s)).$$

Hence we find that

$$\nabla_{\partial_s} = \partial_s + \varphi'(s) \sum_{i=1}^{\ell} f_i + \varphi'(s) \cdot \mathbf{v}(\varphi(s)).$$

In order to bring it back to the form (4.2.2), we need to apply the gauge transformation by $\check{\rho}(\varphi'(s))$, where $\check{\rho} : \mathbb{C}^\times \to H$ is the one-parameter subgroup of H equal to the sum of the fundamental coweights of G, $\check{\rho} = \sum_{i=1}^{\ell} \check{\omega}_i$. (Here we again choose a splitting $\imath : H \to B$ of the homomorphism $B \to H$). Then, considering $\check{\rho}$ as an element of the Lie algebra $\mathfrak{h} = \operatorname{Lie} H$, we have $[\check{\rho}, e_i] = e_i$ and $[\check{\rho}, f_i] = -f_i$ (see Appendix A.3). Therefore we find that

$$\check{\rho}(\varphi'(s)) \cdot \left(\partial_s + \varphi'(s) \sum_{i=1}^{\ell} f_i + \varphi'(s) \cdot \mathbf{v}(\varphi(s)) \right)$$

$$= \partial_s + \sum_{i=1}^{\ell} f_i + \varphi'(s)\check{\rho}(\varphi'(s)) \cdot \mathbf{v}(\varphi(s)) \cdot \check{\rho}(\varphi'(s))^{-1} - \check{\rho}\left(\frac{\varphi''(s)}{\varphi'(s)}\right). \qquad (4.2.3)$$

The above formula defines an action of the group $\operatorname{Aut} \mathcal{O}$ on the space $\operatorname{Op}_G(D)$ of opers on the standard disc,

$$D = \operatorname{Spec} \mathbb{C}[[t]].$$

For a disc D_x around a point x of smooth curve X we may now define $\operatorname{Op}_G(D_x)$ as the twist of $\operatorname{Op}_G(D)$ by the $\operatorname{Aut} \mathcal{O}$-torsor $\mathcal{A}ut_x$ (see Section 3.5.2).

In particular, the above formulas allow us to determine the structure of the H-bundle $\mathcal{F}_H = \mathcal{F}_B \underset{B}{\times} H = \mathcal{F}_B/N$. Let us first describe a general construction of H-bundles which works on any smooth curve X, or a disc, or a punctured disc.

Let $\check{P}$ be the lattice of cocharacters $\mathbb{C}^\times \to H$. Since G is the group of adjoint type associated to G, this lattice is naturally identified with the lattice of integral coweights of the Cartan algebra $\mathfrak{h}$, spanned by $\check{\omega}_i, i = 1, \ldots, \ell$. Let P be the lattice of characters $H \to \mathbb{C}^\times$. We have a natural pairing $\langle \cdot, \cdot \rangle : P \times \check{P} \to \mathbb{Z}$ obtained by composing a character and a cocharacter, which gives us a homomorphism $\mathbb{C}^\times \to \mathbb{C}^\times$ that corresponds to an integer.

Given an H-bundle $\mathcal{F}_H$ on X, for each $\lambda \in P$ we have the associated line bundle $\mathcal{L}_\lambda = \mathcal{F}_H \underset{H}{\times} \mathbb{C}_\lambda$, where $\mathbb{C}_\lambda$ is the one-dimensional representation of H corresponding to λ. The datum of $\mathcal{F}_H$ is equivalent to the data of the line bundles $\{\mathcal{L}_\lambda, \lambda \in P\}$, together with the isomorphisms $\mathcal{L}_\lambda \otimes \mathcal{L}_\mu \simeq \mathcal{L}_{\lambda+\mu}$ satisfying the obvious associativity condition.

Now, given $\check{\mu} \in \check{P}$, we define such data by setting $\mathcal{L}_\lambda = \Omega_X^{\langle \lambda, \check{\mu} \rangle}$. Let us denote the corresponding H-bundle on X by $\Omega^{\check{\mu}}$. It is nothing but the push-forward of the $\mathbb{C}^\times$-bundle $\Omega^\times$ with respect to the homomorphism $\mathbb{C}^\times \to H$ given by the cocharacter $\check{\mu}$.

More concretely, the H-bundle $\Omega^{\check{\mu}}$ may be described as follows: for each choice of a local coordinate t on an open subset of X we have a trivialization of Ω_X generated by the section dt, and hence of each of the line bundles $\mathcal{L}_\lambda$. If we chose a different coordinate s such that $t = \varphi(s)$, then the two trivializations differ by the transition function $\langle \lambda, \check{\mu}(\varphi'(s)) \rangle$. In other words, the transition function for $\Omega^{\check{\mu}}$ is $\check{\mu}(\varphi'(s))$. These transition functions give us an alternative way to define the H-bundle $\Omega^{\check{\mu}}$.

Lemma 4.2.1 *The H-bundle $\mathcal{F}_H = \mathcal{F}_B \underset{B}{\times} H = \mathcal{F}_B/N$ is isomorphic to $\Omega^{\check{\rho}}$.*

Proof. It follows from formula (4.2.3) for the action of the changes of coordinates on opers that if we pass from a coordinate t on D_x to the coordinate s such that $t = \varphi(s)$, then we obtain a new trivialization of the H-bundle $\mathcal{F}_H$, which is related to the old one by the transition function $\check{\rho}(\varphi'(s))$. This precisely means that $\mathcal{F}_H \simeq \Omega^{\check{\rho}}$. □

4.2.4 Canonical representatives

In this section we find canonical representatives in the $N[[t]]$-gauge classes of connections of the form (4.2.2).

In order to do this, we observe that the operator $\operatorname{ad}\check{\rho}$ defines a gradation on $\mathfrak{g}$, called the **principal gradation**, with respect to which we have a direct sum decomposition $\mathfrak{g} = \bigoplus_i \mathfrak{g}_i$. In particular, we have $\mathfrak{b} = \bigoplus_{i\geq 0} \mathfrak{b}_i$, where $\mathfrak{b}_0 = \mathfrak{h}$.

Let now

$$p_{-1} = \sum_{i=1}^{\ell} f_i.$$

The operator $\operatorname{ad} p_{-1}$ acts from $\mathfrak{b}_{i+1}$ to $\mathfrak{b}_i$ injectively for all $i \geq 0$. Hence we can find for each $i \geq 0$ a subspace $V_i \subset \mathfrak{b}_i$, such that $\mathfrak{b}_i = [p_{-1}, \mathfrak{b}_{i+1}] \oplus V_i$. It is well-known that $V_i \neq 0$ if and only if i is an **exponent** of $\mathfrak{g}$, and in that case $\dim V_i$ is equal to the multiplicity of the exponent i. In particular, $V_0 = 0$.

Let $V = \bigoplus_{i\in E} V_i \subset \mathfrak{n}$, where $E = \{d_1, \ldots, d_\ell\}$ is the set of exponents of $\mathfrak{g}$ counted with multiplicity. These are exactly the same exponents that we have encountered previously in Section 2.1.1. They are equal to the orders of the generators of the center of $U(\mathfrak{g})$ minus 1.

We note that the multiplicity of each exponent is equal to 1 in all cases except the case $\mathfrak{g} = D_{2n}, d_n = 2n$, when it is equal to 2.

Lemma 4.2.2 ([DS]) *The action of $N[[t]]$ on $\widetilde{\operatorname{Op}}_G(D)$ is free.*

Proof. We claim that each element of $\partial_t + p_{-1} + \mathbf{v}(t) \in \widetilde{\operatorname{Op}}_G(D)$ can be uniquely represented in the form

$$\partial_t + p_{-1} + \mathbf{v}(t) = \exp(\operatorname{ad} U) \cdot (\partial_t + p_{-1} + \mathbf{c}(t)), \tag{4.2.4}$$

where $U \in \mathfrak{n}[[t]]$ and $\mathbf{c}(t) \in V[[t]]$. To see this, we decompose with respect to the principal gradation: $U = \sum_{j\geq 0} U_j$, $\mathbf{v}(t) = \sum_{j\geq 0} \mathbf{v}_j(t)$, $\mathbf{c}(t) = \sum_{j\in E} \mathbf{c}_j(t)$. Equating the homogeneous components of degree j in both sides of (4.2.4), we obtain that $\mathbf{c}_i + [U_{i+1}, p_{-1}]$ is expressed in terms of $\mathbf{v}_i, \mathbf{c}_j, j < i$, and $U_j, j \leq i$. The injectivity of $\operatorname{ad} p_{-1}$ then allows us to determine uniquely $\mathbf{c}_i$ and U_{i+1}. Hence U and $\mathbf{c}$ satisfying equation (4.2.4) may be found uniquely by induction, and the lemma follows. □

There is a special choice of the transversal subspace $V = \bigoplus_{i \in E} V_i$. Namely, there exists a unique element p_1 in $\mathfrak{n}$, such that $\{p_{-1}, 2\check{\rho}, p_1\}$ is an $\mathfrak{sl}_2$-triple. This means that they have the same relations as the generators $\{e, h, f\}$ of $\mathfrak{sl}_2$ (see Section 2.1.1).

For example, for $\mathfrak{g} = \mathfrak{sl}_n$ (with the rank $\ell = n-1$) we have

$$p_{-1} = \begin{pmatrix} 0 & & \cdots & & \\ 1 & 0 & & & \\ & 1 & 0 & & \\ & & \ddots & \ddots & \\ & & & 1 & 0 \end{pmatrix},$$

$$p_0 = \begin{pmatrix} n-1 & & & & \\ & n-3 & & & \\ & & n-5 & & \\ & & & \ddots & \\ & & & & -n+1 \end{pmatrix},$$

$$p_1 = \begin{pmatrix} 0 & 1(n-1) & & & & \\ & 0 & 2(n-2) & & & \\ & & 0 & 3(n-3) & & \\ & & & \ddots & \ddots & \\ & & & & & (n-1)1 \\ & & & & & 0 \end{pmatrix}.$$

In general, we have $p_1 = \sum_{i=1}^{\ell} m_i e_i$, where the e_i's are generators of $\mathfrak{n}_+$ and the m_i's are certain coefficients uniquely determined by the condition that $\{p_{-1}, 2\check{\rho}, p_1\}$ is an $\mathfrak{sl}_2$-triple.

Let $V^{\mathrm{can}} = \bigoplus_{i \in E} V_i^{\mathrm{can}}$ be the space of $\operatorname{ad} p_1$-invariants in $\mathfrak{n}$. Then p_1 spans V_1^{can}. Let p_j be a linear generator of $V_{d_j}^{\mathrm{can}}$. If the multiplicity of d_j is greater than 1, then we choose linearly independent vectors in $V_{d_j}^{\mathrm{can}}$.

In the case of $\mathfrak{g} = \mathfrak{sl}_n$ we may choose as the elements $p_j, j = 1, \ldots, n-1$, the matrices p_1^j (with respect to the matrix product).

According to Lemma 4.2.2, each G-oper may be represented by a unique operator $\nabla = \partial_t + p_{-1} + \mathbf{v}(t)$, where $\mathbf{v}(t) \in V^{\mathrm{can}}[[t]]$, so that we can write

$$\mathbf{v}(t) = \sum_{j=1}^{\ell} v_j(t) \cdot p_j, \qquad v_j(t) \in \mathbb{C}[[t]].$$

Let us find out how the group of coordinate changes acts on the canonical representatives.

Suppose now that $t = \varphi(s)$, where s is another coordinate on D_x such that $t = \varphi(s)$. With respect to the new coordinate s, ∇ becomes equal to $\partial_s + \widetilde{\mathbf{v}}(s)$, where $\widetilde{\mathbf{v}}(s)$ is expressed via $\mathbf{v}(t)$ and $\varphi(s)$ as in formula (4.2.3). By Lemma 4.2.2, there exists a unique operator $\partial_s + p_{-1} + \overline{\mathbf{v}}(s)$ with $\overline{\mathbf{v}}(s) \in V^{\mathrm{can}}[[s]]$ and $g \in B[[s]]$, such that

$$\partial_s + p_{-1} + \overline{\mathbf{v}}(s) = g \cdot (\partial_s + \widetilde{\mathbf{v}}(s)) . \tag{4.2.5}$$

It is straightforward to find that

$$g = \exp\left(\frac{1}{2}\frac{\varphi''}{\varphi'} \cdot p_1\right) \check{\rho}(\varphi'), \tag{4.2.6}$$

$$\overline{v}_1(s) = v_1(\varphi(s))\,(\varphi')^2 - \frac{1}{2}\{\varphi, s\}, \tag{4.2.7}$$

$$\overline{v}_j(s) = v_j(\varphi(s))\,(\varphi')^{d_j+1}, \qquad j > 1, \tag{4.2.8}$$

where $\{\varphi, s\}$ is the Schwarzian derivative (3.5.12).

Formula (4.2.6) may be used to describe the B-bundle $\mathcal{F}_B$. Namely, in Section 4.1.3 we identified the PGL_2 bundle $\mathcal{F}_{PGL_2}$ underlying all PGL_2-opers and its B_{PGL_2}-reduction $\mathcal{F}_{B_{PGL_2}}$. Let $\imath : PGL_2 \hookrightarrow G$ be the principal embedding corresponding to the principal $\mathfrak{sl}_2$ subalgebra of $\mathfrak{g}$ that we have been using (recall that by our assumption G is of adjoint type). We denote by $\imath_B$ the corresponding embedding $B_{PGL_2} \hookrightarrow B$. Then, according to formula (4.2.6), the G-bundle $\mathcal{F}$ underlying all G-opers and its B-reduction $\mathcal{F}_B$ are isomorphic to the bundles induced from $\mathcal{F}_{PGL_2}$ and $\mathcal{F}_{B_{PGL_2}}$ under the embeddings $\imath$ and $\imath_B$, respectively.

Formulas (4.2.7) and (4.2.8) show that under changes of coordinates, v_1 transforms as a projective connection, and $v_j, j > 1$, transforms as a $(d_j + 1)$-differential on D_x. Thus, we obtain an isomorphism

$$\operatorname{Op}_G(D_x) \simeq \mathcal{P}roj(D_x) \times \bigoplus_{j=2}^{\ell} \Omega_{\mathcal{O}_x}^{\otimes(d_j+1)}, \tag{4.2.9}$$

where $\Omega_{\mathcal{O}_x}^{\otimes n}$ is the space of n-differentials on D_x and $\mathcal{P}roj(D_x)$ is the $\Omega_{\mathcal{O}_x}^{\otimes 2}$-torsor of projective connections on D_x.

This analysis carries over verbatim to the case of the punctured disc $D_x^\times$. In particular, we obtain that for each choice of coordinate t on D_x we have an identification between $\operatorname{Op}_G(D_x^\times)$ with the space of operators $\nabla = \partial_t + p_{-1} + \mathbf{v}(t)$, where now $\mathbf{v}(t) \in V^{\mathrm{can}}((t))$. The action of changes of coordinates is given by the same formula as above, and so we obtain an analogue of the isomorphism (4.2.9):

$$\operatorname{Op}_G(D_x^\times) \simeq \mathcal{P}roj(D_x^\times) \times \bigoplus_{j=2}^{\ell} \Omega_{\mathcal{K}_x}^{\otimes(d_j+1)}. \tag{4.2.10}$$

The natural embedding $\mathrm{Op}_G(D_x) \hookrightarrow \mathrm{Op}_G(D_x^\times)$ is compatible with the natural embeddings of the right hand sides of (4.2.9) and (4.2.10).

In the same way we obtain an identification for a general smooth curve X:

$$\mathrm{Op}_G(X) \simeq \mathcal{P}roj(X) \times \bigoplus_{j=2}^{\ell} \Gamma(X, \Omega_X^{\otimes(d_j+1)}).$$

Just how canonical is this "canonical form"? Suppose we do not wish to make any choices other than that of a Borel subalgebra $\mathfrak{b}$ in $\mathfrak{g}$. The definition of the canonical form involves the choice of the $\mathfrak{sl}_2$-triple $\{p_{-1}, 2\check{\rho}, p_1\}$. However, choosing such an $\mathfrak{sl}_2$-triple is equivalent to choosing a splitting $\mathfrak{h} \hookrightarrow \mathfrak{b}$ of the projection $\mathfrak{b}/\mathfrak{n} \to \mathfrak{h}$ and choosing the generators of the corresponding one-dimensional subspaces $\mathfrak{n}_{-\alpha_i}$. It is easy to see that the group B acts simply transitively on the set of these choices, and hence on the set of $\mathfrak{sl}_2$-triples.

Given an $\mathfrak{sl}_2$-triple $\{p_{-1}, 2\check{\rho}, p_1\}$, we also chose basis elements $p_i, i \in E$, of V^{can}, homogeneous with respect to the grading defined by $\check{\rho}$. This allowed us to make the identification (4.2.9). However, even if the multiplicity of d_i is equal to one, these elements are only well-defined up to a scalar. Therefore, if we want a truly canonical form, we should consider instead of the vector space $\bigoplus_{i=2}^{\ell} \Omega_{\mathcal{O}_x}^{d_i} \cdot p_i$ that we used in (4.2.9), the space

$$V_{>1,x}^{\mathrm{can}} = \Gamma(D_x, \Omega^\times \underset{\mathbb{C}^\times}{\times} V_{>1}^{\mathrm{can}}(1)),$$

where $V_{>1}^{\mathrm{can}}(1)$ is the vector space $\bigoplus_{i>0} V_i^{\mathrm{can}}$ with the grading shifted by 1, which no longer depends on the choice of a basis. Then we obtain a more canonical identification

$$\mathrm{Op}_G(D_x) \simeq \mathcal{P}roj(D_x) \times V_{>1,x}^{\mathrm{can}}, \tag{4.2.11}$$

and similarly for the case of $D_x^\times$ and a more general curve X, which now depends only on the choice of $\mathfrak{sl}_2$-triple. In particular, since $\mathcal{P}roj(D_x)$ is an $\Omega_{\mathcal{O}_x}^{\otimes 2}$-torsor, we find, by identifying $\Omega_{\mathcal{O}_x}^{\otimes 2} = \Omega_{\mathcal{O}_x}^2 \cdot p_1$, that $\mathrm{Op}_G(D_x)$ is a V_x^{can}-torsor, where

$$V_x^{\mathrm{can}} = \Gamma(D_x, \Omega^\times \underset{\mathbb{C}^\times}{\times} V^{\mathrm{can}}(1)).$$

Suppose now that we have another $\mathfrak{sl}_2$-triple $\{\widetilde{p}_{-1}, 2\widetilde{\check{\rho}}, \widetilde{p}_1\}$, and the corresponding space $\widetilde{V}^{\mathrm{can}}$. Then both are obtained from the original ones by the adjoint action of a uniquely determined element $b \in B$. Given an oper ρ, we obtain its two different canonical forms:

$$\nabla = \partial_t + p_{-1} + v_1(t) p_1 + \mathbf{v}_{>1}(t), \qquad \widetilde{\nabla} = \partial_t + \widetilde{p}_{-1} + \widetilde{v}_1(t)\widetilde{p}_1 + \widetilde{\mathbf{v}}_{>1}(t),$$

where $\mathbf{v}(t) \in V_{>1}^{\mathrm{can}}$ and $\widetilde{\mathbf{v}}(t) \in \widetilde{V}_{>1}^{\mathrm{can}}$. But then it follows from the construction that we have $\widetilde{\nabla} = b\nabla b^{-1}$, which means that $\widetilde{v}_1(t) = v_1(t)$ and

$$\widetilde{\mathbf{v}}_{>1}(t) = b\mathbf{v}_{>1}(t)b^{-1}.$$

Since all possible subspaces $V^{\mathrm{can}}_{>1}$ are canonically identified with each other in this way, we may identify all of them with a unique "abstract" graded vector space $V^{\mathrm{abs}}_{<1}$ (note however that $V^{\mathrm{abs}}_{<1}$ cannot be canonically identified with a subspace of $\mathfrak{b}$). We define its twist $V^{\mathrm{abs}}_{>1,x}$ in the same way as above. Then the "true" canonical form of opers is the identification

$$\mathrm{Op}_G(D_x) \simeq \mathcal{P}roj(D_x) \times V^{\mathrm{abs}}_{>1,x}, \tag{4.2.12}$$

which does not depend on any choices (and similarly for the case of $D_x^\times$ and a more general curve X). In particular, we find, as above, that $\mathrm{Op}_G(D_x)$ is a V^{abs}_x-torsor, where V^{abs}_x is the twist of the abstract graded space V^{abs} defined as before.

4.2.5 Alternative choice of representatives for $\mathfrak{sl}_n$

In the case of $\mathfrak{g} = \mathfrak{sl}_n$ there is another choice of representatives of oper gauge classes which is useful in applications. Namely, we choose as the transversal subspace $V = \bigoplus_{i=1}^{n-1} V_i$ the space of traceless matrices with non-zero entries only in the first row (so its ith component V_i is the space spanned by the matrix $E_{1,i+1}$).

The corresponding representatives of PGL_n-opers have the form

$$\partial_t + \begin{pmatrix} 0 & u_1 & u_2 & \cdots & u_{n-1} \\ 1 & 0 & 0 & \cdots & 0 \\ 0 & 1 & 0 & \cdots & 0 \\ \vdots & \ddots & \ddots & \cdots & \vdots \\ 0 & 0 & \cdots & 1 & 0 \end{pmatrix}. \tag{4.2.13}$$

In the same way as in the case of $\mathfrak{sl}_2$ one shows that this space coincides with the space of nth order differential operators

$$\partial_t^n - u_1(t)\partial_t^{n-2} + \ldots + u_{n-2}(t)\partial_t - (-1)^n u_{n-1}(t) \tag{4.2.14}$$

acting from $\Omega^{-(n-1)/2}$ to $\Omega^{(n+1)/2}$. They have principal symbol 1 and subprincipal symbol 0.

The advantage of this form is that it is very explicit and gives a concrete realization of PGL_n-opers in terms of *scalar* differential operators. There are similar realizations for other classical Lie groups (see [DS, BD3]). The disadvantage is that this form is not as canonical, because the "first row" subspace is not canonically defined. In addition, the coefficients $u_i(t)$ of the operator (4.2.14) transform in a complicated way under changes of coordinates, unlike the coefficients $v_i(t)$ of the canonical form from the previous section.

4.3 The center for an arbitrary affine Kac–Moody algebra

We now state one of the main results of this book: a coordinate-independent description of the center $\mathfrak{z}(\widehat{\mathfrak{g}})$ of the vertex algebra $V_{\kappa_c}(\mathfrak{g})$. This description, given in Theorems 4.3.1 and 4.3.2, generalizes the description of $\mathfrak{z}(\widehat{\mathfrak{sl}}_2)$ given in Theorem 3.5.2. We will use it to describe the center $Z(\widehat{\mathfrak{g}})$ of the completed enveloping algebra $\widetilde{U}_{\kappa_c}(\widehat{\mathfrak{g}})$ at the critical level in Theorem 4.3.6. Finally, we will show in Proposition 4.3.9 that the center of $\widetilde{U}_{\kappa}(\widehat{\mathfrak{g}})$ is trivial away from the critical level.

4.3.1 The center of the vertex algebra

Let us recall from Section 3.5.2 that for any smooth point x of a curve X we have the affine Kac–Moody algebra $\widehat{\mathfrak{g}}_x$ associated to x and the algebra $\mathfrak{z}(\widehat{\mathfrak{g}})_x$ defined by formula (3.5.4).

Let ${}^LG$ be the simple Lie group of adjoint type associated to the **Langlands dual Lie algebra** ${}^L\mathfrak{g}$ of $\mathfrak{g}$. By definition, ${}^L\mathfrak{g}$ is the Lie algebra whose Cartan matrix is the transpose of the Cartan matrix of $\mathfrak{g}$ (see Appendix A.3). Thus, ${}^LG$ is the **Langlands dual group** (as defined in Section 1.1.5) of the group G, which is the connected simply-connected simple Lie group with the Lie algebra $\mathfrak{g}$.

The following theorem proved by B. Feigin and myself [FF6] (see also [F4]) is the central result of this book. The detailed proof of this result will be presented in the following chapters.

Theorem 4.3.1 *The algebra $\mathfrak{z}(\widehat{\mathfrak{g}})_x$ is naturally isomorphic to the algebra of functions on the space $\operatorname{Op}_{{}^LG}(D_x)$ of ${}^LG$-opers on the disc D_x.*

Equivalently, we may consider the algebra $\mathfrak{z}(\widehat{\mathfrak{g}})$ corresponding to the standard disc $D = \operatorname{Spec} \mathcal{O}, \mathcal{O} = \mathbb{C}[[t]]$, but keep track of the action of the group $\operatorname{Aut} \mathcal{O}$ and the Lie algebra $\operatorname{Der} \mathcal{O}$. They act on $\mathfrak{z}(\widehat{\mathfrak{g}}) \subset V_{\kappa_c}(\mathfrak{g})$ and on the space $\operatorname{Op}_{{}^LG}(D)$ (and hence on the algebra $\operatorname{Fun} \operatorname{Op}_{{}^LG}(D)$ of functions on it).

Theorem 4.3.2 *The center $\mathfrak{z}(\widehat{\mathfrak{g}})$ is isomorphic to the algebra $\operatorname{Fun} \operatorname{Op}_{{}^LG}(D)$ in a $(\operatorname{Der} O, \operatorname{Aut} \mathcal{O})$-equivariant way.*

Using the isomorphism (4.2.9), we obtain an isomorphism

$$\operatorname{Op}_{{}^LG}(D) \simeq \mathbb{C}[[t]]^{\oplus \ell}$$

(of course, it depends on the coordinate t). Therefore each oper is represented by an ℓ-tuple of formal Taylor series $(v_1(t), \ldots, v_\ell(t))$. Let us write

$$v_i(t) = \sum_{n<0} v_{i,n} t^{-n-1}.$$

Then we obtain an isomorphism

$$\operatorname{Fun} \operatorname{Op}_{^L G}(D) \simeq \mathbb{C}[v_{i,n}]_{i=1,\ldots,\ell; n<0}. \tag{4.3.1}$$

Let $S_i, i = 1, \ldots, \ell$, be the elements of $\mathfrak{z}(\widehat{\mathfrak{g}})$ corresponding to

$$v_{i,-1} \in \operatorname{Fun} \operatorname{Op}_{^L G}(D)$$

under the isomorphism of Theorem 4.3.2. Let us use the fact that it is equivariant under the action of the operator $L_{-1} = -\partial_t \in \operatorname{Der} \mathcal{O}$. On the side of $\operatorname{Fun} \operatorname{Op}_{^L G}(D)$ it is clear that $\frac{1}{m!}(-\partial_t)^m v_{i,-1} = v_{i,-m-1}$. Hence we obtain that under the isomorphism of Theorem 4.3.2 the generators $v_{i,-m-1}$ go to

$$\frac{1}{m!} L_{-1}^m S_i = \frac{1}{m!} T^m S_i,$$

because L_{-1} coincides with the translation operator T on the vertex algebra $V_{\kappa_c}(\mathfrak{g})$. This implies that

$$v_{i,-m-1} \mapsto S_{i,(-m-1)}|0\rangle,$$

where the $S_{i,(n)}$'s are the Fourier coefficients of the vertex operator

$$Y(S_i, z) = \sum_{n<0} S_{i,(n)} z^{-n-1}.$$

Indeed, we recall that $\mathfrak{z}(\widehat{\mathfrak{g}})$ is a commutative vertex algebra, and so we can use formula (2.2.1). Note that the operators $S_{i,(n)}, n \geq 0$, annihilate the vacuum vector $|0\rangle$, and because they commute with the action of $\widehat{\mathfrak{g}}_{\kappa_c}$, their action on $V_{\kappa_c}(\mathfrak{g})$ is identically equal to zero.

Furthermore, we obtain that any polynomial in the $v_{i,n}$'s goes under the isomorphism of Theorem 4.3.2 to the corresponding polynomial in the $S_{i,(n)}$'s, applied to the vacuum vector $|0\rangle \in V_{\kappa_c}(\mathfrak{g})$. Thus, we obtain that

$$\mathfrak{z}(\widehat{\mathfrak{g}}) = \mathbb{C}[S_{i,(n)}]_{i=1,\ldots,\ell; n<0}|0\rangle. \tag{4.3.2}$$

This identification means, in particular, that the map $\operatorname{gr} \mathfrak{z}(\widehat{\mathfrak{g}}) \hookrightarrow \operatorname{Inv} \mathfrak{g}^*[[t]]$ from Lemma 3.3.1 is an isomorphism.†

To see that, we recall the description of the algebra $\operatorname{Inv} \mathfrak{g}^*[[t]]$ from Theorem 3.4.2:

$$\operatorname{Inv} \mathfrak{g}^*[[t]] = \mathbb{C}[\overline{P}_{i,n}]_{i=1,\ldots,\ell; n<0},$$

where $\overline{P}_i, i = 1, \ldots, \ell$, are homogeneous generators of the algebra $\operatorname{Inv} \mathfrak{g}^*$ of degrees $d_i + 1, i = 1, \ldots, \ell$. To compare the two spaces, $\operatorname{gr} \mathfrak{z}(\widehat{\mathfrak{g}})$ and $\operatorname{Inv} \mathfrak{g}^*[[t]]$, we compute their formal characters, i.e., the generating function of their graded dimensions with respect to the grading operator $L_0 = -t\partial_t$.

† We remark that this is, in fact, one of the steps in our proof of Theorem 4.3.2 given below.

If V is a $\mathbb{Z}$-graded vector space $V = \bigoplus_{n \in \mathbb{Z}} V_n$ with finite-dimensional graded components, we define its formal character as

$$\operatorname{ch} V = \sum_{n \in \mathbb{Z}} \dim V_n q^n.$$

Now, it follows from the definition of the generators $\overline{P}_{i,n}$ of $\operatorname{Inv} \mathfrak{g}^*[[t]]$ given in Section 3.3.4 that $\deg \overline{P}_{i,m} = d_i - m$. Therefore we find that

$$\operatorname{ch} \operatorname{Inv} \mathfrak{g}^*[[t]] = \prod_{i=1}^{\ell} \prod_{n_i \geq d_i + 1} (1 - q^{n_i})^{-1}. \tag{4.3.3}$$

On the other hand, from the description of $\operatorname{Op}_{{}^L G}(D)$ given by formula (4.2.9) we obtain the action of $-t\partial_t \in \operatorname{Der} \mathcal{O}$ on the generators $v_{i,m}$. We find that $\deg v_{i,m} = d_i - m$. Hence the character of $\operatorname{Fun} \operatorname{Op}_{{}^L G}(D)$ (and therefore the character of $\mathfrak{z}(\widehat{\mathfrak{g}})$) coincides with the right hand side of formula (4.3.3). Since we have an injective map $\operatorname{gr} \mathfrak{z}(\widehat{\mathfrak{g}}) \hookrightarrow \operatorname{Inv} \mathfrak{g}^*[[t]]$ of two graded spaces whose characters coincide, we obtain the following:

Proposition 4.3.3 *The natural embedding* $\operatorname{gr} \mathfrak{z}(\widehat{\mathfrak{g}}) \hookrightarrow \operatorname{Inv} \mathfrak{g}^*[[t]]$ *is an isomorphism.*

In other words, the center $\mathfrak{z}(\widehat{\mathfrak{g}})$ is "as large as possible." This means that the generators $\overline{P}_{i,-1}$ of the algebra $\operatorname{Inv} \mathfrak{g}^*[[t]]$ may be "lifted" to $\mathfrak{g}[[t]]$-invariant vectors $S_i \in \mathfrak{z}(\widehat{\mathfrak{g}}), i = 1, \ldots, \ell$. In other words, $\overline{P}_{i,-1}$ is the symbol of S_i. It then follows that $\overline{P}_{i,n}$ is the symbol of $S_{i,(n)}|0\rangle \in \mathfrak{z}(\widehat{\mathfrak{g}})$ for all $n < 0$.

In particular, $\overline{P}_{1,-1}$ is the quadratic Casimir generator given by the formula

$$\overline{P}_{1,-1} = \frac{1}{2} \overline{J}^a_{-1} \overline{J}_{a,-1},$$

up to a non-zero scalar. Its lifting to $\mathfrak{z}(\widehat{\mathfrak{g}})$ is the Segal–Sugawara vector S_1 given by formula (3.1.1). Therefore the generators $S_{1,(n)}, n < 0$, are the Segal–Sugawara operators S_{n-1}.

It is natural to ask whether it is possible to find explicit liftings to $\mathfrak{z}(\widehat{\mathfrak{g}})$ of other generators $\overline{P}_{i,-1}, i = 1, \ldots, \ell$, of $\operatorname{Inv} \mathfrak{g}^*[[t]]$. So far, the answer is known only in special cases. For $\mathfrak{g} = \mathfrak{sl}_n$ explicit formulas for $\overline{P}_{i,-1}$ may be constructed using the results of R. Goodman and N. Wallach [GW] which rely on some intricate invariant theory computations.† For $\mathfrak{g}$ of classical types A_ℓ, B_ℓ, C_ℓ another approach was suggested by T. Hayashi in [Ha]: he constructed explicitly the next central elements after the Segal–Sugawara operators, of degree 3 in the case of A_ℓ, and of degree 4 in the case of B_ℓ and C_ℓ. He then generated central elements of higher degrees by using the Poisson structure on the center of $\widetilde{U}_{\kappa_c}(\widehat{\mathfrak{g}})$ discussed in Section 8.3.1 below (which

† Recently, an elegant formula of a different kind for $\mathfrak{g} = \mathfrak{sl}_n$ was obtained in [CT].

Hayashi had essentially introduced for this purpose). Unfortunately, this approach does not not give the entire center when $\mathfrak{g} = D_\ell$ (as the Pfaffian appears to be out of reach) and when $\mathfrak{g}$ is an exceptional Lie algebra. For $\mathfrak{g}$ of types A_ℓ, B_ℓ, C_ℓ one can prove in this way the isomorphism (4.3.2), but not Theorem 4.3.2 identifying the center $\mathfrak{z}(\widehat{\mathfrak{g}})$ with the algebra $\operatorname{Fun}\operatorname{Op}_{^LG}(D)$.

The proof of Theorem 4.3.2 given by B. Feigin and myself [FF6, F4], which is presented in this book, does not rely on explicit formulas for the generators of the center. Instead, we identify $\mathfrak{z}(\widehat{\mathfrak{g}})$ and $\operatorname{Fun}\operatorname{Op}_{^LG}(D)$ as two subalgebras of another algebra, namely, the algebra of functions on the space of connections on a certain ^LH-bundle on the disc, where ^LH is a Cartan subgroup of ^LG.

4.3.2 The center of the enveloping algebra

We now use the above description of the center of $\mathfrak{z}(\widehat{\mathfrak{g}})$ to describe the center $Z(\widehat{\mathfrak{g}})$ of the completed enveloping algebra $\widetilde{U}_{\kappa_c}(\widehat{\mathfrak{g}})$ at the critical level (see Section 2.1.2 for the definition).

Let $B \in \mathfrak{z}(\widehat{\mathfrak{g}}) \subset V_{\kappa_c}(\mathfrak{g})$. Then $\mathfrak{g}[[t]] \cdot B = 0$ and formula (3.2.2) for the commutation relations imply that all elements $B_{[k]} \in \widetilde{U}_{\kappa_c}(\widehat{\mathfrak{g}})$, as defined in Section 3.2.1, commute with the entire affine Kac–Moody algebra $\widehat{\mathfrak{g}}_{\kappa_c}$. Therefore they are central elements of $\widetilde{U}_{\kappa_c}(\widehat{\mathfrak{g}})$. Moreover, any element of $\widetilde{U}(\mathfrak{z}(\widehat{\mathfrak{g}})) \subset \widetilde{U}(V_{\kappa_c}(\mathfrak{g})) = \widetilde{U}_{\kappa_c}(\widehat{\mathfrak{g}})$ is also central (here we use the enveloping algebra functor $\widetilde{U}$ defined in Section 3.2.3 and Lemma 3.2.2). Thus, we obtain a homomorphism $\widetilde{U}(\mathfrak{z}(\widehat{\mathfrak{g}})) \to Z(\widehat{\mathfrak{g}})$.

Now, the algebra $\widetilde{U}(\mathfrak{z}(\widehat{\mathfrak{g}}))$ is a completion of a polynomial algebra. Indeed, recall from the previous section that we have elements $S_i, i = 1, \ldots, \ell$, in $\mathfrak{z}(\widehat{\mathfrak{g}})$ which give us the isomorphism (4.3.2). Let $S_{i,[n]}, n \in \mathbb{Z}$, be the elements of $\widetilde{U}_{\kappa_c}(\widehat{\mathfrak{g}})$ corresponding to $S_i \in V_{\kappa_c}(\mathfrak{g})$ (see Section 3.2.1). It follows from the definition that the topological algebra $\widetilde{U}(\mathfrak{z}(\widehat{\mathfrak{g}}))$ is the completion of the polynomial algebra $\mathbb{C}[S_{i,[n]}]_{i=1,\ldots,\ell;n\in\mathbb{Z}}$ with respect to the topology in which the basis of open neighborhoods of 0 is formed by the ideals generated by $S_{i,[n]}, i = 1, \ldots, \ell; n > N + d_i + 1$, for $N \in \mathbb{Z}_+$. This topology is equivalent to the topology in which the condition $n > N + d_i + 1$ is replaced by the condition $n > N$.

Thus, the completed polynomial algebra $\widetilde{U}(\mathfrak{z}(\widehat{\mathfrak{g}}))$ maps to the center $Z(\widehat{\mathfrak{g}})$ of $\widetilde{U}_{\kappa_c}(\widehat{\mathfrak{g}})$.

Proposition 4.3.4 *This map is an isomorphism, and so $Z(\widehat{\mathfrak{g}})$ is equal to $\widetilde{U}(\mathfrak{z}(\widehat{\mathfrak{g}}))$.*

Proof. We follow the proof given in [BD1], Theorem 3.7.7.

We start by describing the associated graded algebra of $Z(\widehat{\mathfrak{g}})$. The Poincaré–Birkhoff–Witt filtration on $U_{\kappa_c}(\widehat{\mathfrak{g}})$ induces one on the completion $\widetilde{U}_{\kappa_c}(\widehat{\mathfrak{g}})$. The associated graded algebra $\operatorname{gr} U_{\kappa_c}(\widehat{\mathfrak{g}})$ (recall the definition given in Section 3.2.3) is the algebra $\operatorname{Sym}\mathfrak{g}((t)) = \bigoplus_{i\geq 0}\operatorname{Sym}^i\mathfrak{g}((t))$. Let $\overline{I}_N$ be the ideal in $\operatorname{Sym}\mathfrak{g}((t))$ generated by $\mathfrak{g}\otimes t^N\mathbb{C}[[t]]$. The associated graded algebra $\operatorname{gr}\widetilde{U}_{\kappa_c}(\widehat{\mathfrak{g}})$ is the completion $\widetilde{\operatorname{Sym}}\,\mathfrak{g}((t)) = \bigoplus_{i\geq 0}\widetilde{\operatorname{Sym}}^i\mathfrak{g}((t))$ of $\operatorname{Sym}\mathfrak{g}((t))$, where

$$\widetilde{\operatorname{Sym}}^i\mathfrak{g}((t)) = \varprojlim \operatorname{Sym}^i\mathfrak{g}((t))/(\overline{I}_N\cap\operatorname{Sym}^i\mathfrak{g}((t))).$$

In other words, $\widetilde{\operatorname{Sym}}^i\mathfrak{g}((t))$ is the space $\operatorname{Fun}^i(\mathfrak{g}^*\otimes t^{-N}\mathbb{C}[[t]])$ of polynomial functions on $\mathfrak{g}((t))\simeq\mathfrak{g}^*((t))dt\simeq\mathfrak{g}^*((t))$ of degree i. By such a function we mean a function on $\mathfrak{g}^*((t))$ such that for each $N\in\mathbb{Z}_+$ its restriction to $\mathfrak{g}^*\otimes t^{-N}\mathbb{C}[[t]]$ is a polynomial function of degree i (i.e., it comes by pull-back from a polynomial function on a finite-dimensional vector space $\mathfrak{g}^*\otimes(t^{-N}\mathbb{C}[[t]]/t^n\mathbb{C}[[t]])$ for sufficiently large n).

The Lie algebra $\mathfrak{g}((t))$ naturally acts on each space $\widetilde{\operatorname{Sym}}^i\mathfrak{g}((t))$ via the adjoint action, and its Lie subalgebra $\mathfrak{g}[[t]]$ preserves the subspace $\overline{I}_N\cap\operatorname{Sym}^i\mathfrak{g}((t))$ of those functions which vanish on $\mathfrak{g}\otimes t^N\mathbb{C}[[t]]$.

Let $\widetilde{\operatorname{Inv}}\mathfrak{g}^*((t)) = \bigoplus_{i\geq 0}\widetilde{\operatorname{Inv}}^i\mathfrak{g}((t))$ be the subalgebra of $\mathfrak{g}((t))$-invariant elements of $\widetilde{\operatorname{Sym}}\,\mathfrak{g}((t))$. For each $N\in\mathbb{Z}_+$ we have a surjective homomorphism obtained by taking the quotient by $\overline{I}_N$:

$$\widetilde{\operatorname{Sym}}^i\mathfrak{g}((t)) \to \operatorname{Fun}^i(\mathfrak{g}^*\otimes t^{-N}\mathbb{C}[[t]]).$$

The image of $\widetilde{\operatorname{Inv}}^i\mathfrak{g}^*((t))$ under this homomorphism is contained in the space of $\mathfrak{g}[[t]]$-invariants in $\operatorname{Fun}^i(\mathfrak{g}^*\otimes t^{-N}\mathbb{C}[[t]])$.

But we have a natural isomorphism

$$\operatorname{Fun}(\mathfrak{g}^*\otimes t^{-N}\mathbb{C}[[t]]) \simeq \operatorname{Fun}(\mathfrak{g}^*[[t]])$$

(multiplication by t^N) which commutes with the action of $\mathfrak{g}[[t]]$. Since we know the algebra of $\mathfrak{g}[[t]]$-invariant functions on $\mathfrak{g}^*[[t]]$ from Theorem 3.4.2, we find the algebra of $\mathfrak{g}[[t]]$-invariant functions on $\mathfrak{g}^*\otimes t^{-N}\mathbb{C}[[t]]$ by using this isomorphism. To describe it, let $\overline{J}^a_n$ be a linear functional on $\mathfrak{g}^*((t))$ defined as in formula (3.3.4). Note that the restriction of $\overline{J}^a_n, n\geq N$, to $\mathfrak{g}^*\otimes t^{-N}\mathbb{C}[[t]]$ is equal to 0. Let us write

$$\overline{P}_i(z) = \overline{P}_i(\overline{J}^a(z)) = \sum_{n\in\mathbb{Z}}\overline{P}_{i,n}z^{-n-1},$$

where

$$\overline{J}^a(z) = \sum_{n\in\mathbb{Z}}\overline{J}^a_n z^{-n-1}.$$

Now we obtain from the above isomorphism that the algebra of $\mathfrak{g}[[t]]$-invariant functions on $\mathfrak{g}^*\otimes t^{-N}\mathbb{C}[[t]]$ is the free polynomial algebra with the generators $\overline{P}_{i,n}, n<(d_i+1)N$.

However, one checks explicitly that each $\overline{P}_{i,n}$ is not only $\mathfrak{g}[[t]]$-invariant, but $\mathfrak{g}((t))$-invariant. This means that the homomorphism

$$\widetilde{\operatorname{Inv}}\,\mathfrak{g}^*((t)) \to (\operatorname{Fun}(\mathfrak{g}^* \otimes t^{-N}\mathbb{C}[[t]]))^{\mathfrak{g}[[t]]}$$

is surjective for all $N \in \mathbb{Z}_+$. Hence it follows that $\widetilde{\operatorname{Inv}}\,\mathfrak{g}^*((t))$ is the inverse limit of the algebras $(\operatorname{Fun}(\mathfrak{g}^* \otimes t^{-N}\mathbb{C}[[t]]))^{\mathfrak{g}[[t]]}$.

Therefore $\widetilde{\operatorname{Inv}}\,\mathfrak{g}^*((t))$, is the completion of the polynomial algebra

$$\mathbb{C}[\overline{P}_{i,n}]_{i=1,\ldots,\ell;n\in\mathbb{Z}}$$

with respect to the topology in which the basis of open neighborhoods of 0 is formed by the subspaces of the polynomials of fixed degree that lie in the ideals generated by $\overline{P}_{i,n}$, with $n > N$, for $N \in \mathbb{Z}_+$ (using the last condition is equivalent to using the condition $n \geq (d_i+1)N$, which comes from the above analysis).

Now we are ready to prove the proposition. Recall the definition of $\widetilde{U}_{\kappa_c}(\widehat{\mathfrak{g}})$ as the inverse limit of the quotients of $U_{\kappa_c}(\widehat{\mathfrak{g}})/I_N$ by the left ideals I_N generated by $\mathfrak{g} \otimes t^N\mathbb{C}[[t]]$. Therefore

$$Z(\widehat{\mathfrak{g}}) = \varprojlim Z(\widehat{\mathfrak{g}})/Z(\widehat{\mathfrak{g}}) \cap I_N.$$

We have an injective map $Z(\widehat{\mathfrak{g}})/Z(\widehat{\mathfrak{g}}) \cap I_N \hookrightarrow (U_{\kappa_c}(\widehat{\mathfrak{g}})/I_N)^{\mathfrak{g}[[t]]}$. In addition, we have an injective map

$$\operatorname{gr}((U_{\kappa_c}(\widehat{\mathfrak{g}})/I_N)^{\mathfrak{g}[[t]]}) \hookrightarrow (\operatorname{gr} U_{\kappa_c}(\widehat{\mathfrak{g}})/\overline{I}_N)^{\mathfrak{g}[[t]]} = (\operatorname{Fun}(\mathfrak{g}^* \otimes t^{-N}\mathbb{C}[[t]]))^{\mathfrak{g}[[t]]}.$$

We already know that $\mathbb{C}[S_{i,[n]}]_{i=1,\ldots,\ell;n\in\mathbb{Z}}$ is a subalgebra of $Z(\widehat{\mathfrak{g}})$. By construction, the symbol of $S_{i,[n]}$ is equal to $\overline{P}_{i,n}$. Therefore the image of $\mathbb{C}[S_{i,[n]}]_{i=1,\ldots,\ell;n\in\mathbb{Z}}$ under the composition of the above maps is the entire $(\operatorname{Fun}(\mathfrak{g}^* \otimes t^{-N}\mathbb{C}[[t]]))^{\mathfrak{g}[[t]]}$.

This implies that all the intermediate maps are isomorphisms. In particular, we find that for each $N \in \mathbb{Z}_+$ we have

$$Z(\widehat{\mathfrak{g}})/Z(\widehat{\mathfrak{g}}) \cap I_N = \mathbb{C}[S_{i,[n_i]}]_{i=1,\ldots,\ell;n_i \leq N(d_i+1)}.$$

Therefore the center $Z(\mathfrak{g})$ is equal to the completion of the polynomial algebra $\mathbb{C}[S_{i,[n]}]_{i=1,\ldots,\ell;n\in\mathbb{Z}}$ with respect to the topology in which the basis of open neighborhoods of 0 is formed by the ideals generated by $S_{i,[n]}, i = 1, \ldots, \ell; n > N$. But this is precisely $\widetilde{U}(\mathfrak{z}(\widehat{\mathfrak{g}}))$. This completes the proof. □

As a corollary we obtain that $Z(\widehat{\mathfrak{g}})$ is isomorphic to $\widetilde{U}(\operatorname{Fun}\operatorname{Op}_{{}^LG}(D))$.

Lemma 4.3.5 *The algebra* $\widetilde{U}(\operatorname{Fun}\operatorname{Op}_{{}^LG}(D))$ *is canonically isomorphic to the topological algebra* $\operatorname{Fun}\operatorname{Op}_{{}^LG}(D^\times)$ *of functions on the space of* ${}^LG$*-opers on the punctured disc* $D^\times = \operatorname{Spec}\mathbb{C}((t))$.

Proof. As explained in Section 4.3.1, the algebra $\operatorname{Fun}\operatorname{Op}_{^LG}(D)$ is isomorphic to the polynomial algebra $\mathbb{C}[v_{i,n}]_{i=1,\dots,\ell;n<0}$, where the $v_{i,n}$'s are the coefficients of the oper connection

$$\nabla = \partial_t + p_{-1} + \sum_{i=1}^{\ell} v_i(t) p_i, \qquad (4.3.4)$$

where $v_i(t) = \sum_{n<0} v_{i,n} t^{-n-1}$. The construction of the functor $\widetilde{U}$ implies that the algebra $\widetilde{U}(\operatorname{Op}_{^LG}(D)_{\kappa_0^\vee})$ is the completion of the polynomial algebra in the variables $v_{i,n}, i = 1, \dots, \ell; n \in \mathbb{Z}$, with respect to the topology in which the base of open neighborhoods of 0 is formed by the ideals generated by $v_{i,n}, n > N$. But this is precisely the algebra of functions on the space of opers of the form (4.3.4), where $v_i(t) = \sum_{n\in\mathbb{Z}} v_{i,n} t^{-n-1} \in \mathbb{C}((t))$. □

Now Theorem 4.3.2, Proposition 4.3.4 and Lemma 4.3.5 imply the following:

Theorem 4.3.6 *The center $Z(\widehat{\mathfrak{g}})$ is isomorphic to the algebra* $\operatorname{Fun}\operatorname{Op}_{^LG}(D^\times)$ *in a* $(\operatorname{Der} O, \operatorname{Aut} \mathcal{O})$*-equivariant way.*

We will see later that there are other conditions that this isomorphism satisfies, which fix it almost uniquely.

Using this theorem, we can describe the center $Z(\widehat{\mathfrak{g}}_x)$ of the enveloping algebra $\widetilde{U}_{\kappa_c}(\widehat{\mathfrak{g}})_x$ of the Lie algebra $\widehat{\mathfrak{g}}_{\kappa_c,x}$.

Corollary 4.3.7 *The center $Z(\widehat{\mathfrak{g}}_x)$ is isomorphic to the algebra* $\operatorname{Fun}\operatorname{Op}_{^LG}(D_x^\times)$ *of functions on the space of ^LG-opers on $D_x^\times$.*

We also want to record the following useful result, whose proof is borrowed from [BD1], Remark 3.7.11(iii). Note that the group $G((t))$ acts naturally on $\widetilde{U}_{\kappa_c}(\widehat{\mathfrak{g}})$.

Proposition 4.3.8 *The action of $G((t))$ on $Z(\widehat{\mathfrak{g}}) \subset \widetilde{U}_{\kappa_c}(\widehat{\mathfrak{g}})$ is trivial.*

Proof. Let $\widetilde{U}_{\kappa_c}(\widehat{\mathfrak{g}})_{\leq i}$ be the ith term of the filtration on $\widetilde{U}_{\kappa_c}(\widehat{\mathfrak{g}})$ induced by the PBW filtration on $U_{\kappa_c}(\widehat{\mathfrak{g}})$ (see the proof of Proposition 4.3.4). It is clear that the union of $\widetilde{U}_{\kappa_c}(\widehat{\mathfrak{g}})_{\leq i}, i \geq 0$, is dense in $\widetilde{U}_{\kappa_c}(\widehat{\mathfrak{g}})$. Moreover, it follows from the description of the center as a completion of the algebra $\mathbb{C}[S_{i,[n]}]_{i=1,\dots,\ell;n\in\mathbb{Z}}$ that the union of $Z(\widehat{\mathfrak{g}}) \cap \widetilde{U}_{\kappa_c}(\widehat{\mathfrak{g}})_{\leq i}, i \geq 0$, is dense in $Z(\widehat{\mathfrak{g}})$. Therefore it is sufficient to prove that $G((t))$ acts trivially on $Z(\widehat{\mathfrak{g}}) \cap \widetilde{U}_{\kappa_c}(\widehat{\mathfrak{g}})$ (clearly, $G((t))$ preserves each term $\widetilde{U}_{\kappa_c}(\widehat{\mathfrak{g}})_{\leq i}$).

In Proposition 4.3.4 we described the associated graded algebra $\operatorname{gr} Z(\widehat{\mathfrak{g}})$ of $Z(\widehat{\mathfrak{g}})$ with respect to the above filtration. Namely, $\operatorname{gr} Z(\widehat{\mathfrak{g}})$ is a completion of a polynomial algebra. Using explicit formulas for the generators $\overline{P}_{i,n}$ of this algebra presented above, we find that $G((t))$ acts trivially on $\operatorname{gr} Z(\widehat{\mathfrak{g}})$. Therefore the action of $G((t))$ on $Z(\widehat{\mathfrak{g}}) \cap \widetilde{U}_{\kappa_c}(\widehat{\mathfrak{g}})$ factors through a unipotent algebraic group U. But then the corresponding action of its Lie algebra $\mathfrak{g}((t))$ factors through an action of the Lie algebra of U. From our assumption that G is simply-connected we obtain that $G((t))$ is connected. Therefore U is also connected.

Clearly, the differential of a non-trivial action of a connected unipotent Lie group is also non-trivial. Therefore we find that if the action of $G((t))$ on $Z(\widehat{\mathfrak{g}}) \cap \widetilde{U}_{\kappa_c}(\widehat{\mathfrak{g}})$ is non-trivial, then the action of its Lie algebra $\mathfrak{g}((t))$ is also non-trivial. But it is obvious that $\mathfrak{g}((t))$ acts by 0 on $Z(\widehat{\mathfrak{g}})$. Therefore the action of $G((t))$ on $Z(\widehat{\mathfrak{g}}) \cap \widetilde{U}_{\kappa_c}(\widehat{\mathfrak{g}})$, and hence on $Z(\widehat{\mathfrak{g}})$, is trivial. This completes the proof.

Note that if we had not assumed that G were simply-connected, then the fundamental group of G would be a finite group Γ, which would coincide with the group of components of $G((t))$. The above argument shows that the action of $G((t))$ factors through Γ. But a finite group cannot have non-trivial unipotent representations. Hence we obtain that $G((t))$ acts trivially on $Z(\widehat{\mathfrak{g}})$ even without the assumption that G is simply-connected. □

4.3.3 The center away from the critical level

The above results describe the center of the vertex algebra $V_\kappa(\mathfrak{g})$ and the center of the completed enveloping algebra $\widetilde{U}_\kappa(\widehat{\mathfrak{g}})$ at the critical level $\kappa = \kappa_c$. But what happens away from the critical level?

In Proposition 3.3.3 we have answered this question in the case of the vertex algebra: it is trivial, i.e., consists of scalars only. Now, for completeness, we answer it for the completed enveloping algebra $\widetilde{U}_\kappa(\widehat{\mathfrak{g}})$ and show that it is trivial as well. Therefore we are not missing anything, as far as the center of $\widetilde{U}_\kappa(\widehat{\mathfrak{g}})$ is concerned, outside of the critical level.

Proposition 4.3.9 *The center of $\widetilde{U}_\kappa(\widehat{\mathfrak{g}})$ consists of the scalars for $\kappa \neq \kappa_c$.*

Proof. Consider the Lie algebra $\widehat{\mathfrak{g}}'_\kappa = \mathbb{C}d \ltimes \widehat{\mathfrak{g}}_\kappa$, where d is the operator whose commutation relations with $\widehat{\mathfrak{g}}_\kappa$ correspond to the action of the vector field $t\partial_t$. Let $U'_\kappa(\widehat{\mathfrak{g}})$ be the quotient of the universal enveloping algebra of $\widehat{\mathfrak{g}}'_\kappa$ by the ideal generated by $\mathbf{1} - 1$, and $\widetilde{U}'_\kappa(\widehat{\mathfrak{g}})$ its completion defined in the same way as $\widetilde{U}_\kappa(\widehat{\mathfrak{g}})$.

Now, the algebra $\widetilde{U}_\kappa(\widehat{\mathfrak{g}}), \kappa \neq \kappa_c$, contains the operator L_0, which commutes with $\widehat{\mathfrak{g}}_\kappa$ as $-d$. Therefore $L_0 + d$ is a central element of $\widetilde{U}'_\kappa(\widehat{\mathfrak{g}})$, and $\widetilde{U}_\kappa(\widehat{\mathfrak{g}})$ is isomorphic to the quotient of $\widetilde{U}'_\kappa(\widehat{\mathfrak{g}})$ by the ideal generated by $L_0 + d$.

Let $Z_\kappa(\widehat{\mathfrak{g}})$ be the center of $\widetilde{U}_\kappa(\widehat{\mathfrak{g}})$ and $Z'_\kappa(\widehat{\mathfrak{g}})$ the center of $\widetilde{U}'_\kappa(\widehat{\mathfrak{g}})$. We have a natural embedding $\widetilde{U}_\kappa(\widehat{\mathfrak{g}}) \to \widetilde{U}'_\kappa(\widehat{\mathfrak{g}})$. Since any element A of $Z_\kappa(\widehat{\mathfrak{g}})$ must satisfy $[L_0, A] = 0$, we find that the image of A in $\widetilde{U}'_\kappa(\widehat{\mathfrak{g}})$ is also a central element. Therefore we find that $Z'_\kappa(\widehat{\mathfrak{g}}) = Z_\kappa(\widehat{\mathfrak{g}}) \otimes_{\mathbb{C}} \mathbb{C}[L_0 + d]$.

The description of the center $Z'_\kappa(\widehat{\mathfrak{g}})$ follows from the results of V. Kac

[K1]. According to Corollary 1 of [K1], for $\kappa \neq \kappa_c$ there is an isomorphism between $Z'_\kappa(\widehat{\mathfrak{g}})$ and the algebra of invariant functions on the hyperplane $(\widetilde{\mathfrak{h}})^*_\kappa$, corresponding to level κ, in the dual space to the extended Cartan subalgebra $\widetilde{\mathfrak{h}} = (\mathfrak{h} \otimes 1) \oplus \mathbb{C}\mathbf{1} \oplus \mathbb{C}d$, with respect to the ($\rho$-shifted) action of the affine Weyl group. But it is easy to see that the polynomial functions on $(\widetilde{\mathfrak{h}})^*_\kappa$ satisfying this condition are spanned by the powers of the polynomial on $(\widetilde{\mathfrak{h}})^*_\kappa$ obtained by restriction of the invariant inner product on $\widetilde{\mathfrak{h}}^*$. This quadratic polynomial gives rise to the central element $L_0 + d$. Therefore $Z'_\kappa(\widehat{\mathfrak{g}}) = Z_\kappa(\widehat{\mathfrak{g}}) \otimes_{\mathbb{C}} \mathbb{C}[L_0 + d]$, and so $Z_\kappa(\widehat{\mathfrak{g}}) = \mathbb{C}$. □

5
Free field realization

We now set out to prove Theorem 4.3.2 establishing an isomorphism between the center of the vertex algebra $V_{\kappa_c}(\mathfrak{g})$ and the algebra of functions on the space of ${}^L G$-opers on the disc. An essential role in this proof is played by the *Wakimoto modules* over $\widehat{\mathfrak{g}}_{\kappa_c}$. Our immediate goal is to construct these modules and to study their properties. This will be done in this chapter and the next, following [FF4, F4].

5.1 Overview

Our ultimate goal is to prove that the center of the vertex algebra $V_{\kappa_c}(\mathfrak{g})$ is isomorphic to the algebra of functions on the space of ${}^L G$-opers on the disc. It is instructive to look first at the proof of the analogous statement in the finite-dimensional case, which is Theorem 2.1.1. The most direct way to describe the center $Z(\mathfrak{g})$ of $U(\mathfrak{g})$ is to use the so-called Harish-Chandra homomorphism $Z(\mathfrak{g}) \to \operatorname{Fun} \mathfrak{h}^*$, where $\mathfrak{h}$ is the Cartan subalgebra of $\mathfrak{g}$, and to prove that its image is equal to the subalgebra $(\operatorname{Fun} \mathfrak{h}^*)^W$ of W-invariant functions, where W is the Weyl group acting on $\mathfrak{h}^*$ (see Appendix A.3). How can one construct this homomorphism? One possible way is to use geometry; namely, the infinitesimal action of the Lie algebra $\mathfrak{g}$ on the flag manifold $\mathrm{Fl} = G/B_-$, where B_- is the lower Borel subgroup of G. This action preserves the open dense B_+-orbit $\mathcal{U}$ (where B_+ is the upper Borel subgroup of G) in Fl, namely, $B_+ \cdot [1] \simeq N_+ = [B_+, B_+]$. Hence we obtain a Lie algebra homomorphism from $\mathfrak{g}$ to the algebra $\mathcal{D}(N_+)$ of differential operators on the unipotent subgroup $N_+ \subset G$.

One can show that there is in fact a family of such homomorphisms parameterized by $\mathfrak{h}^*$. Thus, we obtain a Lie algebra homomorphism $\mathfrak{g} \to \operatorname{Fun} \mathfrak{h}^* \otimes \mathcal{D}(N_+)$, and hence an algebra homomorphism $U(\mathfrak{g}) \to \operatorname{Fun} \mathfrak{h}^* \otimes \mathcal{D}(N_+)$. Next, one shows that the image of $Z(\mathfrak{g}) \subset U(\mathfrak{g})$ lies entirely in the first factor, which is the commutative subalgebra $\operatorname{Fun} \mathfrak{h}^*$, and this is the sought-after Harish-Chandra homomorphism. One can then prove that the image is invariant under the simple reflections s_i, the generators of W. After this, all that remains to complete the proof is to give an estimate on the "size" of $Z(\mathfrak{g})$.

We will follow the same strategy in the affine case. Instead of an open B_+-orbit of Fl, isomorphic to N_+, we will consider an open orbit of the "loop space" of Fl, which is isomorphic to $N_+((t))$. The infinitesimal action of $\mathfrak{g}((t))$ on $N_+((t))$ gives rise to a homomorphism of Lie algebras $\mathfrak{g}((t)) \to \operatorname{Vect} N_+((t))$. However, there is a new phenomenon which did not exist in the finite-dimensional case: because $N_+((t))$ is infinite-dimensional, we must consider a completion $\widetilde{\mathcal{D}}(N_+((t)))$ of the corresponding algebra of differential operators. It turns out that there is a cohomological obstruction to lifting our homomorphism to a homomorphism $\mathfrak{g}((t)) \to \widetilde{\mathcal{D}}(N_+((t)))$. But we will show, following [FF4, F4], that this obstruction may be partially resolved, giving rise to a homomorphism of the central extension $\widehat{\mathfrak{g}}_{\kappa_c}$ of $\mathfrak{g}((t))$ to $\widetilde{\mathcal{D}}(N_+((t)))$, such that $\mathbf{1} \mapsto 1$. Thus, any module over $\widetilde{\mathcal{D}}(N_+((t)))$ is a $\widehat{\mathfrak{g}}_{\kappa_c}$-module of critical level. This gives us another explanation of the special role of the critical level.

The above homomorphism may be deformed, much like in the finite-dimensional case, to a homomorphism from $\widehat{\mathfrak{g}}_{\kappa_c}$ (and hence from $\widetilde{U}_{\kappa_c}(\widehat{\mathfrak{g}})$) to a (completed) tensor product of $\widetilde{\mathcal{D}}(N((t)))$ and $\operatorname{Fun} \mathfrak{h}^*((t))$. However, we will see that in order to make it $\operatorname{Aut} \mathcal{O}$-equivariant, we need to modify the action of $\operatorname{Aut} \mathcal{O}$ on $\mathfrak{h}^*((t))$: instead of the usual action on $\mathfrak{h}^*((t)) \simeq \mathfrak{h}^* \otimes \Omega_{\mathcal{K}}$ we will need the action corresponding to the $\Omega_{\mathcal{K}}$-torsor $\operatorname{Conn}(\Omega^{-\rho})_{D^\times}$ of connections on the ${}^L H$-bundle $\Omega^{-\rho}$ on the punctured disc. Here ${}^L H$ is the Cartan subgroup of the Langlands dual group (its Lie algebra is identified with $\mathfrak{h}^*$). This is the first inkling of the appearance of connections and of the Langlands dual group.

Next, we show that under this homomorphism the image of the center $Z(\widehat{\mathfrak{g}}) \subset \widetilde{U}_{\kappa_c}(\widehat{\mathfrak{g}})$ is contained in the subalgebra $\operatorname{Fun} \operatorname{Conn}(\Omega^{-\rho})_{D^\times}$. Furthermore, we will show that this image is equal to the centralizer of certain "screening operators" $\overline{V}_i[1], i = 1, \ldots, \ell$ (one may think that these are the analogues of simple reflections from the Weyl group). On the other hand, we will show that this centralizer is also equal to the algebra $\operatorname{Fun} \operatorname{Op}_{{}^L G}(D^\times)$, which is embedded into $\operatorname{Fun} \operatorname{Conn}(\Omega^{-\rho})_{D^\times}$ via the so-called Miura transformation.

Thus, we will obtain the sought-after isomorphism $Z(\widehat{\mathfrak{g}}) \simeq \operatorname{Fun} \operatorname{Op}_{{}^L G}(D^\times)$ of Theorem 4.3.6 identifying the center with the algebra of functions on opers for the Langlands dual group. Actually, in what follows we will work in the setting of vertex algebras and obtain in a similar way an isomorphism $\mathfrak{z}(\widehat{\mathfrak{g}}) \simeq \operatorname{Fun} \operatorname{Op}_{{}^L G}(D)$ of Theorem 4.3.2. (As we saw in Section 4.3.2, this implies the isomorphism of Theorem 4.3.6.)

The homomorphism $\widetilde{U}_{\kappa_c}(\widehat{\mathfrak{g}}) \to \widetilde{\mathcal{D}}(N((t))) \widehat{\otimes} \operatorname{Fun} \mathfrak{h}^*((t))$ and its vertex algebra version are called the free field realization of $\widehat{\mathfrak{g}}_{\kappa_c}$. It may also be deformed away from the critical level, as we will see in the next chapter. This realization was introduced in 1986 by M. Wakimoto [W] in the case of $\widehat{\mathfrak{sl}}_2$ and in 1988 by B. Feigin and the author [FF1] in the general case. It gives rise to a family of $\widehat{\mathfrak{g}}_\kappa$-modules, called Wakimoto modules. They have many applications in representation theory, geometry and conformal field theory.

In this chapter and the next we present the construction of the free field realization and the Wakimoto modules. In the case of $\mathfrak{g} = \mathfrak{sl}_2$ the construction is spelled out in detail in [FB], Ch. 11–12, where the reader is referred for additional motivation and background. Here we explain it in the case of an arbitrary $\mathfrak{g}$. We follow the original approach of [FF4] (with some modifications introduced in [F4]) and prove the existence of the free field realization by cohomological methods. We note that explicit formulas for the Wakimoto realization have been given in [FF1] for $\mathfrak{g} = \mathfrak{sl}_n$ and in [dBF] for general $\mathfrak{g}$. Another proof of the existence of the free field realization has been presented in [FF8]. The construction has also been extended to twisted affine algebras in [Sz].

Here is a more detailed description of the contents of this chapter. We begin in Section 5.2 with the geometric construction of representations of a simple finite-dimensional Lie algebra $\mathfrak{g}$ using an embedding of $\mathfrak{g}$ into a Weyl algebra which is obtained from the infinitesimal action of $\mathfrak{g}$ on the flag manifold. This will serve as a prototype for the construction of Wakimoto modules presented in the subsequent sections. In Section 5.3 we introduce the main ingredients needed for the constructions of Wakimoto modules: the infinite-dimensional Weyl algebra $\mathcal{A}^{\mathfrak{g}}$, the corresponding vertex algebra $M_{\mathfrak{g}}$ and the infinitesimal action of the loop algebra $L\mathfrak{g}$ on the formal loop space $L\mathcal{U}$ of the big cell $\mathcal{U}$ of the flag manifold of $\mathfrak{g}$. We also introduce the local Lie algebra $\mathcal{A}^{\mathfrak{g}}_{\leq 1,\mathfrak{g}}$ of differential operators on $L\mathcal{U}$ of order less than or equal to one. We show that it is a non-trivial extension of the Lie algebra of local vector fields on $L\mathcal{U}$ by local functionals on $L\mathcal{U}$ and compute the corresponding two-cocycle (most of this material has already been presented in [FB], Chapter 12). In Section 5.4 we give a vertex algebra interpretation of this extension. We prove that the embedding of the loop algebra $L\mathfrak{g}$ into the Lie algebra of local vector fields on $L\mathcal{U}$ may be lifted to an embedding of the central extension $\widehat{\mathfrak{g}}$ to $\mathcal{A}^{\mathfrak{g}}_{\leq 1,\mathfrak{g}}$. In order to do that, we need to show that the restriction of the above two-cocycle to $L\mathfrak{g}$ is cohomologically equivalent to the two-cocycle corresponding to its Kac–Moody central extension (of level κ_c). This is achieved in Section 5.5 by replacing the standard cohomological Chevalley complex by a much smaller local subcomplex (where both cocycles belong). In Section 5.6 we compute the cohomology of the latter and prove that the cocycles are indeed cohomologically equivalent.

5.2 Finite-dimensional case

In this section we recall the realization of $\mathfrak{g}$-modules in the space of functions on the big cell of the flag manifold. This will serve as a blueprint for the construction of the free field realization of the affine Kac–Moody algebras in the following sections.

5.2.1 Flag variety

Let again $\mathfrak{g}$ be a simple Lie algebra of rank ℓ with its Cartan decomposition

$$\mathfrak{g} = \mathfrak{n}_+ \oplus \mathfrak{h} \oplus \mathfrak{n}_-, \tag{5.2.1}$$

where $\mathfrak{h}$ is the Cartan subalgebra and $\mathfrak{n}_\pm$ are the upper and lower nilpotent subalgebras. Let

$$\mathfrak{b}_\pm = \mathfrak{h} \oplus \mathfrak{n}_\pm$$

be the upper and lower Borel subalgebras. We will follow the notation of Appendix A.3.

Let G be the connected simply-connected Lie group corresponding to $\mathfrak{g}$, and $N_\pm$ (resp., $B_\pm$) the upper and lower unipotent subgroups (resp., Borel subgroups) of G corresponding to $\mathfrak{n}_\pm$ (resp., $\mathfrak{b}_\pm$).

The homogeneous space $\mathrm{Fl} = G/B_-$ is called the **flag variety** associated to $\mathfrak{g}$. For example, for $G = SL_n$ this is the variety of full flags of subspaces of $\mathbb{C}^n$: $V_1 \subset \ldots \subset V_{n-1} \subset \mathbb{C}^n, \dim V_i = i$. The group SL_n acts transitively on this variety, and the stabilizer of the flag in which $V_i = \mathrm{span}(e_n, \ldots, e_{n-i+1})$ is the subgroup B_- of lower triangular matrices.

The flag variety has a unique open N_+-orbit, the so-called **big cell** $\mathcal{U} = N_+ \cdot [1] \subset G/B_-$, which is isomorphic to N_+. Since N_+ is a unipotent Lie group, the exponential map $\mathfrak{n}_+ \to N_+$ is an isomorphism. Therefore N_+ is isomorphic to the vector space $\mathfrak{n}_+$. Thus, N_+ is isomorphic to the affine space $\mathbb{A}^{|\Delta_+|}$, where Δ_+ is the set of positive roots of $\mathfrak{g}$. Hence the algebra $\mathrm{Fun}\, N_+$ of regular functions on N_+ is a free polynomial algebra. We will call a system of coordinates $\{y_\alpha\}_{\alpha \in \Delta_+}$ on N_+ *homogeneous* if

$$h \cdot y_\alpha = -\alpha(h) y_\alpha, \qquad h \in \mathfrak{h}.$$

In what follows we will consider only homogeneous coordinate systems on N_+.

Note that in order to define $\mathcal{U}$ it is sufficient to choose only a Borel subgroup B_+ of G. Then $N_+ = [B_+, B_+]$ and $\mathcal{U}$ is the open N_+-orbit in the flag manifold defined as the variety of all Borel subgroups of G (so $\mathcal{U}$ is an N_+-torsor). All constructions of this chapter make sense with the choice of B_+ only, i.e., without making the choice of an embedding of $H = B_+/[B_+, B_+]$ into B_+ (which in particular gives the opposite Borel subgroup B_-, see Section 4.2.1). However, to simplify the exposition we will fix a Cartan subgroup $H \subset B_+$ as well. We will see later on that the construction is independent of this choice.

The action of G on G/B_- gives us a map from $\mathfrak{g}$ to the Lie algebra of vector fields on G/B_-, and hence on its open dense subset $\mathcal{U} \simeq N_+$. Thus, we obtain a Lie algebra homomorphism $\mathfrak{g} \to \text{Vect}\, N_+$.

This homomorphism may be described explicitly as follows. Let G° denote the dense open submanifold of G consisting of elements of the form $g_+ g_-, g_+ \in N_+, g_- \in B_-$ (note that such an expression is necessarily unique since $B_- \cap N_+ = 1$). In other words, $G^\circ = p^{-1}(\mathcal{U})$, where p is the projection $G \to G/B_-$. Given $a \in \mathfrak{g}$, consider the one-parameter subgroup $\gamma(\epsilon) = \exp(\epsilon a)$ in G. Since G° is open and dense in G, $\gamma(\epsilon) x \in G^\circ$ for ϵ in the formal neighborhood of 0, so we can write

$$\gamma(\epsilon)^{-1} x = Z_+(\epsilon) Z_-(\epsilon), \qquad Z_+(\epsilon) \in N_+, \quad Z_-(\epsilon) \in B_-.$$

The factor $Z_+(\epsilon)$ just expresses the projection of the subgroup $\gamma(\epsilon)$ onto $N_+ \simeq \mathcal{U} \subset G/B_-$ under the map p. Then the vector field ξ_a (equivalently, a derivation of $\text{Fun}\, N_+$) corresponding to a is given by the formula

$$(\xi_a f)(x) = \left(\frac{d}{d\epsilon} f(Z_+(\epsilon)) \right) \Big|_{\epsilon=0}. \tag{5.2.2}$$

To write a formula for ξ_a in more concrete terms, we choose a faithful representation V of $\mathfrak{g}$ (say, the adjoint representation). Since we only need the ϵ-linear term in our calculation, we can and will assume that $\epsilon^2 = 0$. Considering $x \in N_+$ as a matrix whose entries are polynomials in the coordinates $y_\alpha, \alpha \in \Delta_+$, which expresses a generic element of N_+ in $\text{End}\, V$, we have

$$(1 - \epsilon a) x = Z_+(\epsilon) Z_-(\epsilon). \tag{5.2.3}$$

We find from this formula that $Z_+(\epsilon) = x + \epsilon Z_+^{(1)}$, where $Z_+^{(1)} \in \mathfrak{n}_+$, and $Z_-(\epsilon) = 1 + \epsilon Z_-^{(1)}$, where $Z_-^{(1)} \in \mathfrak{b}_-$. Therefore we obtain from formulas (5.2.2) and (5.2.3) that

$$\xi_a \cdot x = -x(x^{-1} a x)_+, \tag{5.2.4}$$

where z_+ denotes the projection of an element $z \in \mathfrak{g}$ onto $\mathfrak{n}_+$ along $\mathfrak{b}_-$.

For example, let $\mathfrak{g} = \mathfrak{sl}_2$. Then $G/B_- = \mathbb{P}^1$, the variety of lines in $\mathbb{C}^2$. As an open subset of $\mathbb{CP}^1$ we take

$$\mathcal{U} = \left\{ \mathbb{C} \begin{pmatrix} y \\ -1 \end{pmatrix} \right\} \subset \mathbb{CP}^1$$

(the minus sign is chosen here for notational convenience). We obtain a Lie algebra homomorphism $\mathfrak{sl}_2 \to \text{Vect}\, \mathcal{U}$ sending a to ξ_a, which can be calculated explicitly by the formula

$$(v_a \cdot f)(y) = \frac{d}{d\epsilon} f(\exp(-\epsilon a) y)|_{\epsilon=0}.$$

It is easy to find explicit formulas for the vector fields corresponding to elements of the standard basis of $\mathfrak{sl}_2$ (see Section 2.1.1):

$$e \mapsto \frac{\partial}{\partial y}, \qquad h \mapsto -2y\frac{\partial}{\partial y}, \qquad f \mapsto -y^2\frac{\partial}{\partial y}. \tag{5.2.5}$$

5.2.2 The algebra of differential operators

The algebra $\mathcal{D}(\mathcal{U})$ of differential operators on $\mathcal{U}$ is isomorphic to the Weyl algebra with generators $\{y_\alpha, \partial/\partial y_\alpha\}_{\alpha\in\Delta_+}$, and the standard relations

$$\left[\frac{\partial}{\partial y_\alpha}, y_\beta\right] = \delta_{\alpha,\beta}, \qquad \left[\frac{\partial}{\partial y_\alpha}, \frac{\partial}{\partial y_\beta}\right] = [y_\alpha, y_\beta] = 0.$$

The algebra $\mathcal{D}(\mathcal{U})$ has a natural filtration $\{\mathcal{D}_{\leq i}(\mathcal{U})\}$ by the order of the differential operator. In particular, we have an exact sequence

$$0 \to \operatorname{Fun}\mathcal{U} \to \mathcal{D}_{\leq 1}(\mathcal{U}) \to \operatorname{Vect}\mathcal{U} \to 0, \tag{5.2.6}$$

where $\operatorname{Fun}\mathcal{U} \simeq \operatorname{Fun} N_+$ denotes the ring of regular functions on $\mathcal{U}$, and $\operatorname{Vect}\mathcal{U}$ denotes the Lie algebra of vector fields on $\mathcal{U}$. This sequence has a canonical splitting: namely, we lift $\xi \in \operatorname{Vect}\mathcal{U}$ to the unique first order differential operator D_ξ whose symbol equals ξ and which kills the constant functions, i.e., such that $D_\xi \cdot 1 = 0$. Using this splitting, we obtain an embedding $\mathfrak{g} \to \mathcal{D}_{\leq 1}(N_+)$, and hence the structure of a $\mathfrak{g}$-module on the space of functions $\operatorname{Fun} N_+ = \mathbb{C}[y_\alpha]_{\alpha\in\Delta_+}$.

5.2.3 Verma modules and contragredient Verma modules

By construction, the action of $\mathfrak{n}_+$ on $\operatorname{Fun} N_+$ satisfies $e_\alpha \cdot y_\alpha = 1$ and $e_\alpha \cdot y_\beta = 0$ unless α is less than or equal to β with respect to the usual partial ordering on the set of positive roots (for which $\alpha \leq \beta$ if $\beta - \alpha$ is a linear combination of positive simple roots with non-negative coefficients). Therefore, we obtain by induction that for any $A \in \operatorname{Fun} N_+$ there exists $P \in U(\mathfrak{n}_+)$ such that $P \cdot A = 1$.

Consider the pairing

$$U(\mathfrak{n}_+) \times \operatorname{Fun} N_+ \to \mathbb{C},$$

which maps (P, A) to the value of the function $P \cdot A$ at the identity element of N_+. This pairing is $\mathfrak{n}_+$-invariant. Moreover, both $U(\mathfrak{n}_+)$ and $\operatorname{Fun} N_+$ are graded by the positive part Q_+ of the root lattice of $\mathfrak{g}$ (under the action of the Cartan subalgebra $\mathfrak{h}$):

$$U(\mathfrak{n}_+) = \bigoplus_{\gamma\in Q_+} U(\mathfrak{n}_+)_\gamma, \qquad \operatorname{Fun} N_+ = \bigoplus_{\gamma\in Q_+} (\operatorname{Fun} N_+)_{-\gamma},$$

where

$$Q_+ = \left\{ \sum_{i=1}^{\ell} n_i \alpha_i \,\middle|\, n_i \in \mathbb{Z}_+ \right\},$$

and this pairing is homogeneous. This means that the pairing of homogeneous elements is non-zero only if their degrees are opposite.

Let us look at the restriction of the pairing to the finite-dimensional subspaces $U(\mathfrak{n}_+)_\gamma$ and $(\operatorname{Fun} N_+)_{-\gamma}$ for some $\gamma \in Q_+$. Consider he Poincaré–Birkhoff–Witt basis elements

$$\left\{ e_{\alpha(1)} \dots e_{\alpha(k)}, \quad \sum_{i=1}^{k} \alpha(i) = \gamma \right\} \tag{5.2.7}$$

(with respect to some lexicographic ordering) of $U(\mathfrak{n}_+)_\gamma$ and the monomial basis

$$\left\{ y_{\beta(1)} \dots y_{\beta(m)}, \quad \sum_{i=1}^{m} \beta(i) = \gamma \right\}$$

of $\operatorname{Fun} N_+$.

Using formula (5.2.4) we obtain the following formulas for the action of the vector field corresponding to e_α on $\operatorname{Fun} N_+$:

$$e_\alpha \mapsto \frac{\partial}{\partial y_\alpha} + \sum_{\beta \in \Delta_+, \beta > \alpha} P^\alpha_\beta \frac{\partial}{\partial y_\beta}, \tag{5.2.8}$$

where $P^\alpha_\beta \in \operatorname{Fun} N_+$ is a polynomial of degree $\alpha - \beta$, which is a non-zero element of $-Q_+$ (this is what we mean by $\beta > \alpha$ in the above formula).

Let us choose the lexicographic ordering of our monomials (5.2.7) in such a way that $\alpha(i) > \alpha(j)$ only if $i > j$. Then it is easy to see from formula (5.2.8) that the matrix of our pairing, restricted to degree γ, is diagonal with non-zero entries, and so it is non-degenerate.

Therefore we find that the $\mathfrak{n}_+$-module $\operatorname{Fun} N_+$ is isomorphic to the restricted dual $U(\mathfrak{n}_+)^\vee$ of $U(\mathfrak{n}_+)$:

$$U(\mathfrak{n}_+)^\vee \stackrel{\text{def}}{=} \bigoplus_{\gamma \in Q_+} (U(\mathfrak{n}_+)_\gamma)^*.$$

Now we recall the definition of the Verma modules and the contragredient Verma modules.

For each $\chi \in \mathfrak{h}^*$, consider the one-dimensional representation $\mathbb{C}_\chi$ of $\mathfrak{b}_\pm$ on which $\mathfrak{h}$ acts according to χ, and $\mathfrak{n}_\pm$ acts by 0. The **Verma module** M_χ with highest weight $\chi \in \mathfrak{h}^*$ is the induced module

$$M_\chi = \operatorname{Ind}^{\mathfrak{g}}_{\mathfrak{b}_+} \mathbb{C}_\chi \stackrel{\text{def}}{=} U(\mathfrak{g}) \underset{U(\mathfrak{b}_+)}{\otimes} \mathbb{C}_\chi.$$

The Cartan decomposition (5.2.1) gives us an isomorphism of vector spaces $U(\mathfrak{g}) \simeq U(\mathfrak{n}_-) \underset{\mathbb{C}}{\otimes} U(\mathfrak{b}_+)$. Therefore, as an $\mathfrak{n}_-$-module, $M_\chi \simeq U(\mathfrak{n}_-)$.

The **contragredient Verma module** M_χ^* with highest weight $\chi \in \mathfrak{h}^*$ is defined as the (restricted) coinduced module

$$M_\chi^* = \mathrm{Coind}_{\mathfrak{b}_-}^{\mathfrak{g}} \mathbb{C}_\chi \stackrel{\mathrm{def}}{=} \mathrm{Hom}_{U(\mathfrak{b}_-)}^{\mathrm{res}}(U(\mathfrak{g}), \mathbb{C}_\chi). \tag{5.2.9}$$

Here, we consider homomorphisms invariant under the action of $U(\mathfrak{b}_-)$ on $U(\mathfrak{g})$ from the left, and the index "res" means that we consider only those linear maps $U(\mathfrak{g}) \to \mathbb{C}_\chi$ which are finite linear combinations of maps supported on the direct summands $U(\mathfrak{b}_-) \otimes U(\mathfrak{n}_+)_\gamma$ of $U(\mathfrak{g})$ with respect to the isomorphism of vector spaces $U(\mathfrak{g}) \simeq U(\mathfrak{b}_-) \otimes U(\mathfrak{n}_+)$. Then, as an $\mathfrak{n}_+$-module, $M_\chi^* \simeq U(\mathfrak{n}_+)^\vee$.

5.2.4 Identification of Fun N_+ *with* M_χ^*

The module M_0^* is isomorphic to Fun N_+ with its $\mathfrak{g}$-module structure defined above. Indeed, the vector $1 \in \mathrm{Fun}\, N_+$ is annihilated by $\mathfrak{n}_+$ and has weight 0 with respect to h. Hence there is a non-zero homomorphism $\varphi : \mathrm{Fun}\, N_+ \to M_0^*$ sending $1 \in \mathrm{Fun}\, N_+$ to a non-zero vector $v_0^* \in M_\chi^*$ of weight 0. Since Fun N_+ is isomorphic to $U(\mathfrak{n}_+)^\vee$ as an $\mathfrak{n}_+$-module, this homomorphism is injective. Indeed, for any $P \in \mathrm{Fun}\, N_+$ there exists $u \in U(\mathfrak{n}_+)$ such that $U \cdot P = 1$. Therefore $u \cdot \varphi(P) = \varphi(1) = v_0^* \neq 0$, and so $\varphi(P) \neq 0$. But since M_χ^* is also isomorphic to $U(\mathfrak{n}_+)^\vee$ as an $\mathfrak{n}_+$-module, we find that φ is necessarily an isomorphism.

Now we identify the module M_χ^* with an arbitrary weight χ with Fun N_+, where the latter is equipped with a modified action of $\mathfrak{g}$.

Recall that we have a canonical lifting of $\mathfrak{g}$ to $\mathcal{D}_{\leq 1}(N_+)$, $a \mapsto \xi_a$. But this lifting is not unique. We can modify it by adding to each ξ_a a function $\phi_a \in \mathrm{Fun}\, N_+$ so that $\phi_{a+b} = \phi_a + \phi_b$. One readily checks that the modified differential operators $\xi_a + \phi_a$ satisfy the commutation relations of $\mathfrak{g}$ if and only if the linear map $\mathfrak{g} \to \mathrm{Fun}\, N_+$ given by $a \mapsto \phi_a$ is a one-cocycle of $\mathfrak{g}$ with coefficients in Fun N_+.

Each such lifting gives Fun N_+ the structure of a $\mathfrak{g}$-module. Let us impose the extra condition that the modified action of $\mathfrak{h}$ on V remains diagonalizable. This means that ϕ_h is a constant function on N_+ for each $h \in \mathfrak{h}$, and therefore our one-cocycle should be $\mathfrak{h}$-invariant: $\phi_{[h,a]} = \xi_h \cdot \phi_a$, for all $h \in \mathfrak{h}, a \in \mathfrak{g}$. We claim that the space of $\mathfrak{h}$-invariant one-cocycles of $\mathfrak{g}$ with coefficients in $\mathbb{C}[N_+]$ is canonically isomorphic to the first cohomology of $\mathfrak{g}$ with coefficients in $\mathbb{C}[N_+]$, i.e., $H^1(\mathfrak{g}, \mathbb{C}[N_+])$.

Indeed, it is well-known (see, e.g., [Fu]) that if a Lie subalgebra $\mathfrak{h}$ of $\mathfrak{g}$ acts diagonally on $\mathfrak{g}$ and on a $\mathfrak{g}$-module M, then $\mathfrak{h}$ must act by 0 on $H^1(\mathfrak{g}, M)$. Hence $H^1(\mathfrak{g}, \mathbb{C}[N_+])$ is equal to the quotient of the space of $\mathfrak{h}$-invariant one-cocycles by its subspace of $\mathfrak{h}$-invariant coboundaries (i.e., those cocycles which have the form $\phi_b = v_b \cdot f$ for some $f \in \mathbb{C}[N_+]$). But it is clear that the space of $\mathfrak{h}$-invariant coboundaries is equal to 0 in our case, hence the result.

Thus, the set of liftings of $\mathfrak{g}$ to $\mathcal{D}_{\leq 1}(N_+)$ making $\mathbb{C}[N_+]$ into a $\mathfrak{g}$-module with diagonal action of $\mathfrak{h}$ is naturally isomorphic to $H^1(\mathfrak{g}, \mathbb{C}[N_+])$. Since

$$\operatorname{Fun} N_+ = M_0^* = \operatorname{Coind}_{\mathfrak{b}_-}^{\mathfrak{g}} \mathbb{C}_0,$$

we find from the Shapiro lemma (see [Fu], Section 5.4) that

$$H^1(\mathfrak{g}, \operatorname{Fun} N_+) \simeq H^1(\mathfrak{b}_-, \mathbb{C}_0) = (\mathfrak{b}_-/[\mathfrak{b}_-, \mathfrak{b}_-])^* \simeq \mathfrak{h}^*.$$

Thus, for each $\chi \in \mathfrak{h}^*$ we obtain a Lie algebra homomorphism $\rho_\chi : \mathfrak{g} \to \mathcal{D}_{\leq 1}(N_+)$ and hence the structure of an $\mathfrak{h}^*$-graded $\mathfrak{g}$-module on $\operatorname{Fun} N_+$. Let us analyze this $\mathfrak{g}$-module in more detail.

We have $\xi_h \cdot y_\alpha = -\alpha(h) y_\alpha, \alpha \in \Delta_+$, so the weight of any monomial in $\operatorname{Fun} N_+$ is equal to a sum of negative roots. Since our one-cocycle is $\mathfrak{h}$-invariant, we obtain that

$$\xi_h \cdot \phi_{e_\alpha} = \phi_{[h, e_\alpha]} = \alpha(h) \phi_{e_\alpha}, \qquad \alpha \in \Delta_+,$$

so the weight of ϕ_{e_α} has to be equal to the positive root α. Therefore $\phi_{e_\alpha} = 0$ for all $\alpha \in \Delta_+$. Thus, the action of $\mathfrak{n}_+$ on $\operatorname{Fun} N_+$ is not modified. On the other hand, by construction, the action of $h \in \mathfrak{h}$ is modified by $h \mapsto h + \chi(h)$. Therefore the vector $1 \in \operatorname{Fun} N_+$ is still annihilated by $\mathfrak{n}_+$, but now it has weight χ with respect to h. Hence there is a non-zero homomorphism $\operatorname{Fun} N_+ \to M_\chi^*$ sending $1 \in \operatorname{Fun} N_+$ to a non-zero vector $v_\chi^* \in M_\chi^*$ of weight χ. Since both $\operatorname{Fun} N_+$ and M_χ^* are isomorphic to $U(\mathfrak{n}_+)^\vee$ as $\mathfrak{n}_+$-modules, we obtain that this homomorphism is an isomorphism (in the same way as we did at the beginning of this section for $\chi = 0$). Thus, under the modified action obtained via the lifting ρ_χ, the $\mathfrak{g}$-module $\operatorname{Fun} N_+$ is isomorphic to the contragredient Verma module M_χ^*.

To summarize, we have constructed a family of Lie algebra homomorphisms

$$\rho_\chi : \mathfrak{g} \to \mathcal{D}_{\leq 1}(N_+) \to \mathcal{D}(N_+), \qquad \chi \in \mathfrak{h}^*,$$

which give rise to homomorphisms of algebras

$$\widetilde{\rho}_\chi : U(\mathfrak{g}) \to \mathcal{D}(N_+), \qquad \chi \in \mathfrak{h}^*.$$

These homomorphisms combine into a universal homomorphism of algebras

$$\widetilde{\rho} : U(\mathfrak{g}) \to \operatorname{Fun} \mathfrak{h}^* \underset{\mathbb{C}}{\otimes} \mathcal{D}(N_+), \tag{5.2.10}$$

such that for each $\chi \in \mathfrak{h}^*$ we recover ρ_χ as the composition of $\widetilde{\rho}$ and the evaluation at χ homomorphism $\operatorname{Fun} \mathfrak{h}^* \to \mathbb{C}$ along the first factor.

5.2.5 Explicit formulas

Choose a basis $\{J^a\}_{a=1,\ldots,\dim \mathfrak{g}}$ of $\mathfrak{g}$. Under the homomorphism ρ_χ, we have

$$J^a \mapsto P_a\left(y_\alpha, \frac{\partial}{\partial y_\alpha}\right) + f_a(y_\alpha), \tag{5.2.11}$$

where P_a is a polynomial in the y_α's and $\partial/\partial y_\alpha$'s of degree one in the $\partial/\partial y_\alpha$'s, which is independent of χ, and f_a is a polynomial in the y_α's only, which depends on χ.

Let $e_i, h_i, f_i, i = 1, \ldots, \ell$, be the generators of $\mathfrak{g}$. Using formula (5.2.4) we obtain the following formulas:

$$\rho_\chi(e_i) = \frac{\partial}{\partial y_{\alpha_i}} + \sum_{\beta \in \Delta_+} P^i_\beta(y_\alpha) \frac{\partial}{\partial y_\beta}, \tag{5.2.12}$$

$$\rho_\chi(h_i) = -\sum_{\beta \in \Delta_+} \beta(h_i) y_\beta \frac{\partial}{\partial y_\beta} + \chi(h_i), \tag{5.2.13}$$

$$\rho_\chi(f_i) = \sum_{\beta \in \Delta_+} Q^i_\beta(y_\alpha) \frac{\partial}{\partial y_\beta} + \chi(h_i) y_{\alpha_i}, \tag{5.2.14}$$

for some polynomials P^i_β, Q^i_β in $y_\alpha, \alpha \in \Delta_+$.

In addition, we have a Lie algebra anti-homomorphism $\rho^R : \mathfrak{n}_+ \to \mathcal{D}_{\leq 1}(N_+)$, which corresponds to the *right* action of $\mathfrak{n}_+$ on N_+. The differential operators $\rho^R(x), x \in \mathfrak{n}_+$, commute with the differential operators $\rho_\chi(x'), x' \in \mathfrak{n}_+$ (but their commutation relations with $\rho_\chi(x'), x' \notin \mathfrak{n}_+$, are complicated in general). We have

$$\rho^R(e_i) = \frac{\partial}{\partial y_{\alpha_i}} + \sum_{\beta \in \Delta_+} P^{R,i}_\beta(y_\alpha) \frac{\partial}{\partial y_\beta}$$

for some polynomials $P^{R,i}_\beta$ in $y_\alpha, \alpha \in \Delta_+$.

5.3 The case of affine algebras

In this section we develop a similar formalism for the affine Kac–Moody algebras. This will enable us to construct Wakimoto modules, which is an important step in our program.

5.3.1 The infinite-dimensional Weyl algebra

Our goal is to generalize the above construction to the case of affine Kac–Moody algebras. Let again $\mathcal{U}$ be the open N_+-orbit of the flag manifold of G, which we identify with the group N_+ and hence with the Lie algebra $\mathfrak{n}_+$. Consider the formal loop space $L\mathcal{U} = \mathcal{U}((t))$ as a complete topological vector space with the basis of open neighborhoods of $0 \in L\mathcal{U}$ formed by the subspaces $\mathcal{U} \otimes t^N \mathbb{C}[[t]] \subset L\mathcal{U}, N \in \mathbb{Z}$. Thus, $L\mathcal{U}$ is an affine ind-scheme

$$L\mathcal{U} = \varinjlim \mathcal{U} \otimes t^N \mathbb{C}[[t]], \qquad N < 0.$$

Using the coordinates $y_\alpha, \alpha \in \Delta_+$, on $\mathcal{U}$, we can write

$$\mathcal{U} \otimes t^N \mathbb{C}[[t]] \simeq \left\{ \sum_{n \geq N} y_{\alpha,n} t^n \right\}_{\alpha \in \Delta_+} = \operatorname{Spec} \mathbb{C}[y_{\alpha,n}]_{n \geq N}.$$

Therefore we obtain that the ring of functions on $L\mathcal{U}$, denoted by $\operatorname{Fun} L\mathcal{U}$, is the inverse limit of the rings $\mathbb{C}[y_{\alpha,n}]_{\alpha\in\Delta_+, n\geq N}, N < 0$, with respect to the natural surjective homomorphisms

$$s_{N,M} : \mathbb{C}[y_{\alpha,n}]_{\alpha\in\Delta_+,n\geq N} \to \mathbb{C}[y_{\alpha,n}]_{\alpha\in\Delta_+,n\geq M}, \qquad N < M,$$

such that $y_{\alpha,n} \mapsto 0$ for $N \leq n < M$ and $y_{\alpha,n} \mapsto y_{\alpha,n}, n \geq M$. This is a complete topological ring, with the basis of open neighborhoods of 0 given by the ideals generated by $y_{\alpha,n}, n < N$, i.e., the kernels of the homomorphisms $s_{\infty,M} : \operatorname{Fun} L\mathcal{U} \to \operatorname{Fun} \mathcal{U} \otimes t^N \mathbb{C}[[t]]$.

A vector field on $L\mathcal{U}$ is by definition a continuous linear endomorphism ξ of $\operatorname{Fun} L\mathcal{U}$ which satisfies the Leibniz rule: $\xi(fg) = \xi(f)g + f\xi(g)$. In other words, a vector field is a linear endomorphism ξ of $\operatorname{Fun} L\mathcal{U}$ such that for any $M < 0$ there exist $N \leq M$ and a derivation

$$\xi_{N,M} : \mathbb{C}[y_{\alpha,n}]_{\alpha\in\Delta_+,n\geq N} \to \mathbb{C}[y_{\alpha,n}]_{\alpha\in\Delta_+,n\geq M},$$

which satisfies

$$s_{\infty,M}(\xi \cdot f) = \xi_{N,M} \cdot s_{\infty,N}(f)$$

for all $f \in \operatorname{Fun} L\mathcal{U}$. The space of vector fields is naturally a topological Lie algebra, which we denote by $\operatorname{Vect} L\mathcal{U}$.

More concretely, an element of $\operatorname{Fun} L\mathcal{U}$ may be represented as a (possibly infinite) series

$$\sum_{m \leq -M} P_{\alpha,m} y_{\alpha,m},$$

where the $P_{\alpha,m}$'s are arbitrary (finite) polynomials in $y_{\alpha,n}, n \in \mathbb{Z}$. In this formula and in analogous formulas below the summation over $\alpha \in \Delta_+$ is always understood.

The Lie algebra $\operatorname{Vect} L\mathcal{U}$ may also be described as follows. Identify the tangent space $T_0 L\mathcal{U}$ at the origin in $L\mathcal{U}$ with $L\mathcal{U}$, equipped with the structure of a complete topological vector space. Then $\operatorname{Vect} L\mathcal{U}$ is isomorphic to the completed tensor product of $\operatorname{Fun} L\mathcal{U}$ and $L\mathcal{U}$. This means that vector fields on $L\mathcal{U}$ can be described more concretely as series

$$\sum_{n\in\mathbb{Z}} P_{\alpha,n} \frac{\partial}{\partial y_{\alpha,n}},$$

where $P_{\alpha,n} \in \operatorname{Fun} L\mathcal{U}$ satisfies the following property: for each $M \geq 0$, there exists $K \geq M$ such that each $P_n, n \leq -K$, lies in the ideal generated by the $y_{\alpha,m}, \alpha \in \Delta_+, m \leq -M$.

Such an element may be written as follows:

$$\sum_{n \geq N} P_{\alpha,n} \frac{\partial}{\partial y_{\alpha,n}} + \sum_{m \leq -M} y_{\alpha,m} V_{\alpha,m}, \tag{5.3.1}$$

where the $P_{\alpha,n}$'s are polynomials and the $V_{\alpha,m}$'s are polynomial vector fields.

In other words, $\operatorname{Vect} L\mathcal{U}$ is the completion of the Lie algebra of polynomial vector fields in the variables $y_{\alpha,n}, n \in \mathbb{Z}$, with respect to the topology in which the basis of open neighborhoods of 0 is formed by the subspaces $\widetilde{I}_{N,M}$ which consist of the vector fields that are linear combinations of vector fields $P_{\alpha,n}\partial/\partial y_{\alpha,n}, n \geq N$, and $y_{\alpha,m}V_{\alpha,m}, m \leq -M$.

The Lie bracket of vector fields is continuous with respect to this topology. This means that the commutator between two series of the form (5.3.1), computed in the standard way (term by term), is again a series of the above form.

5.3.2 Action of $L\mathfrak{g}$ on $L\mathcal{U}$

From now on we will use the notation $L\mathfrak{g}$ for $\mathfrak{g}((t))$ (L stands for "loops").

We have a natural Lie algebra homomorphism

$$\widehat{\rho} : L\mathfrak{g} \to \operatorname{Vect} L\mathcal{U},$$

which may be described explicitly by the formulas that we obtained in the finite-dimensional case, in which we replace the ordinary variables y_α with the "loop variables" $y_{\alpha,n}$. More precisely, we have the following analogue of formula (5.2.3):

$$(1 - \epsilon A \otimes t^m)x(t) = Z_+(\epsilon)Z_-(\epsilon)$$

where $x \in N_+((t)), Z_+(\epsilon) = x(t) + \epsilon Z_+^{(1)}$, $Z_+^{(1)} \in \mathfrak{n}_+((t))$, and $Z_-(\epsilon) = 1 + \epsilon Z_-^{(1)}$, $Z_-^{(1)} \in \mathfrak{b}_-((t))$. As before, we choose a faithful finite-dimensional representation V of $\mathfrak{g}$ and consider $x(t)$ as a matrix whose entries are Laurent power series in t with coefficients in the ring of polynomials in the coordinates $y_{\alpha,n}, \alpha \in \Delta_+, n \in \mathbb{Z}$, expressing a generic element of $N_+((t))$ in $\operatorname{End} V((t))$. We define $\widehat{\rho}$ by the formula

$$\widehat{\rho}(a \otimes t^m) \cdot x(t) = Z_+^{(1)}.$$

Then we have the following analogue of formula (5.2.4):

$$\widehat{\rho}(a \otimes t^m) \cdot x(t) = -x(t)\left(x(t)^{-1}(a \otimes t^m)x(t)\right)_+, \tag{5.3.2}$$

where z_+ denotes the projection of an element $z \in \mathfrak{g}((t))$ onto $\mathfrak{n}_+((t))$ along $\mathfrak{b}_-((t))$.

This formula implies that for any $a \in \mathfrak{g}$ the series

$$\widehat{\rho}(a(z)) = \sum_{n \in \mathbb{Z}} \widehat{\rho}(a \otimes t^n) z^{-n-1}$$

may be obtained from the formula for $\rho_0(a) = \xi_a$ by the substitution

$$y_\alpha \mapsto \sum_{n \in \mathbb{Z}} y_{\alpha,n} z^n,$$
$$\frac{\partial}{\partial y_\alpha} \mapsto \sum_{n \in \mathbb{Z}} \frac{\partial}{y_{\alpha,n}} z^{-n-1}.$$

5.3.3 The Weyl algebra

Let $\mathcal{A}^{\mathfrak{g}}$ be the Weyl algebra with generators

$$a_{\alpha,n} = \frac{\partial}{\partial y_{\alpha,n}}, \quad a^*_{\alpha,n} = y_{\alpha,-n}, \qquad \alpha \in \Delta_+, n \in \mathbb{Z},$$

and relations

$$[a_{\alpha,n}, a^*_{\beta,m}] = \delta_{\alpha,\beta}\delta_{n,-m}, \qquad [a_{\alpha,n}, a_{\beta,m}] = [a^*_{\alpha,n}, a^*_{\beta,m}] = 0. \tag{5.3.3}$$

The change of sign of n in the definition of $a^*_{\alpha,n}$ is made so as to have $\delta_{n,-m}$ in formula (5.3.3), rather than $\delta_{n,m}$. This will be convenient when we use the vertex algebra formalism.

Introduce the generating functions

$$a_\alpha(z) = \sum_{n \in \mathbb{Z}} a_{\alpha,n} z^{-n-1}, \tag{5.3.4}$$
$$a^*_\alpha(z) = \sum_{n \in \mathbb{Z}} a^*_{\alpha,n} z^{-n}. \tag{5.3.5}$$

Consider a topology on $\mathcal{A}^{\mathfrak{g}}$ in which the basis of open neighborhoods of 0 is formed by the left ideals $I_{N,M}, N, M \in \mathbb{Z}$, generated by $a_{\alpha,n}, \alpha \in \Delta_+, n \geq N$, and $a^*_{\alpha,m}, \alpha \in \Delta_+, m \geq M$. The **completed Weyl algebra** $\widetilde{\mathcal{A}}^{\mathfrak{g}}$ is by definition the completion of $\mathcal{A}^{\mathfrak{g}}$ with respect to this topology. The algebra $\widetilde{\mathcal{A}}^{\mathfrak{g}}$ should be thought of as an analogue of the algebra of differential operators on $L\mathcal{U}$ (see [FB], Chapter 12, for more details).

In concrete terms, elements of $\widetilde{\mathcal{A}}^{\mathfrak{g}}$ may be viewed as arbitrary series of the form

$$\sum_{n \geq N} P_{\alpha,n} a_{\alpha,n} + \sum_{m \geq M} Q_{\alpha,m} a^*_{\alpha,m}, \qquad P_n, Q_m \in \mathcal{A}^{\mathfrak{g}}. \tag{5.3.6}$$

Let $\mathcal{A}^{\mathfrak{g}}_0$ be the (commutative) subalgebra of $\mathcal{A}^{\mathfrak{g}}$ generated by $a^*_{\alpha,n}, \alpha \in \Delta_+, n \in \mathbb{Z}$, and $\widetilde{\mathcal{A}}^{\mathfrak{g}}_0$ its completion in $\widetilde{\mathcal{A}}^{\mathfrak{g}}$. Next, let $\mathcal{A}^{\mathfrak{g}}_{\leq 1}$ be the subspace of $\mathcal{A}^{\mathfrak{g}}$ spanned by the products of elements of $\mathcal{A}^{\mathfrak{g}}_0$ and the generators $a_{\alpha,n}$. Denote by $\widetilde{\mathcal{A}}^{\mathfrak{g}}_{\leq 1}$ its completion in $\widetilde{\mathcal{A}}^{\mathfrak{g}}$. Thus, $\widetilde{\mathcal{A}}^{\mathfrak{g}}_{\leq 1}$ consists of all elements P of $\widetilde{\mathcal{A}}^{\mathfrak{g}}$ with the property that

$$P \bmod I_{N,M} \in \mathcal{A}^{\mathfrak{g}}_{\leq 1} \bmod I_{N,M}, \qquad \forall N, M \in \mathbb{Z}.$$

Here is a more concrete description of $\widetilde{\mathcal{A}}^{\mathfrak{g}}_0$ and $\widetilde{\mathcal{A}}^{\mathfrak{g}}_{\leq 1}$ using the realization of $\widetilde{\mathcal{A}}^{\mathfrak{g}}$ by series of the form (5.3.6). The space $\widetilde{\mathcal{A}}^{\mathfrak{g}}_0$ consists of series of the form (5.3.6), where all $P_{\alpha,n} = 0$ and $Q_{\alpha,m} \in \mathcal{A}^{\mathfrak{g}}_0$. In other words, these are the series which do not contain $a_{\alpha,n}, n \in \mathbb{Z}$. The space $\widetilde{\mathcal{A}}^{\mathfrak{g}}_{\leq 1}$ consists of elements of the form (5.3.6), where $P_{\alpha,n} \in \mathcal{A}^{\mathfrak{g}}_0$ and $Q_{\alpha,m} \in \mathcal{A}^{\mathfrak{g}}_{\leq 1}$.

Proposition 5.3.1 *There is a short exact sequence of Lie algebras*

$$0 \to \operatorname{Fun} L\mathcal{U} \to \widetilde{\mathcal{A}}^{\mathfrak{g}}_{\leq 1} \to \operatorname{Vect} L\mathcal{U} \to 0. \tag{5.3.7}$$

Proof. The statement follows from the following three assertions:

(1) $\widetilde{\mathcal{A}}^{\mathfrak{g}}_{\leq 1}$ is a Lie algebra, and $\widetilde{\mathcal{A}}^{\mathfrak{g}}_0$ is its ideal.
(2) $\widetilde{\mathcal{A}}^{\mathfrak{g}}_0 \simeq \operatorname{Fun} L\mathcal{U}$.
(3) There is a surjective homomorphism of Lie algebras $\widetilde{\mathcal{A}}^{\mathfrak{g}}_{\leq 1} \to \operatorname{Vect} L\mathcal{U}$ whose kernel is $\widetilde{\mathcal{A}}^{\mathfrak{g}}_0$.

(1) follows from the definition of $\widetilde{\mathcal{A}}^{\mathfrak{g}}_{\leq 1}$ and $\widetilde{\mathcal{A}}^{\mathfrak{g}}_0$ as completions of $\mathcal{A}^{\mathfrak{g}}_{\leq 1}$ and $\mathcal{A}^{\mathfrak{g}}_0$, respectively, the fact that $\mathcal{A}^{\mathfrak{g}}_{\leq 1}$ is a Lie algebra containing $\mathcal{A}^{\mathfrak{g}}_0$ as an ideal and the continuity of the Lie bracket.

(2) follows from the definitions.

To prove (3), we construct a surjective linear map $\widetilde{\mathcal{A}}^{\mathfrak{g}}_{\leq 1} \to \operatorname{Vect} L\mathcal{U}$ whose kernel is $\widetilde{\mathcal{A}}^{\mathfrak{g}}_0$ and show that it is a homomorphism of Lie algebras.

A general element of $\widetilde{\mathcal{A}}^{\mathfrak{g}}_{\leq 1}$ may be written in the form

$$\sum_{n \geq N} P_{\alpha,n} a_{\alpha,n} + \sum_{m \geq M_1} \sum_{k \in K_m} Q^{(1)}_{\alpha,\beta,k,m} a_{\beta,k} a^*_{\alpha,m} + \sum_{m \geq M_2} Q^{(2)}_{\alpha,m} a^*_{\alpha,m},$$

for some integers N, M_1, M_2, where $P_{\alpha,n}, Q^{(1)}_{\alpha,\beta,k,m}, Q^{(2)}_{\alpha,m} \in \mathcal{A}^{\mathfrak{g}}_0$, and each $K_m \subset \mathbb{Z}$ is a finite set. We define our map by sending this element to

$$\sum_{n \geq N} P_{\alpha,n} a_{\alpha,n} + \sum_{m \geq M_1} \sum_{k \in K_m} a^*_{\alpha,m} Q^{(1)}_{\alpha,\beta,k,m} a_{\beta,k},$$

which is a well-defined element of $\operatorname{Vect} L\mathcal{U}$ according to its description given at the end of Section 5.3.1.

It is clear that elements of $\widetilde{\mathcal{A}}^{\mathfrak{g}}_{\leq 1}$ which are in the left ideal generated by $a_{\alpha,n}, n \geq N$, and $a^*_{\alpha,m}, n \geq M$, are mapped to the elements of $\operatorname{Vect} L\mathcal{U}$ which are in the sum of the left ideal of $a_{\alpha,n}, n \geq N$, and the right ideal of $a^*_{\alpha,m}, m \geq M$. Therefore we obtain a map of the corresponding quotients. Since elements of these quotients are represented by *finite* linear combinations of elements of $\mathcal{A}^{\mathfrak{g}}$ and $\operatorname{Vect} L\mathcal{U}$, respectively, we obtain that this map of quotients is a Lie algebra homomorphism. The statement that our map is also a Lie algebra homomorphism then follows by continuity of the Lie brackets on $\widetilde{\mathcal{A}}^{\mathfrak{g}}_{\leq 1}$ and $\operatorname{Vect} L\mathcal{U}$ with respect to the topologies defined by the respective ideals. □

Thus, we obtain from Proposition 5.3.1 that $\widetilde{\mathcal{A}}^{\mathfrak{g}}_{\leq 1}$ is an extension of the Lie algebra $\operatorname{Vect} L\mathcal{U}$ by its module $\widetilde{\mathcal{A}}^{\mathfrak{g}}_{0} = \operatorname{Fun} L\mathcal{U}$.

This extension is however different from the standard (split) extension defining the Lie algebra of the usual differential operators on $L\mathcal{U}$ of order less than or equal to 1 (it corresponds to another completion of differential operators defined in [FB], Section 12.1.3). The reason is that the algebra $\widetilde{\mathcal{A}}^{\mathfrak{g}}$ does not act on the space of functions on $L\mathcal{U} = \mathcal{U}((t))$ (see the discussion in [FB], Chapter 12, for more details). It acts instead on the module $M_{\mathfrak{g}}$ defined below. This module may be thought of as the space of "delta-functions" supported on the subspace $L_+\mathcal{U} = \mathcal{U}[[t]]$. Because of that, the trick that we used in the finite-dimensional case, of lifting a vector field to the differential operator annihilating the constant function 1, does not work any more: there is no function annihilated by all the $a_{\alpha,n}$'s in the module $M_{\mathfrak{g}}$! Therefore there is no obvious splitting of the short exact sequence (5.3.7). In fact, we will see below that it is non-split.

Because of that we cannot expect to lift the homomorphism $\widehat{\rho} : L\mathfrak{g} \to \operatorname{Vect} L\mathcal{U}$ to a homomorphism $L\mathfrak{g} \to \widetilde{\mathcal{A}}^{\mathfrak{g}}_{\leq 1}$, as in the finite-dimensional case. Nevertheless, we will show that the homomorphism $\widehat{\rho}$ may be lifted to a homomorphism

$$\widehat{\mathfrak{g}}_{\kappa_c} \to \widetilde{\mathcal{A}}^{\mathfrak{g}}_{\leq 1}, \qquad \mathbf{1} \mapsto 1,$$

where $\widehat{\mathfrak{g}}_{\kappa_c}$ is the affine Kac–Moody algebra at the critical level defined in Section 1.3.6.

Our immediate goal is to prove this assertion. We will begin by showing in the next section that the image of the embedding $L\mathfrak{g} \to \operatorname{Vect} L\mathcal{U}$ belongs to a Lie subalgebra of "local" vector fields $\mathcal{T}^{\mathfrak{g}}_{\mathrm{loc}} \subset \operatorname{Vect} L\mathcal{U}$. This observation will allow us to replace the extension (5.3.7) by its "local" part, which is much smaller and hence more manageable.

5.4 Vertex algebra interpretation

It will be convenient to restrict ourselves to smaller, "local" subalgebras of the Lie algebras appearing in the short exact sequence (5.3.7). These smaller subalgebras will be sufficient for our purposes because, as we will see, the image of $\mathfrak{g}((t))$ lands in the "local" version of $\operatorname{Vect} L\mathcal{U}$. In order to define these subalgebras, we recast everything in the language of vertex algebras.

5.4.1 Heisenberg vertex algebra

Let $M_{\mathfrak{g}}$ be the Fock representation of $\mathcal{A}^{\mathfrak{g}}$ generated by a vector $|0\rangle$ such that

$$a_{\alpha,n}|0\rangle = 0, \quad n \geq 0; \qquad a^*_{\alpha,n}|0\rangle = 0, \quad n > 0.$$

It is clear that the action of $\mathcal{A}^{\mathfrak{g}}$ on $M_{\mathfrak{g}}$ extends to a continuous action of the topological algebra $\widetilde{\mathcal{A}}^{\mathfrak{g}}$ (here we equip $M_{\mathfrak{g}}$ with the discrete topology). Moreover, $M_{\mathfrak{g}}$ carries the structure of a $\mathbb{Z}_+$-graded vertex algebra defined as follows (see Section 2.2.2 for the definition of vertex algebras and the Reconstruction Theorem 2.2.4, which we use to prove that the structures introduced below satisfy the axioms of a vertex algebra):

- $\mathbb{Z}_+$-grading: $\deg a_{\alpha,n} = \deg a^*_{\alpha,n} = -n, \deg|0\rangle = 0$;
- vacuum vector: $|0\rangle$;
- translation operator: $T|0\rangle = 0, \quad [T, a_{\alpha,n}] = -na_{\alpha,n-1}, [T, a^*_{\alpha,n}] = -(n-1)a^*_{\alpha,n-1}$;
- vertex operators:

$$Y(a_{\alpha,-1}|0\rangle, z) = a_\alpha(z), \qquad Y(a^*_{\alpha,0}|0\rangle, z) = a^*_\alpha(z),$$

$$Y(a_{\alpha_1,n_1} \dots a_{\alpha_k,n_k} a^*_{\beta_1,m_1} \dots a^*_{\beta_l,m_l}|0\rangle, z) = \prod_{i=1}^{k} \frac{1}{(-n_i-1)!} \prod_{j=1}^{l} \frac{1}{(-m_j)!} \cdot$$
$$\cdot :\partial_z^{-n_1-1} a_{\alpha_1}(z) \dots \partial_z^{-n_k-1} a_{\alpha_k}(z) \partial_z^{-m_1} a^*_{\beta_1}(z) \dots \partial_z^{-m_l} a^*_{\beta_l}(z):.$$

Here we use the normal ordering operation (denoted by the columns) introduced in Section 2.2.5. In the general case it is defined inductively and so is rather inexplicit, but in the case at hand it can be defined in a more explicit way which we now recall.

Let us call the generators $a_{\alpha,n}, n \geq 0$, and $a^*_{\alpha,m}, m > 0$, **annihilation operators**, and the generators $a_{\alpha,n}, n < 0$, and $a^*_{\alpha,m}, m \leq 0$, **creation operators**. A monomial P in $a_{\alpha,n}$ and $a^*_{\alpha,n}$ is called normally ordered if all factors of P which are annihilation operators stand to the right of all factors of P which are creation operators. Given any monomial P, we define the normally ordered monomial $:P:$ as the monomial obtained by moving all factors of P which are annihilation operators to the right, and all factors of P which are creation operators to the left. For example,

$$:a_{\alpha,4} a^*_{\beta,-5} a^*_{\gamma,1} a^*_{\mu,-3}: = a^*_{\beta,-5} a^*_{\mu,-3} a_{\alpha,4} a^*_{\gamma,1}.$$

Note that since the annihilation operators commute with each other, it does not matter how we order them among themselves. The same is true for the creation operators. This shows that $:P:$ is well-defined by the above conditions.

Given two monomials P and Q, their normally ordered product is by definition the normally ordered monomial $:PQ:$. By linearity, we define the normally ordered product of any number of vertex operators from the vertex algebra $M_{\mathfrak{g}}$ by applying the above definition to each monomial appearing in each Fourier coefficient of the product.

Now let

$$U(M_{\mathfrak{g}}) = (M_{\mathfrak{g}} \otimes \mathbb{C}((t)))/\operatorname{Im}(T \otimes 1 + \operatorname{Id} \otimes \partial_t).$$

be the Lie algebra attached to $M_{\mathfrak{g}}$ as in Section 3.2.1.

As explained in Section 3.2.1, $U(M_{\mathfrak{g}})$ may be viewed as the completion of the span of all Fourier coefficients of vertex operators from $M_{\mathfrak{g}}$. Moreover, we show in the same way as in Proposition 3.2.1 that the map

$$U(M_{\mathfrak{g}}) \to \widetilde{\mathcal{A}}^{\mathfrak{g}}, \qquad A \otimes f(z) \mapsto \operatorname{Res}_{z=0} Y(A,z)f(z)dz$$

is a homomorphism of Lie algebras.

Note that $U(M_{\mathfrak{g}})$ is a Lie algebra, but not an algebra. For instance, it contains the generators $a_{\alpha,n}, a^*_{\alpha,n}$ of the Heisenberg algebra, but does not contain monomials in these generators of degree greater than one. However, we will only need the Lie algebra structure on $U(M_{\mathfrak{g}})$.†

The elements of $\widetilde{\mathcal{A}}^{\mathfrak{g}}_0 = \operatorname{Fun} L\mathcal{U}$ which lie in the image of $U(M_{\mathfrak{g}})$ are usually called **local functionals** on $L\mathcal{U}$. The elements of $\widetilde{\mathcal{A}}^{\mathfrak{g}}$ which belong to $U(M_{\mathfrak{g}})$ are given by (possibly infinite) linear combinations of Fourier coefficients of normally ordered polynomials in $a_\alpha(z), a^*_\alpha(z)$ and their derivatives. We refer to them as **local** elements of $\widetilde{\mathcal{A}}^{\mathfrak{g}}$.

5.4.2 More canonical definition of $M_{\mathfrak{g}}$

The above definition of the vertex algebra $M_{\mathfrak{g}}$ referred to a particular system of coordinates $y_\alpha, \alpha \in \Delta_+$, on the group N_+. If we choose a different coordinate system $y'_\alpha, \alpha \in \Delta_+$, on N_+, we obtain another Heisenberg algebra with generators $a'_{\alpha,n}$ and $a^*_{\alpha,n}{}'$ and a vertex algebra $M'_{\mathfrak{g}}$. However, the vertex algebras $M'_{\mathfrak{g}}$ and $M_{\mathfrak{g}}$ are canonically isomorphic to each other. In particular, it is easy to express the vertex operators $a'_\alpha(z)$ and $a^*_\alpha{}'(z)$ in terms of $a_\alpha(z)$ and $a^*_\alpha(z)$. Namely, if $y'_\alpha = F_\alpha(y_\beta)$, then

$$a'_\alpha(z) \mapsto \sum_{\gamma \in \Delta_+} {:}\partial_{y_\gamma} F_\alpha(a^*_\beta(z))\, a_\gamma(z){:},$$

$$a^*_\alpha{}'(z) \mapsto F_\alpha(a^*_\beta(z)).$$

Note that the homogeneity condition on the coordinate systems severely restricts the possible forms of the functions F_α:

$$F_\alpha = c_\alpha y_\alpha + \sum_{\beta_1+\ldots+\beta_k=\alpha} c_{\beta_1,\ldots,\beta_k} y_{\beta_1} \cdots y_{\beta_k},$$

† As in Lemma 3.2.2 it follows that $\widetilde{\mathcal{A}}^{\mathfrak{g}}$ is isomorphic to $\widetilde{U}(M_{\mathfrak{g}})$, where $\widetilde{U}$ is the functor introduced in Section 3.2.3, but we will not use this fact.

where $c_{\beta_1,\ldots,\beta_k} \in \mathbb{C}$ and $c_\alpha \neq 0$.

It is also possible to define $M_\mathfrak{g}$ without any reference to a coordinate system on N_+. Namely, we may identify $M_\mathfrak{g}$, as an $\mathfrak{n}_+((t))$-module, with

$$M_\mathfrak{g} = \operatorname{Ind}_{\mathfrak{n}_+[[t]]}^{\mathfrak{n}_+((t))} \operatorname{Fun}(N_+[[t]]) \simeq U(\mathfrak{n}_+ \otimes t^{-1}\mathbb{C}[t^{-1}]) \otimes \operatorname{Fun}(N_+[[t]]),$$

where $\operatorname{Fun}(N_+[[t]])$ is the ring of regular functions on the pro-algebraic group $N_+[[t]]$, considered as an $\mathfrak{n}_+[[t]]$-module. If we choose a coordinate system $\{y_\alpha\}_{\alpha\in\Delta_+}$ on N_+, then we obtain a coordinate system $\{y_{\alpha,n}\}_{\alpha\in\Delta_+,n\geq 0}$ on $N_+[[t]]$. Then $u \otimes P(y_{\alpha,n}) \in M_\mathfrak{g}$, where $u \in U(\mathfrak{n}_+ \otimes t^{-1}\mathbb{C}[t^{-1}])$ and $P(y_{\alpha,n}) \in \operatorname{Fun}(N_+[[t]]) = \mathbb{C}[y_{\alpha,n}]_{\alpha\in\Delta_+,n\geq 0}$, corresponds to $u \cdot P(a^*_{\alpha,n})$ in our previous description of $M_\mathfrak{g}$.

It is straightforward to define a vertex algebra structure on $M_\mathfrak{g}$ (see [FF8], Section 2). Namely, the vacuum vector of $M_\mathfrak{g}$ is the vector $1 \otimes 1 \in M_\mathfrak{g}$. The translation operator T is defined as the operator $-\partial_t$, which naturally acts on $\operatorname{Fun}(N_+[[t]])$ as well as on $\mathfrak{n}_+((t))$ preserving $\mathfrak{n}_+[[t]]$. Next, we define the vertex operators corresponding to the elements of $M_\mathfrak{g}$ of the form $x_{-1}|0\rangle$, where $x \in \mathfrak{n}_+$, by the formula

$$Y(x_{-1}|0\rangle, z) = \sum_{n\in\mathbb{Z}} x_n z^{-n-1},$$

where $x_n = x \otimes t^n$, and we consider its action on $M_\mathfrak{g}$ viewed as the induced representation of $\mathfrak{n}_+((t))$. We also need to define the vertex operators

$$Y(P|0\rangle, z) = \sum_{n\in\mathbb{Z}} P_{(n)} z^{-n-1}$$

for $P \in \operatorname{Fun}(N_+[[t]])$. The corresponding linear operators $P_{(n)}$ are completely determined by their action on $|0\rangle$:

$$P_{(n)}|0\rangle = 0, \qquad n \geq 0,$$
$$P_{(n)}|0\rangle = \frac{1}{(-n-1)!} T^{-n-1} P|0\rangle,$$

their mutual commutativity and the following commutation relations with $\mathfrak{n}_+((t))$:

$$[x_m, P_{(k)}] = \sum_{n\geq 0} \binom{m}{n} (x_n \cdot P)_{(m+k-n)}.$$

Using the Reconstruction Theorem 2.2.4, it is easy to prove that these formulas define a vertex algebra structure on $M_\mathfrak{g}$.

In fact, the same definition works if we replace N_+ by any algebraic group G. In the general case, it is natural to consider the central extension $\widehat{\mathfrak{g}}_\kappa$ of the loop algebra $\mathfrak{g}((t))$ corresponding to an invariant inner product κ on $\mathfrak{g}$ defined as in Section 5.3.3. Then we have the induced module

$$\operatorname{Ind}_{\mathfrak{g}[[t]]\oplus\mathbb{C}\mathbf{1}}^{\widehat{\mathfrak{g}}_\kappa} \operatorname{Fun}(G[[t]]),$$

where the central element $\mathbf{1}$ acts on $\mathrm{Fun}(G[[t]])$ as the identity. The corresponding vertex algebra is the algebra of chiral differential operators on G, considered in [GMS, AG]. As shown in [GMS, AG], in addition to the natural (left) action of $\widehat{\mathfrak{g}}_\kappa$ on this vertex algebra, there is another (right) action of $\widehat{\mathfrak{g}}_{-\kappa-\kappa_\mathfrak{g}}$, which commutes with the left action. Here $\kappa_\mathfrak{g}$ is the Killing form on $\mathfrak{g}$, defined by the formula $\kappa_\mathfrak{g}(x,y) = \mathrm{Tr}_\mathfrak{g}(\mathrm{ad}\, x\, \mathrm{ad}\, y)$. In the case when $\mathfrak{g} = \mathfrak{n}_+$, there are no non-zero invariant inner products (in particular, $\kappa_{\mathfrak{n}_+} = 0$), and so we obtain a commuting right action of $\mathfrak{n}_+((t))$ on $M_\mathfrak{g}$. We will use this right action below (see Section 6.1.1).

The above formulas in fact define a canonical vertex algebra structure on

$$\mathrm{Ind}_{\mathfrak{n}_+[[t]]}^{\mathfrak{n}_+((t))} \mathrm{Fun}(\mathcal{U}[[t]]),$$

which is independent of the choice of identification $N_+ \simeq \mathcal{U}$. Recall that in order to define $\mathcal{U}$ we only need to fix a Borel subgroup B_+ of G. Then $\mathcal{U}$ is defined as the open B_+-orbit of the flag manifold and so it is naturally an N_+-torsor. In order to identify $\mathcal{U}$ with N_+ we need to choose a point in $\mathcal{U}$, i.e., an opposite Borel subgroup B_-, or, equivalently, a Cartan subgroup $H = B_+ \cap B_-$ of B_+. But in the above formulas we never used an identification of N_+ and $\mathcal{U}$, only the canonical action of $\mathfrak{n}_+$ on $\mathcal{U}$, which determined a canonical $L\mathfrak{n}_+$-action on $L\mathcal{U}$.

If we do not fix an identification $N_+ \simeq \mathcal{U}$, then the right action of $\mathfrak{n}_+$ on N_+ discussed above becomes an action of the "twisted" Lie algebra $\mathfrak{n}_{+,\mathcal{U}}$, where $\mathfrak{n}_{+,\mathcal{U}} = \mathcal{U} \underset{N_+}{\times} \mathfrak{n}_+$. It is interesting to observe† that unlike $\mathfrak{n}_+$, this twisted Lie algebra $\mathfrak{n}_{+,\mathcal{U}}$ has a canonical decomposition into the one-dimensional subspaces $\mathfrak{n}_{\alpha,\mathcal{U}}$ corresponding to the positive roots $\alpha \in \Delta_+$. Indeed, for any point $u \in \mathcal{U}$ we have an identification $\mathcal{U} \simeq N_+$ and a Cartan subgroup $H_u = B_+ \cap B_u$, where B_u is the stabilizer of u. The subspaces $\mathfrak{n}_{\alpha,\mathcal{U}}$ are defined as the eigenspaces in $\mathfrak{n}_+$ with respect to the adjoint action of H_u. If we choose a different point $u' \in \mathcal{U}$, the identification $\mathcal{U} \simeq N_+$ and H_u will change, but the eigenspaces will remain the same!

Therefore there exist canonical (up to a scalar) generators e_i^R of the right action of $\mathfrak{n}_{\alpha_i,\mathcal{U}}$ on $\mathcal{U}$. We will use these operators below to define the screening operators, and their independence of the choice of the Cartan subgroup in B_+ implies the independence of the kernel of the screening operator from any additional choices.

5.4.3 Local extension

For our purposes we may replace $\widetilde{\mathcal{A}}^\mathfrak{g}$, which is a very large topological algebra, by a relatively small "local part" $U(M_\mathfrak{g})$. Accordingly, we replace $\widetilde{\mathcal{A}}_0^\mathfrak{g}$ and $\widetilde{\mathcal{A}}_{\leq 1}^\mathfrak{g}$ by their local versions $\mathcal{A}_{0,\mathrm{loc}}^\mathfrak{g} = \widetilde{\mathcal{A}}_0^\mathfrak{g} \cap U(M_\mathfrak{g})$ and $\mathcal{A}_{\leq 1,\mathrm{loc}}^\mathfrak{g} = \widetilde{\mathcal{A}}_{\leq 1}^\mathfrak{g} \cap U(M_\mathfrak{g})$.

† I thank D. Gaitsgory for pointing this out.

Let us describe $\mathcal{A}^{\mathfrak{g}}_{0,\mathrm{loc}}$ and $\mathcal{A}^{\mathfrak{g}}_{\leq 1,\mathrm{loc}}$ more explicitly. The space $\mathcal{A}^{\mathfrak{g}}_{0,\mathrm{loc}}$ is spanned (topologically) by the Fourier coefficients of all polynomials in the $\partial_z^n a^*_\alpha(z), n \geq 0$. Note that because the $a^*_{\alpha,n}$'s commute among themselves, these polynomials are automatically normally ordered. The space $\mathcal{A}^{\mathfrak{g}}_{\leq 1,\mathrm{loc}}$ is spanned by the Fourier coefficients of the fields of the form

$$:P(a^*_\alpha(z), \partial_z a^*_\alpha(z), \ldots) a_\beta(z):$$

(the normally ordered product of $P(a^*_\alpha(z), \partial_z a^*_\alpha(z), \ldots)$ and $a_\beta(z)$).

Here we use the fact that the Fourier coefficients of all fields of the form

$$:P(a^*_\alpha(z), \partial_z a^*_\alpha(z), \ldots) \partial_z^m a_\beta(z): , \qquad m > 0,$$

may be expressed as linear combinations of the Fourier coefficients of the fields of the form

$$:P(a^*_\alpha(z), \partial_z a^*_\alpha(z), \ldots) a_\beta(z): . \tag{5.4.1}$$

Further, we define a local version $\mathcal{T}^{\mathfrak{g}}_{\mathrm{loc}}$ of $\mathrm{Vect}\, L\mathcal{U}$ as the subspace which consists of finite linear combinations of Fourier coefficients of the formal power series

$$P(a^*_\alpha(z), \partial_z a^*_\alpha(z), \ldots) a_\beta(z), \tag{5.4.2}$$

where $a_\alpha(z)$ and $a^*_\alpha(z)$ are given by formulas (5.3.4), (5.3.5).

Since $\mathcal{A}^{\mathfrak{g}}_{\leq 1,\mathrm{loc}}$ is the intersection of Lie subalgebras of $\widetilde{\mathcal{A}}^{\mathfrak{g}}$, it is also a Lie subalgebra of $\widetilde{\mathcal{A}}^{\mathfrak{g}}$. By construction, its image in $\mathrm{Vect}\, L\mathcal{U}$ under the homomorphism $\widetilde{\mathcal{A}}^{\mathfrak{g}}_{\leq 1} \to \mathrm{Vect}\, L\mathcal{U}$ equals $\mathcal{T}^{\mathfrak{g}}_{\mathrm{loc}}$. Finally, the kernel of the resulting surjective Lie algebra homomorphism $\mathcal{A}^{\mathfrak{g}}_{\leq 1,\mathrm{loc}} \to \mathcal{T}^{\mathfrak{g}}_{\mathrm{loc}}$ equals $\mathcal{A}^{\mathfrak{g}}_{0,\mathrm{loc}}$. Hence we obtain that the extension (5.3.7) restricts to the "local" extension

$$0 \to \mathcal{A}^{\mathfrak{g}}_{0,\mathrm{loc}} \to \mathcal{A}^{\mathfrak{g}}_{\leq 1,\mathrm{loc}} \to \mathcal{T}^{\mathfrak{g}}_{\mathrm{loc}} \to 0. \tag{5.4.3}$$

This sequence is non-split as will see below. The corresponding two-cocycle will be computed explicitly in Lemma 5.5.2 using the Wick formula (it comes from the "double contractions" of the corresponding vertex operators).

According to Section 5.3.2, the image of $L\mathfrak{g}$ in $\mathrm{Vect}\, L\mathcal{U}$ belongs to $\mathcal{T}^{\mathfrak{g}}_{\mathrm{loc}}$. We will show that the homomorphism $L\mathfrak{g} \to \mathcal{T}^{\mathfrak{g}}_{\mathrm{loc}}$ may be lifted to a homomorphism $\widehat{\mathfrak{g}}_\kappa \to \mathcal{A}^{\mathfrak{g}}_{\leq 1,\mathrm{loc}}$, where $\widehat{\mathfrak{g}}_\kappa$ is the central extension of $L\mathfrak{g}$ defined in Section 5.3.3.

5.5 Computation of the two-cocycle

Recall that an exact sequence of Lie algebras

$$0 \to \mathfrak{h} \to \widetilde{\mathfrak{g}} \to \mathfrak{g} \to 0,$$

where $\mathfrak{h}$ is an abelian ideal, with prescribed $\mathfrak{g}$-module structure, gives rise to a two-cocycle of $\mathfrak{g}$ with coefficients in $\mathfrak{h}$. It is constructed as follows. Choose a splitting $\imath : \mathfrak{g} \to \widetilde{\mathfrak{g}}$ of this sequence (considered as a vector space), and define $\sigma : \bigwedge^2 \mathfrak{g} \to \mathfrak{h}$ by the formula

$$\sigma(a,b) = \imath([a,b]) - [\imath(a), \imath(b)].$$

One checks that σ is a two-cocycle in the Chevalley complex of $\mathfrak{g}$ with coefficients in $\mathfrak{h}$, and that changing the splitting $\imath$ amounts to changing σ by a coboundary.

Conversely, suppose we are given a linear functional $\sigma : \bigwedge^2 \mathfrak{g} \to \mathfrak{h}$. Then we associate to it a Lie algebra structure on the direct sum $\mathfrak{g} \oplus \mathfrak{h}$. Namely, the Lie bracket of any two elements of $\mathfrak{h}$ is equal to 0, $[X,h] = X \cdot h$ for all $X \in \mathfrak{g}, h \in \mathfrak{h}$, and

$$[X,Y] = [X,Y]_{\mathfrak{g}} + \sigma(X,Y), \qquad X,Y \in \mathfrak{g}.$$

These formulas define a Lie algebra structure on $\widetilde{\mathfrak{g}}$ if and only if σ is a two-cocycle in the standard Chevalley complex of $\mathfrak{g}$ with coefficients in $\mathfrak{h}$. Therefore we obtain a bijection between the set of isomorphism classes of extensions of $\mathfrak{g}$ by $\mathfrak{h}$ and the cohomology group $H^2(\mathfrak{g}, \mathfrak{h})$.

5.5.1 Wick formula

Consider the extension (5.4.3). The operation of normal ordering gives us a splitting $\imath$ of this extension as vector space. Namely, $\imath$ maps the nth Fourier coefficient of the series (5.4.2) to the nth Fourier coefficient of the series (5.4.1). To compute the corresponding two-cocycle we have to learn how to compute commutators of Fourier coefficients of generating functions of the form (5.4.1) and (5.4.2). Those may be computed from the operator product expansion (OPE) of the corresponding vertex operators. We now explain how to compute the OPEs of vertex operators using the Wick formula.

In order to state the Wick formula, we have to introduce the notion of contraction of two fields. In order to simplify notation, we will assume that $\mathfrak{g} = \mathfrak{sl}_2$ and suppress the index α in $a_{\alpha,n}$ and $a^*_{\alpha,n}$. The general case is treated in the same way.

From the commutation relations, we obtain the following OPEs:

$$a(z)a^*(w) = \frac{1}{z-w} + {:}a(z)a^*(w){:}\,,$$
$$a^*(z)a(w) = -\frac{1}{z-w} + {:}a^*(z)a(w){:}\,.$$

We view them now as identities on formal power series, in which by $1/(z-w)$ we understand its expansion in positive powers of w/z. Differentiating several times, we obtain

$$\partial_z^n a(z)\partial_w^m a^*(w) = (-1)^n \frac{(n+m)!}{(z-w)^{n+m+1}} + {:}\partial_z^n a(z)\partial_w^m a^*(w){:}\,, \tag{5.5.1}$$

$$\partial_z^m a^*(z)\partial_w^n a(w) = (-1)^{m+1} \frac{(n+m)!}{(z-w)^{n+m+1}} + {:}\partial_z^m a^*(z)\partial_w^n a(w){:} \tag{5.5.2}$$

(here again by $1/(z-w)^n$ we understand its expansion in positive powers of w/z).

Suppose that we are given two normally ordered monomials in $a(z), a^*(z)$ and their derivatives. Denote them by $P(z)$ and $Q(z)$. A single **pairing** between $P(z)$ and $Q(w)$ is by definition either the pairing $(\partial_z^n a(z), \partial_w^m a^*(w))$ of $\partial_z^n a(z)$ occurring in $P(z)$ and $\partial_w^m a^*(w)$ occurring in $Q(w)$, or the pairing $(\partial_z^m a^*(z), \partial_w^n a(w))$ of $\partial_z^m a^*(z)$ occurring in $P(z)$ and $\partial_w^n a(w)$ occurring in $Q(w)$. We attach to it the functions

$$(-1)^n \frac{(n+m)!}{(z-w)^{n+m+1}} \qquad \text{and} \qquad (-1)^{m+1} \frac{(n+m)!}{(z-w)^{n+m+1}},$$

respectively. A multiple pairing B is by definition a disjoint union of single pairings. We attach to it the function $f_B(z,w)$, which is the product of the functions corresponding to the single pairings in B.

Note that the monomials $P(z)$ and $Q(z)$ may well have multiple pairings of the same type. For example, the monomials ${:}a^*(z)^2\partial_z a(z){:}$ and ${:}a(w)\partial_z^2 a^*(w){:}$ have two different pairings of type $(a^*(z), a(w))$; the corresponding function is $-1/(z-w)$. In such a case we say that the multiplicity of this pairing is 2. Note that these two monomials also have a unique pairing $(\partial_z a(z), \partial_w^2 a^*(w))$, and the corresponding function is $-6/(z-w)^4$.

Given a multiple pairing B between $P(z)$ and $Q(w)$, we define $(P(z)Q(w))_B$ as the product of all factors of $P(z)$ and $Q(w)$ which do not belong to the pairing (if there are no factors left, we set $(P(z)Q(w))_B = 1$). The **contraction** of $P(z)Q(w)$ with respect to the pairing B, denoted ${:}P(z)Q(w){:}_B$, is by definition the normally ordered formal power series ${:}(P(z)Q(w))_B{:}$ multiplied by the function $f_B(z,w)$. We extend this definition to the case when B is the empty set by stipulating that

$${:}P(z)Q(w){:}_\emptyset = {:}P(z)Q(w){:}\,.$$

Now we are in a position to state the **Wick formula**, which gives the OPE of two arbitrary normally ordered monomial vertex operators. The proof of this formula is straightforward and is left to the reader.

Lemma 5.5.1 *Let $P(z)$ and $Q(w)$ be two monomials as above. Then the product $P(z)Q(w)$ equals the sum of terms ${:}P(z)Q(w){:}_B$ over all pairings B between P and Q including the empty one, counted with multiplicity.*

Here is an example:

$$:a^*(z)^2\partial_z a(z): :a(w)\partial_z^2 a^*(w): = :a^*(z)^2\partial_z a(z)a(w)\partial_z^2 a^*(w):- \frac{2}{z-w}:a^*(z)\partial_z a(z)\partial_z^2 a^*(w): - \frac{6}{(z-w)^4}:a^*(z)^2 a(w): + \frac{12}{(z-w)^5}a^*(z).$$

5.5.2 Double contractions

Now we can compute our two-cocycle. For this, we need to apply the Wick formula to the fields of the form

$$:R(a^*(z), \partial_z a^*(z), \ldots)a(z): ,$$

whose Fourier coefficients span the preimage of $\mathcal{T}_{\text{loc}}$ in $\mathcal{A}_{\leq 1,\text{loc}}$ under our splitting $\imath$. Two fields of this form may have only single or double pairings, and therefore their OPE can be written quite explicitly.

A field of the above form may be written as $Y(P(a_n^*)a_{-1}, z)$ (or $Y(Pa_{-1}, z)$ for short), where P is a polynomial in the $a_n^*, n \leq 0$ (recall that $a_n^*, n \leq 0$, corresponds to $\partial_z^{-n} a^*(z)/(-n)!$).† Applying the Wick formula, we obtain

Lemma 5.5.2

$$\begin{aligned}Y(Pa_{-1}, z)Y(Qa_{-1}, w) &= :Y(Pa_{-1}, z)Y(Qa_{-1}, w): \\ &+ \sum_{n\geq 0} \frac{1}{(z-w)^{n+1}} :Y(P, z)Y\left(\frac{\partial Q}{\partial a^*_{-n}} a_{-1}, w\right): \\ &- \sum_{n\geq 0} \frac{1}{(z-w)^{n+1}} :Y\left(\frac{\partial P}{\partial a^*_{-n}} a_{-1}, z\right) Y(Q, w): \\ &- \sum_{n,m\geq 0} \frac{1}{(z-w)^{n+m+2}} Y\left(\frac{\partial P}{\partial a^*_{-n}}, z\right) Y\left(\frac{\partial Q}{\partial a^*_{-m}}, w\right).\end{aligned}$$

Note that we do not need to put normal ordering in the last summand.

Using this formula and the commutation relations (3.2.2), we can now easily obtain the commutators of the Fourier coefficients of the fields $Y(Pa_{-1}, z)$ and $Y(Qa_{-1}, w)$.

The first two terms in the right hand side of the formula in Lemma 5.5.2 correspond to single contractions between $Y(Pa_{-1}, z)$ and $Y(Qa_{-1}, w)$. The part in the commutator of the Fourier coefficients induced by these terms will be exactly the same as the commutator of the corresponding vector fields, computed in $\mathcal{T}_{\text{loc}}$. Thus, we see that the discrepancy between the commutators in $\mathcal{A}_{\leq 1,\text{loc}}$ and in $\mathcal{T}_{\text{loc}}$ (as measured by our two-cocycle) is due to the

† In what follows we will write, by abuse of notation, $Y(A, z)$ for the series that, strictly speaking, should be denoted by $Y[A, z]$. The reason is that the homomorphism $U(M_{\mathfrak{g}}) \to \operatorname{End} M_{\mathfrak{g}}, A_{[n]} \mapsto A_{(n)}$ is injective, and so we do not lose anything by considering the image of $U(M_{\mathfrak{g}})$ in $\operatorname{End} M_{\mathfrak{g}}$.

last term in the formula from Lemma 5.5.2, which comes from the **double contractions** between $Y(Pa_{-1}, z)$ and $Y(Qa_{-1}, w)$.

Explicitly, we obtain the following formula for our two-cocycle

$$\omega((Pa_{-1})_{[k]}, (Qa_{-1})_{[s]}) = \tag{5.5.3}$$

$$-\sum_{n,m\geq 0} \mathrm{Res}_{w=0} \frac{1}{(n+m+1)!} \partial_z^{n+m+1} Y\left(\frac{\partial P}{\partial a^*_{-n}}, z\right) Y\left(\frac{\partial Q}{\partial a^*_{-m}}, w\right) z^k w^s \Big|_{z=w} dw.$$

5.5.3 *The extension is non-split*

For example, let us compute the cocycle for the elements $\mathbf{h}_n$ defined by the formula

$$Y(a_0^* a_{-1}, z) = {:}a^*(z)a(z){:} = \sum_{n\in\mathbb{Z}} \mathbf{h}_n z^{-n-1},$$

so that

$$\mathbf{h}_n = \sum_{k\in\mathbb{Z}} {:}a^*_{-k} a_{n+k}{:}\,.$$

According to formula (5.5.3), we have

$$\omega(\mathbf{h}_n, \mathbf{h}_m) = -\,\mathrm{Res}_{w=0}\, n w^{n+m-1} dw = -n\delta_{n,-m}.$$

As the single contraction terms cancel each other, we obtain the following commutation relations in $\mathcal{A}_{\leq 1,\mathrm{loc}}$:

$$[\mathbf{h}_n, \mathbf{h}_m] = -n\delta_{n,-m}.$$

On the other hand, we find that the image of $\mathbf{h}_n$ in $\mathcal{T}_{\mathrm{loc}}$ is the vector field

$$\overline{\mathbf{h}}_n = \sum_{k\in\mathbb{Z}} y_k \frac{\partial}{\partial y_{n+k}}.$$

These vector fields commute with each other.

If the sequence (5.3.7) (or (5.4.3)) were split, then we would be able to find "correction terms" $f_m \in \mathrm{Fun}\,\mathbb{C}((t))$ (or in $\mathcal{A}_{0,\mathrm{loc}}$) such that the lifting $\overline{\mathbf{h}}_n \mapsto \mathbf{h}_n + f_n$ preserves the Lie brackets, that is,

$$[\mathbf{h}_n + f_n, \mathbf{h}_m + f_m] = 0, \qquad \forall n, m \in \mathbb{Z}.$$

But this is equivalent to the formula

$$\overline{\mathbf{h}}_n \cdot f_m - \overline{\mathbf{h}}_m \cdot f_n = -n\delta_{n,-m}.$$

But since $\overline{\mathbf{h}}_n$ is a linear vector field (i.e., linear in the coordinates y_k on $L\mathcal{U}$), the left hand side of this formula cannot be a non-zero constant for any choice of $f_m, n \in \mathbb{Z}$.

Thus, we find that the sequences (5.3.7) and (5.4.3) do not split as exact sequences of Lie algebras.

5.5.4 A reminder on cohomology

Thus, we cannot expect to be able to lift the homomorphism $\widehat{\rho} : L\mathfrak{g} \to \mathcal{T}_{\mathrm{loc}}$ to a homomorphism $L\mathfrak{g} \to \widetilde{\mathcal{A}}^{\mathfrak{g}}_{\leq 1,\mathrm{loc}}$, as in the finite-dimensional case. The next best thing we can hope to accomplish is to lift it to a homomorphism from the central extension $\widehat{\mathfrak{g}}_{\kappa}$ of $L\mathfrak{g}$ to $\widetilde{\mathcal{A}}^{\mathfrak{g}}_{\leq 1,\mathrm{loc}}$. Let us see what cohomological condition corresponds to the existence of such a lifting.

So let again

$$0 \to \mathfrak{h} \to \widetilde{\mathfrak{l}} \to \mathfrak{l} \to 0$$

be an extension of Lie algebras, where $\mathfrak{h}$ is an abelian Lie subalgebra and an ideal in $\widetilde{\mathfrak{l}}$. Choosing a splitting $\imath$ of this sequence considered as a vector space we define a two-cocycle of $\mathfrak{l}$ with coefficients in $\mathfrak{h}$ as in Section 5.5. Suppose that we are given a Lie algebra homomorphism $\alpha : \mathfrak{g} \to \mathfrak{l}$ for some Lie algebra $\mathfrak{g}$. Pulling back our two-cocycle under α we obtain a two-cocycle of $\mathfrak{g}$ with coefficients in $\mathfrak{h}$.

On the other hand, given a homomorphism $\mathfrak{h}' \xrightarrow{i} \mathfrak{h}$ of $\mathfrak{g}$-modules we obtain a map i_* between the spaces of two-cocycles of $\mathfrak{g}$ with coefficients in $\mathfrak{h}$ and $\mathfrak{h}'$. The corresponding map of the cohomology groups $H^2(\mathfrak{g}, \mathfrak{h}') \to H^2(\mathfrak{g}, \mathfrak{h})$ will also be denoted by i_*.

Lemma 5.5.3 *Suppose that we are given a two-cocycle σ of $\mathfrak{g}$ with coefficients in $\mathfrak{h}'$ such that the cohomology classes of $i_*(\sigma)$ and ω are equal in $H^2(\mathfrak{g}, \mathfrak{h}')$. Denote by $\widetilde{\mathfrak{g}}$ the extension of $\mathfrak{g}$ by $\mathfrak{h}'$ corresponding to σ defined as above. Then the map $\mathfrak{g} \to \mathfrak{l}$ may be augmented to a map of commutative diagrams*

$$\begin{array}{ccccccccc} 0 & \longrightarrow & \mathfrak{h} & \longrightarrow & \widetilde{\mathfrak{l}} & \longrightarrow & \mathfrak{l} & \longrightarrow & 0 \\ & & \uparrow & & \uparrow & & \uparrow & & \\ 0 & \longrightarrow & \mathfrak{h}' & \longrightarrow & \widetilde{\mathfrak{g}} & \longrightarrow & \mathfrak{g} & \longrightarrow & 0. \end{array} \tag{5.5.4}$$

Moreover, the set of isomorphism classes of such diagrams is a torsor over $H^1(\mathfrak{g}, \mathfrak{h})$.

Proof. If the cohomology classes of $i_*(\sigma)$ and ω coincide, then $i_*(\sigma) + d\gamma = \omega$, where γ is a one-cochain, i.e., a linear functional $\mathfrak{g} \to \mathfrak{h}$. Define a linear map $\beta : \widetilde{\mathfrak{g}} \to \widetilde{\mathfrak{l}}$ as follows. By definition, we have a splitting $\widetilde{\mathfrak{g}} = \mathfrak{g} \oplus \mathfrak{h}'$ as a vector space. We set $\beta(X) = \imath(\alpha(X)) + \gamma(X)$ for all $X \in \mathfrak{g}$ and $\beta(h) = i(h)$ for all $h \in \mathfrak{h}$. Then the above equality of cocycles implies that β is a Lie algebra homomorphism which makes the diagram (5.5.4) commutative. However, the choice of γ is not unique as we may modify it by adding to it an arbitrary one-cocycle γ'. But the homomorphisms corresponding to γ and to $\gamma + \gamma'$, where γ' is a coboundary, lead to isomorphic diagrams. This implies that the set of isomorphism classes of such diagrams is a torsor over $H^1(\mathfrak{g}, \mathfrak{h})$. $\square$

5.5.5 Two cocycles

Restricting the two-cocycle ω of $\mathcal{T}_{\mathrm{loc}}$ with coefficients in $\mathcal{A}^{\mathfrak{g}}_{0,\mathrm{loc}}$ corresponding to the extension (5.4.3) to $L\mathfrak{g} \subset \mathcal{T}_{\mathrm{loc}}$, we obtain a two-cocycle of $L\mathfrak{g}$ with coefficients in $\mathcal{A}^{\mathfrak{g}}_{0,\mathrm{loc}}$. We also denote it by ω. The $L\mathfrak{g}$-module $\mathcal{A}^{\mathfrak{g}}_{0,\mathrm{loc}}$ contains the trivial subrepresentation $\mathbb{C}$, i.e., the span of $|0\rangle_{[-1]}$ (which we view as the constant function on $L\mathcal{U}$), and the inclusion $\mathbb{C} \xrightarrow{i} \mathcal{A}^{\mathfrak{g}}_{0,\mathrm{loc}}$ induces a map i_* of the corresponding spaces of two-cocycles and the cohomology groups $H^2(L\mathfrak{g}, \mathbb{C}) \to H^2(L\mathfrak{g}, \mathcal{A}^{\mathfrak{g}}_{0,\mathrm{loc}})$.

As we discussed in Section 5.3.3, the cohomology group $H^2(L\mathfrak{g}, \mathbb{C})$ is one-dimensional and is isomorphic to the space of invariant inner products on $\mathfrak{g}$. We denote by σ the class corresponding to the inner product $\kappa_c = -\frac{1}{2}\kappa_{\mathfrak{g}}$, where $\kappa_{\mathfrak{g}}$ denotes the Killing form on $\mathfrak{g}$. Thus, by definition,

$$\kappa_c(x, y) = -\frac{1}{2} \operatorname{Tr}(\operatorname{ad} x \operatorname{ad} y).$$

The cocycle σ is then given by the formula

$$\sigma(J^a_n, J^b_m) = n\delta_{n,-m}\kappa_c(J^a, J^b) \tag{5.5.5}$$

(see Section 5.3.3).

In the next section we will show that the cohomology classes of $i_*(\sigma)$ and ω are equal. Lemma 5.5.3 will then imply that there exists a family of Lie algebra homomorphisms $\widehat{\mathfrak{g}}_{\kappa_c} \to \mathcal{A}^{\mathfrak{g}}_{\leq 1,\mathrm{loc}}$ such that $\mathbf{1} \mapsto 1$.

5.6 Comparison of cohomology classes

Unfortunately, the Chevalley complex that calculates $H^2(L\mathfrak{g}, \mathcal{A}^{\mathfrak{g}}_{0,\mathrm{loc}})$ is unmanageably large. So as the first step we will show in the next section that ω and σ actually both belong to a much smaller subcomplex, where they are more easily compared.

5.6.1 Clifford algebras

Choose a basis $\{J^a\}_{a=1,\dots,\dim\mathfrak{g}}$ of $\mathfrak{g}$, and set $J^a_n = J^a \otimes t^n$. Introduce the Clifford algebra with generators $\psi_{a,n}, \psi^*_{a,m}, a = 1, \dots, \dim\mathfrak{g}; m, n \in \mathbb{Z}$, with anti-commutation relations

$$[\psi_{a,n}, \psi_{b,n}]_+ = [\psi^*_{a,n}, \psi^*_{b,m}]_+ = 0, \qquad [\psi_{a,n}, \psi^*_{b,m}]_+ = \delta_{a,b}\delta_{n,-m}.$$

Let $\bigwedge_{\mathfrak{g}}$ be the module over this Clifford algebra generated by a vector $|0\rangle$ such that

$$\psi_{a,n}|0\rangle = 0, \quad n > 0, \qquad \psi^*_{a,n}|0\rangle = 0, \quad n \geq 0.$$

Then $\bigwedge_{\mathfrak{g}}$ carries the following structure of a $\mathbb{Z}_+$-graded vertex superalgebra (see [FB], Section 15.1.1):

- $\mathbb{Z}_+$-grading: $\deg \psi_{a,n} = \deg \psi^*_{a,n} = -n, \deg |0\rangle = 0$;
- vacuum vector: $|0\rangle$;
- translation operator: $T|0\rangle = 0, [T, \psi_{a,n}] = -n\psi_{a,n-1}, [T, \psi^*_{a,n}] = -(n-1)\psi^*_{a,n-1}$;
- vertex operators:

$$Y(\psi_{a,-1}|0\rangle, z) = \psi_a(z) = \sum_{n\in\mathbb{Z}} \psi_{a,n} z^{-n-1},$$
$$Y(\psi^*_{a,0}|0\rangle, z) = \psi^*_a(z) = \sum_{n\in\mathbb{Z}} \psi^*_{a,n} z^{-n},$$

$$Y(\psi_{a_1,n_1} \dots \psi_{a_k,n_k} \psi^*_{b_1,m_1} \dots \psi^*_{b_l,m_l}|0\rangle, z) = \prod_{i=1}^{k} \frac{1}{(-n_i-1)!} \prod_{j=1}^{l} \frac{1}{(-m_j)!} \cdot$$
$$:\partial_z^{-n_1-1}\psi_{a_1}(z)\dots\partial_z^{-n_k-1}\psi_{a_k}(z)\partial_z^{-m_1}\psi^*_{b_1}(z)\dots\partial_z^{-m_l}\psi^*_{b_l}(z):.$$

The tensor product of two vertex superalgebras is naturally a vertex superalgebra (see Lemma 1.3.6 of [FB]), and so $M_\mathfrak{g} \otimes \bigwedge_\mathfrak{g}$ is a vertex superalgebra.

5.6.2 The local Chevalley complex

The ordinary Chevalley complex computing the cohomology $H^\bullet(L\mathfrak{g}, \mathcal{A}^\mathfrak{g}_{0,\mathrm{loc}})$ is $C^\bullet(L\mathfrak{g}, \mathcal{A}^\mathfrak{g}_{0,\mathrm{loc}}) = \bigoplus_{i\geq 0} C^i(L\mathfrak{g}, \mathcal{A}^\mathfrak{g}_{0,\mathrm{loc}})$, where

$$C^i(L\mathfrak{g}, \mathcal{A}^\mathfrak{g}_{0,\mathrm{loc}}) = \mathrm{Hom}_{\mathrm{cont}}(\textstyle\bigwedge^i(L\mathfrak{g}), \mathcal{A}^\mathfrak{g}_{0,\mathrm{loc}}),$$

where $\bigwedge^i(L\mathfrak{g})$ stands for the natural completion of the ordinary ith exterior power of the topological vector space $L\mathfrak{g}$, and we consider all continuous linear maps. For $f \in \mathcal{A}^\mathfrak{g}_{0,\mathrm{loc}}$ we denote by $\psi^*_{b_1,-k_1} \dots \psi^*_{b_i,-k_i} f$ the linear functional $\phi \in \mathrm{Hom}_{\mathrm{cont}}(\bigwedge^i(L\mathfrak{g}), \mathcal{A}^\mathfrak{g}_{0,\mathrm{loc}})$ defined by the formula

$$\phi(J^{a_1}_{m_1} \wedge \dots \wedge J^{a_i}_{m_i}) = \begin{cases} (-1)^{l(\tau)} f, & ((a_1, m_1), \dots, (a_i, m_i)) = \tau((b_1, k_1), \dots, (b_i, k_i)), \\ 0, & ((a_1, m_1), \dots, (a_i, m_i)) \neq \tau((b_1, k_1), \dots, (b_i, k_i)), \end{cases} \tag{5.6.1}$$

where τ runs over the symmetric group on i letters and $l(\tau)$ is the length of τ. Then any element of the space $C^i(L\mathfrak{g}, \mathcal{A}^\mathfrak{g}_{0,\mathrm{loc}})$ may be written as a (possibly infinite) linear combination of terms of this form.

The differential $d : C^i(L\mathfrak{g}, \mathcal{A}^\mathfrak{g}_{0,\mathrm{loc}}) \to C^{i+1}(L\mathfrak{g}, \mathcal{A}^\mathfrak{g}_{0,\mathrm{loc}})$ is given by the formula

$$(d\phi)(X_1, \dots, X_{i+1}) = \sum_{j=1}^{i+1} (-1)^{j+1} X_j \phi(X_1, \dots, \widehat{X_j}, \dots, X_{i+1})$$
$$+ \sum_{j<k} (-1)^{j+k+1} \phi([X_j, X_k], X_1, \dots, \widehat{X_j}, \dots, \widehat{X_k}, \dots, X_{i+1}).$$

It follows from the definition of the vertex operators that the linear maps

$$\int Y(\psi^*_{a_1,n_1}\cdots\psi^*_{a_i,n_i}a^*_{\alpha_1,m_1}\cdots a^*_{\alpha_j,m_j}|0\rangle, z)\, dz, \qquad n_p \leq 0, m_p \leq 0. \quad (5.6.2)$$

from $\bigwedge^i(L\mathfrak{g})$ to $\mathcal{A}^{\mathfrak{g}}_{0,\mathrm{loc}}$ are continuous. Here and below we will use the notation

$$\int f(z)dz = \mathrm{Res}_{z=0}\, f(z)dz.$$

Let $C^i_{\mathrm{loc}}(L\mathfrak{g}, \mathcal{A}^{\mathfrak{g}}_{0,\mathrm{loc}})$ be the subspace of the space $\mathrm{Hom}_{\mathrm{cont}}(\bigwedge^i(L\mathfrak{g}), \mathcal{A}^{\mathfrak{g}}_{0,\mathrm{loc}})$, spanned by all linear maps of the form (5.6.2).

Lemma 5.6.1 *The Chevalley differential maps the subspace* $C^i_{\mathrm{loc}}(L\mathfrak{g}, \mathcal{A}^{\mathfrak{g}}_{0,\mathrm{loc}}) \subset C^i(L\mathfrak{g}, \mathcal{A}^{\mathfrak{g}}_{0,\mathrm{loc}})$ *to* $C^{i+1}_{\mathrm{loc}}(L\mathfrak{g}, \mathcal{A}^{\mathfrak{g}}_{0,\mathrm{loc}}) \subset C^{i+1}(L\mathfrak{g}, \mathcal{A}^{\mathfrak{g}}_{0,\mathrm{loc}})$, *and so*

$$C^\bullet_{\mathrm{loc}}(L\mathfrak{g}, \mathcal{A}^{\mathfrak{g}}_{0,\mathrm{loc}}) = \bigoplus_{i\geq 0} C^i_{\mathrm{loc}}(L\mathfrak{g}, \mathcal{A}^{\mathfrak{g}}_{0,\mathrm{loc}})$$

is a subcomplex of $C^\bullet(L\mathfrak{g}, \mathcal{A}^{\mathfrak{g}}_{0,\mathrm{loc}})$.

Proof. The action of the differential d on $\int Y(A,z)dz$, where A is of the form (5.6.2), may be written as the commutator

$$\left[\int Q(z)dz, \int Y(A,z)dz\right],$$

where $Q(z)$ is the field

$$Q(z) = \sum_a J^a(z)\psi^*_a(z) - \frac{1}{2}\sum_{a,b,c}\mu^{ab}_c{:}\psi^*_a(z)\psi^*_b(z)\psi_c(z){:}\,, \qquad (5.6.3)$$

where μ^{ab}_c denotes the structure constants of $\mathfrak{g}$. Therefore it is equal to $Y\left(\int Q(z)dz \cdot A, z\right)$, which is of the form (5.6.2). $\square$

We call $C^i_{\mathrm{loc}}(L\mathfrak{g}, \mathcal{A}^{\mathfrak{g}}_{0,\mathrm{loc}})$ the **local** subcomplex of $C^\bullet(L\mathfrak{g}, \mathcal{A}^{\mathfrak{g}}_{0,\mathrm{loc}})$.

Lemma 5.6.2 *Both* ω *and* $i_*(\sigma)$ *belong to* $C^2_{\mathrm{loc}}(L\mathfrak{g}, \mathcal{A}^{\mathfrak{g}}_{0,\mathrm{loc}})$.

Proof. We begin with σ, which is given by formula (5.5.5). Therefore the cocycle $i_*(\sigma)$ is equal to the following element of $C^2_{\mathrm{loc}}(L\mathfrak{g}, \mathcal{A}^{\mathfrak{g}}_{0,\mathrm{loc}})$:

$$\begin{aligned} i_*(\sigma) &= \sum_{a\leq b}\sum_{n\in\mathbb{Z}} \kappa_c(J^a, J^b)n\psi^*_{a,-n}\psi^*_{b,n} \\ &= \sum_{a\leq b}\kappa_c(J^a, J^b)\int Y(\psi^*_{a,-1}\psi^*_{b,0}|0\rangle, z)\, dz. \end{aligned} \qquad (5.6.4)$$

Next we consider ω. Combining the discussion of Section 5.3.2 with formulas of Section 5.2.5, we obtain that

$$\imath(J^a_n) = \sum_{\beta\in\Delta_+}\int Y(R^\beta_a(a^*_{\alpha,0})a_{\beta,-1}|0\rangle, z)\, z^n dz,$$

where R_a^β is a polynomial. Then formula (5.5.3) implies that

$$\omega(J_n^a, J_m^b) = - \sum_{\alpha,\beta\in\Delta_+} \int \left(Y\left(T\frac{\partial R_a^\alpha}{\partial a_{\beta,0}^*} \cdot \frac{\partial R_b^\beta}{\partial a_{\alpha,0}^*}, w \right) w^{n+m} \right.$$
$$\left. + nY\left(\frac{\partial R_a^\alpha}{\partial a_{\beta,0}^*} \frac{\partial R_b^\beta}{\partial a_{\alpha,0}^*}, w \right) w^{n+m-1} \right) dw.$$

Therefore

$$\omega = - \sum_{a\leq b;\alpha,\beta\in\Delta_+} \int \left(Y\left(\psi_{a,0}^*\psi_{b,0}^* T\frac{\partial R_a^\alpha}{\partial a_{\beta,0}^*} \cdot \frac{\partial R_b^\beta}{\partial a_{\alpha,0}^*}, z \right) \right. \tag{5.6.5}$$
$$\left. + Y\left(\psi_{a,-1}^*\psi_{b,0}^* \frac{\partial R_a^\alpha}{\partial a_{\beta,0}^*} \frac{\partial R_b^\beta}{\partial a_{\alpha,0}^*}, z \right) \right) dz$$

(in the last two formulas we have omitted $|0\rangle$). Hence it belongs to the space $C^2_{\rm loc}(L\mathfrak{g}, \mathcal{A}^{\mathfrak{g}}_{0,{\rm loc}})$. □

We need to show that the cocycles $i_*(\sigma)$ and ω represent the same cohomology class in the local complex $C^\bullet_{\rm loc}(L\mathfrak{g}, \mathcal{A}^{\mathfrak{g}}_{0,{\rm loc}})$. We will show that this is equivalent to checking that the restrictions of these cocycles to the Lie subalgebra $L\mathfrak{h} \subset L\mathfrak{g}$ are the same. These restrictions can be easily computed and we indeed obtain that they coincide. The passage to $L\mathfrak{h}$ is achieved by a version of the Shapiro lemma, as we explain in the next section.

5.6.3 Another complex

Given a Lie algebra $\mathfrak{l}$, we denote the Lie subalgebra $\mathfrak{l}[[t]]$ of $L\mathfrak{l} = \mathfrak{l}((t))$ by $L_+\mathfrak{l}$.

Observe that the Lie algebra $L_+\mathfrak{g}$ acts naturally on the space

$$M_{\mathfrak{g},+} = \mathbb{C}[a_{\alpha,n}^*]_{\alpha\in\Delta_+,n\leq 0} \simeq \mathbb{C}[y_{\alpha,n}]_{\alpha\in\Delta_+,n\geq 0},$$

which is identified with the ring of functions on the space $\mathcal{U}[[t]] \simeq N_+[[t]]$. We identify the standard Chevalley complex

$$C^\bullet(L_+\mathfrak{g}, M_{\mathfrak{g},+}) = {\rm Hom}_{\rm cont}(\textstyle\bigwedge^\bullet L_+\mathfrak{g}, M_{\mathfrak{g},+})$$

with the tensor product $M_{\mathfrak{g},+} \otimes \bigwedge^\bullet_{\mathfrak{g},+}$, where

$$\textstyle\bigwedge^\bullet_{\mathfrak{g},+} = \bigwedge(\psi_{a,n}^*)_{n\leq 0}.$$

Introduce a superderivation T on $C^\bullet(L_+\mathfrak{g}, M_{\mathfrak{g},+})$ acting by the formulas

$$T \cdot a_{a,n}^* = -(n-1)a_{a,n-1}^*, \qquad T \cdot \psi_{a,n}^* = -(n-1)\psi_{a,n-1}^*.$$

We have a linear map

$$\int : C^\bullet(L_+\mathfrak{g}, M_{\mathfrak{g},+}) \to C^\bullet_{\rm loc}(L\mathfrak{g}, \mathcal{A}^{\mathfrak{g}}_{0,{\rm loc}})$$

sending $A \in C^\bullet(L_+\mathfrak{g}, M_{\mathfrak{g},+})$ to $\int Y(A,z)dz$ (recall that $\int$ picks out the (-1)st Fourier coefficient of a formal power series).

Recall that in any vertex algebra V we have the identity $Y(TA,z) = \partial_z Y(A,z)$. Hence if $A \in \operatorname{Im} T$, then $\int Y(A,z)dz = 0$.

Lemma 5.6.3 *The map* $\int$ *defines an isomorphism*

$$C^\bullet_{\mathrm{loc}}(L\mathfrak{g}, \mathcal{A}^{\mathfrak{g}}_{0,\mathrm{loc}}) \simeq C^\bullet(L_+\mathfrak{g}, M_{\mathfrak{g},+})/(\operatorname{Im} T + \mathbb{C}),$$

and $\operatorname{Ker} T = \mathbb{C}$. *Moreover, the following diagram is commutative:*

$$\begin{array}{ccccc} C^\bullet(L_+\mathfrak{g}, M_{\mathfrak{g},+}) & \xrightarrow{T} & C^\bullet(L_+\mathfrak{g}, M_{\mathfrak{g},+}) & \xrightarrow{\int} & C^\bullet_{\mathrm{loc}}(L\mathfrak{g}, \mathcal{A}^{\mathfrak{g}}_{0,\mathrm{loc}}) \\ {\scriptstyle d}\uparrow & & {\scriptstyle d}\uparrow & & {\scriptstyle d}\uparrow \\ C^\bullet(L_+\mathfrak{g}, M_{\mathfrak{g},+}) & \xrightarrow{T} & C^\bullet(L_+\mathfrak{g}, M_{\mathfrak{g},+}) & \xrightarrow{\int} & C^\bullet_{\mathrm{loc}}(L\mathfrak{g}, \mathcal{A}^{\mathfrak{g}}_{0,\mathrm{loc}}). \end{array}$$

Proof. It is easy to see that the differential of the Chevalley complex $C^\bullet(L_+\mathfrak{g}, M_{\mathfrak{g},+})$ acts by the formula $A \mapsto \int Q(z)dz \cdot A$, where $Q(z)$ is given by formula (5.6.3). The lemma now follows from the argument used in the proof of Lemma 5.6.1. □

Consider the double complex

$$\begin{array}{ccccccc} \mathbb{C} & \longrightarrow & C^\bullet(L_+\mathfrak{g}, M_{\mathfrak{g},+}) & \xrightarrow{T} & C^\bullet(L_+\mathfrak{g}, M_{\mathfrak{g},+}) & \longrightarrow & \mathbb{C} \\ & & {\scriptstyle d}\uparrow & & {\scriptstyle d}\uparrow & & \\ \mathbb{C} & \longrightarrow & C^\bullet(L_+\mathfrak{g}, M_{\mathfrak{g},+}) & \xrightarrow{T} & C^\bullet(L_+\mathfrak{g}, M_{\mathfrak{g},+}) & \longrightarrow & \mathbb{C}. \end{array}$$

According to the lemma, the cohomology of the complex $C^\bullet_{\mathrm{loc}}(L\mathfrak{g}, \mathcal{A}^{\mathfrak{g}}_{0,\mathrm{loc}})$ is given by the second term of the spectral sequence, in which the zeroth differential is vertical.

We start by computing the first term of this spectral sequence. Let us observe that the Lie algebra $L_+\mathfrak{n}_+$ has a $Q \times \mathbb{Z}_+$ grading, where Q is the root lattice of $\mathfrak{g}$, defined by the formulas $\deg e_{\alpha,n} = (\alpha, n)$. Therefore the universal enveloping algebra $U(L_+\mathfrak{n}_+)$ is also $\mathbb{Z}_+ \times Q$-graded. Moreover, all homogeneous components $U(L_+\mathfrak{n}_+)_{(\gamma,n)}$, for $\gamma \in Q, n \in \mathbb{Z}_+$, of $U(L_+\mathfrak{n}_+)$ are finite-dimensional. The corresponding restricted dual space is

$$U(L_+\mathfrak{n}_+)^\vee = \bigoplus_{(\gamma,n)\in Q\times\mathbb{Z}_+} (U(L_+\mathfrak{n}_+)_{(\gamma,n)})^*.$$

It carries a natural structure of $L_+\mathfrak{n}_+$-module.

Now let

$$\operatorname{Coind}_{L_+\mathfrak{b}_-}^{L_+\mathfrak{g}} \mathbb{C} \stackrel{\text{def}}{=} \operatorname{Hom}^{\mathrm{res}}_{U(L_+\mathfrak{b}_-)}(U(L_+\mathfrak{g}), \mathbb{C})$$

(compare with formula (5.2.9)). The right hand side consists of all "restricted" linear functionals $U(L_+\mathfrak{g}) \to \mathbb{C}$ invariant under the action of $U(\mathfrak{b}_-)$ on $U(\mathfrak{g})$

from the left, where "restricted" means that we consider only those functionals which are *finite* linear combinations of maps supported on the direct summands $U(\mathfrak{b}_-) \otimes U(\mathfrak{n}_+)_{(\gamma,n)}$ of $U(\mathfrak{g})$, with respect to the isomorphism of vector spaces $U(\mathfrak{g}) \simeq U(\mathfrak{b}_-) \otimes U(\mathfrak{n}_+)$. In other words, as an $L_+\mathfrak{n}_+$-module, $\mathrm{Coind}_{L_+\mathfrak{b}_-}^{L_+\mathfrak{g}} \mathbb{C} \simeq U(L_+\mathfrak{n}_+)^\vee$.

Lemma 5.6.4 *The $L_+\mathfrak{g}$-module $M_{\mathfrak{g},+}$ is isomorphic to the coinduced module* $\mathrm{Coind}_{L_+\mathfrak{b}_-}^{L_+\mathfrak{g}} \mathbb{C}$.

Proof. We apply verbatim the argument used in our proof given in Sections 5.2.3 and 5.2.4 that $\mathrm{Fun}\, N_+$ is isomorphic to $\mathrm{Coind}_{\mathfrak{b}_-}^{\mathfrak{g}} \mathbb{C}$ as a $\mathfrak{g}$-modules. First we prove that $M_{\mathfrak{g},+} = \mathrm{Fun}\, L_+N_+$ is isomorphic to $U(L_+\mathfrak{n}_+)^\vee$ as an $L_+\mathfrak{n}_+$-module by showing that the monomial basis in $M_{\mathfrak{g},+}$ is dual to a PBW monomial basis in $U(L_+\mathfrak{n}_+)$, with respect to an $L_+\mathfrak{n}_+$-invariant pairing of the two modules. We then use this fact to prove that the natural homomorphism $M_{\mathfrak{g},+} \to \mathrm{Coind}_{L_+\mathfrak{b}_-}^{L_+\mathfrak{g}} \mathbb{C}$ is an isomorphism. □

Define a map of complexes

$$\mu' : C^\bullet(L_+\mathfrak{g}, M_{\mathfrak{g},+}) \to C^\bullet(L_+\mathfrak{b}_-, \mathbb{C})$$

as follows. If γ is an i-cochain in the complex

$$C^\bullet(L_+\mathfrak{g}, M_{\mathfrak{g},+}) = \mathrm{Hom}_{\mathrm{cont}}(\textstyle\bigwedge^i L_+\mathfrak{g}, M_{\mathfrak{g},+}),$$

then $\mu'(\gamma)$ is by definition the restriction of γ to $\bigwedge^i L_+\mathfrak{b}_-$ composed with the natural projection $M_{\mathfrak{g},+} = \mathrm{Coind}_{L_+\mathfrak{b}_-}^{L_+\mathfrak{g}} \mathbb{C} \to \mathbb{C}$. It is clear that μ' is a morphism of complexes. The following is an example of the Shapiro lemma (see [Fu], Section 5.4, for the proof).

Lemma 5.6.5 *The map μ' induces an isomorphism at the level of cohomologies, i.e.,*

$$H^\bullet(L_+\mathfrak{g}, M_{\mathfrak{g},+}) \simeq H^\bullet(L_+\mathfrak{g}, \mathrm{Coind}_{L_+\mathfrak{b}_-}^{L_+\mathfrak{g}} \mathbb{C}) \simeq H^\bullet(L_+\mathfrak{b}_-, \mathbb{C}). \tag{5.6.6}$$

Now we compute the right hand side of (5.6.6). Since $L_+\mathfrak{n}_- \subset L_+\mathfrak{b}_-$ is an ideal, and $L_+\mathfrak{b}_-/L_+\mathfrak{n}_- \simeq L_+\mathfrak{h}$, we obtain from the Serre–Hochschild spectral sequence (see [Fu], Section 5.1) that

$$H^n(L_+\mathfrak{b}_-, \mathbb{C}) = \bigoplus_{p+q=n} H^p(L_+\mathfrak{h}, H^q(L_+\mathfrak{n}_-, \mathbb{C})).$$

But $\mathfrak{h} \otimes 1 \in L_+\mathfrak{h}$ acts diagonally on $H^\bullet(L_+\mathfrak{n}_-, \mathbb{C})$, inducing an inner grading. According to [Fu], Section 5.2,

$$H^p(L_+\mathfrak{h}, H^q(L_+\mathfrak{n}_-, \mathbb{C})) = H^p(L_+\mathfrak{h}, H^q(L_+\mathfrak{n}_-, \mathbb{C})_0),$$

where $H^q(L_+\mathfrak{n}_-, \mathbb{C})_0$ is the subspace where $\mathfrak{h} \otimes 1$ acts by 0. Clearly, the space $H^0(L_+\mathfrak{n}_-, \mathbb{C})_0$ is the one-dimensional subspace of the scalars $\mathbb{C} \subset$

$H^0(L_+\mathfrak{n}_-, \mathbb{C})$, and for $q \neq 0$ we have $H^q(L_+\mathfrak{n}_-, \mathbb{C})_0 = 0$. Thus, we find that

$$H^p(L_+\mathfrak{h}, H^q(L_+\mathfrak{n}_-, \mathbb{C})) = H^p(L_+\mathfrak{h}, \mathbb{C}).$$

Furthermore, we have the following result. Define a map of complexes

$$\mu : C^\bullet(L_+\mathfrak{g}, M_{\mathfrak{g},+}) \to C^\bullet(L_+\mathfrak{h}, \mathbb{C}) \tag{5.6.7}$$

as follows. If γ is an i-cochain in the complex

$$C^\bullet(L_+\mathfrak{g}, M_{\mathfrak{g},+}) = \mathrm{Hom}_{\mathrm{cont}}(\textstyle\bigwedge^i L_+\mathfrak{g}, M_{\mathfrak{g},+}),$$

then $\mu(\gamma)$ is by definition the restriction of γ to $\bigwedge^i L_+\mathfrak{h}$ composed with the natural projection

$$p : M_{\mathfrak{g},+} = \mathrm{Coind}_{L_+\mathfrak{b}_-}^{L_+\mathfrak{g}} \mathbb{C} \to \mathbb{C}. \tag{5.6.8}$$

Lemma 5.6.6 *The map μ induces an isomorphism at the level of cohomologies, i.e.,*

$$H^\bullet(L_+\mathfrak{g}, M_{\mathfrak{g},+}) \simeq H^\bullet(L_+\mathfrak{h}, \mathbb{C}).$$

In particular, the cohomology class of a cocycle in the complex $C^i(L_+\mathfrak{g}, M_{\mathfrak{g},+})$ is uniquely determined by its restriction to $\bigwedge^i L_+\mathfrak{h}$.

Proof. It follows from the construction of the Serre–Hochschild spectral sequence (see [Fu], Section 5.1) that the restriction map $C^\bullet(L_+\mathfrak{b}_-, \mathbb{C}) \to C^\bullet(L_+\mathfrak{h}, \mathbb{C})$ induces an isomorphism at the level of cohomologies. The statement of the lemma now follows by combining this with Lemma 5.6.5. □

Since $L_+\mathfrak{h}$ is abelian, we have

$$H^\bullet(L_+\mathfrak{h}, \mathbb{C}) = \textstyle\bigwedge^\bullet(L_+\mathfrak{h})^*,$$

and so

$$H^\bullet(L_+\mathfrak{g}, M_{\mathfrak{g},+}) \simeq \textstyle\bigwedge^\bullet(L_+\mathfrak{h})^*.$$

It is clear that this isomorphism is compatible with the action of T on both sides. The kernel of T acting on the right hand side is equal to the subspace of scalars. Therefore we obtain

$$H^\bullet_{\mathrm{loc}}(L\mathfrak{g}, \mathcal{A}^{\mathfrak{g}}_{0,\mathrm{loc}}) \simeq \textstyle\bigwedge^\bullet(L_+\mathfrak{h})/(\mathrm{Im}\, T + \mathbb{C}).$$

5.6.4 Restricting the cocycles

Any cocycle in $C^i_{\mathrm{loc}}(L\mathfrak{g}, \mathcal{A}^{\mathfrak{g}}_{0,\mathrm{loc}})$ is a cocycle in

$$C^i(L\mathfrak{g}, \mathcal{A}^{\mathfrak{g}}_{0,\mathrm{loc}}) = \mathrm{Hom}_{\mathrm{cont}}(\textstyle\bigwedge^i(L\mathfrak{g}), \mathcal{A}^{\mathfrak{g}}_{0,\mathrm{loc}}),$$

and as such it may be restricted to $\bigwedge^i(L\mathfrak{h})$.

The following lemma is the crucial result, which will enable us to show that ω and $i_*(\sigma)$ define the same cohomology class.

Lemma 5.6.7 *Any two cocycles in $C^i_{\mathrm{loc}}(L\mathfrak{g}, \mathcal{A}^{\mathfrak{g}}_{0,\mathrm{loc}})$, whose restrictions to $\bigwedge^i(L\mathfrak{h})$ coincide, represent the same cohomology class.*

Proof. We need to show that any cocycle ϕ in $C^i_{\mathrm{loc}}(L\mathfrak{g}, \mathcal{A}^{\mathfrak{g}}_{0,\mathrm{loc}})$, whose restriction to $\bigwedge^i(L\mathfrak{h})$ is equal to zero, is equivalent to the zero cocycle. According to the above computation, ϕ may be written as $\int Y(A,z)dz$, where A is a cocycle in $C^i(L_+\mathfrak{g}, M_{\mathfrak{g},+}) = \mathrm{Hom}_{\mathrm{cont}}(\bigwedge^i L_+\mathfrak{g}, M_{\mathfrak{g},+})$. But then the restriction of A to $\bigwedge^i L_+\mathfrak{h} \subset \bigwedge^i L_+\mathfrak{g}$, denoted by $\overline{A}$, must be in the image of the operator T, and hence so is $p(\overline{A}) = \mu(A) \in \bigwedge^i(L_+\mathfrak{h})^*$ (here p is the projection defined in (5.6.8) and μ is the map defined in (5.6.7)). Thus, $\mu(A) = T(h)$ for some $h \in \bigwedge^i(L_+\mathfrak{h})^*$. Since T commutes with the differential and the kernel of T on $\bigwedge^i(L_+\mathfrak{h})^*$ consists of the scalars, we obtain that h is also a cocycle in $C^i(L_+\mathfrak{h}, \mathbb{C}) = \bigwedge^i(L_+\mathfrak{h})^*$.

According to Lemma 5.6.6, the map μ induces an isomorphism on the cohomologies. Hence h is equal to $\mu(B)$ for some cocycle B in $C^i(L_+\mathfrak{g}, M_{\mathfrak{g},+})$. It is clear from the definition that μ commutes with the action of the translation operator T. Therefore it follows that the cocycles A and $T(B)$ are equivalent in $C^i(L_+\mathfrak{g}, M_{\mathfrak{g},+})$. But then ϕ is equivalent to the zero cocycle. □

Now we are ready to prove the main result of this chapter.

Theorem 5.6.8 *The cocycles ω and $i_*(\sigma)$ represent the same cohomology class in $H^2_{\mathrm{loc}}(L\mathfrak{g}, \mathcal{A}^{\mathfrak{g}}_{0,\mathrm{loc}})$. Therefore there exists a lifting of the homomorphism $L\mathfrak{g} \to \mathcal{T}^{\mathfrak{g}}_{\mathrm{loc}}$ to a homomorphism $\widehat{\mathfrak{g}}_{\kappa_c} \to \mathcal{A}^{\mathfrak{g}}_{\leq 1,\mathrm{loc}}$ such that $\mathbf{1} \mapsto 1$. Moreover, this homomorphism may be chosen in such a way that*

$$J^a(z) \mapsto Y(P_a(a^*_{\alpha,0}, a_{\beta,-1})|0\rangle, z) + Y(B_a, z), \tag{5.6.9}$$

where P_a is the polynomial introduced in formula (5.2.11) *and B_a is a polynomial in $a^*_{\alpha,n}, n \leq 0$, of degree* 1 *(with respect to the assignment* $\deg a^*_{\alpha,n} = -n$*).*

Proof. By Lemma 5.6.7, it suffices to check that the restrictions of ω and $i_*(\sigma)$ to $\bigwedge^2(L\mathfrak{h})$ coincide. We have

$$i_*(\sigma)(h_n, h'_m) = n\kappa_c(h,h')\delta_{n,-m}$$

for all $h, h' \in \mathfrak{h}$. Now let us compute the restriction of ω. We find from the formulas given in Section 5.2.5 and Section 5.3.2 that

$$\imath(h(z)) = -\sum_{\alpha\in\Delta_+} \alpha(h) :a^*_\alpha(z) a_\alpha(z): \tag{5.6.10}$$

(recall that $h(z) = \sum_{n\in\mathbb{Z}} h_n z^{-n-1}$). Therefore we find, in the same way as in the computation at the end of Section 5.5.3, that

$$\omega(h_n, h'_m) = -n\delta_{n,-m} \sum_{\alpha \in \Delta_+} \alpha(h)\alpha(h').$$

Now the key observation is that

$$\sum_{\alpha \in \Delta_+} \alpha(h)\alpha(h') = \kappa_c(h, h'), \tag{5.6.11}$$

because by definition $\kappa_c(\cdot, \cdot) = -\frac{1}{2}\kappa_{\mathfrak{g}}(\cdot, \cdot)$ and for the Killing form $\kappa_{\mathfrak{g}}$ we have

$$\kappa_{\mathfrak{g}}(h, h') = 2 \sum_{\alpha \in \Delta_+} \alpha(h)\alpha(h').$$

Therefore we find that

$$\omega(h_n, h'_m) = n\kappa_c(h, h')\delta_{n,-m},$$

and so the restrictions of the cocycles ω and $i_*(\sigma)$ to $\bigwedge^2(L\mathfrak{h})$ coincide. By Lemma 5.6.7, they represent the same class in $H^2_{\text{loc}}(L\mathfrak{g}, \mathcal{A}^{\mathfrak{g}}_{0,\text{loc}})$.

Hence there exists $\gamma \in C^1_{\text{loc}}(L\mathfrak{g}, \mathcal{A}^{\mathfrak{g}}_{0,\text{loc}})$ such that $\omega + d\gamma = i_*(\sigma)$. We may write γ as

$$\gamma = \sum_a \int \psi^*_a(z) Y(B_a, z)\, dz, \qquad B_a \in M_{\mathfrak{g},+}.$$

It follows from the computations made in the proof of Lemma 5.6.2 that both ω and $i_*(\sigma)$ are homogeneous elements of $C^2_{\text{loc}}(L\mathfrak{g}, \mathcal{A}^{\mathfrak{g}}_{0,\text{loc}})$ of degree 0. Therefore γ may be chosen to be of degree 0, i.e., B_a may be chosen to be of degree 1.

Then by Lemma 5.5.3 formulas (5.6.9) define a homomorphism of Lie algebras $\widehat{\mathfrak{g}}_{\kappa_c} \to \mathcal{A}^{\mathfrak{g}}_{\leq 1,\text{loc}}$ of the form (5.6.9) such that $\mathbf{1} \mapsto 1$. This completes the proof. □

5.6.5 Obstruction for other flag varieties

The key fact that we used in the proof above is that the inner product on $\mathfrak{h}$ defined by the left hand side of formula (5.6.11) is the restriction to $\mathfrak{h}$ of an invariant inner product on $\mathfrak{g}$. We then find that this invariant inner product is the critical inner product κ_c.

Let us observe that the necessary and sufficient condition for an inner product on $\mathfrak{h}$ to be equal to the restriction to $\mathfrak{h}$ of an invariant inner product on $\mathfrak{g}$ is that it is *W-invariant*, where W is the Weyl group of $\mathfrak{g}$ acting on $\mathfrak{h}$. In the case of the inner product defined by the left hand side of formula (5.6.11) this is obvious as the set Δ_+ is W-invariant. But this observation also singles out the "full" flag variety G/B_- as the only possible flag variety of G for which the cohomological obstruction may be overcome.

Indeed, consider a more general flag variety G/P, where P is a parabolic subgroup. For example, in the case of $\mathfrak{g} = \mathfrak{sl}_n$ conjugacy classes of parabolic subgroups correspond to non-trivial partitions of n. The standard (lower) parabolic subgroup corresponding to a partition $n = n_1 + \ldots + n_k$ consists of "block lower triangular" matrices with the blocks of sizes $n_1, \ldots, n_k$. The corresponding flag variety is the variety of flags $V_1 \subset \ldots \subset V_{k-1} \subset \mathbb{C}^n$, where $\dim V_i = n_1 + \ldots + n_i$.

Let P be a parabolic subgroup that is not a Borel subgroup and $\mathfrak{p}$ the corresponding Lie subalgebra of $\mathfrak{g}$. We may assume without loss of generality that $\mathfrak{p}$ contains $\mathfrak{b}_-$ and is invariant under the adjoint action of H. Then we have a decomposition $\mathfrak{g} = \mathfrak{p} \oplus \mathfrak{n}$, where $\mathfrak{n}$ is a Lie subalgebra of $\mathfrak{n}_+$, spanned by generators e_α, where α runs over a proper subset $\Delta_P \subset \Delta_+$. We can develop the same formalism for G/P as for G/B_-. The Lie subalgebra $\mathfrak{n}$ will now play the role of $\mathfrak{n}_+$: its Lie group is isomorphic to an open dense subset of G/P, etc. In particular, formula (5.6.10) is modified: instead of the summation over Δ_+ we have the summation only over Δ_S. Therefore in the corresponding cocycle we obtain the inner product on $\mathfrak{h}$ given by the formula

$$\nu_P(h, h') = \sum_{\alpha \in \Delta_P} \alpha(h)\alpha(h').$$

But no proper subset of Δ_+ is W-invariant. Therefore the inner product ν_P is *not* W-invariant and hence cannot be obtained as the restriction to $\mathfrak{h}$ of an invariant inner product on $\mathfrak{g}$. Therefore our argument breaks down, and we see that in the case of a parabolic subgroup P other than the Borel we cannot lift the homomorphism from $L\mathfrak{g}$ to the corresponding Lie algebra of vector fields to a homomorphism from a central extension $\widehat{\mathfrak{g}}_\kappa$ to the algebra of differential operators.

This illustrates the special role played by the full flag variety G/B_-. It is closely related to the fact that the first Pontryagin class of G/B_- vanishes while it is non-zero for other flag varieties, which leads to an "anomaly." Nevertheless, we will show in Section 6.3 how to overcome this anomaly in the general case. The basic idea is to add additional representations of the affinization of the Levi subgroup $\mathfrak{m}$ of $\mathfrak{p}$. By judiciously choosing the levels of these representations, we cancel the anomaly.

To summarize the results of this chapter, we have now obtained a "free field realization" homomorphism from $\widehat{\mathfrak{g}}_{\kappa_c}$ to the Lie algebra $\mathcal{A}^{\mathfrak{g}}_{\leq 1,\mathrm{loc}}$ of Fourier coefficients of vertex operators built from "free fields" $a_\alpha(z)$ and $a^*_\alpha(z)$ corresponding to differential operators on the flag variety G/B_-. In the next chapter we will interpret this homomorphism in the vertex algebra language.

6
Wakimoto modules

In this chapter we extend the free field realization constructed above to non-critical levels and recast it in the vertex algebra language. In these terms the free field realization amounts to a homomorphism from the vertex algebra $V_\kappa(\mathfrak{g})$ associated to $\widehat{\mathfrak{g}}_\kappa$ to a "free field" vertex algebra $M_\mathfrak{g} \otimes \pi_0^{\kappa-\kappa_c}$ associated to an infinite-dimensional Heisenberg Lie algebra.

Modules over the vertex algebra $M_\mathfrak{g} \otimes \pi_0^{\kappa-\kappa_c}$ now become $\widehat{\mathfrak{g}}_\kappa$-modules, which we call the Wakimoto modules. They were first constructed by M. Wakimoto [W] in the case $\mathfrak{g} = \widehat{\mathfrak{sl}}_2$ and by B. Feigin and the author [FF1, FF4] for general $\mathfrak{g}$. The Wakimoto modules may be viewed as representations of $\widehat{\mathfrak{g}}$ which are "semi-infinitely induced" from representations of its Heisenberg subalgebra $\widehat{\mathfrak{h}}$. In contrast to the usual induction, however, the level of the $\widehat{\mathfrak{h}}$-module gets shifted by the critical value κ_c. In particular, if we start with an $\widehat{\mathfrak{h}}$-module of level zero (e.g., a one-dimensional module corresponding to a character of the abelian Lie algebra $L\mathfrak{h}$), then the resulting $\widehat{\mathfrak{g}}$-module will be at the critical level. This is the reason why it is the critical level, and not the zero level, as one might naively expect, that is the "middle point" among all levels.

The Wakimoto modules (of critical and non-critical levels) will be crucial in achieving our ultimate goal: describing the center of $V_{\kappa_c}(\mathfrak{g})$ (see Chapter 8).

Here is a more detailed description of the material of this chapter. We start in Section 6.1 with the case of critical level. Our main result is the existence of a vertex algebra homomorphism $V_{\kappa_c}(\mathfrak{g}) \to M_\mathfrak{g} \otimes \pi_0$, where π_0 is the commutative vertex algebra associated to $L\mathfrak{h}$. Using this homomorphism, we construct a $\widehat{\mathfrak{g}}$-module structure of critical level on the tensor product $M_\mathfrak{g} \otimes N$, where N is an arbitrary (smooth) $L\mathfrak{h}$-module. These are the Wakimoto modules of critical level (we will see below that they are irreducible for generic values of the parameters).

In Section 6.2 we generalize this construction to arbitrary levels. We prove the existence of a homomorphism of vertex algebras

$$w_\kappa : V_\kappa(\mathfrak{g}) \to M_\mathfrak{g} \otimes \pi_0^{\kappa-\kappa_c},$$

where $\pi_0^{\kappa-\kappa_c}$ is the vertex algebra associated to the Heisenberg Lie algebra $\widehat{\mathfrak{h}}_{\kappa-\kappa_c}$. As in our earlier discussion of the center, we pay special attention to the action of the group $\operatorname{Aut}\mathcal{O}$ of coordinate changes on the Wakimoto modules. This is necessary to achieve a coordinate-independent construction of these modules, which we will need to obtain a coordinate-independent description of the center of $V_{\kappa_c}(\mathfrak{g})$. To do this, we describe the (quasi)conformal structure on $M_{\mathfrak{g}} \otimes \pi_0^{\kappa-\kappa_c}$ corresponding to the Segal–Sugawara (quasi)conformal structure on $V_\kappa(\mathfrak{g})$.

Next, we extend the construction of the Wakimoto modules in Section 6.3 to a more general context in which the Lie subalgebra $\widehat{\mathfrak{h}}$ is replaced by a central extension of the loop algebra of the Levi subalgebra of an arbitrary parabolic Lie subalgebra of $\mathfrak{g}$ following the ideas of [FF4]. Thus, we establish a "semi-infinite parabolic induction" pattern for representations of affine Kac–Moody algebras, similar to the parabolic induction for reductive groups over local non-archimedian fields.

6.1 Wakimoto modules of critical level

In this section we will recast the free field realization obtained in the previous chapter as a homomorphism of vertex algebras $V_{\kappa_c}(\mathfrak{g}) \to M_{\mathfrak{g}} \otimes \pi_0$, where π_0 is the commutative vertex algebra associated to $L\mathfrak{h}$.

6.1.1 Homomorphism of vertex algebras

Recall that to the affine Kac–Moody algebra $\widehat{\mathfrak{g}}_\kappa$ we associate the vacuum $\widehat{\mathfrak{g}}_\kappa$-module $V_\kappa(\mathfrak{g})$, as defined in Section 2.2.4. We have shown in Theorem 2.2.2 that $V_\kappa(\mathfrak{g})$ is a $\mathbb{Z}$-graded vertex algebra.

We wish to interpret a Lie algebra homomorphism $\widehat{\mathfrak{g}}_{\kappa_c} \to \mathcal{A}^{\mathfrak{g}}_{\leq 1,\mathrm{loc}}$ in terms of a homomorphism of vertex algebras.

Lemma 6.1.1 *Defining a homomorphism of $\mathbb{Z}$-graded vertex algebras $V_\kappa(\mathfrak{g}) \to V$ is equivalent to choosing vectors $\widetilde{J}^a_{-1}|0\rangle_V \in V, a = 1, \ldots, \dim\mathfrak{g}$, of degree 1 such that the Fourier coefficients $\widetilde{J}^a_n$ of the vertex operators*

$$Y(\widetilde{J}^a_{-1}|0\rangle_V, z) = \sum_{n\in\mathbb{Z}} \widetilde{J}^a_n z^{-n-1},$$

satisfy the relations (1.3.4) *with* $\mathbf{1} = 1$.

Proof. Given a homomorphism $\rho : V_\kappa(\mathfrak{g}) \to V$, set $\widetilde{J}^a_{-1}|0\rangle_V = \rho(J^a_{-1}|0\rangle)$. The fact that ρ is a homomorphism of vertex algebras implies that the OPEs of the vertex operators $Y(\widetilde{J}^a_{-1}|0\rangle_V, z)$ will be the same as those of the vertex operators $J^a(z)$, and hence the commutation relations of their Fourier coefficients $\widetilde{J}^a_n$ are the same as those of the J^a_n's.

Conversely, suppose that we are given vectors $\widetilde{J}^a_{-1}|0\rangle_V$ satisfying the condition of the lemma. Define a linear map $\rho : V_\kappa(\mathfrak{g}) \to V$ by the formula

$$J^{a_1}_{n_1} \ldots J^{a_m}_{n_m}|0\rangle \mapsto \widetilde{J}^{a_1}_{n_1} \ldots \widetilde{J}^{a_m}_{n_m}|0\rangle.$$

It is easy to check that this map is a homomorphism of $\mathbb{Z}$-graded vertex algebras. □

The vertex algebra $M_{\mathfrak{g}}$ is $\mathbb{Z}$-graded (see Section 5.4.1). The following is a corollary of Theorem 5.6.8.

Corollary 6.1.2 *There exists a homomorphism of $\mathbb{Z}$-graded vertex algebras $V_{\kappa_c}(\mathfrak{g}) \to M_{\mathfrak{g}}$.*

Proof. According to Theorem 5.6.8, there exists a homomorphism $\widehat{\mathfrak{g}}_{\kappa_c} \to U(M_{\mathfrak{g}})$ such that $\mathbf{1} \mapsto 1$ and

$$J^a(z) \mapsto Y(P_a(a^*_{\alpha,0}, a_{\beta,-1})|0\rangle, z) + Y(B_a, z),$$

where the vectors P_a and B_a have degree 1. Therefore by Lemma 6.1.1 we obtain a homomorphism of vertex algebras $V_{\kappa_c}(\mathfrak{g}) \to M_{\mathfrak{g}}$ such that

$$J^a_{-1}|0\rangle \mapsto P_a(a^*_{\alpha,0}, a_{\beta,-1})|0\rangle + B_a.$$

□

The complex $C^\bullet_{\mathrm{loc}}(L\mathfrak{g}, \mathcal{A}^{\mathfrak{g}}_{0,\mathrm{loc}})$ which we used to prove Theorem 5.6.8 carries a gradation with respect to the root lattice of $\mathfrak{g}$ defined by the formulas

$$\mathrm{wt}\, a^*_{\alpha,n} = -\alpha, \qquad \mathrm{wt}\, \psi^*_{a,n} = -\,\mathrm{wt}\, J^a, \qquad \mathrm{wt}\,|0\rangle = 0$$

so that $\mathrm{wt} \int Y(A,z)dz = \mathrm{wt}\, A$. The differential preserves this gradation, and it is clear that the cocycles ω and $i_*(\sigma)$ are homogeneous with respect to it. Therefore the element γ introduced in the proof of Theorem 5.6.8 may be chosen in such a way that it is also homogeneous with respect to the weight gradation. This means that $B_a \in M_{\mathfrak{g},+}$ may be chosen in such a way that $\mathrm{wt}\, B_a = \mathrm{wt}\, J^a$, in addition to the condition $\deg B_a = 1$.

We then necessarily have $B_a = 0$ for all $J^a \in \mathfrak{h} \oplus \mathfrak{n}_+$, because there are no elements of such weights in $M_{\mathfrak{g},+}$. Furthermore, we find that the term B_a corresponding to $J^a = f_i$ must be proportional to $a^*_{\alpha_i,-1}|0\rangle \in M_{\mathfrak{g},+}$. Using the formulas of Section 5.2.5 and the discussion of Section 5.3.2, we therefore obtain a more explicit description of the homomorphism $V_{\kappa_c}(\mathfrak{g}) \to M_{\mathfrak{g}}$:

Theorem 6.1.3 *There exist constants $c_i \in \mathbb{C}$ such that the Fourier coefficients* *of the vertex operators*

$$e_i(z) = a_{\alpha_i}(z) + \sum_{\beta\in\Delta_+} :P^i_\beta(a^*_\alpha(z))a_\beta(z):,$$
$$h_i(z) = -\sum_{\beta\in\Delta_+} \beta(h_i):a^*_\beta(z)a_\beta(z):,$$
$$f_i(z) = \sum_{\beta\in\Delta_+} :Q^i_\beta(a^*_\alpha(z))a_\beta(z): + c_i\partial_z a^*_{\alpha_i}(z),$$

where the polynomials P^i_β, Q^i_β are introduced in formulas (5.2.12)–(5.2.14), *generate an action of $\widehat{\mathfrak{g}}_{\kappa_c}$ on $M_{\mathfrak{g}}$.*

In addition to the above homomorphism of Lie algebras $w_{\kappa_c} : \widehat{\mathfrak{g}}_{\kappa_c} \to \mathcal{A}^{\mathfrak{g}}_{\leq 1,\mathrm{loc}}$, there is also a Lie algebra anti-homomorphism

$$w^R : L\mathfrak{n}_+ \to \mathcal{A}^{\mathfrak{g}}_{\leq 1,\mathrm{loc}}$$

which is induced by the right action of $\mathfrak{n}_+$ on N_+ (see Section 5.2.5). By construction, the images of $L\mathfrak{n}_+$ under w_{κ_c} and w^R commute. We have

$$w^R(e_i(z)) = a_{\alpha_i}(z) + \sum_{\beta\in\Delta_+} P^{R,i}_\beta(a^*_\gamma(z))a_\beta(z), \tag{6.1.1}$$

where the polynomials $P^{R,i}_\beta$ were defined in Section 5.2.5. More generally, we have

$$w_{\kappa_c}(e_\alpha(z)) = a_\alpha(z) + \sum_{\beta\in\Delta_+;\beta>\alpha} P^\alpha_\beta(a^*_\gamma(z))a_\beta(z), \tag{6.1.2}$$
$$w^R(e_\alpha(z)) = a_\alpha(z) + \sum_{\beta\in\Delta_+;\beta>\alpha} P^{R,\alpha}_\beta(a^*_\gamma(z))a_\beta(z), \tag{6.1.3}$$

for some polynomials P^α_β and $P^{R,\alpha}_\beta$, where the condition $\beta > \alpha$ comes from the commutation relations

$$[h, e_\alpha] = \alpha(h)e_\alpha, \qquad [h, e^R_\alpha] = \alpha(h)e^R_\alpha, \qquad h \in \mathfrak{h}.$$

Note that there is no need to put normal ordering in the above three formulas, because P^α_β and $P^{R,\alpha}_\beta$ cannot contain a^*_β, for the same reason.

6.1.2 Other $\widehat{\mathfrak{g}}$-module structures on $M_{\mathfrak{g}}$

In Theorem 6.1.3 we constructed the structure of a $\widehat{\mathfrak{g}}_{\kappa_c}$-module on $M_{\mathfrak{g}}$. To obtain other $\widehat{\mathfrak{g}}_{\kappa_c}$-module structures of critical level on $M_{\mathfrak{g}}$, we need to consider other homomorphisms $\widehat{\mathfrak{g}}_{\kappa_c} \to \mathcal{A}^{\mathfrak{g}}_{\leq 1,\mathrm{loc}}$ lifting the homomorphism $L\mathfrak{g} \to \mathcal{T}^{\mathfrak{g}}_{\mathrm{loc}}$. According to Lemma 5.5.3, the set of isomorphism classes of such liftings is a torsor over $H^1(L\mathfrak{g}, \mathcal{A}^{\mathfrak{g}}_{0,\mathrm{loc}})$, which is the first cohomology of the "big" Chevalley complex

$$C^i(L\mathfrak{g}, \mathcal{A}^{\mathfrak{g}}_{0,\mathrm{loc}}) = \mathrm{Hom}_{\mathrm{cont}}(\textstyle\bigwedge^i(L\mathfrak{g}), \mathcal{A}^{\mathfrak{g}}_{0,\mathrm{loc}}).$$

Recall that our complex $C^\bullet_{\mathrm{loc}}(L\mathfrak{g}, \mathcal{A}^{\mathfrak{g}}_{0,\mathrm{loc}})$ has a weight gradation, and our cocycle ω has weight 0. Therefore among all liftings we consider those which have weight 0. The set of such liftings is in bijection with the weight 0 homogeneous component of $H^1(L\mathfrak{g}, \mathcal{A}^{\mathfrak{g}}_{0,\mathrm{loc}})$.

Lemma 6.1.4 *The weight* 0 *component of* $H^1(L\mathfrak{g}, \mathcal{A}^{\mathfrak{g}}_{0,\mathrm{loc}})$ *is isomorphic to the (topological) dual space* $(L\mathfrak{h})^*$ *to* $L\mathfrak{h}$.

Proof. First we show that the weight 0 component of $H^1(L\mathfrak{g}, \mathcal{A}^{\mathfrak{g}}_{0,\mathrm{loc}})$ is isomorphic to the space of weight 0 cocycles in $C^1(L\mathfrak{g}, \mathcal{A}^{\mathfrak{g}}_{0,\mathrm{loc}})$. Indeed, since $\mathrm{wt}\, a^*_{\alpha,n} = -\alpha$, the weight 0 part of $C^0(L\mathfrak{g}, \mathcal{A}^{\mathfrak{g}}_{0,\mathrm{loc}}) = \mathcal{A}^{\mathfrak{g}}_{0,\mathrm{loc}}$ is one-dimensional and consists of the constants. If we apply the differential to a constant, we obtain 0, and so there are no coboundaries of weight 0 in $C^1(L\mathfrak{g}, \mathcal{A}^{\mathfrak{g}}_{0,\mathrm{loc}})$.

Next, we show that any weight 0 one-cocycle ϕ is uniquely determined by its restriction to $L\mathfrak{h} \subset L\mathfrak{g}$, which may be an arbitrary (continuous) linear functional on $L\mathfrak{h}$. Indeed, since the weights occurring in $\mathcal{A}^{\mathfrak{g}}_{0,\mathrm{loc}}$ are less than or equal to 0, the restriction of ϕ to $L_+\mathfrak{n}_-$ is equal to 0, and the restriction to $L\mathfrak{h}$ takes values in the constants $\mathbb{C} \subset \mathcal{A}^{\mathfrak{g}}_{0,\mathrm{loc}}$. Now let us fix $\phi|_{L\mathfrak{h}}$. We identify $(L\mathfrak{h})^*$ with $\mathfrak{h}^*((t))dt$ using the residue pairing, and write $\phi|_{L\mathfrak{h}}$ as $\chi(t)dt$ using this identification. Here

$$\chi(t) = \sum_{n\in\mathbb{Z}} \chi_n t^{-n-1}, \qquad \chi_n \in \mathfrak{h}^*,$$

where $\chi_n(h) = \phi(h_n)$. We denote $\langle \chi(t), h_i \rangle$ by $\chi_i(t)$.

We claim that for any $\chi(t)dt \in \mathfrak{h}^*((t))dt$, there is a unique one-cocycle ϕ of weight 0 in $C^1(L\mathfrak{g}, \mathcal{A}^{\mathfrak{g}}_{0,\mathrm{loc}})$ such that

$$\phi(e_i(z)) = 0, \tag{6.1.4}$$

$$\phi(h_i(z)) = \chi_i(z), \tag{6.1.5}$$

$$\phi(f_i(z)) = \chi_i(z) a^*_{\alpha,n}(z). \tag{6.1.6}$$

Indeed, having fixed $\phi(e_i(z))$ and $\phi(h_i(z))$ as in (6.1.4) and (6.1.5), we obtain using the formula

$$\phi\left([e_{i,n}, f_{j,m}]\right) = e_{i,n} \cdot \phi(f_{j,m}) - f_{j,m} \cdot \phi(e_{i,n})$$

that

$$\delta_{i,j} \chi_{i,n+m} = e_{i,n} \cdot \phi(f_{j,m}).$$

This equation on $\phi(f_{j,m})$ has a unique solution in $\mathcal{A}^{\mathfrak{g}}_{0,\mathrm{loc}}$ of weight $-\alpha_i$, namely, the one given by formula (6.1.6). The cocycle ϕ, if it exists, is uniquely determined once we fix its values on $e_{i,n}, h_{i,n}$, and $f_{i,n}$. Let us show that it exists. This is equivalent to showing that the Fourier coefficients of the fields

$$e_i(z) = a_{\alpha_i}(z) + \sum_{\beta \in \Delta_+} {:}P^i_\beta(a^*_\alpha(z))a_\beta(z){:}, \tag{6.1.7}$$

$$h_i(z) = -\sum_{\beta \in \Delta_+} \beta(h_i){:}a^*_\beta(z)a^*_\beta(z){:} + \chi_i(z), \tag{6.1.8}$$

$$f_i(z) = \sum_{\beta \in \Delta_+} {:}Q^i_\beta(a^*_\alpha(z))a_\beta(z){:} + c_i\partial_z a^*_{\alpha_i}(z) + \chi_i(z)a^*_{\alpha,n}(z), \tag{6.1.9}$$

satisfy the relations of $\widehat{\mathfrak{g}}_{\kappa_c}$ with $\mathbf{1} = 1$.

Let us remove the normal ordering and set $c_i = 0, i = 1, \ldots, \ell$. Then the corresponding Fourier coefficients are no longer well-defined as linear operators on $M_{\mathfrak{g}}$. But they are well-defined linear operators on the space $\operatorname{Fun} L\mathcal{U}$. The resulting $L\mathfrak{g}$-module structure is easy to describe. Indeed, the Lie algebra $L\mathfrak{g}$ acts on $\operatorname{Fun} L\mathcal{U}$ by vector fields. More generally, for any $L\mathfrak{b}_-$-module R we obtain a natural action of $L\mathfrak{g}$ on the tensor product $\operatorname{Fun} L\mathcal{U} \widehat{\otimes} R$ (this is just the topological $L\mathfrak{g}$-module induced from the $L\mathfrak{b}_-$-module R). If we choose as R the one-dimensional representation on which all $f_{i,n}$ act by 0 and $h_{i,n}$ acts by multiplication by $\chi_{i,n}$ for all $i = 1, \ldots, \ell$ and $n \in \mathbb{Z}$, then the corresponding $L\mathfrak{g}$-action on $\operatorname{Fun} L\mathcal{U} \widehat{\otimes} R \simeq \operatorname{Fun} L\mathcal{U}$ is given by the Fourier coefficients of the above formulas, but with the normal ordering removed and $c_i = 0$. Hence if we remove the normal ordering and set $c_i = 0$, then these Fourier coefficients do satisfy the commutation relations of $L\mathfrak{g}$.

When we restore the normal ordering, these commutation relations may in general be distorted, due to the double contractions, as explained in Section 5.5.2. But we know from Theorem 6.1.3 that when we restore normal ordering and set all $\chi_{i,n} = 0$, then there exist the numbers c_i such that these Fourier coefficients satisfy the commutation relations of $\widehat{\mathfrak{g}}_{\kappa_c}$ with $\mathbf{1} = 1$. The terms (6.1.4)–(6.1.6) will not generate any new double contractions in the commutators. Therefore the commutation relations of $\widehat{\mathfrak{g}}_{\kappa_c}$ are satisfied for all non-zero values of $\chi_{i,n}$. This completes the proof. □

Corollary 6.1.5 *For each $\chi(t) \in \mathfrak{h}^*((t))$ there is a $\widehat{\mathfrak{g}}$-module structure of critical level on $M_{\mathfrak{g}}$, with the action given by formulas* (6.1.7)–(6.1.9).

We call these modules the **Wakimoto modules of critical level** and denote them by $W_{\chi(t)}$.

Let π_0 be the commutative algebra $\mathbb{C}[b_{i,n}]_{i=1,\ldots,\ell;n<0}$ with the derivation T given by the formula

$$T \cdot b_{i_1,n_1} \ldots b_{i_m,n_m} = -\sum_{j=1}^{m} n_j b_{i_1,n_1} \ldots b_{i_j,n_j-1} \cdots b_{i_m,n_m}.$$

Then π_0 is naturally a commutative vertex algebra (see Section 2.2.2). In particular, we have

$$Y(b_{i,-1}, z) = b_i(z) = \sum_{n<0} b_{i,n} z^{-n-1}.$$

Using the same argument as in the proof of Lemma 6.1.4, we now obtain a stronger version of Corollary 6.1.2.

Theorem 6.1.6 *There exists a homomorphism of vertex algebras*

$$w_{\kappa_c} : V_{\kappa_c}(\mathfrak{g}) \to M_{\mathfrak{g}} \otimes \pi_0$$

such that

$$e_i(z) \mapsto a_{\alpha_i}(z) + \sum_{\beta \in \Delta_+} :P^i_\beta(a^*_\alpha(z)) a_\beta(z):,$$

$$h_i(z) \mapsto - \sum_{\beta \in \Delta_+} \beta(h_i) :a^*_\beta(z) a_\beta(z): + b_i(z),$$

$$f_i(z) \mapsto \sum_{\beta \in \Delta_+} :Q^i_\beta(a^*_\alpha(z)) a_\beta(z): + c_i \partial_z a^*_{\alpha_i}(z) + b_i(z) a^*_{\alpha_i}(z),$$

where the polynomials P^i_β, Q^i_β are introduced in formulas (5.2.12)–(5.2.14).

Thus, any module over the vertex algebra $M_{\mathfrak{g}} \otimes \pi_0$ becomes a $V_{\kappa_c}(\mathfrak{g})$-module, and hence a $\widehat{\mathfrak{g}}$-module of critical level. We will not require that the module is necessarily $\mathbb{Z}$-graded. In particular, for any $\chi(t) \in \mathfrak{h}^*((t))$ we have a one-dimensional π_0-module $\mathbb{C}_{\chi(t)}$, on which $b_{i,n}$ acts by multiplication by $\chi_{i,n}$. The corresponding $\widehat{\mathfrak{g}}_{\kappa_c}$-module is the Wakimoto module $W_{\chi(t)}$ introduced above.

6.2 Deforming to other levels

In this section we extend the construction of Wakimoto modules to non-critical levels. We prove the existence of a homomorphism of vertex algebras

$$w_\kappa : V_\kappa(\mathfrak{g}) \to M_{\mathfrak{g}} \otimes \pi_0^{\kappa - \kappa_c},$$

where $\pi_0^{\kappa-\kappa_c}$ is the vertex algebra associated to the Heisenberg Lie algebra $\widehat{\mathfrak{h}}_{\kappa-\kappa_c}$, and analyze in detail the compatibility of this homomorphism with the action of the group $\operatorname{Aut} \mathcal{O}$ of changes of coordinate on the disc. This will enable us to give a coordinate-independent interpretation of the free field realization.

6.2.1 Homomorphism of vertex algebras

As before, we denote by $\mathfrak{h}$ the Cartan subalgebra of $\mathfrak{g}$. Let $\widehat{\mathfrak{h}}_\kappa$ be the one-dimensional central extension of the loop algebra $L\mathfrak{h} = \mathfrak{h}((t))$ with the two-cocycle obtained by restriction of the two-cocycle on $L\mathfrak{g}$ corresponding to the inner product κ. Then according to formula (1.3.4), $\widehat{\mathfrak{h}}_\kappa$ is a Heisenberg Lie algebra. We will consider a copy of this Lie algebra with generators $b_{i,n}, i = 1, \ldots, \ell, n \in \mathbb{Z}$, and $\mathbf{1}$ with the commutation relations

$$[b_{i,n}, b_{j,m}] = n\kappa(h_i, h_j)\delta_{n,-m}\mathbf{1}. \tag{6.2.1}$$

Thus, the $b_{i,n}$'s satisfy the same relations as the $h_{i,n}$'s. Let π_0^κ denote the $\widehat{\mathfrak{h}}_\kappa$-module induced from the one-dimensional representation of the abelian Lie subalgebra of $\widehat{\mathfrak{h}}_\kappa$ spanned by $b_{i,n}, i = 1, \ldots, \ell, n \geq 0$, and $\mathbf{1}$, on which $\mathbf{1}$ acts as the identity and all other generators act by 0. We denote by $|0\rangle$ the generating vector of this module. It satisfies $b_{i,n}|0\rangle = 0, n \geq 0$. Then π_0^κ has the following structure of a $\mathbb{Z}_+$-graded vertex algebra (see Theorem 2.3.7 of [FB]):

- $\mathbb{Z}_+$-grading: $\deg b_{i_1,n_1} \ldots b_{i_m,n_m}|0\rangle = -\sum_{i=1}^m n_i$;
- vacuum vector: $|0\rangle$;
- translation operator: $T|0\rangle = 0$, $[T, b_{i,n}] = -nb_{i,n-1}$;
- vertex operators:

$$Y(b_{i,-1}|0\rangle, z) = b_i(z) = \sum_{n\in\mathbb{Z}} b_{i,n}z^{-n-1},$$

$$Y(b_{i_1,n_1} \ldots b_{i_m,n_m}|0\rangle, z) = \prod_{j=1}^{n} \frac{1}{(-n_j-1)!} :\partial_z^{-n_1-1}b_{i_1}(z) \ldots \partial_z^{-n_m-1}b_{i_m}(z): .$$

The tensor product $M_\mathfrak{g} \otimes \pi_0^{\kappa-\kappa_c}$ also acquires a vertex algebra structure.

The following theorem extends the result of Theorem 6.1.6 away from the critical level.

Theorem 6.2.1 *There exists a homomorphism of vertex algebras*

$$w_\kappa : V_\kappa(\mathfrak{g}) \to M_\mathfrak{g} \otimes \pi_0^{\kappa-\kappa_c}$$

such that

$$\begin{aligned}
e_i(z) &\mapsto a_{\alpha_i}(z) + \sum_{\beta\in\Delta_+} :P^i_\beta(a^*_\alpha(z))a_\beta(z): , \\
h_i(z) &\mapsto -\sum_{\beta\in\Delta_+} \beta(h_i):a^*_\beta(z)a_\beta(z): + b_i(z), \qquad (6.2.2)\\
f_i(z) &\mapsto \sum_{\beta\in\Delta_+} :Q^i_\beta(a^*_\alpha(z))a_\beta(z): + (c_i + (\kappa-\kappa_c)(e_i, f_i))\,\partial_z a^*_{\alpha_i}(z) + b_i(z)a^*_{\alpha_i}(z),
\end{aligned}$$

where the polynomials P^i_β, Q^i_β *were introduced in formulas* (5.2.12)–(5.2.14).

Proof. Denote by $\widetilde{\mathcal{A}}^\mathfrak{g}_{\mathrm{loc}}$ the Lie algebra $U(M_\mathfrak{g} \otimes \pi_0^{\kappa-\kappa_c})$. By Lemma 6.1.1, in order to prove the theorem, we need to show that formulas (6.2.2) define a homomorphism of Lie algebras $\widehat{\mathfrak{g}}_\kappa \to \widetilde{\mathcal{A}}^{\mathfrak{g}}_{\mathrm{loc}}$ sending the central element $\mathbf{1}$ to the identity.

Formulas (6.2.2) certainly define a linear map $\overline{w}_\kappa : L\mathfrak{g} \to \widetilde{\mathcal{A}}^{\mathfrak{g}}_{\mathrm{loc}}$. Denote by ω_κ the linear map $\bigwedge^2 L\mathfrak{g} \to \widetilde{\mathcal{A}}^{\mathfrak{g}}_{\mathrm{loc}}$ defined by the formula

$$\omega_\kappa(f,g) = [\overline{w}_\kappa(f), \overline{w}_\kappa(g)] - \overline{w}_\kappa([f,g]).$$

Evaluating it explicitly in the same way as in the proof of Lemma 5.6.2, we find that ω_κ takes values in $\mathcal{A}^{\mathfrak{g}}_{0,\mathrm{loc}} \subset \widetilde{\mathcal{A}}^{\mathfrak{g}}_{\mathrm{loc}}$. Furthermore, by construction of $\overline{w}_\kappa$, for any $X \in \mathcal{A}^{\mathfrak{g}}_{0,\mathrm{loc}}$ and $f \in L\mathfrak{g}$ we have $[\overline{w}_\kappa(f), X] = f \cdot X$, where in the right hand side we consider the action of f on the $L\mathfrak{g}$-module $\mathcal{A}^{\mathfrak{g}}_{0,\mathrm{loc}}$. This immediately implies that ω_κ is a two-cocycle of $L\mathfrak{g}$ with coefficients in $\mathcal{A}^{\mathfrak{g}}_{0,\mathrm{loc}}$. By construction, ω_κ is local, i.e., belongs to $C^2_{\mathrm{loc}}(L\mathfrak{g}, \mathcal{A}^{\mathfrak{g}}_{0,\mathrm{loc}})$.

Let us compute the restriction of ω_κ to $\bigwedge^2 L\mathfrak{h}$. The calculations made in the proof of Lemma 5.6.8 imply that

$$\omega_\kappa(h_n, h'_m) = n(\kappa_c(h,h') + (\kappa - \kappa_c)(h,h')) = n\kappa(h,h').$$

Therefore this restriction is equal to the restriction of the Kac–Moody two-cocycle σ_κ on $L\mathfrak{g}$ corresponding to κ. Now Lemma 5.6.7 implies that the two-cocycle ω_κ is cohomologically equivalent to $i_*(\sigma_\kappa)$. We claim that it is actually equal to $i_*(\sigma_\kappa)$.

Indeed, the difference between these cocycles is the coboundary of some element $\gamma \in C^1_{\mathrm{loc}}(L\mathfrak{g}, \mathcal{A}^{\mathfrak{g}}_{0,\mathrm{loc}})$. The discussion before Theorem 6.1.3 implies that $\gamma(e_i(z)) = \gamma(h_i(z)) = 0$ and $\gamma(f_i(z)) = c'_i \partial_z a^*_{\alpha_i}(z)$ for some constants $c'_i \in \mathbb{C}$. In order to find the constants c'_i we compute the value of the corresponding two-cocycle $\omega_\kappa + d\gamma$ on $e_{i,n}$ and $f_{i,-n}$. We find that it is equal to $n\sigma_\kappa(e_{i,n}, f_{i,-n}) + c'_i n$. On the other hand, the commutation relations in $\widehat{\mathfrak{g}}_\kappa$ require that it be equal to $n\sigma_\kappa(e_{i,n}, f_{i,-n})$. Therefore $c'_i = 0$ for all $i = 1, \ldots, \ell$, and so $\gamma = 0$. Hence we have $\omega_\kappa = i_*(\sigma_\kappa)$. This implies that formulas (6.2.2) indeed define a homomorphism of Lie algebras $\widehat{\mathfrak{g}}_\kappa \to \mathcal{A}^{\mathfrak{g}}_{\leq 1,\mathrm{loc}}$ sending the central element $\mathbf{1}$ to the identity. This completes the proof. □

Now any module over the vertex algebra $M_{\mathfrak{g}} \otimes \pi_0^{\kappa-\kappa_c}$ becomes a $V_\kappa(\mathfrak{g})$-module and hence a $\widehat{\mathfrak{g}}_\kappa$-module (with K acting as 1). For $\lambda \in \mathfrak{h}^*$, let $\pi_\lambda^{\kappa-\kappa_c}$ be the Fock representation of $\widehat{\mathfrak{h}}_{\kappa-\kappa_c}$ generated by a vector $|\lambda\rangle$ satisfying

$$b_{i,n}|\lambda\rangle = 0, \quad n > 0, \qquad b_{i,0}|\lambda\rangle = \lambda(h_i)|\lambda\rangle, \qquad \mathbf{1}|\lambda\rangle = |\lambda\rangle.$$

Then

$$W_{\lambda,\kappa} \stackrel{\mathrm{def}}{=} M_{\mathfrak{g}} \otimes \pi_\lambda^{\kappa-\kappa_c}$$

is an $M_{\mathfrak{g}} \otimes \pi_0^{\kappa-\kappa_c}$-module, and hence a $\widehat{\mathfrak{g}}_\kappa$-module. We call it the **Wakimoto module of level κ and highest weight λ.**

6.2.2 Wakimoto modules over $\widehat{\mathfrak{sl}}_2$

In this section we describe explicitly the Wakimoto modules over $\widehat{\mathfrak{sl}}_2$.

Let $\{e, h, f\}$ be the standard basis of the Lie algebra $\mathfrak{sl}_2$. Let κ_0 be the invariant inner product on $\mathfrak{sl}_2$ normalized in such a way that $\kappa_0(h,h) = 2$. We will write an arbitrary invariant inner product κ on $\mathfrak{sl}_2$ as $k\kappa_0$, where $k \in \mathbb{C}$, and will use k in place of κ in our notation. In particular, κ_c corresponds to $k = -2$. The set Δ_+ consists of one element in the case of $\mathfrak{sl}_2$, so we will drop the index α in $a_\alpha(z)$ and $a^*_\alpha(z)$. Likewise, we will drop the index i in $b_i(z)$, etc., and will write M for $M_{\mathfrak{sl}_2}$. We will also identify the dual space to the Cartan subalgebra $\mathfrak{h}^*$ with $\mathbb{C}$ by sending $\chi \in \mathfrak{h}^*$ to $\chi(h)$.

The Weyl algebra $\mathcal{A}_{\mathfrak{sl}_2}$ has generators $a_n, a^*_n, n \in \mathbb{Z}$, with the commutation relations

$$[a_n, a^*_m] = \delta_{n,-m}.$$

Its Fock representation is denoted by M. The Heisenberg Lie algebra $\widehat{\mathfrak{h}}_k$ has generators $b_n, n \in \mathbb{Z}$, and $\mathbf{1}$, with the commutation relations

$$[b_n, b_m] = 2kn\delta_{n,-m}\mathbf{1},$$

and π^k_λ is its Fock representation generated by a vector $|\lambda\rangle$ such that

$$b_n|\lambda\rangle = 0, \quad n > 0; \qquad b_0|\lambda\rangle = \lambda|\lambda\rangle; \qquad \mathbf{1}|\lambda\rangle = |\lambda\rangle.$$

The module π^k_0 and the tensor product $M \otimes \pi^k_0$ are vertex algebras.

The homomorphism $w_k : V_k(\mathfrak{sl}_2) \to M \otimes \pi^{k+2}_0$ of vertex algebras is given by the following formulas:

$$\begin{aligned} e(z) &\mapsto a(z) \\ h(z) &\mapsto -2{:}a^*(z)a(z){:} + b(z) \\ f(z) &\mapsto -{:}a^*(z)^2a(z){:} + k\partial_z a^*(z) + a^*(z)b(z). \end{aligned} \tag{6.2.3}$$

The Wakimoto module $M \otimes \pi^{k+2}_\lambda$ will be denoted by $W_{\lambda,k}$ and its highest weight vector will be denoted by $|\lambda\rangle$.

We will use these explicit formulas in Chapter 7 in order to construct intertwining operators between Wakimoto modules.

6.2.3 Conformal structures at non-critical levels

In this section we show that the homomorphism w_κ of Theorem 6.2.1 is a homomorphism of conformal vertex algebras when $\kappa \neq \kappa_c$ (see Section 3.1.4 for the definition of conformal vertex algebras). This will allow us to obtain a coordinate-independent version of this homomorphism. By taking the limit $\kappa \to \kappa_c$, we will also obtain a coordinate-independent version of the homomorphism w_{κ_c}.

The vertex algebra $V_\kappa(\mathfrak{g}), \kappa \neq \kappa_c$, has the structure of a conformal vertex algebra given by the Segal–Sugawara vector

$$\mathbf{S}_\kappa = \frac{1}{2}\sum_a J^a_{-1} J_{a,-1}|0\rangle, \tag{6.2.4}$$

where $\{J_a\}$ is the basis of $\mathfrak{g}$ dual to the basis $\{J^a\}$ with respect to the inner product $\kappa - \kappa_c$ (see Section 3.1.4, where this vector was denoted by $\widetilde{S}$). We need to calculate the image of $\mathbf{S}_\kappa$ under w_κ.

Proposition 6.2.2 *The image of* $\mathbf{S}_\kappa$ *under* w_κ *is equal to*

$$\left(\sum_{\alpha\in\Delta_+} a_{\alpha,-1}a^*_{\alpha,-1} + \frac{1}{2}\sum_{i=1}^{\ell} b_{i,-1}b^i_{-1} - \rho_{-2}\right)|0\rangle, \tag{6.2.5}$$

where $\{b^i\}$ *is a dual basis to* $\{b_i\}$ *and* ρ *is the element of* $\mathfrak{h}$ *corresponding to* $\rho \in \mathfrak{h}^*$ *under the isomorphism induced by the inner product* $(\kappa - \kappa_c)|_{\mathfrak{h}}$.

The vector $w_\kappa(\mathbf{S}_\kappa)$ *defines the structure of a conformal algebra and hence an action of the Virasoro algebra on* $\mathcal{M}_\mathfrak{g} \otimes \pi_0^{\kappa-\kappa_0}$. *The homomorphism* w_κ *intertwines the corresponding action of* $(\operatorname{Der}\mathcal{O}, \operatorname{Aut}\mathcal{O})$ *on* $\mathcal{M}_\mathfrak{g} \otimes \pi_0^{\kappa-\kappa_0}$ *and the natural action of* $(\operatorname{Der}\mathcal{O}, \operatorname{Aut}\mathcal{O})$ *on* $V_\kappa(\mathfrak{g})$.

Proof. The vertex algebras $V_\kappa(\mathfrak{g})$ and $M_\mathfrak{g} \otimes \pi_0^{\kappa-\kappa_c}$ carry $\mathbb{Z}$-gradings and also weight gradings by the root lattice of $\mathfrak{g}$ coming from the action of $\mathfrak{h}$ (note that $\operatorname{wt} a_{\alpha,n} = -\operatorname{wt} a^*_{\alpha,n} = \alpha, \operatorname{wt} b_{i,n} = 0$). The homomorphism w_κ preserves these gradings. The vector $\mathbf{S}_\kappa \in V_\kappa(\mathfrak{g})$ is of degree 2 with respect to the $\mathbb{Z}$-grading and of weight 0 with respect to the root lattice grading. Hence the same is true for $w_\kappa(\mathbf{S}_\kappa)$.

The basis in the corresponding homogeneous subspace of $M_\mathfrak{g} \otimes \pi_0^{\kappa-\kappa_c}$ is formed by the monomials of the form

$$b_{i,-1}b_{j,-1}, \qquad b_{i,-2}, \qquad a_{\alpha,-1}a^*_{\alpha,-1}, \tag{6.2.6}$$

$$a_{\alpha,-1}a_{\beta,-1}a^*_{\alpha,0}a^*_{\beta,0}, \qquad a_{\alpha,-1}a^*_{\alpha,0}b_{i,-1}, \tag{6.2.7}$$

$$a_{\alpha,-2}a^*_{\alpha,0}, \qquad a_{\alpha+\beta,-2}a^*_{\alpha,0}a^*_{\beta,0}, \qquad a_{\alpha+\beta,-1}a^*_{\alpha,-1}a^*_{\beta,0}, \tag{6.2.8}$$

applied to the vacuum vector $|0\rangle$. We want to show that $w_\kappa(\mathbf{S}_\kappa)$ is a linear combination of the monomials (6.2.6).

By construction, the Fourier coefficients $L_n, n \in \mathbb{Z}$, of the vertex operator $w_\kappa(\mathbf{S}_\kappa)$ preserve the weight grading on $M_\mathfrak{g} \otimes \pi_0^{\kappa-\kappa_c}$, and we have $\deg L_n = -n$ with respect to the vertex algebra grading on $M_\mathfrak{g} \otimes \pi_0^{\kappa-\kappa_c}$. We claim that any vector of the form $P(a^*_{\alpha,0})|0\rangle$ is annihilated by $L_n, n \geq 0$. This is clear for $n > 0$ for degree reasons.

To see that the same is true for L_0, observe that

$$L_0 \cdot P(a^*_{\alpha,0})|0\rangle = \frac{1}{2} \sum_a w_\kappa(J_{a,0}) w_\kappa(J^a_0) \cdot P(a^*_{\alpha,0})|0\rangle.$$

According to the formulas for the homomorphism w_κ given in Theorem 6.2.1, the action of the constant subalgebra $\mathfrak{g} \subset \widehat{\mathfrak{g}}_\kappa$ on $\mathbb{C}[a^*_{\alpha,0}]_{\alpha\in\Delta_+}|0\rangle$, obtained via w_κ, coincides with the natural action of $\mathfrak{g}$ on $\mathbb{C}[y_\alpha]_{\alpha\in\Delta_+} = \operatorname{Fun} N_+$, if we substitute $a^*_{\alpha,0} \mapsto y_\alpha$. Therefore the action of L_0 on $P(a^*_{\alpha,0})|0\rangle$ coincides with the action of the Casimir operator $\frac{1}{2}\sum_a J_a J^a$ on $\mathbb{C}[y_\alpha]_{\alpha\in\Delta_+}$, under this substitution. But as a $\mathfrak{g}$-module, the latter is the contragredient Verma module M^*_0 as we showed in Section 5.2.3. Therefore the action of the Casimir operator on it is equal to 0.

The fact that all vectors of the form $P(a^*_{\alpha,0})|0\rangle$ are annihilated by $L_n, n \geq 0$, precludes the monomials (6.2.7) and the monomials (6.2.8), except for the last one, from appearing in $w_\kappa(\mathbf{S}_\kappa)$. In order to eliminate the last monomial in (6.2.8) we will prove that

$$L_n \cdot a_{\alpha,-1}|0\rangle = 0, \quad n > 0 \qquad L_0 \cdot a_{\alpha,-1}|0\rangle = a_{\alpha,-1}|0\rangle, \qquad \alpha \in \Delta_+.$$

The first formula holds for degree reasons. To show the second formula, let $\{e_\alpha\}_{\alpha\in\Delta_+}$ be a root basis of $\mathfrak{n}_+ \subset \mathfrak{g}$ such that $e_{\alpha_i} = e_i$. Then the vectors $e_{\alpha,-1}|0\rangle \in V_\kappa(\mathfrak{g})$, and hence $w_\kappa(e_{\alpha,-1}|0\rangle) \in M_\mathfrak{g} \otimes \pi_0^{\kappa-\kappa_c}$ are annihilated by $L_n, n > 0$, and are eigenvectors of L_0 with eigenvalue 1. We have

$$w_\kappa(e_{\alpha,-1}|0\rangle) = a_{\alpha,-1}|0\rangle + \sum_{\beta\in\Delta_+;\beta>\alpha} P^\alpha_\beta(a^*_{\alpha,0}) a_{\beta,-1}|0\rangle, \tag{6.2.9}$$

where the polynomials P^α_β are found from the following formula for the action of e_α on $\mathcal{U}$:

$$e_\alpha = \frac{\partial}{\partial y_\alpha} + \sum_{\beta\in\Delta_+;\beta>\alpha} P^\alpha_\beta(y_\alpha) \frac{\partial}{\partial y_\beta}$$

(see formula (6.1.2)).

Starting from the maximal root $\alpha_{\max}$ and proceeding by induction on decreasing heights of the roots, we derive from these formulas that $L_0 \cdot a_{\alpha,-1}|0\rangle = a_{\alpha,-1}|0\rangle$. This eliminates the last monomial in (6.2.8) and gives us the formula

$$w_\kappa(\mathbf{S}_\kappa) = \sum_{\alpha\in\Delta_+} a_{\alpha,-1} a^*_{\alpha,-1}|0\rangle \tag{6.2.10}$$

plus the sum of the first two types of monomials (6.2.6). It remains to determine the coefficients with which they enter the formula.

In order to do that we use the following formula for $w_\kappa(h_{i,-1}|0\rangle)$, which follows from Theorem 6.2.1:

$$w_\kappa(h_{i,-1}|0\rangle) = \left(- \sum_{\beta\in\Delta_+} \beta(h_i) a^*_{0,\beta} a_{\beta,-1} + b_{i,-1} \right) |0\rangle. \tag{6.2.11}$$

We find from it the action of L_0 and L_1 on the first summand of $w_\kappa(h_{i,-1}|0\rangle)$:

$$L_0 \cdot -\sum_{\beta\in\Delta_+} \beta(h_i) a^*_{0,\beta} a_{\beta,-1}|0\rangle = -\sum_{\beta\in\Delta_+} \beta(h_i) a^*_{0,\beta} a_{\beta,-1}|0\rangle,$$

$$L_1 \cdot -\sum_{\beta\in\Delta_+} \beta(h_i) a^*_{0,\beta} a_{\beta,-1}|0\rangle = -\sum_{\beta\in\Delta_+} \beta(h_i)|0\rangle = -2\rho(h_i)|0\rangle.$$

In addition, $L_n, n > 1$, act by 0 on it.

On the other hand, we know that $w_\kappa(h_{i,-1}|0\rangle)$ has to be annihilated by $L_n, n > 0$, and is an eigenvector of L_0 with eigenvalue 1. There is a unique combination of the first two types of monomials (6.2.6) which, when added to (6.2.10), satisfies this condition; namely,

$$\frac{1}{2}\sum_{i=1}^{\ell} b_{i,-1} b^i_{-1} - \rho_{-2}. \tag{6.2.12}$$

This proves formula (6.2.5). Now we verify directly, using Lemma 3.1.2, that the vector $w_\kappa(\mathbf{S}_\kappa)$ given by this formula satisfies the axioms of a conformal vector.

This completes the proof. □

6.2.4 Quasi-conformal structures at the critical level

Now we use this lemma to obtain additional information about the homomorphism w_{κ_c}. Denote by O the complete topological ring $\mathbb{C}[[t]]$ and by $\operatorname{Der}\mathcal{O}$ the Lie algebra of its continuous derivations. Note that $\operatorname{Der}\mathcal{O} \simeq \mathbb{C}[[t]]\partial_t$.

Recall that a vertex algebra V is called **quasi-conformal** if it carries an action of the Lie algebra $\operatorname{Der}\mathcal{O}$ satisfying the following conditions:

- the formula

$$[L_m, A_{(k)}] = \sum_{n\geq -1} \binom{m+1}{n+1} (L_n \cdot A)_{(m+k-n)}$$

  holds for all $A \in V$;
- the element $L_{-1} = -\partial_t$ acts as the translation operator T;
- the element $L_0 = -t\partial_t$ acts semi-simply with integral eigenvalues;
- the Lie subalgebra $\operatorname{Der}_+\mathcal{O}$ acts locally nilpotently

(see Definition 6.3.4 of [FB]). In particular, a conformal vertex algebra is automatically quasi-conformal (with the $\operatorname{Der}\mathcal{O}$-action coming from the Virasoro action).

The Lie algebra $\operatorname{Der}\mathcal{O}$ acts naturally on $\widehat{\mathfrak{g}}_{\kappa_c}$ preserving $\mathfrak{g}[[t]]$, and hence it acts on $V_{\kappa_c}(\mathfrak{g})$. The $\operatorname{Der}\mathcal{O}$-action on $V_{\kappa_c}(\mathfrak{g})$ coincides with the limit $\kappa \to \kappa_c$ of the $\operatorname{Der}\mathcal{O}$-action on $V_\kappa(\mathfrak{g}), \kappa \neq \kappa_c$, obtained from the Sugawara conformal structure. Therefore this action defines the structure of a quasi-conformal vertex algebra on $V_{\kappa_c}(\mathfrak{g})$.

Next, we define the structure of a quasi-conformal algebra on $M_{\mathfrak{g}} \otimes \pi_0$ as follows. The vertex algebra $M_{\mathfrak{g}}$ is conformal with the conformal vector (6.2.10), and hence it is also quasi-conformal. The commutative vertex algebra π_0 is the $\kappa \to \kappa_c$ limit of the family of conformal vertex algebras $\pi_0^{\kappa-\kappa_c}$ with the conformal vector (6.2.12). The induced action of the Lie algebra $\operatorname{Der} \mathcal{O}$ on $\pi_0^{\kappa-\kappa_c}$ is well-defined in the limit $\kappa \to \kappa_c$ and so it induces a $\operatorname{Der} \mathcal{O}$-action on π_0. Therefore it gives rise to the structure of a quasi-conformal vertex algebra on π_0. The $\operatorname{Der} \mathcal{O}$-action is in fact given by derivations of the algebra structure on $\pi_0 \simeq \mathbb{C}[b_{i,n}]$, and hence by Lemma 6.3.5 of [FB] it defines the structure of a quasi-conformal vertex algebra on π_0. Explicitly, the action of the basis elements $L_n = -t^{n+1}\partial_t, n \geq -1$, of $\operatorname{Der} \mathcal{O}$ on π_0 is determined by the following formulas:

$$\begin{aligned} L_n \cdot b_{i,m} &= -m b_{i,n+m}, \qquad -1 \leq n < -m, \\ L_n \cdot b_{i,-n} &= n(n+1), \qquad n > 0, \\ L_n \cdot b_{i,m} &= 0, \qquad n > -m \end{aligned} \tag{6.2.13}$$

(note that $\langle \rho, h_i \rangle = 1$ for all i). Now we obtain a quasi-conformal vertex algebra structure on $M_{\mathfrak{g}} \otimes \pi_0$ by taking the sum of the above $\operatorname{Der} \mathcal{O}$-actions.

Since the quasi-conformal structures on $V_{\kappa_c}(\mathfrak{g})$ and $M_{\mathfrak{g}} \otimes \pi_0$ both arose as the limits of conformal structures as $\kappa \to \kappa_c$, we obtain the following corollary of Proposition 6.2.2:

Corollary 6.2.3 *The homomorphism $w_{\kappa_c} : V_{\kappa_c} \to M_{\mathfrak{g}} \otimes \pi_0$ preserves quasi-conformal structures. In particular, it intertwines the actions of* $\operatorname{Der} \mathcal{O}$ *and* $\operatorname{Aut} \mathcal{O}$ *on both sides.*

6.2.5 Transformation formulas for the fields

We can now obtain the transformation formulas for the fields $a_\alpha(z), a^*_\alpha(z)$ and $b_i(z)$ and the modules $M_{\mathfrak{g}}$ and π_0.

From the explicit formula (6.2.5) for the conformal vector $w_\kappa(\mathbf{S}_\kappa)$ we find the following commutation relations:

$$[L_n, a_{\alpha,m}] = -m a_{m+n}, \qquad [L_n, a^*_{\alpha,m}] = -(m-1) a^*_{\alpha,m-1}.$$

In the same way as in Section 3.5.5 we derive from this that $a_\alpha(z)$ transforms as a one-form on the punctured disc $D^\times = \operatorname{Spec} \mathbb{C}((z))$, while $a^*_\alpha(z)$ transforms as a function on $D^\times$. In particular, we obtain the following description of the module $M_{\mathfrak{g}}$.

Let us identify $\mathcal{U}$ with $\mathfrak{n}_+$. Consider the Heisenberg Lie algebra Γ, which is a central extension of the commutative Lie algebra $\mathfrak{n}_+ \otimes \mathcal{K} \oplus \mathfrak{n}^*_+ \otimes \Omega_{\mathcal{K}}$ with the cocycle given by the formula

$$f(t), g(t)dt \mapsto \int \langle f(t), g(t) \rangle dt.$$

This cocycle is coordinate-independent, and therefore Γ carries natural actions of $\operatorname{Der} \mathcal{O}$, which preserve the Lie subalgebra $\Gamma_+ = \mathfrak{n}_+ \otimes \mathcal{O} \oplus \mathfrak{n}_+^* \otimes \Omega_{\mathcal{O}}$. We identify the completed Weyl algebra $\widetilde{\mathcal{A}}^{\mathfrak{g}}$ with a completion of $U(\Gamma)/(\mathbf{1}-1)$, where $\mathbf{1}$ is the central element. The module $M_{\mathfrak{g}}$ is then identified with the Γ-module induced from the one-dimensional representation of $\Gamma_+ \oplus \mathbb{C}\mathbf{1}$, on which Γ_+ acts by 0, and $\mathbf{1}$ acts as the identity. The $\operatorname{Der} \mathcal{O}$-action on $M_{\mathfrak{g}}$ considered above is nothing but the natural action on the induced module.

Now we consider the fields $b_i(z) = \sum_{n<0} b_{i,n} z^{-n-1}$ and the module π_0. Formulas (6.2.13) describe the action of $\operatorname{Der} \mathcal{O}$ on the $b_{i,n}$'s and hence on the series $b_i(z)$. In fact, these formulas imply that $\partial_z + b_i(z)$ transforms as a connection on the line bundle $\Omega^{-\langle \rho, h_i \rangle}$.

More precisely, let ${}^L H$ be the dual group to H, i.e., it is the complex torus that is determined by the property that its lattice of characters ${}^L H \to \mathbb{C}^\times$ is the lattice of cocharacters $\mathbb{C}^\times \to H$, and the lattice of cocharacters of ${}^L H$ is the lattice of characters of H.† The Lie algebra ${}^L\mathfrak{h}$ of ${}^L H$ is then canonically identified with $\mathfrak{h}^*$. Denote by $\Omega^{-\rho}$ the unique principal ${}^L H$-bundle on the disc D such that the line bundle associated to any character $\check{\lambda} : {}^L H \to \mathbb{C}^\times$ (equivalently, a cocharacter of H) is $\Omega^{-\langle \rho, \check{\lambda} \rangle}$ (see Section 4.2.3). Denote by $\operatorname{Conn}(\Omega^{-\rho})_D$ the space of all connections on this ${}^L H$-bundle. This is a torsor over $\mathfrak{h}^* \otimes \Omega_{\mathcal{O}}$.

The above statement about $b_i(z)$ may be reformulated as follows: consider the $\mathfrak{h}^*$-valued field $\mathbf{b}(z) = \sum_{i=1}^{\ell} b_i(z) \omega_i$ such that $\langle \mathbf{b}(z), h_i \rangle = b_i(z)$. Then the operator $\partial_z + \mathbf{b}(z)$ transforms as a connection on the ${}^L H$-bundle $\Omega^{-\rho}$ over D. Equivalently, $\partial_z + b_i(z)$ transforms as a connection on the line bundle $\Omega^{-\langle \rho, h_i \rangle}$ over D.

To see that, let w be a new coordinate such that $z = \varphi(w)$; then the same connection will appear as $\partial_w + \widetilde{\mathbf{b}}(w)$, where

$$\widetilde{\mathbf{b}}(w) = \varphi' \cdot \mathbf{b}(\varphi(w)) + \rho \frac{\varphi''}{\varphi'}. \tag{6.2.14}$$

It is straightforward to check that formula (6.2.14) is equivalent to formula (6.2.13).

This implies that π_0 is isomorphic to the algebra $\operatorname{Fun}(\operatorname{Conn}(\Omega^{-\rho})_D)$ of functions on the space $\operatorname{Conn}(\Omega^{-\rho})_D$. If we choose a coordinate z on D, then we identify $\operatorname{Conn}(\Omega^{-\rho})_D$ with $\mathfrak{h}^* \otimes \Omega_{\mathcal{O}}$ this algebra with the free polynomial algebra $\mathbb{C}[b_{i,n}]_{i=1,\ldots,\ell; n<0}$, whose generators $b_{i,n}$ are the following linear functionals

† Thus, ${}^L H$ is the Cartan subgroup of the Langlands dual group ${}^L G$ of G introduced in Section 1.1.5.

on $\mathfrak{h}^* \otimes \Omega_\mathcal{O} \simeq \mathfrak{h}^*[[z]]dz$:

$$b_{i,n}(\chi(z)dz) = \mathrm{Res}_{z=0}\langle \chi(z), h_i \rangle z^n dz.$$

6.2.6 Coordinate-independent version

Up to now, we have considered Wakimoto modules as representations of the Lie algebra $\widehat{\mathfrak{g}}_{\kappa_c}$, which is the central extension of $L\mathfrak{g} = \mathfrak{g}((t))$. As explained in Section 3.5.2, it is important to develop a theory which applies to the central extension $\widehat{\mathfrak{g}}_{\kappa,x}$ of the Lie algebra $\mathfrak{g}(\mathcal{K}_x) = \mathfrak{g} \otimes \mathcal{K}_x$, where $\mathcal{K}_x$ is the algebra of functions on the punctured disc around a point x of a smooth curve. In other words, $\mathcal{K}_x$ is the completion of the field of functions on X corresponding to x. It is a topological algebra which is isomorphic to $\mathbb{C}((t))$, but non-canonically. If we choose a formal coordinate t at x, we may identify $\mathcal{K}_x$ with $\mathbb{C}((t))$, but this identification is non-canonical as there is usually no preferred choice of coordinate t at x. However, as we saw in Section 3.5.2, we have a canonical central extension $\widehat{\mathfrak{g}}_{\kappa_c,x}$ of $\mathfrak{g}(\mathcal{K}_x)$ and the vacuum $\widehat{\mathfrak{g}}_{\kappa_c,x}$-module

$$V_\kappa(\mathfrak{g})_x = \mathrm{Ind}_{\mathfrak{g}\otimes\mathcal{O}_x \oplus \mathbb{C}K}^{\widehat{\mathfrak{g}}_{\kappa,x}} \mathbb{C}.$$

where, as before, $\mathcal{O}_x$ denotes the ring of integers in $\mathcal{K}_x$ which is isomorphic to $\mathbb{C}[[t]]$.

As explained in Section 3.5.4, $V_\kappa(\mathfrak{g})_x$ may also be obtained as the twist of $V_\kappa(\mathfrak{g})$ by the $\mathrm{Aut}\,\mathcal{O}$-torsor $\mathcal{A}ut_x$ of formal coordinates at x:

$$V_\kappa(\mathfrak{g})_x = \mathcal{A}ut_x \underset{\mathrm{Aut}\,\mathcal{O}}{\times} V_\kappa(\mathfrak{g}).$$

Now we wish to recast the free field realization homomorphism

$$w_\kappa : V_\kappa(\mathfrak{g}) \to M_\mathfrak{g} \otimes \pi_0^{\kappa-\kappa_c}$$

of Theorem 6.2.1 in a coordinate-independent way.

We have an analogue of the Heisenberg Lie algebra Γ introduced in the previous section, attached to a point x. By definition, this Lie algebra, denoted by Γ_x, is the central extension of $(\mathfrak{n}_+ \otimes \mathcal{K}_x) \oplus (\mathfrak{n}_+^* \otimes \Omega_{\mathcal{K}_x})$. Let $\Gamma_{+,x}$ be its commutative Lie subalgebra $(\mathfrak{n}_+ \otimes \mathcal{O}_x) \oplus (\mathfrak{n}_+^* \otimes \Omega_{\mathcal{O}_x})$. Let $M_{\mathfrak{g},x}$ be the Γ_x-module

$$M_{\mathfrak{g},x} = \mathrm{Ind}_{\Gamma_{+,x}\oplus\mathbb{C}\mathbf{1}}^{\Gamma_x} \mathbb{C}.$$

As in the case of $V_\kappa(\mathfrak{g})_x$, we have

$$M_{\mathfrak{g},x} = \mathcal{A}ut_x \underset{\mathrm{Aut}\,\mathcal{O}}{\times} M_\mathfrak{g},$$

where the action of $\mathrm{Aut}\,\mathcal{O}$ on $M_\mathfrak{g}$ comes from the action of coordinate changes on $M_\mathfrak{g}$ induced by its action on the Lie algebra Γ, as described in the previous section.

Next, we need a version of the Heisenberg Lie algebra $\mathfrak{h}_\nu$ and its module π_0^ν attached to the point x.

Consider the vector space $\mathrm{Conn}_{\{\lambda\}}(\Omega^{-\rho})_{D_x^\times}$ of λ-connections on the ${}^L H$-bundle $\Omega^{-\rho}$ on $D_x^\times = \mathrm{Spec}\,\mathcal{K}_x$, for all possible complex values of λ. If we choose an isomorphism $\mathcal{K}_x \simeq \mathbb{C}((t))$, then a λ-connection is an operator $\nabla = \lambda\partial_t + \chi(t)$, where $\chi(t) \in \mathfrak{h}^*((t))$. We have an exact sequence

$$0 \to \mathfrak{h}^* \otimes \Omega_{\mathcal{K}_x} \to \mathrm{Conn}_{\{\lambda\}}(\Omega^{-\rho})_{D_x^\times} \to \mathbb{C}\partial_t \to 0,$$

where the penultimate map sends ∇ as above to $\lambda\partial_t$.

Let $\widehat{\mathfrak{h}}_{\nu,x}$ be the topological dual vector space to $\mathrm{Conn}_{\{\lambda\}}(\Omega^{-\rho})_{D_x^\times}$. It fits into an exact sequence (here we use the residue pairing between $\Omega_{\mathcal{K}_x}$ and $\mathcal{K}_x$)

$$0 \to \mathbb{C}\mathbf{1} \to \widehat{\mathfrak{h}}_{\nu,x} \to \mathfrak{h} \otimes \mathcal{K}_x \to 0,$$

where $\mathbf{1}$ is the element dual to ∂_x. The Lie bracket is given by the old formula (see Section 6.2.1); it is easy to see that this formula (which depends on ν) is coordinate-independent.

Note that this sequence does not have a natural coordinate-independent splitting. However, there is a natural splitting $\mathfrak{h} \otimes \mathcal{O}_x \to \widehat{\mathfrak{h}}_{\nu,x}$ (the image is the orthogonal complement to the space of λ-connections on the disc D_x). Therefore we define an $\widehat{\mathfrak{h}}_{\nu,x}$-module

$$\pi^\nu_{\lambda,x} = \mathrm{Ind}^{\widehat{\mathfrak{h}}_{\nu,x}}_{\mathfrak{h}\otimes\mathcal{O}_x \oplus \mathbb{C}\mathbf{1}} \lambda.$$

From the description of the action of $\mathrm{Aut}\,\mathcal{O}$ on the Lie algebra $\widehat{\mathfrak{h}}_\nu$ obtained in the previous section it follows immediately that

$$\pi^\nu_{\lambda,x} = \mathcal{A}ut_x \underset{\mathrm{Aut}\,\mathcal{O}}{\times} \pi^\nu_\lambda.$$

Now we can state a coordinate-independent version of the free field homomorphism w_κ.

Proposition 6.2.4 *For any level κ and any point x there is a natural homomorphism of $\widehat{\mathfrak{g}}_{\kappa,x}$-modules $V_\kappa(\mathfrak{g})_x \to M_{\mathfrak{g},x} \otimes \pi_{0,x}^{\kappa-\kappa_c}$. Furthermore, for any highest weight $\lambda \in \mathfrak{h}^*$,*

$$W_{\lambda,\kappa,x} = M_{\mathfrak{g},x} \otimes \pi_{0,x}^{\kappa-\kappa_c}$$

carries a natural structure of $\widehat{\mathfrak{g}}_{\kappa,x}$-module.

Proof. According to Proposition 6.2.2 and Corollary 6.2.3, the map w_κ, considered as a homomorphism of $\widehat{\mathfrak{g}}_{\kappa,x}$-modules, is compatible with the action of $\mathrm{Aut}\,\mathcal{O}$ on both sides. Therefore we may twist it by the torsor $\mathcal{A}ut_x$. According to the above discussion, we obtain the first assertion of the proposition. Likewise, twisting the action of $\widehat{\mathfrak{g}}$ on $W_{\lambda,\kappa}$ by $\mathcal{A}ut_x$, we obtain the second assertion. □

In particular, for $\nu = 0$ the Lie algebra $\widehat{\mathfrak{h}}_{0,x}$ is commutative. Any connection ∇ on $\Omega^{-\rho}$ over the punctured disc $D_x^\times$ defines a linear functional on $\widehat{\mathfrak{h}}_{0,x}$, and hence a one-dimensional representation $\mathbb{C}_\nabla$ of $\widehat{\mathfrak{h}}_{0,x}$, considered as a commutative Lie algebra. If we choose an isomorphism $\mathcal{K}_x \simeq \mathbb{C}((t))$, then the connection is given by the formula $\nabla = \partial_t + \chi(t)$, where $\chi(t) \in \mathfrak{h}^*((t))$. The action of the generators $b_{i,n}$ on $\mathbb{C}_\nabla$ is then given by the formula

$$b_{i,n} \mapsto \int \langle \chi(z), h_i \rangle z^n dz.$$

By Theorem 6.1.6, there exists a homomorphism of vertex algebras $w_{\kappa_c} : V_{\kappa_c}(\mathfrak{g}) \to M_{\mathfrak{g}} \otimes \pi_0$. According to Corollary 6.2.3, it commutes with the action of $\operatorname{Aut} \mathcal{O}$ on both sides. Therefore the corresponding homomorphism of Lie algebras $\widehat{\mathfrak{g}} \to U(M_{\mathfrak{g}} \otimes \pi_0)$ also commutes with the action of $\operatorname{Aut} \mathcal{O}$. Hence we may twist this homomorphism with the $\operatorname{Aut} \mathcal{O}$-torsor $\mathcal{A}ut_x$. Then we obtain a homomorphism of Lie algebras

$$\widehat{\mathfrak{g}}_{\kappa_c,x} \to U(M_{\mathfrak{g}} \otimes \pi_0)_x \stackrel{\text{def}}{=} \mathcal{A}ut_x \underset{\operatorname{Aut} \mathcal{O}}{\times} U(M_{\mathfrak{g}} \otimes \pi_0).$$

Let us call a $\Gamma_x \oplus \widehat{\mathfrak{h}}_{0,x}$-module **smooth** if any vector in this module is annihilated by the Lie subalgebra

$$(\mathfrak{n}_+ \otimes \mathfrak{m}_x^N) \oplus (\mathfrak{n}_+^* \otimes \mathfrak{m}_x^N \Omega_{\mathcal{O}}) \oplus (\mathfrak{h} \otimes \mathfrak{m}_x^N),$$

where $\mathfrak{m}$ is the maximal ideal of $\mathcal{O}_x$, for sufficiently large $N \in \mathbb{Z}_+$. Clearly, any smooth $\Gamma \oplus \widehat{\mathfrak{h}}_0$-module is automatically a $U(M_{\mathfrak{g}} \otimes \pi_0)$-module. Hence any smooth $\Gamma_x \oplus \widehat{\mathfrak{h}}_{0,x}$-module is automatically a $U(M_{\mathfrak{g}} \otimes \pi_0)_x$-module and hence a $\widehat{\mathfrak{g}}_{\kappa_c,x}$-module.

Proposition 6.2.5 *For any connection on the ${}^L H$-bundle $\Omega^{-\rho}$ over the punctured disc $D_x^\times = \operatorname{Spec} \mathcal{K}_x$ there is a canonical $\widehat{\mathfrak{g}}_{\kappa_c,x}$-module structure on $M_{\mathfrak{g},x}$.*

Proof. Note that $M_{\mathfrak{g},x}$ is a smooth Γ_x-module, and $\mathbb{C}_\nabla$ is a smooth $\widehat{\mathfrak{h}}_{0,x}$-module for any connection ∇ on the ${}^L H$-bundle $\Omega^{-\rho}$ over the punctured disc $D_x^\times$. Taking the tensor product of these two modules we obtain a $\widehat{\mathfrak{g}}_{\kappa_c,x}$-module, which is isomorphic to $M_{\mathfrak{g},x}$ as a vector space. □

Thus, we obtain a family of $\widehat{\mathfrak{g}}_{\kappa_c,x}$-modules parameterized by the connections on the ${}^L H$-bundle $\Omega^{-\rho}$ over the punctured disc $D_x^\times$.

If we choose an isomorphism $\mathcal{K}_x \simeq \mathbb{C}((t))$ and write a connection ∇ as $\nabla = \partial_t + \chi(t)$, then the module corresponding to ∇ is nothing but the Wakimoto module $W_{\chi(z)}$ introduced in Corollary 6.1.5.

6.3 Semi-infinite parabolic induction

In this section we generalize the construction of Wakimoto modules by considering an arbitrary parabolic subalgebra of $\mathfrak{g}$ instead of a Borel subalgebra.

6.3.1 Wakimoto modules as induced representations

The construction of the Wakimoto modules presented above may be summarized as follows: for each representation N of the Heisenberg Lie algebra $\widehat{\mathfrak{h}}_\kappa$, we have constructed a $\widehat{\mathfrak{g}}_{\kappa+\kappa_c}$-module structure on $M_\mathfrak{g} \otimes N$. The procedure consists of extending the $\widehat{\mathfrak{h}}_\kappa$-module by 0 to $\widehat{\mathfrak{b}}_{-,\kappa}$, followed by what may be viewed as a semi-infinite analogue of induction functor from $\widehat{\mathfrak{b}}_{-,\kappa}$-modules to $\widehat{\mathfrak{g}}_{\kappa+\kappa_c}$-modules. An important feature of this construction, as opposed to the ordinary induction, is that the level gets shifted by κ_c. In particular, if we start with an $\widehat{\mathfrak{h}}_0$-module, or, equivalently, a representation of the commutative Lie algebra $L\mathfrak{h}$, then we obtain a $\widehat{\mathfrak{g}}_{\kappa_c}$-module of critical level, rather than of level 0. For instance, we can apply this construction to irreducible smooth representations of the commutative Lie algebra $L\mathfrak{h}$. These are one-dimensional and are in one-to-one correspondence with the elements $\chi(t)$ of the (topological) dual space $(L\mathfrak{h})^* \simeq \mathfrak{h}^*((t))dt$. As the result we obtain the Wakimoto modules $W_{\chi(t)}$ of critical level introduced above. Looking at the transformation properties of $\chi(t)$ under the action of the group $\operatorname{Aut}\mathcal{O}$ of changes of the coordinate t, we find that $\chi(t)$ actually transforms not as a one-form, but as a connection on a specific ${}^LH$-bundle. This "anomaly" is a typical feature of "semi-infinite" constructions.

In contrast, if $\kappa \neq 0$, the irreducible smooth $\widehat{\mathfrak{h}}_\kappa$-modules are just the Fock representations π_λ^κ. To each of them we attach a $\widehat{\mathfrak{g}}_{\kappa+\kappa_c}$-module $W_{\lambda,\kappa+\kappa_c}$.

Now we want to generalize this construction by replacing the Borel subalgebra $\mathfrak{b}_-$ and its Levi quotient $\mathfrak{h}$ by an arbitrary parabolic subalgebra $\mathfrak{p}$ and its Levi quotient $\mathfrak{m}$. Then we wish to attach to a module over a central extension of the loop algebra $L\mathfrak{m}$ a $\widehat{\mathfrak{g}}$-module. It turns out that this is indeed possible provided that we pick a suitable central extension of $L\mathfrak{m}$. We call the resulting $\widehat{\mathfrak{g}}$-modules the generalized Wakimoto modules corresponding to $\mathfrak{p}$. Thus, we obtain a functor from the category of smooth $\widehat{\mathfrak{m}}$-modules to the category of smooth $\widehat{\mathfrak{g}}$-modules. It is natural to call it the functor of **semi-infinite parabolic induction** (by analogy with a similar construction for representations of reductive groups).

6.3.2 The main result

Let $\mathfrak{p}$ be a parabolic Lie subalgebra of $\mathfrak{g}$. We will assume that $\mathfrak{p}$ contains the lower Borel subalgebra $\mathfrak{b}_-$ (and so, in particular, $\mathfrak{p}$ contains $\mathfrak{h} \subset \mathfrak{b}_-$). Let

$$\mathfrak{p} = \mathfrak{m} \oplus \mathfrak{r} \tag{6.3.1}$$

be a Levi decomposition of $\mathfrak{p}$, where $\mathfrak{m}$ is a Levi subgroup containing $\mathfrak{h}$ and $\mathfrak{r}$ is the nilpotent radical of $\mathfrak{p}$. Further, let

$$\mathfrak{m} = \bigoplus_{i=1}^{s} \mathfrak{m}_i \oplus \mathfrak{m}_0$$

be the direct sum decomposition of $\mathfrak{m}$ into the direct sum of simple Lie subalgebras $\mathfrak{m}_i, i = 1, \ldots, s$, and an abelian subalgebra $\mathfrak{m}_0$ such that these direct summands are mutually orthogonal with respect to the inner product on $\mathfrak{g}$. We denote by $\kappa_{i,c}$ the critical inner product on $\mathfrak{m}_i, i = 1, \ldots, s$, defined as in Section 5.3.3. We also set $\kappa_{0,c} = 0$.

Given a set of invariant inner products κ_i on $\mathfrak{m}_i, 0 = 1, \ldots, s$, we obtain an invariant inner product on $\mathfrak{m}$. Let $\widehat{\mathfrak{m}}_{(\kappa_i)}$ be the corresponding affine Kac–Moody algebra, i.e., the one-dimensional central extension of $L\mathfrak{m}$ with the commutation relations given by formula (1.3.4). We denote by $V_{\kappa_i}(\mathfrak{m}_i), i = 1, \ldots, s$, the vacuum module over $\widehat{\mathfrak{m}}_{i,\kappa_i}$ with the vertex algebra structure defined as in Section 6.1.1. We also denote by $V_{\kappa_0}(\mathfrak{m}_0)$ the Fock representation $\pi_0^{\kappa_0}$ of the Heisenberg Lie algebra $\mathfrak{m}_{\kappa_0}$ with its vertex algebra structure defined as in Section 6.2.1. Let

$$V_{(\kappa_i)}(\mathfrak{m}) \stackrel{\text{def}}{=} \bigotimes_{i=0}^{s} V_{\kappa_i}(\mathfrak{m}_i)$$

be the vacuum module over $\widehat{\mathfrak{m}}_{(\kappa_i)}$ with the tensor product vertex algebra structure.

Denote by Δ'_+ the set of positive roots of $\mathfrak{g}$ which do not occur in the root space decomposition of $\mathfrak{p}$ (note that by our assumption that $\mathfrak{b}_- \subset \mathfrak{p}$, all negative roots do occur). Let $\mathcal{A}^{\mathfrak{g},\mathfrak{p}}$ be the Weyl algebra with generators $a_{\alpha,n}, a^*_{\alpha,n}, \alpha \in \Delta'_+, n \in \mathbb{Z}$, and relations (5.3.3). Let $M_{\mathfrak{g},\mathfrak{p}}$ be the Fock representation of $\mathcal{A}^{\mathfrak{g},\mathfrak{p}}$ generated by a vector $|0\rangle$ such that

$$a_{\alpha,n}|0\rangle = 0, \quad n \geq 0; \qquad a^*_{\alpha,n}|0\rangle = 0, \quad n > 0.$$

Then $M_{\mathfrak{g},\mathfrak{p}}$ carries a vertex algebra structure defined as in Section 5.4.1.

We have the following analogue of Theorem 6.2.1.

Theorem 6.3.1 *Suppose that $\kappa_i, i = 0, \ldots, s$, is a set of inner products such that there exists an inner product κ on $\mathfrak{g}$ whose restriction to $\mathfrak{m}_i$ equals $\kappa_i - \kappa_{i,c}$ for all $i = 0, \ldots, s$. Then there exists a homomorphism of vertex algebras*

$$w_\kappa^{\mathfrak{p}} : V_{\kappa+\kappa_c}(\mathfrak{g}) \to M_{\mathfrak{g},\mathfrak{p}} \otimes V_{(\kappa_i)}(\mathfrak{m}).$$

Proof. The proof is a generalization of the proof of Theorem 6.2.1 (in fact, Theorem 6.2.1 is a special case of Theorem 6.3.1 when $\mathfrak{p} = \mathfrak{b}_-$). Let P be the Lie subgroup of G corresponding to $\mathfrak{p}$, and consider the homogeneous space G/P. It has an open dense subset $\mathcal{U}_\mathfrak{p} = N_\mathfrak{p} \cdot [1]$, where $N_\mathfrak{p}$ is the subgroup of N_+ corresponding to the subset $\Delta'_+ \subset \Delta_+$. We identify $\mathcal{U}_\mathfrak{p}$ with $N_\mathfrak{p}$ and with its Lie algebra $\mathfrak{n}_\mathfrak{p}$ using the exponential map.

Set $L\mathcal{U}_\mathfrak{p} = \mathcal{U}_\mathfrak{p}((t))$. We define functions and vector fields on $L\mathcal{U}_\mathfrak{p}$, denoted by $\operatorname{Fun} L\mathcal{U}_\mathfrak{p}$ and $\operatorname{Vect} L\mathcal{U}_\mathfrak{p}$, respectively, in the same way as in Section 5.3.1. The action of $L\mathfrak{g}$ on $\mathcal{U}_\mathfrak{p}((t))$ gives rise to a Lie algebra homomorphism $L\mathfrak{g} \to$

Vect $L\mathcal{U}_{\mathfrak{p}}$, in the same way as before. We generalize this homomorphism as follows.

Consider the quotient G/R, where R is the Lie subgroup of G corresponding to the nilpotent Lie algebra $\mathfrak{r}$ appearing in the Levi decomposition (6.3.1). We have a natural projection $G/R \to G/P$, which is an M-bundle, where $M = P/R$ is the Levi subgroup of G corresponding to $\mathfrak{m}$. Over $\mathcal{U}_{\mathfrak{p}} \subset G/P$ this bundle may be trivialized, and so it is isomorphic to $\mathcal{U}_{\mathfrak{p}} \times M$. The Lie algebra $\mathfrak{g}$ acts on this bundle, and hence on $\mathcal{U}_{\mathfrak{p}} \times M$. As the result, we obtain a Lie algebra homomorphism $\mathfrak{g} \to \mathrm{Vect}(\mathcal{U}_{\mathfrak{p}} \times M)$. It is easy to see that it factors through homomorphisms

$$\mathfrak{g} \to (\mathrm{Vect}\,\mathcal{U}_{\mathfrak{p}} \otimes 1) \oplus (\mathrm{Fun}\,\mathcal{U}_{\mathfrak{p}} \otimes \mathfrak{m})$$

and $\mathfrak{m} \to \mathrm{Vect}\, M$.

The loop version of this construction gives rise to a (continuous) homomorphism of (topological) Lie algebras

$$L\mathfrak{g} \to \mathrm{Vect}\, L\mathcal{U}_{\mathfrak{p}} \oplus \mathrm{Fun}\,\mathcal{U}_{\mathfrak{p}} \widehat{\otimes} L\mathfrak{m}.$$

Moreover, the image of this homomorphism is contained in the "local part," i.e., the direct sum of the local part $\mathcal{T}^{\mathfrak{g},\mathfrak{p}}_{\mathrm{loc}}$ of Vect $L\mathcal{U}_{\mathfrak{p}}$ defined as in Section 5.4.3 and the local part $\mathcal{I}^{\mathfrak{g},\mathfrak{p}}_{\mathrm{loc}}$ of $\mathrm{Fun}\,\mathcal{U}_{\mathfrak{p}} \widehat{\otimes} L\mathfrak{m}$. By definition, $\mathcal{I}^{\mathfrak{g},\mathfrak{p}}_{\mathrm{loc}}$ is the span of the Fourier coefficients of the formal power series $P(\partial_z^n a^*_\alpha(z))J^a(z)$, where P is a differential polynomial in $a^*_\alpha(z), \alpha \in \Delta'_+$, and $J^a \in \mathfrak{m}$.

Let $\mathcal{A}^{\mathfrak{g},\mathfrak{p}}_{0,\mathrm{loc}}$ and $\mathcal{A}^{\mathfrak{g},\mathfrak{p}}_{\leq 1,\mathrm{loc}}$ be the zeroth and the first terms of the natural filtration on the local completion of the Weyl algebra $\mathcal{A}^{\mathfrak{g},\mathfrak{p}}$, defined as in Section 5.3.3. We have a non-split exact sequence of Lie algebras

$$0 \to \mathcal{A}^{\mathfrak{g},\mathfrak{p}}_{0,\mathrm{loc}} \to \mathcal{A}^{\mathfrak{g},\mathfrak{p}}_{\leq 1,\mathrm{loc}} \to \mathcal{T}^{\mathfrak{g},\mathfrak{p}}_{\mathrm{loc}} \to 0. \tag{6.3.2}$$

Set

$$\mathcal{J}^{\mathfrak{g},\mathfrak{p}}_{\mathrm{loc}} \stackrel{\mathrm{def}}{=} \mathcal{A}^{\mathfrak{g},\mathfrak{p}}_{\leq 1,\mathrm{loc}} \oplus \mathcal{I}^{\mathfrak{g},\mathfrak{p}}_{\mathrm{loc}}, \tag{6.3.3}$$

and note that $\mathcal{J}^{\mathfrak{g},\mathfrak{p}}_{\mathrm{loc}}$ is naturally a Lie subalgebra of the local Lie algebra $U(M_{\mathfrak{g},\mathfrak{p}} \otimes V_{(\kappa_i)}(\mathfrak{m}))$. Using the splitting of the sequence (6.3.2) as a vector space via the normal ordering, we obtain a linear map $\overline{w}_{(\kappa_i)} : L\mathfrak{g} \to \mathcal{J}^{\mathfrak{g},\mathfrak{p}}_{\mathrm{loc}}$.

We need to compute the failure of $\overline{w}_{(\kappa_i)}$ to be a Lie algebra homomorphism. Thus, we consider the corresponding linear map $\omega_{(\kappa_i)} : \bigwedge^2 L\mathfrak{g} \to \mathcal{J}^{\mathfrak{g},\mathfrak{p}}_{\mathrm{loc}}$ defined by the formula

$$\omega_{(\kappa_i)}(f,g) = [\overline{w}_\kappa(f), \overline{w}_\kappa(g)] - \overline{w}_\kappa([f,g]).$$

Evaluating it explicitly in the same way as in the proof of Lemma 5.6.2, we find that $\omega_{(\kappa_i)}$ takes values in $\mathcal{A}^{\mathfrak{g},\mathfrak{p}}_{0,\mathrm{loc}} \subset \mathcal{J}^{\mathfrak{g},\mathfrak{p}}_{\mathrm{loc}}$. Furthermore, $\mathcal{A}^{\mathfrak{g},\mathfrak{p}}_{0,\mathrm{loc}}$ is naturally an $L\mathfrak{g}$-module, and by construction of $\overline{w}_{(\kappa_i)}$, for any $X \in \mathcal{A}^{\mathfrak{g},\mathfrak{p}}_{0,\mathrm{loc}}$ and $f \in L\mathfrak{g}$ we have $[\overline{w}_{(\kappa_i)}(f), X] = f \cdot X$. This implies that $\omega_{(\kappa_i)}$ is a two-cocycle of

$L\mathfrak{g}$ with coefficients in $\mathcal{A}^{\mathfrak{g},\mathfrak{p}}_{0,\mathrm{loc}}$. By construction, it is local, i.e., belongs to $C^2_{\mathrm{loc}}(L\mathfrak{g}, \mathcal{A}^{\mathfrak{g},\mathfrak{p}}_{0,\mathrm{loc}})$ (defined as in Section 5.6.2).

Following the argument used in the proof of Lemma 5.6.7, we show that any two cocycles in $C^2_{\mathrm{loc}}(L\mathfrak{g}, \mathcal{A}^{\mathfrak{g},\mathfrak{p}}_{0,\mathrm{loc}})$, whose restrictions to $\bigwedge^2(L\mathfrak{m})$ coincide, represent the same cohomology class.

Let us compute the restriction of $\omega_{(\kappa_i)}$ to $\bigwedge^2 L\mathfrak{m}$. For this we evaluate $\overline{w}_{(\kappa_i)}$ on elements of $L\mathfrak{m}$. Let $\{J^\alpha\}_{\alpha\in\Delta'_+}$ be a basis of $\mathfrak{n}_\mathfrak{p}$. The adjoint action of the Lie algebra $\mathfrak{m}$ on $\mathfrak{g}$ preserves $\mathfrak{n}_\mathfrak{p}$, and so we obtain a representation $\rho_{\mathfrak{n}_\mathfrak{p}}$ of $\mathfrak{m}$ on $\mathfrak{n}_\mathfrak{p}$. For any element $A \in \mathfrak{m}$ we have

$$\rho_{\mathfrak{n}_\mathfrak{p}}(A) \cdot J^\alpha = \sum_{\beta\in\Delta'_+} c^\alpha_\beta(A) J^\beta$$

for some $c^\alpha_\beta(A) \in \mathbb{C}$. Therefore we obtain the following formula:

$$\overline{w}_{(\kappa_i)}(A(z)) = -\sum_{\beta\in\Delta'_+} c^\alpha_\beta(A){:}a^*_\beta(z)a_\beta(z){:} + \widetilde{A}(z), \qquad A \in \mathfrak{m}, \tag{6.3.4}$$

where $\widetilde{A}(z) = \sum_{n\in\mathbb{Z}}(A \otimes t^n)z^{-n-1}$, considered as a generating series of elements of $\mathfrak{m} \subset \mathcal{I}^{\mathfrak{g},\mathfrak{p}}_{\mathrm{loc}}$.

Let $\kappa_{\mathfrak{n}_\mathfrak{p}}$ be the inner product on $\mathfrak{m}$ defined by the formula

$$\kappa_{\mathfrak{n}_\mathfrak{p}}(A, B) = \mathrm{Tr}_{\mathfrak{n}_\mathfrak{p}}\, \rho_{\mathfrak{n}_\mathfrak{p}}(A)\rho_{\mathfrak{n}_\mathfrak{p}}(B).$$

Computing directly the commutation relations between the coefficients of the series (6.3.4), we find that for $A \in \mathfrak{m}_i, B \in \mathfrak{m}_j$ we have the following formulas:

$$\omega_{(\kappa_i)}(A_n, B_m) = -n\kappa_{\mathfrak{n}_\mathfrak{p}}(A, B)\delta_{n,-m}, \tag{6.3.5}$$

if $i \neq j$, and

$$\omega_{(\kappa_i)}(A_n, B_m) = n(-\kappa_{\mathfrak{n}_\mathfrak{p}}(A, B) + \kappa_i(A, B))\delta_{n,-m}, \tag{6.3.6}$$

if $i = j$. Thus, the restriction of $\omega_{(\kappa_i)}$ to $\bigwedge^2 L\mathfrak{m}$ takes values in the subspace of constants $\mathbb{C} \subset \mathcal{A}^{\mathfrak{g},\mathfrak{p}}_{0,\mathrm{loc}}$.

Let $\kappa_\mathfrak{g}$ be the Killing form on $\mathfrak{g}$ and $\kappa_{\mathfrak{m}_i}$ the Killing form on $\mathfrak{m}_i$ (in particular, $\kappa_{\mathfrak{m}_0} = 0$). Then we have

$$\begin{aligned} \kappa_\mathfrak{g}(A, B) &= \kappa_{\mathfrak{m}_i}(A, B) + 2\kappa_{\mathfrak{n}_\mathfrak{p}}(A, B), \qquad \text{if} \quad i = j, \\ \kappa_\mathfrak{g}(A, B) &= 2\kappa_{\mathfrak{n}_\mathfrak{p}}(A, B), \qquad \text{if} \quad i \neq j. \end{aligned} \tag{6.3.7}$$

The factor of 2 is due to the fact that we have to include both positive and negative roots. Recall our assumption that $\mathfrak{m}_i$ and $\mathfrak{m}_j$ are orthogonal with respect to $\kappa_\mathfrak{g}$ for all $i \neq j$. This implies that $\kappa_{\mathfrak{n}_\mathfrak{p}}(A, B) = 0$, if $i \neq j$.

Recall that by definition $\kappa_c = -\frac{1}{2}\kappa_\mathfrak{g}$, and $\kappa_{i,c} = -\frac{1}{2}\kappa_{\mathfrak{m}_i}$. Hence formula (6.3.7) implies that

$$\kappa_{\mathfrak{n}_\mathfrak{p}}|_{\mathfrak{m}_i} = -\kappa_c + \kappa_{i,c}.$$

Therefore we find that if $\kappa_i = \kappa|_{\mathfrak{m}_i} + \kappa_{i,c}$ for some invariant inner product κ on $\mathfrak{g}$, then

$$-\kappa_{\mathfrak{n}_\mathfrak{p}}|_{\mathfrak{m}_i} + \kappa_i = (\kappa + \kappa_c)|_{\mathfrak{m}_i}.$$

By inspection of formulas (6.3.5) and (6.3.6), we now find that if κ is an invariant inner product on $\mathfrak{g}$ whose restriction to $\mathfrak{m}_i$ equals $\kappa_i - \kappa_{i,c}$ for all $i = 0, \ldots, s$, then the restriction of the two-cocycle $\omega_{(\kappa_i)}$ to $\bigwedge^2(L\mathfrak{m})$ is equal to the restriction to $\bigwedge^2(L\mathfrak{m})$ of the two-cocycle $\sigma_{\kappa+\kappa_c}$ on $L\mathfrak{g}$ (this is the cocycle representing the one-dimensional central extension of $\mathfrak{g}$ corresponding to the inner product $\kappa + \kappa_c$ on $\mathfrak{g}$).

Applying the argument used in the proof of Lemma 5.6.7, we find that under the above conditions, which are precisely the conditions stated in the theorem, the two-cocycle $\omega_{(\kappa_i)}$ on $L\mathfrak{g}$ is equivalent to the two-cocycle $\sigma_{\kappa+\kappa_c}$.

Therefore we obtain, in the same way as in the proof of Theorem 5.6.8, that under these conditions the linear map $\overline{w}_{(\kappa_i)} : L\mathfrak{g} \to \mathcal{J}^{\mathfrak{g},\mathfrak{p}}_{\mathrm{loc}}$ may be modified by the addition of an element of $C^1_{\mathrm{loc}}(L\mathfrak{g}, \mathcal{A}^{\mathfrak{g},\mathfrak{p}}_{0,\mathrm{loc}})$ to give us a Lie algebra homomorphism

$$\widehat{\mathfrak{g}}_{\kappa+\kappa_c} \to \mathcal{J}^{\mathfrak{g},\mathfrak{p}}_{\mathrm{loc}} \subset U(M_{\mathfrak{g},\mathfrak{p}} \otimes V_{(\kappa_i)}(\mathfrak{m})).$$

Now Lemma 6.1.1 implies that there exists a homomorphism of vertex algebras

$$w^{\mathfrak{p}}_\kappa : V_{\kappa+\kappa_c}(\mathfrak{g}) \to M_{\mathfrak{g},\mathfrak{p}} \otimes V_{(\kappa_i)}(\mathfrak{m}).$$

This completes the proof. □

Let us call an $\widehat{\mathfrak{m}}_{(\kappa_i)}$-module **smooth** if any vector in it is annihilated by the Lie subalgebra $\mathfrak{m} \otimes t^N\mathbb{C}[[t]]$ for sufficiently large N.

Corollary 6.3.2 *For any smooth $\widehat{\mathfrak{m}}_{(\kappa_i)}$-module R with the κ_i's satisfying the conditions of Theorem 6.3.1, the tensor product $M_{\mathfrak{g},\mathfrak{p}} \otimes R$ is naturally a smooth $\widehat{\mathfrak{g}}_{\kappa+\kappa_c}$-module. There is a functor from the category of smooth $\widehat{\mathfrak{m}}_{(\kappa_i)}$-modules to the category of smooth $\widehat{\mathfrak{g}}_{\kappa+\kappa_c}$-modules sending a module R to $M_{\mathfrak{g},\mathfrak{p}} \otimes R$ and $\widehat{\mathfrak{m}}_{(\kappa_i)}$-homomorphism $R_1 \to R_2$ to the $\widehat{\mathfrak{g}}_{\kappa+\kappa_c}$-homomorphism $M_{\mathfrak{g},\mathfrak{p}} \otimes R_1 \to M_{\mathfrak{g},\mathfrak{p}} \otimes R_2$.*

We call the $\widehat{\mathfrak{g}}_{\kappa+\kappa_c}$-module $M_{\mathfrak{g},\mathfrak{p}} \otimes R$ the **generalized Wakimoto module corresponding to** R.

Consider the special case when R is the tensor product of the Wakimoto modules W_{λ_i,κ_i} over $\widehat{\mathfrak{m}}_i, i = 1, \ldots, s$, and the Fock representation $\pi^{\kappa_0}_{\lambda_0}$ over the Heisenberg Lie algebra $\widehat{\mathfrak{m}}_0$. In this case it follows from the construction that the corresponding $\widehat{\mathfrak{g}}_{\kappa+\kappa_c}$-module $M_{\mathfrak{g},\mathfrak{p}} \otimes R$ is isomorphic to the Wakimoto module $W_{\lambda,\kappa+\kappa_c}$ over $\widehat{\mathfrak{g}}_{\kappa+\kappa_c}$, where $\lambda = (\lambda_i)$.

Finally, let us suppose that the κ_i's are chosen in such a way that the conditions of Theorem 6.3.1 are not satisfied. Then the two-cocycle $\omega_{(\kappa_i)}$ on $L\mathfrak{g}$ with coefficients in $\mathcal{A}^{\mathfrak{g},\mathfrak{p}}_{0,\mathrm{loc}}$ defined in the proof of Theorem 6.3.1, restricted to $L\mathfrak{m}$, still gives rise to a two-cocycle of $L\mathfrak{m}$ with coefficients in $\mathbb{C}$. However, this two-cocycle is no longer equivalent to the restriction to $L\mathfrak{m}$ of any two-cocycle σ on $L\mathfrak{g}$ (those can be represented by the cocycles σ_ν corresponding to invariant inner products ν on $\mathfrak{g}$). Using the same argument as in the proof of Lemma 5.6.7, we obtain that the two-cocycle $\omega_{(\kappa_i)}$ on $L\mathfrak{g}$ with coefficients in $\mathcal{A}^{\mathfrak{g},\mathfrak{p}}_{0,\mathrm{loc}}$ cannot be equivalent to a two-cocycle on $L\mathfrak{g}$ with coefficients in $\mathbb{C} \subset \mathcal{A}^{\mathfrak{g},\mathfrak{p}}_{0,\mathrm{loc}}$. Therefore the map $\overline{w}_{(\kappa_i)} : L\mathfrak{g} \to \mathcal{J}^{\mathfrak{g},\mathfrak{p}}_{\mathrm{loc}}$ cannot be lifted to a Lie algebra homomorphism $\widehat{\mathfrak{g}}_\nu \to \mathcal{J}^{\mathfrak{g},\mathfrak{p}}_{\mathrm{loc}}$ (for any ν) in this case. In other words, the conditions of Theorem 6.3.1 are the necessary and sufficient conditions for the existence of such a homomorphism.

6.3.3 General parabolic subalgebras

So far we have worked under the assumption that the parabolic subalgebra $\mathfrak{p}$ contains $\mathfrak{b}_-$. It is also possible to construct Wakimoto modules associated to other parabolic subalgebras. Let us explain how to do this in the case when $\mathfrak{p} = \mathfrak{b}_+$.

Let N be any module over the vertex algebra $M_\mathfrak{g} \otimes \pi_0^{\kappa-\kappa_c}$. There is an involution of $\mathfrak{g}$ sending e_i to f_i and h_i to $-h_i$. Under this involution $\mathfrak{b}_-$ goes to $\mathfrak{b}_+$. This involution induces an involution on $\widehat{\mathfrak{g}}_\kappa$. Then Theorem 6.2.1 implies that the following formulas define a $\widehat{\mathfrak{g}}_\kappa$-structure on N (with $\mathbf{1}$ acting as the identity):

$$f_i(z) \mapsto a_{\alpha_i}(z) + \sum_{\beta\in\Delta_+} :P^i_\beta(a^*_\alpha(z))a_\beta(z): ,$$

$$h_i(z) \mapsto \sum_{\beta\in\Delta_+} \beta(h_i):a^*_\beta(z)a_\beta(z): - b_i(z),$$

$$e_i(z) \mapsto \sum_{\beta\in\Delta_+} :Q^i_\beta(a^*_\alpha(z))a_\beta(z): + (c_i + (\kappa - \kappa_c)(e_i, f_i))\, \partial_z a^*_{\alpha_i}(z) + b_i(z)a^*_{\alpha_i}(z),$$

where the polynomials P^i_β, Q^i_β were introduced in formulas (5.2.12)–(5.2.14).

For the resulting $\widehat{\mathfrak{g}}_\kappa$-module to be a module with highest weight, we choose N as follows. Let $M'_\mathfrak{g}$ be the Fock representation of the Weyl algebra $\mathcal{A}^\mathfrak{g}$ generated by a vector $|0\rangle'$ such that

$$a_{\alpha,n}|0\rangle' = 0, \quad n > 0; \qquad a^*_{\alpha,n}|0\rangle' = 0, \quad n \geq 0.$$

We take as N the module $M'_\mathfrak{g} \otimes \pi^{\kappa-\kappa_c}_{-2\rho-\lambda}$, where $\pi^{\kappa-\kappa_c}_{-2\rho-\lambda}$ is the $\pi_0^{\kappa-\kappa_c}$-module defined in Section 6.2.1. We denote this module by $W^+_{\lambda,\kappa}$. This is the generalized Wakimoto module corresponding to the parabolic subalgebra $\mathfrak{b}_+$. We will use the same notation $|0\rangle'$ for the vector

$$|0\rangle' \otimes |-2\rho-\lambda\rangle \in M'_\mathfrak{g} \otimes \pi^{\kappa-\kappa_c}_{-2\rho-\lambda} = W^+_{\lambda,\kappa}. \tag{6.3.8}$$

The following result, which identifies a particular Wakimoto module with a Verma module, will be used in Section 8.1.1.

Consider the Lie algebra $\widehat{\mathfrak{n}}_+ = (\mathfrak{g} \otimes t\mathbb{C}[[t]]) \oplus (\mathfrak{n}_+ \otimes 1)$. For $\lambda \in \mathfrak{h}^*$, let $\mathbb{C}_\lambda$ be the one-dimensional representation of $\widehat{\mathfrak{n}}_+ \oplus (\mathfrak{h} \otimes 1) \oplus \mathbb{C}\mathbf{1}$, on which $\widehat{\mathfrak{n}}_+ = (\mathfrak{g} \otimes t\mathbb{C}[[t]]) \oplus (\mathfrak{n}_+ \otimes 1)$ acts by 0, $\mathfrak{h} \otimes 1$ acts according to λ, and $\mathbf{1}$ acts as the identity. Define the **Verma module** $\mathbb{M}_{\lambda,\kappa}$ of level κ and highest weight λ as the corresponding induced $\widehat{\mathfrak{g}}_\kappa$ module:

$$\mathbb{M}_{\lambda,\kappa} = \operatorname{Ind}_{\widehat{\mathfrak{n}}_+ \oplus (\mathfrak{h} \otimes 1) \oplus \mathbf{1}}^{\widehat{\mathfrak{g}}_\kappa} \mathbb{C}_\lambda. \tag{6.3.9}$$

The image in $\mathbb{M}_{\lambda,\kappa}$ of the vector $1 \otimes 1$ of this tensor product is the **highest weight vector** of $\mathbb{M}_{\lambda,\kappa}$. We denote it by $v_{\lambda,\kappa}$.

Proposition 6.3.3 *The Wakimoto module W^+_{0,κ_c} is isomorphic to the Verma module $\mathbb{M}_{0,\kappa_c}$.*

Proof. The vector $|0\rangle' \in W^+_{0,\kappa_c}$ given by formula (6.3.8) satisfies the same properties as the highest weight vector $v_{\lambda,\kappa} \in \mathbb{M}_{0,\kappa_c}$: it is annihilated by $\widehat{\mathfrak{n}}_+$, the Lie algebra $\mathfrak{h} \otimes 1$ acts via the functional λ, and $\mathbf{1}$ acts as the identity. Therefore there is a non-zero homomorphism $\mathbb{M}_{0,\kappa_c} \to W^+_{0,\kappa_c}$ sending the highest weight vector of $\mathbb{M}_{0,\kappa_c}$ to $|0\rangle' \in W^+_{0,\kappa_c}$.

We start by showing that the characters of the modules $\mathbb{M}_{0,\kappa_c}$ and W^+_{0,κ_c} are equal. This will reduce the problem to showing that the above homomorphism is surjective.

Let us recall the notion of character of a $\widehat{\mathfrak{g}}_\kappa$-module. Suppose that we have a $\widehat{\mathfrak{g}}_\kappa$-module M equipped with an action of the grading operator $L_0 = -t\partial_t$, compatible with its action on $\widehat{\mathfrak{g}}_\kappa$.†

Suppose in addition that L_0 and $\mathfrak{h} \otimes 1 \subset \widehat{\mathfrak{g}}_\kappa$ act diagonally on M with finite-dimensional common eigenspaces. Then we define the character of M as the formal series

$$\operatorname{ch} M = \sum_{\widehat{\lambda} \in (\mathfrak{h} \oplus \mathbb{C}L_0)^*} \dim M(\widehat{\lambda})\, e^{\widehat{\lambda}}, \tag{6.3.10}$$

where $M(\widehat{\lambda})$ is the generalized eigenspace of L_0 and $\mathfrak{h} \otimes 1$ corresponding to $\widehat{\lambda} : (\mathfrak{h} \oplus \mathbb{C}L_0)^* \to \mathbb{C}$.

The direct sum $(\mathfrak{h} \otimes 1) \oplus \mathbb{C}L_0 \oplus \mathbb{C}\mathbf{1}$ is in fact the Cartan subalgebra of the extended Kac–Moody algebra $\widehat{\mathfrak{g}}'_\kappa = \mathbb{C}L_0 \ltimes \widehat{\mathfrak{g}}_\kappa$ (see [K2]). Elements of the dual space to this Cartan subalgebra are called weights. We will consider the weights occurring in modules on which the central element $\mathbf{1}$ acts as the identity. Therefore without loss of generality we may view these weights as elements of the dual space to $\widetilde{\mathfrak{h}} = (\mathfrak{h} \otimes 1) \oplus \mathbb{C}L_0$, and hence as pairs (λ, ϕ), where $\lambda \in \mathfrak{h}^*$ and ϕ is the value of $-L_0 = t\partial_t$. We will use the standard notation $\delta = (0, 1)$.

† Note that if $\kappa \neq \kappa_c$, then any smooth $\widehat{\mathfrak{g}}_\kappa$-module carries an action of the Virasoro algebra obtained via the Segal–Sugawara construction, and so in particular an L_0 action. However, general $\widehat{\mathfrak{g}}_{\kappa_c}$-modules do not necessarily carry an L_0 action.

The set of positive roots of $\widehat{\mathfrak{g}}$ is naturally a subset of $\widetilde{\mathfrak{h}}^*$:

$$\widehat{\Delta}_+ = \{\alpha + n\delta \,|\, \alpha \in \Delta_+, n \geq 0\} \sqcup \{-\alpha + n\delta \,|\, \alpha \in \Delta_+, n > 0\} \sqcup \{n\delta \,|\, n > 0\}.$$

The roots of the first two types are real roots; they have multiplicity 1. The roots of the last type are imaginary; they have multiplicity ℓ.

We have a natural partial order on the set $\widetilde{\mathfrak{h}}^*$ of weights: $\widehat{\lambda} > \widehat{\mu}$ if $\widehat{\lambda} - \widehat{\mu} = \sum_i \widehat{\beta}_i$, where the $\widehat{\beta}_i$'s are positive roots.

Let $\mathbb{M}_{\lambda,\kappa}$ be the Verma module over $\widehat{\mathfrak{g}}_\kappa$ defined above. There is a unique way to extend the action of $\widehat{\mathfrak{g}}_\kappa$ to $\widehat{\mathfrak{g}}'_\kappa$ by setting $L_0 \cdot v_{\lambda,\kappa_c} = 0$ and using the commutation relations $[L_0, A_n] = -nA_n$ between L_0 and $\widehat{\mathfrak{g}}_\kappa$ to define the action of L_0 on the rest of $\mathbb{M}_{\lambda,\kappa}$. The resulting module is the Verma module over $\widehat{\mathfrak{g}}'_{\kappa_c}$ with highest weight $\widehat{\lambda} = (\lambda, 0)$, which we will denote by $\mathbb{M}_{\widehat{\lambda},\kappa}$.

By the Poincaré–Birkhoff–Witt theorem, as a vector space $\mathbb{M}_{\widehat{\lambda},\kappa}$ is isomorphic to $U(\widehat{\mathfrak{n}}_-)$, where $\widehat{\mathfrak{n}}_- = (\mathfrak{g} \otimes t^{-1}\mathbb{C}[t^{-1}]) \oplus (\mathfrak{n}_- \otimes 1)$. Therefore we obtain the following formula for the character of $\mathbb{M}_{\widehat{\lambda},\kappa}$:

$$\operatorname{ch} \mathbb{M}_{\widehat{\lambda},\kappa} = e^{\widehat{\lambda}} \prod_{\widehat{\alpha} \in \widehat{\Delta}_+} (1 - e^{-\widehat{\alpha}})^{-\operatorname{mult} \widehat{\alpha}}, \tag{6.3.11}$$

where $\widehat{\Delta}_+$ is the set of positive roots of $\widehat{\mathfrak{g}}_\kappa$.

On the other hand, we have an action of $\operatorname{Der} \mathcal{O}$, and in particular of L_0, on W^+_{0,κ_c} coming from the quasi-conformal vertex algebra structure on $M_\mathfrak{g} \otimes \pi_0$ described in Section 6.2.4. It follows from the formulas obtained in Section 6.2.5 that we have the following commutation relations:

$$[L_0, a_{\alpha,n}] = -na_{\alpha,n}, \quad [L_0, a^*_{\alpha,n}] = -na^*_{\alpha,n}, \quad [L_0, b_{i,n}] = -nb_{i,n}.$$

These formulas, together with the requirement that $L_0|0\rangle' = 0$, uniquely determine the action of L_0 on W^+_{0,κ_c}. Next, we have an action of the Cartan subalgebra $\mathfrak{h} \otimes 1 \subset \widehat{\mathfrak{g}}_{\kappa_c}$ on W^+_{0,κ_c} such that

$$[h, a_{\alpha,n}] = \alpha(h)a_{\alpha,n}, \quad [h, a^*_{\alpha,n}] = \alpha(h)a^*_{\alpha,n}, \quad [h, b_{i,n}] = 0$$

for $h \in \mathfrak{h}$. Both L_0 and $\mathfrak{h} \otimes 1$ act by 0 on $|0\rangle \in W^+_{0,\kappa_c}$. Since W^+_{0,κ_c} has a basis of monomials in $a_{\alpha_n}, \alpha \in \Delta_+, n < 0$; $a^*_{\alpha,n}, \alpha \in \Delta_+, n \leq 0$; and $b_{i,n}, i = 1, \ldots, \ell, n < 0$, we find that the character of W^+_{0,κ_c} is equal to the character of $\mathbb{M}_{0,\kappa_c}$ given by formula (6.3.11).

The homomorphism $\mathbb{M}_{0,\kappa_c} \to W^+_{0,\kappa_c}$ intertwines the action of $\widehat{\mathfrak{g}}'_{\kappa_c} = \mathbb{C}L_0 \ltimes \widehat{\mathfrak{g}}_{\kappa_c}$ on both modules. Therefore our proposition will follow if we show that the homomorphism $\mathbb{M}_{0,\kappa_c} \to W^+_{0,\kappa_c}$ is surjective, or, equivalently, that W^+_{0,κ_c} is generated by the vector $|0\rangle'$ given by formula (6.3.8).

Suppose that W^+_{0,κ_c} is not generated by $|0\rangle'$. Then the space of coinvariants of W^+_{0,κ_c} with respect to the Lie algebra

$$\widehat{\mathfrak{n}}_- = (\mathfrak{n}_- \otimes 1) \oplus (\mathfrak{g} \otimes t^{-1}\mathbb{C}[t^{-1}])$$

has dimension greater than 1, because it must include some vectors in addition to the one-dimensional subspace spanned by the image of the highest weight vector. This means that there exists a homogeneous linear functional on W^+_{0,κ_c}, whose weight is less than the highest weight $(0,0)$ and which is $\widehat{\mathfrak{n}}_-$-invariant.

Then it is in particular invariant under the Lie subalgebra

$$L_-\mathfrak{b}_- = \mathfrak{n}_- \otimes \mathbb{C}[t^{-1}] \oplus \mathfrak{h} \otimes t^{-1}\mathbb{C}[t^{-1}].$$

Therefore this functional necessarily factors through the space of coinvariants of W^+_{0,κ_c} by $L_-\mathfrak{b}_-$.

However, it follows from the construction of W^+_{0,κ_c} that $L_-\mathfrak{b}_-$ acts freely on W^+_{0,κ_c}, and the space of coinvariants with respect to this action is isomorphic to the subspace

$$\mathbb{C}[a^*_{\alpha,n}]_{\alpha\in\Delta_+,n<0} \subset W^+_{0,\kappa_c}.$$

Indeed, it follows from the explicit formulas (6.1.2) and (6.2.2) for the action of $e_\alpha(z)$ and $h_i(z)$ (which become $f_\alpha(z)$ and $-h_i(z)$ after we apply our involution) that the lexicographically ordered monomials

$$\prod_{l_a<0} h_{i_a,l_a} \prod_{m_b\leq 0} f_{\alpha_b,m_b} \prod_{n_c<0} a^*_{\beta_c,n_b}|0\rangle'$$

form a basis of W^+_{0,κ_c}.

Hence we obtain that any $L_-\mathfrak{b}_-$-invariant functional on W^+_{0,κ_c} is completely determined by its restriction to the subspace $\mathbb{C}[a^*_{\alpha,n}]_{\alpha\in\Delta_+,n<0}$. Thus, a non-zero $L_-\mathfrak{b}_-$-invariant functional on W^+_{0,κ_c} of weight strictly less than the highest weight $(0,0)$ necessarily takes a non-zero value on a homogeneous subspace of $\mathbb{C}[a^*_{\alpha,n}]_{\alpha\in\Delta_+,n<0}$ of non-zero weight. But the weights of these subspaces are of the form

$$-\sum_j (n_j\delta - \beta_j), \qquad n_j > 0, \quad \beta_j \in \Delta_+. \tag{6.3.12}$$

Since the weight of our subspace is supposed to be less than the highest weight by our assumption, the number of summands in this formula has to be non-zero.

This implies that W^+_{0,κ_c} must have an irreducible subquotient of highest weight of this form. Since the characters of W^+_{0,κ_c} and $\mathbb{M}_{0,\kappa_c}$ coincide, and the characters of irreducible highest weight representations are linearly independent (see [KK]), we find that $\mathbb{M}_{0,\kappa_c}$ also has an irreducible subquotient of highest weight of the form (6.3.12).

Now recall the Kac–Kazhdan theorem [KK] describing the set of highest weights of irreducible subquotients of Verma modules. In the case at hand the statement is as follows. A weight $\widehat{\mu} = (\mu, n)$ appears as the highest weight of an irreducible subquotient $\mathbb{M}_{\widehat{\lambda},\kappa_c}$, where $\widehat{\lambda} = (\lambda, 0)$, if and only if $n \leq 0$

and either $\mu = \lambda$ or there exists a finite sequence of weights $\mu_0, \ldots, \mu_m \in \mathfrak{h}^*$ such that $\mu_0 = \mu, \mu_m = 0$, $\mu_{i+1} = \mu_i \pm m_i \beta_i$ for some positive roots β_i and positive integers m_i which satisfy

$$2(\mu_i + \rho, \beta_i) = m_i(\beta_i, \beta_i) \tag{6.3.13}$$

(here $(\cdot, \cdot)$ is the inner product on $\mathfrak{h}^*$ induced by an arbitrary non-degenerate invariant inner product on $\mathfrak{g}$).

Now observe that the equations (6.3.13) coincide with the equations appearing in the analysis of irreducible subquotients of the Verma modules over $\mathfrak{g}$ of highest weights in the orbit of λ under the ρ-shifted action of the Weyl group. In other words, a weight $\widehat{\mu} = (\mu, n)$ appears in the decomposition of $\mathbb{M}_{0,\kappa_c}$ if and only $n \leq 0$ and $\mu = w(\rho) - \rho$ for some element w of the Weyl group of $\mathfrak{g}$. But for any w, the weight $w(\rho) - \rho$ equals the sum of negative simple roots of $\mathfrak{g}$. Hence the weight of any irreducible subquotient of $\mathbb{M}_{0,\kappa_c}$ has the form $-n\delta - \sum_i m_i \alpha_i, n \geq 0, m_i \geq 0$. Such a weight cannot be of the form (6.3.12).

Therefore W^+_{0,κ_c} is generated by the highest weight vector. Therefore the homomorphism $\mathbb{M}_{0,\kappa_c} \to W^+_{0,\kappa_c}$ is surjective. Since the characters of the two modules coincide, we find that W^+_{0,κ_c} is isomorphic to $\mathbb{M}_{0,\kappa_c}$. □

In Proposition 9.5.1 we will generalize this result to the case of Verma modules of other highest weights.

In Section 8.1.1 we will need one more result on the structure of W^+_{0,κ_c}. Consider the Lie algebra $\widetilde{\mathfrak{b}}_+ = (\mathfrak{b}_+ \otimes 1) \oplus (\mathfrak{g} \otimes t\mathbb{C}[[t]])$.

Lemma 6.3.4 *The space of $\widetilde{\mathfrak{b}}_+$-invariants of W^+_{0,κ_c} is equal to $\pi_0 \subset W^+_{0,\kappa_c}$.*

Proof. It follows from the formulas for the action of $\widehat{\mathfrak{g}}_{\kappa_c}$ on W^+_{0,κ_c} given at the beginning of this section that all vectors in π_0 are annihilated by $\widetilde{\mathfrak{b}}_+$. Let us show that there are no other $\widetilde{\mathfrak{b}}_+$-invariant vectors in W^+_{0,κ_c}.

A $\widetilde{\mathfrak{b}}_+$-invariant vector is in particular annihilated by the Lie subalgebra $L_+\mathfrak{n}_- = \mathfrak{n}_- \otimes t\mathbb{C}[[t]]$. In formula (6.1.3) we defined the operators $e^R_{\alpha,n}, \alpha \in \Delta_+, n \in \mathbb{Z}$. These operators generate the right action of the Lie algebra $L\mathfrak{n}_+$ on $M_{\mathfrak{g}}$, which commutes with the (left) action of $L\mathfrak{n}_+$ (which is part of the free field realization of $\widehat{\mathfrak{g}}_{\kappa_c}$). These operators now act on W^+_{0,κ_c}, but because we have applied the involution exchanging $\mathfrak{n}_+$ and $\mathfrak{n}_-$, we will now denote them by $f^R_{\alpha,n}$. They generate an action of the Lie algebra $L\mathfrak{n}_-((t))$ which commutes with the action of $L\mathfrak{n}_-$ which is part of the action of $\widehat{\mathfrak{g}}_{\kappa_c}$ on W^+_{0,κ_c} (see the formula at the beginning of this section).

It is easy to see from the explicit formulas for these operators that the lexicographically ordered monomials of the form

$$\prod_{l_a<0} b_{i_a,l_a} \prod_{m_b \leq 0} f^R_{\alpha_b,m_b} \prod_{n_c<0} a^*_{\alpha_c,n_c} |0\rangle'$$

form a basis in W^+_{0,κ_c}. Thus, we have a tensor product decomposition

$$W^+_{0,\kappa_c} = \overline{W}^+_{0,\kappa_c} \otimes W^{+,*}_{0,\kappa_c},$$

where $W^{+,*}_{0,\kappa_c}$ (resp., $\overline{W}^+_{0,\kappa_c}$) is the span of monomials in $a^*_{\alpha,n}$ only (resp., in $f^R_{\alpha,m}$ and $b_{i,l}$ only). Moreover, because the action of $L_+\mathfrak{n}_-$ commutes with $f^R_{\alpha,m}$ and $b_{i,l}$, we find that $L_+\mathfrak{n}_-$ acts only along the second factor of this tensor product decomposition.

Next, we prove, in the same way as in Lemma 5.6.4 that, as an $L_+\mathfrak{n}_-$-module, $W^{+,*}_{0,\kappa_c}$ is isomorphic to the restricted dual of the free module with one generator. Therefore the space of $L_+\mathfrak{n}_-$-invariants in $W^{+,*}_{0,\kappa_c}$ is one-dimensional, spanned by the highest weight vector. We conclude that the space of $L_+\mathfrak{n}_-$-invariants of W^+_{0,κ_c} is equal to the subspace $\overline{W}^+_{0,\kappa_c}$.

Now suppose that we have a $\widetilde{\mathfrak{b}}_+$-invariant vector in W^+_{0,κ_c}. Then it necessarily belongs to $\overline{W}^+_{0,\kappa_c}$. But it is also annihilated by other elements of $\widetilde{\mathfrak{b}}_+$, in particular, by $\mathfrak{h} \otimes 1 \subset \widetilde{\mathfrak{b}}_+$. Since

$$[h \otimes 1, a^*_{\alpha,n}] = \alpha(h) a^*_{\alpha,n}, \qquad h \in \mathfrak{h},$$

we find that a vector in $\overline{W}^+_{0,\kappa_c}$ is annihilated by $\mathfrak{h} \otimes 1$ if only if it belongs to π_0 (in which case it is annihilated by the entire Lie subalgebra $\widetilde{\mathfrak{b}}_+$). Hence the space of $\widetilde{\mathfrak{b}}_+$-invariants of W^+_{0,κ_c} is equal to π_0. □

6.4 Appendix: Proof of the Kac–Kazhdan conjecture

As an application of the Wakimoto modules, we give a proof of the Kac–Kazhdan conjecture from [KK], following [F1, F4].

We will use the extended affine Kac–Moody algebra, $\widehat{\mathfrak{g}}'_\kappa = \mathbb{C}L_0 \ltimes \widehat{\mathfrak{g}}_\kappa$ and the weights of the extended Cartan subalgebra $\widetilde{\mathfrak{h}} = \mathbb{C}L_0 \oplus (\mathfrak{h} \otimes 1)$, as described in the proof of Proposition 6.3.3. It is known that the Verma module $\mathbb{M}_{\widehat{\lambda},\kappa}$ over $\widetilde{\mathfrak{g}}_\kappa$ has a unique irreducible quotient, which we denote by $L_{\widehat{\lambda},\kappa}$.

Let us recall that in [KK] a certain subset $H^\kappa_{\beta,m} \in \widetilde{\mathfrak{h}}^*$ is defined for any pair (β, m), where β is a positive root of $\widetilde{\mathfrak{g}}_\kappa$ and m is a positive integer. If β is a real root, then $H^\kappa_{\beta,m}$ is a hyperplane in $\widetilde{\mathfrak{h}}^*$ and if β is an imaginary root, then $H^{\kappa_c}_{\beta,m} = \widetilde{\mathfrak{h}}^*$ and $H^\kappa_{\beta,m} = \emptyset$ for $\kappa \neq \kappa_c$. It is shown in [KK] that $L_{\widehat{\lambda},\kappa}$ is a subquotient of $\mathbb{M}_{\widehat{\mu},\kappa}$ if and only if the following condition is satisfied: there exists a finite sequence of weights $\widehat{\mu}_0, \ldots, \widehat{\mu}_n$ such that $\widehat{\mu}_0 = \widehat{\lambda}, \widehat{\mu}_n = \widehat{\mu}$, $\widehat{\mu}_{i+1} = \widehat{\mu}_i - m_i\beta_i$ for some positive roots β_i and positive integers m_i, and $\widehat{\mu}_i \in H^\kappa_{\beta_i,m_i}$ for all $i = 1, \ldots, n$.

Denote by $\widehat{\Delta}^{\text{re}}_+$ the set of positive real roots of $\widehat{\mathfrak{g}}_\kappa$ (see the proof of Proposition 6.3.3). Let us call a weight $\widehat{\lambda}$ a **generic weight of critical level** if $\widehat{\lambda}$ does not belong to any of the hyperplanes $H^\kappa_{\beta,m}, \beta \in \widehat{\Delta}^{\text{re}}_+$. It is easy to see from the above condition that $\widehat{\lambda}$ is a generic weight of critical level if and only if the only irreducible subquotients of $\mathbb{M}_{\widehat{\lambda},\kappa_c}$ have highest weights $\widehat{\lambda}-n\delta$, where n is a non-negative integer (i.e., their $\mathfrak{h}^*$ components are equal to the $\mathfrak{h}^*$ component of $\widehat{\lambda}$). The following assertion is the Kac–Kazhdan conjecture for the untwisted affine Kac–Moody algebras.

Theorem 6.4.1 *For generic weight $\widehat{\lambda}$ of critical level*

$$\operatorname{ch} L_{\widehat{\lambda},\kappa_c} = e^{\widehat{\lambda}} \prod_{\alpha\in\widehat{\Delta}_+^{\mathrm{re}}} (1-e^{-\alpha})^{-1}.$$

Proof. Without loss of generality, we may assume that $\widehat{\lambda} = (\lambda, 0)$.

Introduce the gradation operator L_0 on the Wakimoto module $W_{\chi(t)}$ by using the vertex algebra gradation on $M_{\mathfrak{g}}$. It is clear from the formulas defining the $\widehat{\mathfrak{g}}_{\kappa_c}$-action on $W_{\chi(t)}$ given in Theorem 6.1.6 that this action is compatible with the gradation if and only if $\chi(t) = \lambda/t$, where $\lambda \in \mathfrak{h}^*$. In that case

$$\operatorname{ch} W_{\lambda/t} = e^{\widehat{\lambda}} \prod_{\alpha\in\widehat{\Delta}_+^{\mathrm{re}}} (1-e^{-\alpha})^{-1},$$

where $\widehat{\lambda} = (\lambda, 0)$. Thus, in order to prove the theorem we need to show that if $\widehat{\lambda}$ is a generic weight of critical level, then $W_{\lambda/t}$ is irreducible. Suppose that this is not so. Then either $W_{\lambda/t}$ contains a singular vector, i.e., a vector annihilated by the Lie subalgebra $\widehat{\mathfrak{n}}_+ = (\mathfrak{g}\otimes t\mathbb{C}[[t]])\oplus(\mathfrak{n}_+\otimes 1)$, other than the multiples of the highest weight vector, or $W_{\lambda/t}$ is not generated by its highest weight vector.

Suppose that $W_{\lambda/t}$ contains a singular vector other than a multiple of the highest weight vector. Such a vector must then be annihilated by the Lie subalgebra $L_+\mathfrak{n}_+ = \mathfrak{n}_+[[t]]$. We have introduced in Remark 6.1.1 the right action of $\mathfrak{n}_+((t))$ on $M_{\mathfrak{g}}$, which commutes with the left action. It is clear from formula (6.1.3) that the monomials

$$\prod_{n_a<0} e^R_{\alpha_a,n_a} \prod_{m_b\le 0} a^*_{\alpha_b,m_b}|0\rangle \tag{6.4.1}$$

form a basis of $M_{\mathfrak{g}}$. We show in the same way as in the proof of Lemma 6.3.4 that the space of $L_+\mathfrak{n}_+$-invariants of W_{0,κ_c} is equal to the subspace $\overline{W}_{0,\kappa_c}$ of W_{0,κ_c} spanned by all monomials (6.4.1) not containing $a^*_{\alpha,m}$'s.

In particular, we find that the weight of any singular vector of $W_{\lambda/t}$ which is not equal to the highest weight vector has the form $(\lambda,0)-\sum_j(n_j\delta-\beta_j)$, where $n_j>0$ and each β_j is a positive root of $\mathfrak{g}$. But then $W_{\lambda/t}$ contains an irreducible subquotient of such a weight.

Now observe that

$$\operatorname{ch} \mathbb{M}_{(\lambda,0),\kappa_c} = \prod_{n>0}(1-q^n)^{-\ell}\cdot \operatorname{ch} W_{\lambda/t},$$

where $q = e^{-\delta}$ (see the proof of Proposition 6.3.3). If an irreducible module $L_{\widehat{\mu},\kappa_c}$ appears as a subquotient of $W_{\lambda/t}$, then it appears in the decomposition of $\operatorname{ch} W_{\lambda/t}$ into the sum of characters of irreducible representations and hence in the decomposition of $\operatorname{ch} \mathbb{M}_{(\lambda,0),\kappa_c}$. Since the characters of irreducible representations are linearly independent (see [KK]), this implies that $L_{\widehat{\mu},\kappa_c}$ is an irreducible subquotient of $\mathbb{M}_{(\lambda,0),\kappa_c}$. But this contradicts our assumption that $(\lambda, 0)$ is a generic weight of critical level. Therefore we conclude that $W_{\lambda/t}$ does not contain any singular vectors other than the multiples of the highest weight vector.

Next, suppose that $W_{\lambda/t}$ is not generated by its highest weight vector. But then there exists a homogeneous linear functional on $W_{\lambda/t}$, whose weight is less than the highest weight and which is invariant under $\widehat{\mathfrak{n}}_- = (\mathfrak{g} \otimes t^{-1}\mathbb{C}[t^{-1}]) \oplus (\mathfrak{n}_- \otimes 1)$, and in particular, under its Lie subalgebra $L_-\mathfrak{n}_+ = \mathfrak{n}_+ \otimes t^{-1}\mathbb{C}[t^{-1}]$. Therefore this functional factors through the space of coinvariants of $W_{\lambda/t}$ by $L_-\mathfrak{n}_+$. But $L_-\mathfrak{n}_+$ acts freely on $W_{\lambda/t}$, and the space of coinvariants is isomorphic to the subspace $\mathbb{C}[a^*_{\alpha,n}]_{\alpha\in\Delta_+,n\leq 0}$ of $W_{\lambda/t}$. Hence we obtain that the weight of this functional has the form $(\lambda, 0) - \sum_j (n_j\delta + \beta_j)$, where $n_j \geq 0$ and each β_j is a positive root of $\mathfrak{g}$. In the same way as above, it follows that this contradicts our assumption that λ is a generic weight. Therefore $W_{\lambda/t}$ is generated by its highest weight vector. We also know that it does not contain any singular vectors other than the highest weight vector. Hence $W_{\lambda/t}$ is irreducible. This completes the proof. □

7
Intertwining operators

We are now ready to prove Theorem 4.3.2. The proof is presented in this chapter and the next, following [F4]. The theorem was originally proved in [FF6, F1] (we note that a closely related statement, Theorem 8.3.1, was conjectured by V. Drinfeld).

At the beginning of this chapter we outline the overall strategy of the proof (see Section 7.1) and then make the first two steps in the proof. We then develop the theory of intertwining operators between Wakimoto modules, which is an important step of the proof of Theorem 4.3.2. First, we do it in Section 7.2 in the case of $\widehat{\mathfrak{sl}}_2$. We construct explicitly intertwining operators, which we call the screening operators of the first and second kind. In Section 7.3 we use these operators and the functor of parabolic induction to construct intertwining operators between Wakimoto modules over an arbitrary affine Kac–Moody algebra.

7.1 Strategy of the proof

Our strategy in proving Theorem 4.3.2 will be as follows. In Theorem 6.1.6 we constructed a free field realization homomorphism of vertex algebras

$$w_{\kappa_c} : V_{\kappa_c}(\mathfrak{g}) \to M_{\mathfrak{g}} \otimes \pi_0. \tag{7.1.1}$$

Step 1. We will show that the homomorphism (7.1.1) is injective.

Step 2. We will show that the image of $\mathfrak{z}(\widehat{\mathfrak{g}}) \subset V_{\kappa_c}(\mathfrak{g})$ under w_{κ_c} is contained in $\pi_0 \subset M_{\mathfrak{g}} \otimes \pi_0$.

Thus, we need to describe the image of $\mathfrak{z}(\widehat{\mathfrak{g}})$ in π_0.

Step 3. We will construct the **screening operators** $\overline{S}_i, i = 1, \ldots, \ell$, from $W_{0,\kappa_c} = M_{\mathfrak{g}} \otimes \pi_0$ to some other modules, which commute with the action of $\widehat{\mathfrak{g}}_{\kappa_c}$.

Step 4. We will show that the image of $V_{\kappa_c}(\mathfrak{g})$ under w_{κ_c} is contained in

$\bigcap_{i=1}^{\ell} \operatorname{Ker} \overline{S}_i$. This implies that the image of $\mathfrak{z}(\widehat{\mathfrak{g}})$ is contained in $\bigcap_{i=1}^{\ell} \operatorname{Ker} \overline{V}_i[1]$, where $\overline{V}_i[1]$ is the restriction of $\overline{S}_i$ to π_0.

Step 5. By using the associated graded of our modules and the isomorphism $W^+_{0,\kappa_c} \simeq \mathbb{M}_{0,\kappa_c}$ from Proposition 6.3.3, we will find the character of $\mathfrak{z}(\widehat{\mathfrak{g}})$. We will show that it is equal to the character of $\bigcap_{i=1}^{\ell} \operatorname{Ker} \overline{V}_i[1]$. Therefore we will obtain that

$$\mathfrak{z}(\widehat{\mathfrak{g}}) = \bigcap_{i=1}^{\ell} \operatorname{Ker} \overline{V}_i[1].$$

Step 6. By using **Miura opers**, we will show that there is a natural isomorphism

$$\operatorname{Fun} \operatorname{Op}_{{}^L G}(D) \simeq \bigcap_{i=1}^{\ell} \operatorname{Ker} \overline{V}_i[1].$$

Therefore we obtain that

$$\mathfrak{z}(\widehat{\mathfrak{g}}) \simeq \operatorname{Fun} \operatorname{Op}_{{}^L G}(D).$$

This will give us the proof of Theorem 6.1.6 because we will show that the above identifications preserve the natural actions of the group $\operatorname{Aut} \mathcal{O}$. We will also show that this isomorphism satisfies various other compatibilities.

The proof of Theorem 6.1.6 presented in this book follows closely the paper [F4]. This proof is different from the original proof from [FF6, F1] in two respects. First of all, we use the screening operators of the second kind rather than the first kind; their $\kappa \to \kappa_c$ limits are easier to study. Second, we use the isomorphism between the Verma module of critical level with highest weight 0 and a certain Wakimoto module and the computation of the associated graded of the spaces of singular vectors to estimate the character of the center.

We now perform Steps 1 and 2 of the above plan. Then we will take up Steps 3 and 4 in the rest of this chapter.

7.1.1 Finite-dimensional case

We start with the statement of Step 1. In fact, we will prove a more general result that applies to an arbitrary κ, and not only to κ_c.

Before presenting the proof of this statement, it is instructive to consider its analogue in the finite-dimensional case.

Recall that in Section 5.2 we constructed the homomorphism (5.2.10),

$$\widetilde{\rho} : U(\mathfrak{g}) \to \operatorname{Fun} \mathfrak{h}^* \underset{\mathbb{C}}{\otimes} \mathcal{D}(N_+). \tag{7.1.2}$$

We wish to prove that this homomorphism is injective.

Consider the Poincaré–Birkhoff–Witt filtration on $U(\mathfrak{g})$, and the filtration on $\operatorname{Fun}\mathfrak{h}^* \underset{\mathbb{C}}{\otimes} \mathcal{D}(N_+)$ defined as follows: its nth term is the direct sum

$$\bigoplus_{i=0}^{n} (\operatorname{Fun}\mathfrak{h}^*)_{\leq i} \otimes \mathcal{D}(N_+)_{\leq (n-i)}.$$

Here $(\operatorname{Fun}\mathfrak{h}^*)_{\leq i}$ denotes the space of polynomials of degrees less than or equal to i, and $\mathcal{D}(N_+)_{\leq (n-i)}$ is the space of differential operators of order less than or equal to $n-i$.

It is clear from the formulas presented in Section 5.2 that the homomorphism $\widetilde{\rho}$ preserves these filtrations. Consider the corresponding homomorphism of the associated graded algebras. Recall that $\operatorname{gr} U(\mathfrak{g}) = \operatorname{Fun}\mathfrak{g}^*$ and $\operatorname{gr}\mathcal{D}(N_+) = \operatorname{Fun} T^*N_+$. Obviously, $\operatorname{gr}\operatorname{Fun}\mathfrak{h}^* = \operatorname{Fun}\mathfrak{h}^*$. Therefore we obtain a homomorphism of commutative algebras

$$\operatorname{Fun}\mathfrak{g}^* \to \operatorname{Fun}\mathfrak{h}^* \underset{\mathbb{C}}{\otimes} \operatorname{Fun} T^*N_+. \tag{7.1.3}$$

It corresponds to a morphism of affine algebraic varieties

$$\mathfrak{h}^* \times T^*N_+ \to \mathfrak{g}^*.$$

Observe that the cotangent bundle to N_+ is identified with the trivial vector bundle over N_+ with the fiber $\mathfrak{n}_+$. Since $\mathfrak{b}_+ = \mathfrak{h} \oplus \mathfrak{n}$, we find that $\mathfrak{h}^* \oplus \mathfrak{n}_+^* = \mathfrak{b}_+^*$. Hence the above morphism is equivalent to a morphism

$$\mathfrak{b}_+^* \times N_+ \to \mathfrak{g}^*.$$

By using a non-degenerate invariant inner product κ_0 on $\mathfrak{g}$, we obtain a morphism

$$p : \mathfrak{b}_- \times N_+ \to \mathfrak{g}. \tag{7.1.4}$$

The latter morphism is easy to describe explicitly. It follows from the definitions that it sends

$$(x, g) \in \mathfrak{b}_- \times N_+ \mapsto gxg^{-1} \in \mathfrak{g}.$$

Therefore the image of p in $\mathfrak{g}$ consists of all elements of $\mathfrak{g}$ which belong to a Borel subalgebra $\mathfrak{b}$ such that the corresponding point of $\mathrm{Fl} = G/B_-$ is in the open N_+-orbit $\mathcal{U} = N_+ \cdot [1]$ (we say that such $\mathfrak{b}$ is in **generic relative position** with $\mathfrak{b}_-$). Therefore the image of p is open and dense in $\mathfrak{g}$. In other words, p is dominant. Moreover, a generic element in the image is contained in a unique such Borel subalgebra $\mathfrak{b}$, so p is generically one-to-one. Therefore we find that the homomorphism (7.1.3) is injective, which implies that the homomorphism $\widetilde{\rho}$ of (7.1.2) is also injective.

Note that the morphism p may be recast in the context of the **Grothendieck alteration**. Let $\widetilde{\mathfrak{g}}$ be the variety of pairs $(\mathfrak{b}, x)$, where $\mathfrak{b}$ is a Borel subalgebra in $\mathfrak{g}$ and $x \in \mathfrak{b}$. There is a natural morphism $\widetilde{\mathfrak{g}} \to \mathrm{Fl}$, mapping $(\mathfrak{b}, x)$ to $\mathfrak{b} \in \mathrm{Fl}$.

This map identifies $\widetilde{\mathfrak{g}}$ with a vector bundle over the flag variety Fl, whose fiber over $\mathfrak{b} \in \mathrm{Fl}$ is the vector space $\mathfrak{b}$. There is also a morphism $\widetilde{\mathfrak{g}} \to \mathfrak{g}$ sending $(\mathfrak{b}, x)$ to x. Now let $\mathcal{U}$ be the big cell, i.e., the open N_+-orbit of Fl, and $\widetilde{\mathcal{U}}$ its preimage in $\widetilde{\mathfrak{g}}$. In other words, $\widetilde{\mathcal{U}}$ consists of those pairs $(\mathfrak{b}, x)$ for which $\mathfrak{b}$ is in generic relative position with $\mathfrak{b}_-$. Then $\widetilde{\mathcal{U}}$ is naturally isomorphic to $\mathfrak{b} \times N_+$ and the restriction of the morphism $\widetilde{\mathfrak{g}} \to \mathfrak{g}$ to $\widetilde{\mathcal{U}}$ is the above morphism p.

7.1.2 Injectivity

Now we are ready to consider the analogous question for the affine Lie algebras, which corresponds to Step 1 of our plan.

Proposition 7.1.1 *The homomorphism w_κ of Theorem 6.2.1 is injective for any κ.*

Proof. We apply the same argument as in the finite-dimensional case. Namely, we introduce filtrations on $V_\kappa(\mathfrak{g})$ and $W_{0,\kappa} = M_{\mathfrak{g}} \otimes \pi_0^{\kappa-\kappa_c}$ which are preserved by w_κ, and then show that the induced map $\operatorname{gr} w_\kappa : \operatorname{gr} V_\kappa(\mathfrak{g}) \to \operatorname{gr} W_{0,\kappa}$ (which turns out to be independent of κ) is injective.

The Poincaré–Birkhoff–Witt filtration on $U(\widehat{\mathfrak{g}}_\kappa)$ induces a filtration on $V_\kappa(\mathfrak{g})$ as explained in Section 2.2.5.

Now we define a filtration $\{W_{0,\kappa}^{\leq p}\}$ on $W_{0,\kappa}$ By definition, $W_{0,\kappa}^{\leq p}$ is the span of monomials in the $a_{\alpha,n}$'s, $a^*_{\alpha,n}$'s and $b_{i,n}$'s whose combined degree in the $a_{\alpha,n}$'s and $b_{i,n}$'s is less than or equal to p (this is analogous to the filtration used above in the finite-dimensional case). It is clear from the construction of the homomorphism w_κ that it preserves these filtrations. Moreover, these are filtrations of vertex algebras, and the associated graded spaces are commutative vertex algebras, and in particular commutative algebras. The associated graded $\operatorname{gr} w_\kappa$ of w_κ is therefore a homomorphism of these commutative algebras.

Now we describe the corresponding commutative algebras $\operatorname{gr} V_\kappa(\mathfrak{g})$ and $\operatorname{gr} W_{0,\kappa}$ and the homomorphism $\operatorname{gr} w_\kappa : \operatorname{gr} V_\kappa(\mathfrak{g}) \to \operatorname{gr} W_{0,\kappa}$. We will identify $\mathfrak{g}$ with $\mathfrak{g}^*$ using a non-degenerate invariant inner product κ_0.

Let $J(\mathfrak{b}_- \times N_+) = J\mathfrak{b}_- \times JN_+$ and $J\mathfrak{g}$ be the infinite jet schemes of $\mathfrak{b}_- \times N_+$ and $\mathfrak{g}$, defined as in Section 3.4.2. Then $\operatorname{gr} V_\kappa(\mathfrak{g}) = \operatorname{Fun} J\mathfrak{g}$ and $\operatorname{gr} W_{0,\kappa} = \operatorname{Fun} J(\mathfrak{b}_- \times N_+)$. The homomorphism

$$\operatorname{gr} w_\kappa : \operatorname{Fun} J\mathfrak{g} \to \operatorname{Fun} J(\mathfrak{b}_- \times N_+) \tag{7.1.5}$$

corresponds to a morphism of jet schemes

$$J(\mathfrak{b}_- \times N_+) \to J\mathfrak{g}.$$

It is clear from the construction that it is nothing but the morphism Jp corresponding, by functoriality of J, to the morphism p given by formula (7.1.4).

But p is dominant and generically one-to one. Let $\mathfrak{g}_{\text{gen}}$ be the locus in the image of p in $\mathfrak{g}$ over which p is one-to-one. Then $\mathfrak{g}_{\text{gen}}$ is open and dense in $\mathfrak{g}$. Therefore $J\mathfrak{g}_{\text{gen}}$ is open and dense in $J\mathfrak{g}$, and $J\mathfrak{g}_{\text{gen}}$ is clearly in the image of Jp. Hence we find that Jp is dominant and generically one-to-one. Therefore the homomorphism of rings of functions (7.1.5) is injective. This implies that w_κ is also injective. □

Now we specialize to the critical level. The vertex algebra $V_{\kappa_c}(\mathfrak{g})$ contains the commutative subalgebra $\mathfrak{z}(\widehat{\mathfrak{g}})$, its center. Recall that $\mathfrak{z}(\widehat{\mathfrak{g}})$ is the space of $\mathfrak{g}[[t]]$-invariant vectors in $V_{\kappa_c}(\mathfrak{g})$. On the other hand, $W_{0,\kappa_c} = M_{\mathfrak{g}} \otimes \pi_0$ contains the commutative subalgebra π_0, which is its center.

Lemma 7.1.2 *The image of $\mathfrak{z}(\widehat{\mathfrak{g}}) \subset V_{\kappa_c}(\mathfrak{g})$ in W_{0,κ_c} under w_{κ_c} is contained in $\pi_0 \subset W_{0,\kappa_c}$.*

Proof. We use the same argument as in the proof of Lemma 6.3.4.
Let us observe that the lexicographically ordered monomials of the form

$$\prod_{l_a<0} b_{i_a,l_a} \prod_{m_b<0} e^R_{\alpha_b,m_b} \prod_{n_c\leq 0} a^*_{\alpha_c,n_c}|0\rangle, \tag{7.1.6}$$

where the $e^R_{\alpha,n}$'s are given by formula (6.1.3), form a basis of W_{0,κ_c}. Thus, we have a tensor product decomposition

$$W_{0,\kappa_c} = \overline{W}_{0,\kappa_c} \otimes M_{\mathfrak{g},+},$$

where $M_{\mathfrak{g},+}$ (resp., $\overline{W}_{0,\kappa_c}$) is the span of monomials in $a^*_{\alpha,n}$ only (resp., in $e^R_{\alpha,m}$ and $b_{i,l}$ only).

The image of any element of $\mathfrak{z}(\widehat{\mathfrak{g}})$ in W_{0,κ_c} is an $L_+\mathfrak{g}$-invariant vector, where, as before, we use the notation $L_+\mathfrak{g} = \mathfrak{g}[[t]]$. In particular, it is annihilated by $L_+\mathfrak{n}_+$ and by $\mathfrak{h}$. Since $L_+\mathfrak{n}_+$ commutes with $e^R_{\alpha,n}$ and $b_{i,n}$, it acts only along the second factor $M_{\mathfrak{g},+}$ of the above tensor product decomposition. According to Lemma 5.6.4, $M_{\mathfrak{g},+} \simeq \operatorname{Coind}_{L_+\mathfrak{b}_-}^{L_+\mathfrak{g}} \mathbb{C}$ as an $L_+\mathfrak{n}_+$-module. Therefore the space of $L_+\mathfrak{n}_+$-invariants in $M_{\mathfrak{g},+}$ is one-dimensional, spanned by constants.

Thus, we obtain that the space of $L_+\mathfrak{n}_+$-invariants in W_{0,κ_c} is equal to $\overline{W}_{0,\kappa_c}$. However, the weight of the monomial (7.1.6) without the second factor is equal to the sum of positive roots corresponding to the factors $e^R_{\alpha,m}$. Such a monomial is $\mathfrak{h}$-invariant if and only if there are no such factors present. Therefore the subspace of $(\mathfrak{h} \oplus L_+\mathfrak{n}_+)$-invariants of W_{0,κ_c} is the span of the monomials (7.1.6), which only involve the $b_{i,l}$'s. This is precisely the subspace $\pi_0 \subset W_{0,\kappa_c}$. □

We are now done with Steps 1 and 2 of the proof of Theorem 4.3.2.

7.2 The case of $\mathfrak{sl}_2$

Now we perform Steps 3 and 4 of the plan outlined in Section 7.1. This will enable us to describe the image of $\mathfrak{z}(\widehat{\mathfrak{g}})$ inside π_0 under w_{κ_c} as the intersection of kernels of certain operators.

We start by giving a uniform construction of intertwining operators between Wakimoto modules, the so-called screening operators (of two kinds) following [FF2, FF3, F1, FFR, F4]. First, we analyze in detail the case of $\mathfrak{g} = \mathfrak{sl}_2$.

7.2.1 Vertex operators associated to a module over a vertex algebra

We need to recall some general results on the vertex operators associated to a module over a vertex algebra, following [FHL], Section 5.1. Let V be a conformal vertex algebra V (see the definition in Section 3.1.4) and M a V-module, i.e., a vector space together with a linear map

$$Y_M : V \to \operatorname{End} M[[z^{\pm 1}]]$$

satisfying the axioms of Definition 5.1.1 of [FB]. In particular, the Fourier coefficients of

$$Y_M(\omega, z) = \sum_{n \in \mathbb{Z}} L_n^M z^{-n-2},$$

where ω is the conformal vector of V, define an action of the Virasoro algebra on M. We denote L_{-1}^M by T.

Define a linear map

$$Y_{V,M} : M \to \operatorname{Hom}(V, M)[[z^{\pm 1}]]$$

by the formula

$$Y_{V,M}(A, z)B = e^{zT} Y_M(B, -z)A, \qquad A \in M, B \in V \tag{7.2.1}$$

(compare with the skew-symmetry property of Proposition 2.3.2).

This is an example of **intertwining operators** introduced in [FHL]. By Proposition 5.1.2 of [FHL], this map satisfies the following property: for any $A \in V, B \in M, C \in V$, there exists an element

$$f \in M[[z, w]][z^{-1}, w^{-1}, (z - w)^{-1}]$$

such that the formal power series

$$Y_M(A, z) Y_{V,M}(B, w) C, \qquad Y_{V,M}(B, w) Y(A, z) C,$$

$$Y_{V,M}(Y_{V,M}(B, w - z)A, z)C, \qquad Y_{V,M}(Y(A, z - w)B, w)C$$

are expansions of f in

$$M((z))((w)), \qquad M((w))((z)), \qquad M((z))((z - w)), \qquad M((w))((z - w)),$$

respectively (compare with Corollary 3.2.3 of [FB]). Abusing notation, we will write

$$Y_M(A,z)Y_{V,M}(B,w) = Y_{V,M}(Y(A,z-w)B,w),$$

and call this formula the operator product expansion (OPE), as in the case $M = V$ when $Y_{V,M} = Y$ (see Section 2.3.3).

In the formulas below we will use the same notation $Y(A,z)$ for $Y(A,z)$ and $Y_M(A,z)$. The following commutation relations between the Fourier coefficients of $Y(A,z)$ and $Y_{V,M}(B,w)$ are proved in exactly the same way as in the case $M = V$ (see [FB], Section 3.3.6). If we write

$$Y(A,z) = \sum_{n\in\mathbb{Z}} A_{(n)}z^{-n-1}, \qquad Y_{V,M}(B,w) = \sum_{n\in\mathbb{Z}} B_{(n)}w^{-n-1},$$

then we have

$$[B_{(m)}, A_{(k)}] = \sum_{n\geq 0} \binom{m}{n} (B_{(n)}A)_{(m+k-n)}. \tag{7.2.2}$$

In particular, we obtain that

$$\left[\int Y_{V,M}(B,z)dz, Y(A,w)\right] = Y_{V,M}\left(\int Y_{V,M}(B,z)dz \cdot A, w\right). \tag{7.2.3}$$

Here, as before, $\int$ denotes the residue of at $z = 0$.

Another property that we will need is the following analogue of formula (2.3.5):

$$Y_{V,M}(TA,z) = \partial_z Y_{V,M}(A,z). \tag{7.2.4}$$

It is proved as follows:

$$Y_{V,M}(TA,z)B = e^{zT}Y_M(B,-z)TA =$$

$$e^{zT}TY_M(B,-z)A - e^{zT}[T, Y_M(B,-z)]A = \partial_z(Y_{V,M}(A,z)B),$$

where we use the identity

$$[T, Y_M(B,z)] = \partial_z Y_M(B,z),$$

which follows from [FB], Proposition 5.1.2.

7.2.2 The screening operator

Let us apply the results of Section 7.2.1 in the case of the vertex algebra $W_{0,k}$ and its module $W_{-2,k}$ for $k \neq -2$. As before, we denote by $|0\rangle$ and $|-2\rangle$ the highest weight vectors of these modules.

Recall from Section 6.2.2 that under the homomorphism w_k the generators e_n of $\widehat{\mathfrak{sl}}_2$ are mapped to a_n (from now on, by abuse of notation, we will identify the elements of $\widehat{\mathfrak{sl}}_2$ with their images under w_k). The commutation relations of $\widehat{\mathfrak{sl}}_2$ imply the following formulas:

$$[e_n, a_{-1}] = 0, \qquad [h_n, a_{-1}] = 2a_{n-1}, \quad [f_n, a_{-1}] = -h_{n-1} + k\delta_{n,1}.$$

In addition, it follows from formulas (6.2.3) that

$$e_n|-2\rangle = a_n|-2\rangle = 0, \quad n \geq 0; \qquad h_n|-2\rangle = f_n|-2\rangle = 0, \quad n > 0,$$

$$h_0|-2\rangle = -2|-2\rangle.$$

Therefore

$$e_n \cdot a_{-1}|-2\rangle = h_n \cdot a_{-1}|-2\rangle = 0, \qquad n \geq 0,$$

and

$$f_n \cdot a_{-1}|-2\rangle = 0, \qquad n > 1.$$

We also find that

$$f_1 \cdot a_{-1}|-2\rangle = (k+2)|-2\rangle,$$

and

$$\begin{aligned} f_0 \cdot a_{-1}|-2\rangle &= a_{-1}f_0|-2\rangle - h_0|-2\rangle \\ &= -b_{-1}|-2\rangle = (k+2)T|-2\rangle, \end{aligned} \tag{7.2.5}$$

where T is the translation operator. The last equality follows from the formula for the conformal vector in $W_{0,k}$ given in Proposition 6.2.2, which implies that the action of T on $\pi_0 \subset W_{0,k}$ is given by

$$T = \frac{1}{2(k+2)} \sum_{n \in \mathbb{Z}} b_n b_{-n-1}.$$

We wish to write down an explicit formula for the operator

$$S_k(z) \stackrel{\text{def}}{=} Y_{W_{0,k},W_{-2,k}}(a_{-1}|-2\rangle) : \; W_{0,k} \to W_{-2,k}$$

and use the above properties to show that its residue is an intertwining operator, i.e., it commutes with the action of $\widehat{\mathfrak{sl}}_{2,k}$. Since the vertex subalgebras $M_{\mathfrak{sl}_2}$ and π_0^{k+2} of $W_{0,k}$ commute with each other, we find that

$$S_k(z) = Y_{M_{\mathfrak{sl}_2}}(a_{-1}|0\rangle, z) Y_{\pi_0^{k+2},\pi_{-2}^{k+2}}(|-2\rangle, z) = a(z) Y_{\pi_0^{k+2},\pi_{-2}^{k+2}}(|-2\rangle, z).$$

It remains to determine

$$V_{-2}(z) \stackrel{\text{def}}{=} Y_{\pi_0^{k+2},\pi_{-2}^{k+2}}(|-2\rangle, z) : \; \pi_0^{k+2} \to \pi_{-2}^{k+2}.$$

The identity (7.2.2) specialized to the case $A = b_{-1}|0\rangle$ and $B = |-2\rangle$ implies the following commutation relations:

$$[b_n, V_{-2}(z)] = -2z^n V_{-2}(z). \tag{7.2.6}$$

In addition, we obtain from formulas (7.2.4) and (7.2.5) that

$$(k+2)\partial_z V_{-2}(z) = -{:}b(z)V_{-2}(z){:}. \tag{7.2.7}$$

It is easy to see that formulas (7.2.6) and (7.2.7) determine $V_{-2}(z)$ uniquely (this is explained in detail in [FB], Section 5.2.6). As the result, we obtain the following explicit formula:

$$V_{-2}(z) = T_{-2} \exp\left(\frac{1}{k+2}\sum_{n<0}\frac{b_n}{n}z^{-n}\right)\exp\left(\frac{1}{k+2}\sum_{n>0}\frac{b_n}{n}z^{-n}\right). \tag{7.2.8}$$

Here we denote by T_{-2} the translation operator $\pi_0^{k+2} \to \pi_{-2}^{k+2}$ sending the highest weight vector $|0\rangle$ to the highest weight vector $|-2\rangle$ and commuting with all $b_n, n \neq 0$.

Now, using formula (7.2.2) we obtain that the operator $S_k(z)$ has the following OPEs:

$$e(z)S_k(w) = \text{reg.}, \qquad h(z)S_k(w) = \text{reg.},$$
$$f(z)S_k(w) = \frac{(k+2)V_{-2}(w)}{(z-w)^2} + \frac{(k+2)\partial_w V_{-2}(w)}{z-w} + \text{reg.}$$
$$= (k+2)\partial_w \frac{V_{-2}(w)}{z-w} + \text{reg.}$$

Since the residue of a total derivative is equal to 0, this implies the following:

Proposition 7.2.1 *The residue $S_k = \int S_k(w)dw$ is an intertwining operator between the $\widehat{\mathfrak{sl}}_2$-modules $W_{0,k}$ and $W_{-2,k}$.*

We call S_k the **screening operator of the first kind** for $\widehat{\mathfrak{sl}}_2$.

Proposition 7.2.2 *For $k \notin -2 + \mathbb{Q}_{\geq 0}$ the sequence*

$$0 \to V_k(\mathfrak{sl}_2) \to W_{0,k} \xrightarrow{S_k} W_{-2,k} \to 0$$

is exact.

Proof. By Proposition 7.1.1, $V_k(\mathfrak{sl}_2)$ is naturally an $\widehat{\mathfrak{sl}}_2$-submodule of $W_{0,k}$ for any value of k. The module $V_k(\mathfrak{sl}_2)$ is generated from the vacuum vector $|0\rangle$, whose image in $W_{0,k}$ is the highest weight vector. We have

$$S_k = \sum_{n\in\mathbb{Z}} a_n V_{-2,-n},$$

where $V_{-2,-n}$ is the coefficient in front of z^n in $V_{-2}(z)$. It is clear from formula (7.2.8) that $V_{-2,m}|0\rangle = 0$ for all $m > 0$. We also have $a_n|0\rangle = 0$ for all $n \geq 0$. Therefore $|0\rangle$ belongs to the kernel of S_k. But since S_k commutes with the action of $\widehat{\mathfrak{sl}}_2$, this implies that the entire submodule $V_k(\mathfrak{sl}_2)$ lies in the kernel of S_k.

In order to prove that $V_k(\mathfrak{sl}_2)$ coincides with the kernel of S_k, we compare their characters (see the proof of Proposition 6.3.3 for the definition of the characters). We will use the notation $q = e^{-\delta}, u = e^{\alpha}$.

Since $V_k(\mathfrak{sl}_2)$ is isomorphic to the universal enveloping algebra of the Lie algebra spanned by e_n, h_n and f_n with $n < 0$, we find that

$$\operatorname{ch} V_k(\mathfrak{g}) = \prod_{n>0} (1-q^n)^{-1}(1-uq^n)^{-1}(1-u^{-1}q^n)^{-1}. \tag{7.2.9}$$

Similarly, we obtain that

$$\operatorname{ch} W_{\lambda,k} = u^{\lambda/2} \prod_{n>0} (1-q^n)^{-1}(1-uq^n)^{-1}(1-u^{-1}q^{n-1})^{-1}.$$

Thus, $\operatorname{ch} W_{\lambda,k} = \operatorname{ch} \mathbb{M}_{\lambda,k}$, where $\mathbb{M}_{\lambda,k}$ is the Verma module over $\widehat{\mathfrak{sl}}_2$ with highest weight (λ, k). Therefore if $\mathbb{M}_{\lambda,k}$ is irreducible, then so is $W_{\lambda,k}$. The set of values (λ, k) for which $\mathbb{M}_{\lambda,k}$ is irreducible is described in [KK]. It follows from this description that if $k \notin -2 + \mathbb{Q}_{\geq 0}$, then $\mathbb{M}_{-2,k}$, and hence $W_{-2,k}$, is irreducible. It is easy to check that $S_k(a_0^*|0\rangle) = |-2\rangle$, so that S_k is a non-zero homomorphism. Hence it is surjective for such values of k. Therefore the character of its kernel for such k is equal to $\operatorname{ch} W_{0,k} - \operatorname{ch} W_{-2,k} = \operatorname{ch} V_k(\mathfrak{sl}_2)$. This completes the proof. □

Next, we will describe the second screening operator for $\widehat{\mathfrak{sl}}_2$. For this we need to recall the Friedan–Martinec–Shenker bosonization.

7.2.3 Friedan–Martinec–Shenker bosonization

Consider the Heisenberg Lie algebra with the generators $p_n, q_n, n \in \mathbb{Z}$, and the central element $\mathbf{1}$ with the commutation relations

$$[p_n, p_m] = n\delta_{n,-m}\mathbf{1}, \qquad [q_n, q_m] = -n\delta_{n,-m}\mathbf{1}, \qquad [p_n, q_m] = 0.$$

We set

$$p(z) = \sum_{n\in\mathbb{Z}} p_n z^{-n-1}, \qquad q(z) = \sum_{n\in\mathbb{Z}} q_n z^{-n-1}.$$

For $\lambda \in \mathbb{C}, \mu \in \mathbb{C}$, let $\Pi_{\lambda,\mu}$ be the Fock representation of this Lie algebra generated by a highest weight vector $|\lambda, \mu\rangle$ such that

$$p_n|\lambda,\mu\rangle = \lambda\delta_{n,0}|\lambda,\mu\rangle, \quad q_n|\lambda,\mu\rangle = \mu\delta_{n,0}|\lambda,\mu\rangle, \quad n \geq 0; \quad \mathbf{1}|\lambda,\mu\rangle = |\lambda,\mu\rangle.$$

Consider the vertex operators $V_{\lambda,\mu}(z) : \Pi_{\lambda',\mu'} \to \Pi_{\lambda+\lambda',\mu+\mu'}$ given by the formula

$$V_{\lambda,\mu}(z) = $$

$$T_{\lambda,\mu} z^{\lambda\lambda'-\mu\mu'} \exp\left(-\sum_{n<0} \frac{\lambda p_n + \mu q_n}{n} z^{-n}\right) \exp\left(-\sum_{n>0} \frac{\lambda p_n + \mu q_n}{n} z^{-n}\right),$$

where $T_{\lambda,\mu}$ is the translation operator $\Pi_{0,0} \to \Pi_{\lambda,\mu}$ sending the highest weight vector to the highest weight vector and commuting with all $p_n, q_n, n \neq 0$.

Abusing notation, we will write these operators as $e^{\lambda u + \mu v}$, where $u(z)$ and $v(z)$ stand for the anti-derivatives of $p(z)$ and $q(z)$, respectively, i.e., $p(z) = \partial_z u(z), q(z) = \partial_z v(z)$.

For $\gamma \in \mathbb{C}$, set

$$\Pi_\gamma = \bigoplus_{n \in \mathbb{Z}} \Pi_{n+\gamma, n+\gamma}.$$

Using the vertex operators, one defines a vertex algebra structure on the direct sum Π_0 (with the vacuum vector $|0,0\rangle$) as in [FB], Section 5.2.6.

Moreover, Π_γ is a module over the vertex algebra Π_0 for any $\gamma \in \mathbb{C}$.

The following realization of the vertex algebra M (also known as the "$\beta\gamma$-system") in terms of the vertex algebra Π_0 is due to Friedan, Martinec, and Shenker [FMS]. (The isomorphism between the image of M in Π_0 and the kernel of $\int e^u dz$ stated in the theorem was established in [FF5].)

Theorem 7.2.3 *There is a (unique) embedding of vertex algebras $M \hookrightarrow \Pi_0$ under which the fields $a(z)$ and $a^*(z)$ are mapped to the fields*

$$\widetilde{a}(z) = e^{u+v}, \qquad \widetilde{a}^*(z) = (\partial_z e^{-u})e^{-v} = -{:}p(z)e^{-u-v}{:}\,.$$

Further, the image of M in Π_0 is equal to the kernel of the operator $\int e^u dz$.

Equivalently, Π_0 may be described as the localization of M with respect to a_{-1}, i.e.,

$$\Pi_0 \simeq M[(a_{-1})^{-1}] = \mathbb{C}[a_n]_{n<-1} \otimes \mathbb{C}[a_n^*]_{n\leq 0} \otimes \mathbb{C}[(a_{-1})^{\pm 1}]. \qquad (7.2.10)$$

The vertex algebra structure on Π_0 is obtained by a natural extension of the vertex algebra structure on M.

Under the embedding $M \hookrightarrow \Pi_0$ the Virasoro field $T(z) = {:}\partial_z a^*(z)a(z){:}$ of M described in Section 6.2.3 is mapped to the following field in Π_0:

$$\frac{1}{2}{:}p(z)^2{:} - \frac{1}{2}\partial_z p(z) - \frac{1}{2}{:}q(z)^2{:} + \frac{1}{2}\partial_z q(z).$$

Thus, the map $M \hookrightarrow \Pi_0$ becomes a homomorphism of conformal vertex algebras with respect to the conformal structures corresponding to these fields.

The reason why the FMS bosonization is useful to us is that it allows us to make sense of the field $a(z)^\gamma$, where γ is an arbitrary complex number. Namely, we replace $a(z)^\gamma$ with the field

$$\widetilde{a}(z)^\gamma = e^{\gamma(u+v)} : \Pi_0 \to \Pi_\gamma,$$

which is well-defined.

Now we take the tensor product $\Pi_0 \otimes \pi_0^{k+2}$, where we again assume that $k \neq -2$. This is a vertex algebra which contains $M \otimes \pi_0^{k+2}$, and hence $V_k(\mathfrak{sl}_2)$, as vertex subalgebras. In particular, for any $\gamma, \lambda \in \mathbb{C}$, the tensor product $\Pi_\gamma \otimes \pi_\lambda^{k+2}$ is a module over $V_k(\mathfrak{sl}_2)$ and hence over $\widehat{\mathfrak{sl}}_2$. We denote it by $\widetilde{W}_{\gamma,\lambda,k}$. In addition, we introduce the bosonic vertex operator

$$V_{2(k+2)}(z) = T_{2(k+2)} \exp\left(-\sum_{n<0} \frac{b_n}{n} z^{-n}\right) \exp\left(-\sum_{n>0} \frac{b_n}{n} z^{-n}\right). \tag{7.2.11}$$

Let us set

$$\widetilde{S}_k(z) = \widetilde{a}(z)^{-(k+2)} V_{2(k+2)}(z). \tag{7.2.12}$$

A straightforward computation similar to the one performed in Section 7.2.2 yields:

Proposition 7.2.4 *The residue*

$$\widetilde{S}_k = \int \widetilde{S}_k(z) dz : \widetilde{W}_{0,0,k} \to \widetilde{W}_{-(k+2),2(k+2),k}$$

is an intertwining operator between the $\widehat{\mathfrak{sl}}_2$*-modules* $\widetilde{W}_{0,0,k}$ *and* $\widetilde{W}_{-(k+2),2(k+2),k}$.

We call $\widetilde{S}_k$, or its restriction to $W_{0,k} \subset \widetilde{W}_{0,0,k}$, the **screening operator of the second kind** for $\widehat{\mathfrak{sl}}_2$. This operator was first introduced by V. Dotsenko [Do].

Proposition 7.2.5 *For generic* k *the* $\widehat{\mathfrak{sl}}_2$*-submodule* $V_\kappa(\mathfrak{sl}_2) \subset W_{0,k}$ *is equal to the kernel of* $\widetilde{S}_k : W_{0,k} \to \widetilde{W}_{-(k+2),2(k+2),k}$.

Proof. We will show that for generic k the kernel of $\widetilde{S}_k : W_{0,k} \to \widetilde{W}_{-(k+2),2(k+2),k}$ coincides with the kernel of the screening operator of the first kind, $S_k : W_{0,k} \to W_{-2,k}$. This, together with Proposition 7.2.2, will imply the statement of the proposition.

Note that the operator S_k has an obvious extension to an operator $\widetilde{W}_{0,0,k} \to \widetilde{W}_{0,-2,k}$ defined by the same formula. The kernel of $\widetilde{S}_k$ (resp., S_k) in $W_{0,k}$ is equal to the intersection of $W_{0,k} \subset \widetilde{W}_{0,0,k}$ and the kernel of $\widetilde{S}_k$ (resp., S_k) acting from $\widetilde{W}_{0,0,k}$ to $\widetilde{W}_{-(k+2),2(k+2),k}$ (resp., $\widetilde{W}_{0,-2,k}$). Therefore it is sufficient to show that the kernels of $\widetilde{S}_k$ and S_k on $\widetilde{W}_{0,0,k}$ are equal for generic k.

Now let $\phi(z)$ be the anti-derivative of $b(z)$, i.e., $b(z) = \partial_z \phi(z)$. By abusing notation we will write $V_{-2}(z) = e^{-(k+2)^{-1}\phi}$ and $V_{2(k+2)}(z) = e^{\phi}$. Then the screening currents $S_k(z)$ and $\widetilde{S}_k(z)$ become

$$S_k(z) = e^{u+v-(k+2)^{-1}\phi(z)}, \qquad \widetilde{S}_k(z) = e^{-(k+2)u-(k+2)v+\phi}.$$

Consider a more general situation: let $\mathfrak{h}$ be an abelian Lie algebra with a non-degenerate inner product κ. Using this inner product, we identify $\mathfrak{h}$ with

$\mathfrak{h}^*$. Let $\widehat{\mathfrak{h}}_\kappa$ be the Heisenberg Lie algebra and $\pi_\lambda^\kappa, \lambda \in \mathfrak{h}^* \simeq \mathfrak{h}$ be its Fock representations, defined as in Section 6.2.1. Then for any $\chi \in \mathfrak{h}^*$ there is a vertex operator $V_\chi^\kappa(z) : \pi_0^\kappa \to \pi_\chi^\kappa$ given by the formula

$$V_\chi^\kappa(z) = T_\chi \exp\left(-\sum_{n<0} \frac{\chi_n}{n} z^{-n}\right) \exp\left(-\sum_{n>0} \frac{\chi_n}{n} z^{-n}\right), \tag{7.2.13}$$

where the $\chi_n = \chi \otimes t^n$ are the elements of the Heisenberg Lie algebra $\widehat{\mathfrak{h}}_\kappa$ corresponding to $\chi \in \mathfrak{h}^*$, which we identify with $\mathfrak{h}$ using the inner product κ.

Suppose that $\kappa(\chi, \chi) \neq 0$, and denote by $\check{\chi}$ the element of $\mathfrak{h}$ equal to $-2\chi/\kappa(\chi,\chi)$. We claim that if χ is generic (i.e., away from countably many hypersurfaces in $\mathfrak{h}$), then the kernels of $\int V_\chi^\kappa(z)dz$ and $\int V_{\check{\chi}}^\kappa(z)dz$ in π_0^κ coincide. Indeed, we have a decomposition of $\widehat{\mathfrak{h}}_\kappa$ into a direct sum of the Heisenberg Lie subalgebra $\widehat{\mathfrak{h}}_\kappa^\chi$ generated by $\chi_n, n \in \mathbb{Z}$, and the Heisenberg Lie subalgebra $\widehat{\mathfrak{h}}_\kappa^\perp$ which is the centralizer of $\chi_n, n \in \mathbb{Z}$, in $\widehat{\mathfrak{h}}_\kappa$ (this is a Heisenberg Lie subalgebra of $\widehat{\mathfrak{h}}_\kappa$ corresponding to the orthogonal complement of χ in $\mathfrak{h}$; note that by our assumption on χ, this orthogonal complement does not contain χ). The Fock representation π_0^κ decomposes into a tensor product of Fock representations of these two Lie subalgebras. The operators $\int V_\chi^\kappa(z)dz$ and $\int V_{\check{\chi}}^\kappa(z)dz$ commute with $\widehat{\mathfrak{h}}_\kappa^\perp$. Therefore the kernel of each of these operators in π_0^κ is equal to the tensor product of the Fock representation of $\widehat{\mathfrak{h}}_\kappa^\perp$ and the kernel of this operator on the Fock representation of $\widehat{\mathfrak{h}}_\kappa^\chi$.

In other words, the kernel is determined by the corresponding kernel on the Fock representation of the Heisenberg Lie algebra $\widehat{\mathfrak{h}}_\kappa^\chi$. The latter kernels for $\int V_\chi^\kappa(z)dz$ and $\int V_{\check{\chi}}^\kappa(z)dz$ coincide for generic values of $\kappa(\chi,\chi)$, as shown in [FB], Section 15.4.15. Hence the kernels of $\int V_\chi^\kappa(z)dz$ and $\int V_{\check{\chi}}^\kappa(z)dz$ also coincide generically.

Now we apply this result in our situation, which corresponds to the three-dimensional Lie algebra $\mathfrak{h}$ with a basis $\overline{u}, \overline{v}, \overline{\phi}$ and the following non-zero inner products of the basis elements:

$$\kappa(\overline{u}, \overline{u}) = -\kappa(\overline{v}, \overline{v}) = 1, \qquad \kappa(\overline{\phi}, \overline{\phi}) = 2(k+2).$$

Our screening currents $S_k(z)$ and $\widetilde{S}_k(z)$ are equal to $V_\chi^\kappa(z)$ and $V_{\check{\chi}}^\kappa(z)$, where

$$\chi = \overline{u} + \overline{v} - (k+2)^{-1}\overline{\phi}, \qquad \check{\chi} = -(k+2)\chi = -(k+2)\overline{u} - (k+2)\overline{v} + \overline{\phi}.$$

Therefore for generic k the kernels of the screening operators S_k and $\widetilde{S}_k$ coincide. $\square$

7.3 Screening operators for an arbitrary $\mathfrak{g}$

In this section we construct screening operators between Wakimoto modules over $\widehat{\mathfrak{g}}_\kappa$ for an arbitrary simple Lie algebra $\mathfrak{g}$ and use them to characterize $V_\kappa(\mathfrak{g})$ inside $W_{0,\kappa}$.

7.3.1 *Parabolic induction*

Denote by $\mathfrak{sl}_2^{(i)}$ the Lie subalgebra of $\mathfrak{g}$, isomorphic to $\mathfrak{sl}_2$, which is generated by e_i, h_i, and f_i. Let $\mathfrak{p}^{(i)}$ be the parabolic subalgebra of $\mathfrak{g}$ spanned by $\mathfrak{b}_-$ and e_i, and $\mathfrak{m}^{(i)}$ its Levi subalgebra. Thus, $\mathfrak{m}^{(i)}$ is equal to the direct sum of $\mathfrak{sl}_2^{(i)}$ and the orthogonal complement $\mathfrak{h}_i^\perp$ of h_i in $\mathfrak{h}$.

We apply to $\mathfrak{p}^{(i)}$ the results on semi-infinite parabolic induction of Section 6.3. According to Corollary 6.3.2, we obtain a functor from the category of smooth representations of $\widehat{\mathfrak{sl}}_{2,k} \oplus \widehat{\mathfrak{h}}_{i,\kappa_0}^\perp$, with k and κ_0 satisfying the conditions of Theorem 6.3.1, to the category of smooth $\widehat{\mathfrak{g}}_{\kappa+\kappa_c}$-modules. The condition on k and κ_0 is that the inner products on $\mathfrak{sl}_2^{(i)}$ corresponding to $(k+2)$ and κ_0 are both restrictions of an invariant inner product $(\kappa-\kappa_c)$ on $\mathfrak{g}$. In other words, $(\kappa-\kappa_c)(h_i,h_i) = 2(k+2)$ and $\kappa_0 = \kappa|_{\mathfrak{h}_i^\perp}$. By abuse of notation we will write κ for κ_0. If k and κ satisfy this condition, then for any smooth $\widehat{\mathfrak{sl}}_2$-module R of level k and any smooth $\widehat{\mathfrak{h}}_\kappa^\perp$-module L the tensor product $M_{\mathfrak{g},\mathfrak{p}^{(i)}} \otimes R \otimes L$ is a smooth $\widehat{\mathfrak{g}}_{\kappa+\kappa_c}$-module.

Note that the above condition means that

$$(\kappa-\kappa_c)(h_i,h_j) = \beta a_{ji}, \tag{7.3.1}$$

where a_{ji} is the (ji)th entry of the Cartan matrix of $\mathfrak{g}$.

In particular, if we choose R to be the Wakimoto module $W_{\lambda,k}$ over $\mathfrak{sl}_2$, and L to be the Fock representation $\pi_{\lambda_0}^\kappa$, the corresponding $\widehat{\mathfrak{g}}_\kappa$-module will be isomorphic to the Wakimoto module $W_{(\lambda,\lambda_0),\kappa+\kappa_c}$, where (λ,λ_0) is the weight of $\mathfrak{g}$ built from λ and λ_0. Under this isomorphism the generators $a_{\alpha_i,n}, n \in \mathbb{Z}$, will have a special meaning: they correspond to the *right* action of the elements $e_{i,n}$ of $\widehat{\mathfrak{g}}$, which was defined in Section 6.1.1. In other words, making the above identification of modules forces us to choose a system of coordinates $\{y_\alpha\}_{\alpha\in\Delta_+}$ on N_+ such that $\rho^R(e_i) = \partial/\partial y_{\alpha_i}$ (in the notation of Section 5.2.5), and so $w^R(e_i(z)) = a_{\alpha_i}(z)$ (in the notation of Section 6.1.1). From now on we will denote $w^R(e_i(z))$ by $e_i^R(z)$. For a general coordinate system on N_+ we have

$$e_i^R(z) = a_{\alpha_i}(z) + \sum_{\beta\in\Delta_+} P_\beta^{R,i}(a_\alpha^*(z))a_\beta(z) \tag{7.3.2}$$

(see formula (6.1.1)).

Now any intertwining operator between Wakimoto modules $W_{\lambda_1,k}$ and $W_{\lambda_2,k}$ over $\widehat{\mathfrak{sl}}_2$ gives rise to an intertwining operator between the Wakimoto modules $W_{(\lambda_1,\lambda_0),\kappa+\kappa_c}$ and $W_{(\lambda_2,\lambda_0),\kappa+\kappa_c}$ over $\widehat{\mathfrak{g}}_{\kappa+\kappa_c}$ for any weight λ_0 of $\mathfrak{h}_i^\perp$. We will use this fact and the $\widehat{\mathfrak{sl}}_2$ screening operators introduced in the previous section to construct intertwining operators between Wakimoto modules over $\widehat{\mathfrak{g}}_\kappa$.

7.3.2 Screening operators of the first kind

Let κ be a non-zero invariant inner product on $\mathfrak{g}$. We will use the same notation for the restriction of κ to $\mathfrak{h}$. Now to any $\chi \in \mathfrak{h}$ we associate a vertex operator $V^\kappa_\chi(z) : \pi^\kappa_0 \to \pi^\kappa_\chi$ given by formula (7.2.13). For $\kappa \neq \kappa_c$, we set

$$S_{i,\kappa}(z) \stackrel{\text{def}}{=} e^R_i(z) V^{\kappa-\kappa_c}_{-\alpha_i}(z) : W_{0,\kappa} \to W_{-\alpha_i,\kappa},$$

where $e^R_i(z)$ is given by formula (6.1.1). Note that

$$S_{i,\kappa}(z) = Y_{W_{0,\kappa},W_{-\alpha_i,\kappa}}(e^R_{i,-1}|-\alpha_i\rangle, z) \tag{7.3.3}$$

in the notation of Section 7.2.1.

According to Proposition 7.2.1 and the above discussion, the operator

$$S_{i,\kappa} = \int S_{i,\kappa}(z) dz : W_{0,\kappa} \to W_{-\alpha_i,\kappa} \tag{7.3.4}$$

is induced by the screening operator of the first kind S_k for the ith $\widehat{\mathfrak{sl}}_2$ subalgebra, where k is determined from the formula $(\kappa - \kappa_c)(h_i, h_i) = 2(k+2)$. Hence Proposition 7.2.1 implies:

Proposition 7.3.1 *The operator $S_{i,\kappa}$ is an intertwining operator between the $\widehat{\mathfrak{g}}_\kappa$-modules $W_{0,\kappa}$ and $W_{-\alpha_i,\kappa}$ for each $i = 1, \ldots, \ell$.*

We call $S_{i,\kappa}$ the ith screening operator of the first kind for $\widehat{\mathfrak{g}}_\kappa$.

Recall that by Proposition 7.1.1 $V_\kappa(\mathfrak{g})$ is naturally a $\widehat{\mathfrak{g}}_\kappa$-submodule and a vertex subalgebra of $W_{0,\kappa}$. On the other hand, the intersection of the kernels of $S_{i,\kappa}, i = 1, \ldots, \ell$, is a $\widehat{\mathfrak{g}}_\kappa$-submodule of $W_{0,\kappa}$, by Proposition 7.3.1, and a vertex subalgebra of $W_{0,\kappa}$, due to formula (7.3.3) and the commutation relations (7.2.3). The following proposition is proved in [FF8] (we will not use it here).

Proposition 7.3.2 *For generic κ, $V_\kappa(\mathfrak{g})$ is equal to the intersection of the kernels of the screening operators $S_{i,\kappa}, i = 1, \ldots, \ell$.*

Furthermore, in [FF8], Section 3, a complex $C^\bullet_\kappa(\mathfrak{g})$ of $\widehat{\mathfrak{g}}_\kappa$-modules is constructed for generic κ. Its ith degree term is the direct sum of the Wakimoto modules $W_{w(\rho)-\rho,\kappa}$, where w runs over all elements of the Weyl group of $\mathfrak{g}$ of length i. Its zeroth cohomology is isomorphic to $V_\kappa(\mathfrak{g})$, and all other cohomologies vanish. For $\mathfrak{g} = \mathfrak{sl}_2$ the complex $C^\bullet_\kappa(\mathfrak{sl}_2)$ has length 1 and coincides with the one appearing in Proposition 7.2.2. In general, the degree 0 term $C^0_\kappa(\mathfrak{g})$ of the complex is $W_{0,\kappa}$, the degree 1 term $C^1_\kappa(\mathfrak{g})$ is $\bigoplus_{i=1}^\ell W_{-\alpha_i,\kappa}$. The zeroth differential is the sum of the screening operators $S_{i,\kappa}$.

In [FF8] it is also explained how to construct other intertwining operators as compositions of the screening operators $S_{i,\kappa}$ using the Bernstein–Gelfand–Gelfand resolution of the trivial representation of the quantum group $U_q(\mathfrak{g})$. Roughly speaking, the screening operators $S_{i,\kappa}$ satisfy the q-Serre relations, i.e., the defining relations of the quantized enveloping algebra $U_q(\mathfrak{n}_+)$ with appropriate parameter q. Then for generic κ we attach to a singular vector of weight μ in the Verma module M_λ over $U_q(\mathfrak{g})$ an intertwining operator $W_{\lambda,\kappa} \to W_{\mu,\kappa}$. This operator is equal to the integral of a product of the screening currents $S_{i,\kappa}(z)$ over a certain cycle on the configuration space with coefficients in a local system that is naturally attached to the above singular vector.

The simplest operators correspond to the singular vectors $f_i v_0$ of weight $-\alpha_i$ in the Verma module M_0 over $U_q(\mathfrak{g})$. The corresponding intertwining operators are nothing but our screening operators (7.3.4).

7.3.3 Screening operators of the second kind

In order to define the screening operators of the second kind, we need to make sense of the series $(e_i^R(z))^\gamma$ for complex values of γ. So we choose a system of coordinates on N_+ in such a way that $e_i^R(z) = a_{\alpha_i}(z)$ (note that this cannot be achieved for all $i = 1, \ldots, \ell$ simultaneously). This is automatically so if we define the Wakimoto modules over $\widehat{\mathfrak{g}}$ via the semi-infinite parabolic induction from Wakimoto modules over the ith subalgebra $\widehat{\mathfrak{sl}}_2$ (see Section 7.3.1).

Having chosen such a coordinate system, we define the series $a_{\alpha_i}(z)^\gamma$ using the Friedan–Martinec–Shenker bosonization of the Weyl algebra generated by $a_{\alpha_i,n}, a^*_{\alpha_i,n}$, $n \in \mathbb{Z}$, as explained in Section 7.2.3. Namely, we have a vertex algebra

$$\Pi_0^{(i)} = \mathbb{C}[a_{\alpha_i,n}]_{n<-1} \otimes \mathbb{C}[a^*_{\alpha_i,n}]_{n\leq 0} \otimes \mathbb{C}[a^{\pm 1}_{\alpha_i,-1}]$$

containing

$$M_{\mathfrak{g}}^{(i)} = \mathbb{C}[a_{\alpha_i,n}]_{n\leq -1} \otimes \mathbb{C}[a^*_{\alpha_i,n}]_{n\leq 0}$$

and a $\Pi_0^{(i)}$-module $\Pi_\gamma^{(i)}$ defined as in Section 7.2.3. We then set

$$\widetilde{W}^{(i)}_{\gamma,\lambda,\kappa} = W_{\lambda,\kappa} \underset{M_{\mathfrak{g}}^{(i)}}{\otimes} \Pi_\gamma^{(i)}.$$

This is a $\widehat{\mathfrak{g}}_\kappa$-module, which contains $W_{\lambda,\kappa}$ if $\gamma = 0$. Note that $\widetilde{W}^{(i)}_{0,0,\kappa_c}$ is the $\widehat{\mathfrak{g}}$-module obtained by the semi-infinite parabolic induction from the $\widehat{\mathfrak{sl}}_2$-module $\widetilde{W}_{0,0,-2}$.

Now let $\beta = \frac{1}{2}(\kappa - \kappa_c)(h_i, h_i)$ and define the field

$$\widetilde{S}_{\kappa,i}(z) \stackrel{\text{def}}{=} (e_i^R(z))^{-\beta} V_{\check{\alpha}_i}(z) : \widetilde{W}^{(i)}_{0,0,\kappa} \to \widetilde{W}^{(i)}_{-\beta,\beta\check{\alpha}_i,\kappa}. \qquad (7.3.5)$$

Here $\check{\alpha}_i = h_i \in \mathfrak{h}$ denotes the ith coroot of $\mathfrak{g}$. Then the operator

$$\widetilde{S}_{\kappa,i} = \int \widetilde{S}_{\kappa,i}(z) dz$$

is induced by the screening operator of the second kind $\widetilde{S}_k$ for the ith $\widehat{\mathfrak{sl}}_2$ subalgebra, where k is determined from the formula $(\kappa - \kappa_c)(h_i, h_i) = 2(k+2)$. Hence Proposition 7.2.1 implies (a similar result was also obtained in [PRY]):

Proposition 7.3.3 *The operator $\widetilde{S}_{\kappa,i}$ is an intertwining operator between the $\widehat{\mathfrak{g}}_\kappa$-modules $\widetilde{W}^{(i)}_{0,0,\kappa}$ and $\widetilde{W}^{(i)}_{-\beta,\beta\check{\alpha}_i,\kappa}$.*

Note that $\widetilde{S}_{i,\kappa}$ is the residue of $Y_{V,M}((e^R_{i,-1})^{-\beta}|\check{\alpha}_i\rangle, z)$, where $V = \widetilde{W}^{(i)}_{0,0,\kappa}$ and $M = \widetilde{W}^{(i)}_{-\beta,\beta\check{\alpha}_i,\kappa}$ (see Section 7.2.1). Therefore, according to the commutation relations (7.2.3), the intersection of kernels of $\widetilde{S}_{i,\kappa}, i = 1, \ldots, \ell$, is naturally a vertex subalgebra of $\widetilde{W}^{(i)}_{0,0,\kappa}$ or $W_{0,\kappa}$. Combining Proposition 7.3.2 and Proposition 7.2.5, we obtain:

Proposition 7.3.4 *For generic κ, $V_\kappa(\mathfrak{g})$ is isomorphic, as a $\widehat{\mathfrak{g}}_\kappa$-module and as a vertex algebra, to the intersection of the kernels of the screening operators*

$$\widetilde{S}_{i,\kappa} : W_{0,\kappa} \to \widetilde{W}^{(i)}_{-\beta,\beta\check{\alpha}_i,\kappa}, \qquad i = 1, \ldots, \ell.$$

We remark that one can use the screening operators $\widetilde{S}_{i,\kappa}$ to construct more general intertwining operators following the procedure of [FF8].

7.3.4 Screening operators of second kind at the critical level

We would like to use the screening operators $\widetilde{S}_{i,\kappa}$ to characterize the image of the homomorphism $V_{\kappa_c}(\mathfrak{g}) \to W_{0,\kappa_c}$ (implementing Step 3 of our plan from Section 7.1). First, we need to define their limits as $\kappa \to \kappa_c$.

We start with the case when $\mathfrak{g} = \mathfrak{sl}_2$. In order to define the limit of $\widetilde{S}_k$ as $k \to -2$ we make $W_{0,k}$ and $\widetilde{W}_{-(k+2),2(k+2),k}$ into free modules over $\mathbb{C}[\beta]$, where β is a formal variable representing $k+2$, and then consider the quotient of these modules by the ideal generated by β.

More precisely, let $\pi_0[\beta]$ (resp., $\pi_{2\beta}[\beta]$) be the free $\mathbb{C}[\beta]$-module spanned by the monomials in $b_n, n < 0$, applied to a vector $|0\rangle$ (resp., $|2\beta\rangle$). We define the structure of vertex algebra over $\mathbb{C}[\beta]$ on $\pi_0[\beta]$ as in Section 6.2.1. Then $\pi_{2\beta}[\beta]$ is a module over $\pi_0[\beta]$. The dependence on β comes from the commutation relations

$$[b_n, b_m] = 2\beta n \delta_{n,-m}$$

and the fact that b_0 acts on $\pi_{2\beta}[\beta]$ by multiplication by 2β. Taking the quotient of $\pi_0[\beta]$ (resp., $\pi_{2\beta}[\beta]$) by the ideal generated by $(\beta - k), k \in \mathbb{C}$, we obtain the vertex algebra π_0^k (resp., the module π_{2k}^k over π_0^k) introduced in Section 6.2.2.

We define free $\mathbb{C}[\beta]$-modules $W_0[\beta]$ and $\widetilde{W}_{0,0}[\beta]$ as the tensor products

$$M \otimes_{\mathbb{C}} \pi_0[\beta] \qquad \text{and} \qquad \Pi_0 \otimes_{\mathbb{C}} \pi_0[\beta],$$

respectively. These are vertex algebras over $\mathbb{C}[\beta]$, and their quotients by the ideals generated by $(\beta - k), k \in \mathbb{C}$, are the vertex algebras $W_{0,k}$ and $\widetilde{W}_{0,0,k}$, respectively.

Next, let $\Pi_{-\beta+n,-\beta+n}$ be the free $\mathbb{C}[\beta]$-module spanned by the lexicographically ordered monomials in $p_n, q_n, n < 0$, applied to a vector $|-\beta+n, -\beta+n\rangle$. Set

$$\Pi_{-\beta} = \bigoplus_{n \in \mathbb{Z}} \Pi_{-\beta+n,-\beta+n},$$

and

$$\widetilde{W}_{-\beta,2\beta} = \Pi_{-\beta} \otimes_{\mathbb{C}[\beta]} \pi_{2\beta}[\beta].$$

Each Fourier coefficient of the formal power series

$$V_{2\beta}(z) : \pi_0[\beta] \to \pi_{2\beta}[\beta] \otimes_{\mathbb{C}} \mathbb{C}[\beta], \qquad \widetilde{a}(z)^{-\beta} : \Pi_0 \otimes_{\mathbb{C}} \mathbb{C}[\beta] \to \Pi_{-\beta},$$

given by the formulas above, is a well-defined linear operator commuting with the action of $\mathbb{C}[\beta]$. Hence the Fourier coefficients of their product are also well-defined linear operators from $\widetilde{W}_0[\beta]$ to $\widetilde{W}_{-\beta,2\beta}$. The corresponding operators on the quotients by the ideals generated by $(\beta - k), k \in \mathbb{C}$, coincide with the operators introduced above. We need to compute explicitly the leading term in the β-expansion of the residue $\int \widetilde{a}(z)^{-\beta} V_{2\beta}(z) dz$ and its restriction to $W_0[\beta]$. We will use this leading term as the screening operator at the critical level.

Consider first the expansion of $V_{2\beta}(z)$ in powers of β. Let us write $V_{2\beta}(z) = \sum_{n \in \mathbb{Z}} V_{2\beta}[n] z^{-n}$. Introduce the operators $\overline{V}[n], n \leq 0$, via the formal power series

$$\sum_{n \leq 0} \overline{V}[n] z^{-n} = \exp\left(\sum_{m>0} \frac{b_{-m}}{m} z^m \right). \tag{7.3.6}$$

Using formula (7.2.11), we obtain the following expansion of $V_{2\beta}[n]$:

$$V_{2\beta}[n] = \begin{cases} \overline{V}[n] + \beta(\ldots), & n \leq 0, \\ \beta\overline{V}[n] + \beta^2(\ldots), & n > 0, \end{cases}$$

where

$$\overline{V}[n] = -2 \sum_{m \leq 0} \overline{V}[m] \frac{\partial}{\partial b_{m-n}}, \qquad n > 0. \tag{7.3.7}$$

Next, we consider the expansion of $\widetilde{a}(z)^{-\beta} = e^{-\beta(u+v)}$ in powers of β. Let us write $\widetilde{a}(z)^{-\beta} = \sum_{n\in\mathbb{Z}} \widetilde{a}(z)^{-\beta}_{[n]} z^{-n}$. We will identify $\Pi_{-\beta}$ with $\Pi_0 \otimes_{\mathbb{C}} \mathbb{C}[\beta]$, as $\mathbb{C}[\beta]$-modules. Then we find that

$$\widetilde{a}(z)^{-\beta}_{[n]} = \begin{cases} 1 + \beta(\ldots), & n = 0, \\ \beta \dfrac{p_n + q_n}{n} + \beta^2(\ldots), & n \neq 0. \end{cases}$$

The above formulas imply the following expansion of the screening operator $\widetilde{S}_{\beta-2}$:

$$\int \widetilde{a}(z)^{-\beta} V_{2\beta}(z) dz = \beta \left(\overline{V}[1] + \sum_{n>0} \frac{1}{n} \overline{V}[-n+1](p_n + q_n) \right) + \beta^2(\ldots).$$

Therefore we define the limit of the screening operator $\widetilde{S}_{\beta-2}$ at $\beta = 0$ (corresponding to $k = -2$) as the operator

$$\overline{S} \stackrel{\text{def}}{=} \overline{V}[1] + \sum_{n>0} \frac{1}{n} \overline{V}[-n+1](p_n + q_n), \tag{7.3.8}$$

acting from $W_{0,-2} = M_{\mathfrak{sl}_2} \otimes \pi_0$ to $\widetilde{W}_{0,0,-2} = \Pi_0 \otimes \pi_0$. By construction, $\overline{S}$ is equal to the leading term in the β-expansion of the screening operator $\widetilde{S}_{\beta-2}$. Hence $\overline{S}$ commutes with the $\widehat{\mathfrak{sl}}_2$-action on $W_{0,-2}$ and $\widetilde{W}_{0,0,-2}$.

It is possible to express the operators

$$\frac{p_n + q_n}{n} = -(u_n + v_n), \qquad n > 0,$$

in terms of the Heisenberg algebra generated by $a_m, a^*_m, m \in \mathbb{Z}$. Namely, from the definition $\widetilde{a}(z) = e^{u+v}$ it follows that $u(z) + v(z) = \log \widetilde{a}(z)$, so that the field $u(z) + v(z)$ commutes with $a(w)$, and we have the following OPE with $a^*(w)$:

$$(u(z) + v(z)) a^*(w) = \frac{1}{z-w} \widetilde{a}(w)^{-1} + \text{reg}.$$

Therefore, writing $\widetilde{a}(z)^{-1} = \sum_{n\in\mathbb{Z}} \widetilde{a}(z)^{-1}_{[n]} z^{-n}$, we obtain the following commutation relations:

$$\left[\frac{p_n + q_n}{n}, a^*_m \right] = -\widetilde{a}(z)^{-1}_{[n+m-1]}, \qquad n > 0. \tag{7.3.9}$$

Using the realization (7.2.10) of Π_0, the series $\widetilde{a}(z)^{-1}$ is expressed as follows:

$$\widetilde{a}(z)^{-1} = (a_{-1})^{-1} \left(1 + (a_{-1})^{-1} \sum_{n\neq-1} a_n z^{-n-1} \right)^{-1}, \tag{7.3.10}$$

where the right hand side is expanded as a formal power series in positive powers of $(a_{-1})^{-1}$. It is easy to see that each Fourier coefficient of this power series is well-defined as a linear operator on Π_0.

The above formulas completely determine the action of $(p_n + q_n)/n, n > 0$, and hence of $\overline{S}$, on any vector in $W_{0,-2} = M \otimes \pi_0 \subset \widetilde{W}_{0,0,-2}$. Namely, we use the commutation relations (7.3.9) to move $(p_n + q_n)/n$ through the a^*_m's. As the result, we obtain Fourier coefficients of $\widetilde{a}(z)^{-1}$, which are given by formula (7.3.10). Applying each of them to any vector in $W_{0,-2}$, we always obtain a finite sum.

This completes the construction of the limit $\overline{S}$ of the screening operator $\widetilde{S}_k$ as $k \to -2$ in the case of $\mathfrak{sl}_2$. Now we consider the case of an arbitrary $\mathfrak{g}$.

The limit of $\widetilde{S}_{i,\kappa}$ as $\kappa \to \kappa_c$ is by definition the operator

$$\overline{S}_i : W_{0,\kappa_c} \to \widetilde{W}^{(i)}_{0,0,\kappa_c},$$

obtained from $\overline{S}$ via the functor of semi-infinite parabolic induction. Therefore it is given by the formula

$$\overline{S}_i = \overline{V}_i[1] + \sum_{n>0} \frac{1}{n} \overline{V}_i[-n+1](p_{i,n} + q_{i,n}). \tag{7.3.11}$$

Here $\overline{V}_i[n] : \pi_0 \to \pi_0$ are the linear operators given by the formulas

$$\sum_{n \leq 0} \overline{V}_i[n] z^{-n} = \exp\left(\sum_{m>0} \frac{b_{i,-m}}{m} z^m \right), \tag{7.3.12}$$

$$\overline{V}_i[1] = -\sum_{m \leq 0} \overline{V}_i[m]\, \mathbf{D}_{b_{i,m-1}}, \tag{7.3.13}$$

where $\mathbf{D}_{b_{i,m}}$ denotes the derivative in the direction of $b_{i,m}$ given by the formula

$$\mathbf{D}_{b_{i,m}} \cdot b_{j,n} = a_{ji} \delta_{n,m}, \tag{7.3.14}$$

and (a_{kl}) is the Cartan matrix of $\mathfrak{g}$ (it is normalized so that we have $\mathbf{D}_{b_{i,m}} \cdot b_{i,n} = 2\delta_{n,m}$, as in the case of $\mathfrak{sl}_2$). This follows from the commutation relations (6.2.1) between the $b_{i,n}$'s and formula (7.3.1).

The operators $(p_{i,n} + q_{i,n})/n$ acting on $\Pi^{(i)}_0$ are defined in the same way as above.

Thus, we obtain well-defined linear operators $\overline{S}_i : W_{0,\kappa_c} \to \widetilde{W}^{(i)}_{0,0,\kappa_c}$. By construction, they commute with the action of $\widehat{\mathfrak{g}}_{\kappa_c}$ on both modules. It is clear that the operators $\overline{S}_i, i = 1, \ldots, \ell$, annihilate the highest weight vector of W_{0,κ_c}. Therefore they annihilate all vectors obtained from the highest weight vector under the action of $\widehat{\mathfrak{g}}_{\kappa_c}$, i.e., all vectors in $V_{\kappa_c}(\mathfrak{g}) \subset W_{0,\kappa_c}$. Thus we obtain

Proposition 7.3.5 *The image of the vacuum module $V_{\kappa_c}(\mathfrak{g})$ under w_{κ_c} is contained in the intersection of the kernels of the operators $\overline{S}_i : W_{0,\kappa_c} \to \widetilde{W}^{(i)}_{0,0,\kappa_c}, i = 1, \ldots, \ell$.*

It follows from Proposition 7.3.5, Proposition 7.1.1 and Lemma 7.1.2 that the image of $\mathfrak{z}(\widehat{\mathfrak{g}})$ under the homomorphism $w_{\kappa_c} : V_{\kappa_c}(\mathfrak{g}) \hookrightarrow W_{0,\kappa_c}$ is contained in the intersection of the kernels of the operators $\overline{S}_i, i = 1, \ldots, \ell$, restricted to $\pi_0 \subset W_{0,\kappa_c}$. But according to formula (7.3.11), the restriction of $\overline{S}_i$ to π_0 is nothing but the operator $\overline{V}_i[1] : \pi_0 \to \pi_0$ given by formula (7.3.13). Therefore we obtain the following:

Proposition 7.3.6 *The center $\mathfrak{z}(\widehat{\mathfrak{g}})$ of $V_{\kappa_c}(\mathfrak{g})$ is contained in the intersection of the kernels of the operators $\overline{V}_i[1], i = 1, \ldots, \ell$, in π_0.*

This completes Step 4 of the plan outlined in Section 7.1. In the next chapter we will use this result to describe the center $\mathfrak{z}(\widehat{\mathfrak{g}})$ and identify it with the algebra of functions on the space of opers.

7.3.5 Other Fourier components of the screening curents

In Section 9.6 below we will need other Fourier components of the screening currents of the second kind constructed above (analogous results may be obtained for the screening currents of the first kind along the same lines).

We start with the case of $\mathfrak{g} = \mathfrak{sl}_2$. Consider the screening current

$$\widetilde{S}_k(z) = \widetilde{a}(z)^{-(k+2)} V_{2(k+2)}(z),$$

introduced in formula (7.2.12). In Proposition 7.2.4 we considered its residue as a linear operator $\widetilde{W}_{0,0,k} \to \widetilde{W}_{-(k+2),2(k+2),k}$. In particular $V_{2(k+2)}(z)$ was defined in formula (7.2.11) as a current acting from $\pi_0^{(k+2)}$ to $\pi_{2(k+2)}^{k+2}$.

We may also consider an action of this current on other Fock modules $\pi_\lambda^{k+2}, \lambda \in \mathbb{C}$, over the Heisenberg algebra. However, in that case we also need to multiply the current given by formula (7.2.11) by the factor $z^{b_0} = z^\lambda$. Only then will the relations between this current and $b(z)$ obtained above be valid. (If $\lambda = 0$, then $z^\lambda = 1$, and that is why we did not include it in formula (7.2.11)).) From now on we will denote by $V_{2(k+2)}(z)$ the current (7.2.11) multiplied by $z^{b_0} = z^\lambda$. This is a formal linear combination of z^α, where $\alpha \in \lambda + \mathbb{Z}$. Therefore for non-integer values of λ we cannot take its residue. But if $\lambda \in \mathbb{Z}$, then we obtain a well-defined linear operator

$$\widetilde{S}_k^{(\lambda)} \stackrel{\text{def}}{=} \int \widetilde{S}_k(z) dz : \quad \widetilde{W}_{\lambda,k} \to \widetilde{W}_{-(k+2),\lambda+2(k+2),k}, \tag{7.3.15}$$

which again commutes with the action of $\widehat{\mathfrak{sl}}_2$.

Next, we consider the limit of this operator when $k = \beta - 2$ and $\beta \to 0$. Suppose that $\lambda \in \mathbb{Z}_+$. Then, in the same way as in Section 7.3.4, we find that $\widetilde{S}_{\beta-2}^{(\lambda)} \to \beta \overline{S}^{(\lambda)}$, where

$$\overline{S}^{(\lambda)} = \overline{V}[\lambda + 1] + \sum_{n>0} \frac{1}{n} \overline{V}[-n+1](p_{n+\lambda} + q_{n+\lambda}), \tag{7.3.16}$$

acting from $W_{\lambda,-2} = M_{\mathfrak{sl}_2} \otimes \pi_\lambda$ to $\widetilde{W}_{0,\lambda,-2} = \Pi_0 \otimes \pi_\lambda$. The operators $\overline{V}[n]$

acting on π_λ are defined by formulas (7.3.6) and (7.3.7), and the operators $p_n + q_n$ acting from $M_{\mathfrak{sl}_2}$ to Π_0 are defined by formulas (7.3.9) and (7.3.10).

Since $\overline{S}^{(\lambda)}$ is obtained as a limit of an intertwining operator, we obtain that $\overline{S}^{(\lambda)}$ is also an intertwining operator (i.e., it commutes with the action of $\widehat{\mathfrak{sl}}_2$).

Now we construct intertwining operators for an arbitrary $\mathfrak{g}$ by applying the semi-infinite parabolic induction to the operators (7.3.15) and (7.3.16), as in Section 7.3.4. First, for any weight $\lambda \in \mathfrak{h}^*$ such that $\lambda_i = \langle \lambda, \check{\alpha}_i \rangle \in \mathbb{Z}$ we obtain the intertwining operator

$$\widetilde{S}^{(\lambda_i)}_{\kappa,i} = \int \widetilde{S}_{\kappa,i}(z)dz : \quad \widetilde{W}^{(i)}_{\lambda,\kappa} \to \widetilde{W}^{(i)}_{-\beta,\lambda+\beta\check{\alpha}_i,\kappa},$$

where $\widetilde{S}_{\kappa,i}(z)$ is the current given by formula (7.3.5) multiplied by the factor z^{λ_i}.

Next, we consider the $\kappa \to \kappa_c$ limit of this operator for $\lambda_i \in \mathbb{Z}_+$. As before, this corresponds to the limit $\beta \to 0$, where $\beta = \frac{1}{2}(\kappa - \kappa_c)(h_i, h_i)$. The limiting operator is given by the formula

$$\overline{S}^{(\lambda_i)}_i = \overline{V}_i[\lambda_i + 1] + \sum_{n>0} \frac{1}{n} \overline{V}_i[-n+1](p_{i,n+\lambda_i} + q_{i,n+\lambda_i}), \tag{7.3.17}$$

where $\overline{V}_i[n], n \leq 0$, are given by formula (7.3.12), and

$$\overline{V}_i[\lambda_i + 1] = -\sum_{m \leq \lambda_i} \overline{V}_i[m - \lambda_i] \, \mathbf{D}_{b_{i,m-1}}, \tag{7.3.18}$$

where, as before, $\mathbf{D}_{b_{i,m}}$ is the derivation on $\pi_\lambda \simeq \mathbb{C}[b_{i,n}]_{i=1,\ldots,\ell;n<0}$ such that

$$\mathbf{D}_{b_{i,m}} \cdot b_{j,n} = a_{ji} \delta_{n,m}.$$

The operator $\overline{S}^{(\lambda_i)}_i$ acts from W_{λ,κ_c} to $\widetilde{W}_{0,\lambda,\kappa_c}$. In particular, we find that the restriction of $\overline{S}_i$ to $\pi_\lambda \subset W_{\lambda,\kappa_c}$ is equal to $\overline{V}_i[\lambda_i + 1]$, given by formula (7.3.18).

We will use this result in the proof of Proposition 9.6.7 below, which generalizes Proposition 7.3.6 to arbitrary dominant integral weights.

8
Identification of the center with functions on opers

In this chapter we use the above description of the center of the vertex algebra $V_{\kappa_c}(\mathfrak{g})$ in terms of the kernels of the screening operators in order to complete the plan outlined in Section 7.1. We follow the argument of [F4]. We first show that this ceter is canonically isomorphic to the so-called classical $\mathcal{W}$-algebra associated to the Langlands dual Lie algebra ${}^L\mathfrak{g}$. Next, we prove that the classical $\mathcal{W}$-algebra is in turn isomorphic to the algebra of functions on the space of ${}^LG$-opers on the formal disc D. The algebra $\operatorname{Fun}\operatorname{Op}_{^LG}(D)$ of functions on the space of ${}^LG$-opers on D carries a vertex Poisson structure. We show that our isomorphism between the center of $V_{\kappa_c}(\mathfrak{g})$ (with its canonical vertex Poisson structure coming from the deformation of the level) and $\operatorname{Fun}\operatorname{Op}_{^LG}(D)$ preserves vertex Poisson structures. Using this isomorphism, we show that the center of the completed enveloping algebra of $\widehat{\mathfrak{g}}$ at the critical level is isomorphic, as a Poisson algebra, to the algebra $\operatorname{Fun}\operatorname{Op}_{^LG}(D^\times)$ of functions on the space of opers on the punctured disc $D^\times$. The latter isomorphism was conjectured by V. Drinfeld.

Here is a more detailed description of the material of this chapter.

In Section 8.1 we identify the center of the vertex algebra $V_{\kappa_c}(\mathfrak{g})$ with the intersection of the kernels of certain operators, which we identify in turn with the classical $\mathcal{W}$-algebra associated to the Langlands dual Lie algebra ${}^L\mathfrak{g}$. We also show that the corresponding vertex Poisson algebra structures coincide. In Section 8.2.1 we define the Miura opers and explain their relationship to the connections on a certain H-bundle. We then show in Section 8.2 that this classical $\mathcal{W}$-algebra (resp., the commutative vertex algebra π_0) is nothing but the algebra of functions on the space of ${}^LG$-opers on the disc (resp., Miura ${}^LG$-opers on the disc). Furthermore, the embedding of the classical $\mathcal{W}$-algebra into π_0 coincides with the Miura map between the two algebras of functions. Finally, in Section 8.3 we consider the center of the completed universal enveloping algebra of $\widehat{\mathfrak{g}}$ at the critical level. We identify it with the algebra of functions on the space of ${}^LG$-opers on the punctured disc. We then prove that this identification satisfies various compatibilities. In particular, we show that an affine analogue of the Harish-Chandra homomorphism obtained by evaluating central elements on the Wakimoto modules is nothing but the Miura transformation from Miura ${}^LG$-opers to ${}^LG$-opers on the punctured disc.

8.1 Description of the center of the vertex algebra

In this section we use Proposition 7.3.6 to describe the center $\mathfrak{z}(\widehat{\mathfrak{g}})$ of $V_{\kappa_c}(\mathfrak{g})$ as defined in Section 3.3.2.

According to Proposition 7.3.6, $\mathfrak{z}(\widehat{\mathfrak{g}})$ is identified with a subspace in the intersection of the kernels of the operators $\overline{V}_i[1], i = 1, \ldots, \ell$, in π_0. We now show that $\mathfrak{z}(\widehat{\mathfrak{g}})$ is actually equal to this intersection. This is Step 5 of the plan described in Section 7.1.

8.1.1 Computation of the character of $\mathfrak{z}(\widehat{\mathfrak{g}})$

As explained in Section 3.3.3, the $\widehat{\mathfrak{g}}_{\kappa_c}$-module $V_{\kappa_c}(\mathfrak{g})$ has a natural filtration induced by the Poincaré–Birkhoff–Witt filtration on the universal enveloping algebra $U(\widehat{\mathfrak{g}}_{\kappa_c})$, and the associated graded space $\operatorname{gr} V_{\kappa_c}(\mathfrak{g})$ is isomorphic to

$$\operatorname{Sym} \mathfrak{g}((t))/\mathfrak{g}[[t]] \simeq \operatorname{Fun} \mathfrak{g}^*[[t]],$$

where we use the following (coordinate-dependent) pairing

$$\langle A \otimes f(t), B \otimes g(t) \rangle = \langle A, B \rangle \operatorname{Res}_{t=0} f(t)g(t)dt \tag{8.1.1}$$

for $A \in \mathfrak{g}^*$ and $B \in \mathfrak{g}$.

Recall that $\mathfrak{z}(\widehat{\mathfrak{g}})$ is equal to the space of $\mathfrak{g}[[t]]$-invariants in $V_{\kappa_c}(\mathfrak{g})$. The symbol of a $\mathfrak{g}[[t]]$-invariant vector in $V_{\kappa_c}(\mathfrak{g})$ is a $\mathfrak{g}[[t]]$-invariant vector in $\operatorname{gr} V_{\kappa_c}(\mathfrak{g})$, i.e., an element of the space of $\mathfrak{g}[[t]]$-invariants in $\operatorname{Fun} \mathfrak{g}^*[[t]]$. Hence we obtain an injective map

$$\operatorname{gr} \mathfrak{z}(\widehat{\mathfrak{g}}) \hookrightarrow (\operatorname{Fun} \mathfrak{g}^*[[t]])^{\mathfrak{g}[[t]]} = \operatorname{Inv}\, \mathfrak{g}^*[[t]] \tag{8.1.2}$$

(see Lemma 3.3.1).

According to Theorem 3.4.2, the algebra $\operatorname{Inv} \mathfrak{g}^*[[t]]$ is the free polynomial algebra in the generators $\overline{P}_{i,n}, i = 1, \ldots, \ell; n < 0$ (see Section 3.3.4 for the definition of the polynomials $\overline{P}_{i,n}$).

We have an action of the operator $L_0 = -t\partial_t$ on $\operatorname{Fun} \mathfrak{g}^*[[t]]$. It defines a $\mathbb{Z}$-gradation on $\operatorname{Fun} \mathfrak{g}^*[[t]]$ such that $\deg \overline{J}^a_n = -n$. Then $\deg \overline{P}_{i,n} = d_i - n$, and according to formula (4.3.3), the character of $\operatorname{Inv} \mathfrak{g}^*[[t]]$ is equal to

$$\operatorname{ch} \operatorname{Inv} \mathfrak{g}^*[[t]] = \prod_{i=1}^{\ell} \prod_{n_i \geq d_i+1} (1 - q^{n_i})^{-1}. \tag{8.1.3}$$

We now show that the map (8.1.2) is an isomorphism using Proposition 6.3.3. Consider the Lie subalgebra $\widetilde{\mathfrak{b}}_+ = (\mathfrak{b}_+ \otimes 1) \oplus (\mathfrak{g} \otimes t\mathbb{C}[[t]])$ of $\mathfrak{g}[[t]]$. The natural surjective homomorphism $\mathbb{M}_{0,\kappa_c} \to V_{\kappa_c}(\mathfrak{g})$ gives rise to a map of the corresponding subspaces of $\widetilde{\mathfrak{b}}_+$-invariants

$$\phi : (\mathbb{M}_{0,\kappa_c})^{\widetilde{\mathfrak{b}}_+} \to V_{\kappa_c}(\mathfrak{g})^{\widetilde{\mathfrak{b}}_+}.$$

Both $\mathbb{M}_{0,\kappa_c}$ and $V_{\kappa_c}(\mathfrak{g})$ have natural filtrations (induced by the PBW filtration on $U_{\kappa_c}(\widehat{\mathfrak{g}})$) which are preserved by the homomorphism between them. Therefore we have the corresponding map of associated graded

$$\phi_{\mathrm{cl}} : (\mathrm{gr}\, \mathbb{M}_{0,\kappa_c})^{\widetilde{\mathfrak{b}}_+} \to (\mathrm{gr}\, V_{\kappa_c}(\mathfrak{g}))^{\widetilde{\mathfrak{b}}_+}.$$

Since $V_{\kappa_c}(\mathfrak{g})$ is a direct sum of finite-dimensional representations of the constant subalgebra $\mathfrak{g} \otimes 1$ of $\mathfrak{g}[[t]]$, we find that any $\widetilde{\mathfrak{b}}_+$-invariant in $V_{\kappa_c}(\mathfrak{g})$ of $\mathrm{gr}\, V_{\kappa_c}(\mathfrak{g})$ is automatically a $\mathfrak{g}[[t]]$-invariant. Therefore we obtain that

$$V_{\kappa_c}(\mathfrak{g})^{\widetilde{\mathfrak{b}}_+} = V_{\kappa_c}(\mathfrak{g})^{\mathfrak{g}[[t]]},$$
$$(\mathrm{gr}\, V_{\kappa_c}(\mathfrak{g}))^{\widetilde{\mathfrak{b}}_+} = (\mathrm{gr}\, V_{\kappa_c}(\mathfrak{g}))^{\mathfrak{g}[[t]]} = \mathbb{C}[\overline{P}_{i,m}]_{i=1,\ldots,\ell;m<0}.$$

We need to describe $(\mathrm{gr}\, \mathbb{M}_{0,\kappa_c})^{\widetilde{\mathfrak{b}}_+}$. First, observe that

$$\mathrm{gr}\, \mathbb{M}_{0,\kappa_c} = \mathrm{Sym}\, \mathfrak{g}(\!(t)\!)/\widetilde{\mathfrak{b}}_+ \simeq \mathrm{Fun}\, \mathfrak{g}^*[[t]]_{(-1)},$$

where

$$\mathfrak{g}^*[[t]]_{(-1)} = ((\mathfrak{n}_-)^* \otimes t^{-1}) \oplus \mathfrak{g}^*[[t]] \simeq (\mathfrak{g}(\!(t)\!)/\widetilde{\mathfrak{b}}_+)^*,$$

and we use the pairing (8.1.1) between $\mathfrak{g}(\!(t)\!)$ and $\mathfrak{g}^*(\!(t)\!)$. In terms of this identification, the map ϕ_{cl} becomes a ring homomorphism

$$\mathrm{Fun}\, \mathfrak{g}^*[[t]]_{(-1)} \to \mathrm{Fun}\, \mathfrak{g}^*[[t]]$$

induced by the natural embedding $\mathfrak{g}^*[[t]] \to \mathfrak{g}^*[[t]]_{(-1)}$.

Suppose that our basis $\{J^a\}$ of $\mathfrak{g}$ is chosen in such a way that it is the union of two subsets, which constitute bases in $\mathfrak{b}_+$ and in $\mathfrak{n}_-$. Let $\overline{J}^a_n$ be the polynomial function on $\widehat{\mathfrak{n}}_+^{(-1)}$ defined by the formula

$$\overline{J}^a_n(A(t)) = \mathrm{Res}_{t=0}\langle A(t), J^a\rangle t^n dt. \tag{8.1.4}$$

Then, as an algebra, $\mathrm{Fun}\, \mathfrak{g}^*[[t]]_{(-1)}$ is generated by these linear functionals $\overline{J}^a_n$, where $n < 0$, or $n = 0$ and $J^a \in \mathfrak{n}_-$.

Next, we construct $\widetilde{\mathfrak{b}}_+$-invariant functions on $\mathfrak{g}^*[[t]]_{(-1)}$ in the same way as in Section 3.3.4, by substituting the generating functions

$$\overline{J}^a(z) = \sum_n \overline{J}^a_n z^{-n-1}$$

(with the summation over $n < 0$ or $n \leq 0$ depending on whether $J^a \in \mathfrak{b}_+$ or $\mathfrak{n}_-$) into the invariant polynomials $\overline{P}_i, i = 1, \ldots, \ell$ on $\mathfrak{g}$. But since now $\overline{J}^a(z)$ has non-zero z^{-1} coefficients if $J^a \in \mathfrak{n}_-$, the resulting series

$$\overline{P}_i(\overline{J}^a(z)) = \sum_{m\in\mathbb{Z}} \overline{P}_{i,m} z^{-m-1}$$

will have non-zero coefficients $\overline{P}_{i,m}$ for all $m < d_i$. Thus, we obtain a natural homomorphism

$$\mathbb{C}[P_{i,m_i}]_{i=1,\dots,\ell;m_i<d_i} \to (\operatorname{Fun}\mathfrak{g}^*[[t]]_{(-1)})^{\widetilde{\mathfrak{b}}_+}.$$

Lemma 8.1.1 *This homomorphism is an isomorphism.*

Proof. Let

$$\mathfrak{g}^*[[t]]_{(0)} = ((\mathfrak{n}_-)^* \otimes 1) \oplus (\mathfrak{g}^* \otimes t\mathbb{C}[[t]]) = t\mathfrak{g}^*[[t]]_{(-1)}.$$

Clearly, the spaces of $\widetilde{\mathfrak{b}}_+$-invariant functions on $\mathfrak{g}^*[[t]]_{(-1)}$ and $\mathfrak{g}^*[[t]]_{(0)}$ are isomorphic (albeit the $\mathbb{Z}$-gradings are different), so we will consider the latter space.

Denote by $\mathfrak{g}^*[[t]]^{\mathrm{reg}}_{(0)}$ the intersection of $\mathfrak{g}^*[[t]]_{(0)}$ and

$$J\mathfrak{g}^*_{\mathrm{reg}} = \mathfrak{g}^*_{\mathrm{reg}} \times (\mathfrak{g}^* \otimes t\mathbb{C}[[t]]).$$

Thus, $\mathfrak{g}^*[[t]]^{\mathrm{reg}}_{(0)} = ((\mathfrak{n}_-)^{*,\mathrm{reg}} \otimes 1) \oplus (\mathfrak{g}^* \otimes t\mathbb{C}[[t]])$, where $(\mathfrak{n}_-)^{*,\mathrm{reg}} = (\mathfrak{n}_-)^* \cap \mathfrak{g}^*_{\mathrm{reg}}$ is an open dense subset of $(\mathfrak{n}_-)^*$, so that $\widehat{\mathfrak{n}}^{\mathrm{reg}}_+$ is open and dense in $\widehat{\mathfrak{n}}_+$.

Recall the morphism $Jp : J\mathfrak{g}^*_{\mathrm{reg}} \to J\mathcal{P}$ introduced in the proof of Theorem 3.4.2. It was shown there that the group $JG = G[[t]]$ acts transitively along the fibers of Jp. Let $\widetilde{B}_+$ be the subgroup of $G[[t]]$ corresponding to the Lie algebra $\widetilde{\mathfrak{b}}_+ \subset \mathfrak{g}[[t]]$. Note that for any $x \in \mathfrak{g}^*[[t]]^{\mathrm{reg}}_{(0)}$, the group $\widetilde{B}_+$ is equal to the subgroup of all elements g of $G[[t]]$ such that $g \cdot x \in \mathfrak{g}^*[[t]]^{\mathrm{reg}}_{(0)}$. Therefore $\widetilde{B}_+$ acts transitively along the fibers of the restriction of the morphism Jp to $\mathfrak{g}^*[[t]]^{\mathrm{reg}}_{(0)}$.

This implies, in the same way as in the proof of Theorem 3.4.2, that the ring of $\widetilde{B}_+$-invariant (equivalently, $\widetilde{\mathfrak{b}}_+$-invariant) polynomials on $\mathfrak{g}^*[[t]]^{\mathrm{reg}}_{(0)}$ is the ring of functions on the image of $\mathfrak{g}^*[[t]]^{\mathrm{reg}}_{(0)}$ in $J\mathcal{P}$ under the map Jp. But it follows from the construction that the image of $\mathfrak{g}^*[[t]]^{\mathrm{reg}}_{(0)}$ in $J\mathcal{P}$ is the subspace determined by the equations $\overline{P}_{i,m} = 0, i = 1, \dots, \ell; m = -1$. Hence the ring of $\widetilde{\mathfrak{b}}_+$-invariant polynomials on $\mathfrak{g}^*[[t]]^{\mathrm{reg}}_{(0)}$ is equal to $\mathbb{C}[\overline{P}_{i,m_i}]_{i=1,\dots,\ell;m_i<-1}$. Since $\mathfrak{g}^*[[t]]^{\mathrm{reg}}_{(0)}$ is dense in $\mathfrak{g}^*[[t]]_{(0)}$, we obtain that this is also the ring of invariant polynomials on $\widehat{\mathfrak{n}}_+$.

When we pass from $\mathfrak{g}^*[[t]]_{(0)}$ to $\mathfrak{g}^*[[t]]_{(-1)}$, we need to take into account the shifting of the indices $\overline{P}_{i,m_i} \mapsto \overline{P}_{i,m_i+d_i+1}$ of invariant polynomials corresponding to the shift $\overline{J}^a_n \mapsto \overline{J}^a_{n+1}$. Then we obtain the statement of the lemma. □

Corollary 8.1.2 *The map ϕ_{cl} is surjective.*

Proof. The map ϕ_{cl} corresponds to taking the quotient of the free polynomial algebra on $P_{i,m_i}, i = 1, \dots, \ell; m_i < d_i$, by the ideal generated by $P_{i,m_i}, i = 1, \dots, \ell; 0 \le m_i < d_i$. □

It follows from the construction that $\deg \overline{P}_{i,m} = m - d_i$. Hence we obtain the following formula for the character of $(\operatorname{gr} \mathbb{M}_{0,\kappa_c})^{\widetilde{\mathfrak{b}}_+} = (\operatorname{Fun} \widehat{\mathfrak{n}}_+^{(-1)})^{\widetilde{\mathfrak{b}}_+}$:

$$\operatorname{ch} (\operatorname{gr} \mathbb{M}_{0,\kappa_c})^{\widetilde{\mathfrak{b}}_+} = \prod_{m>0} (1 - q^m)^{-\ell}.$$

Now recall that by Theorem 6.3.3 the Verma module $\mathbb{M}_{0,\kappa_c}$ is isomorphic to the Wakimoto module W^+_{0,κ_c}. Hence $(\mathbb{M}_{0,\kappa_c})^{\widetilde{\mathfrak{b}}_+} = (W^+_{0,\kappa_c})^{\widetilde{\mathfrak{b}}_+}$. In addition, according to Lemma 6.3.4 we have $(W^+_{0,\kappa_c})^{\widetilde{\mathfrak{b}}_+} = \pi_0$, and so its character is also equal to $\prod_{m>0} (1 - q^m)^{-\ell}$. Therefore we find that the natural embedding

$$\operatorname{gr}(\mathbb{M}_{0,\kappa_c}^{\widetilde{\mathfrak{b}}_+}) \hookrightarrow (\operatorname{gr} \mathbb{M}_{0,\kappa_c})^{\widetilde{\mathfrak{b}}_+}$$

is an isomorphism.

Consider the commutative diagram

$$\begin{array}{ccc} \operatorname{gr}(\mathbb{M}_{0,\kappa_c}^{\widetilde{\mathfrak{b}}_+}) & \longrightarrow & \operatorname{gr}(V_{\kappa_c}(\mathfrak{g})^{\mathfrak{g}[[t]]}) \\ \downarrow & & \downarrow \\ (\operatorname{gr} \mathbb{M}_{0,\kappa_c})^{\widetilde{\mathfrak{b}}_+} & \longrightarrow & (\operatorname{gr} V_{\kappa_c}(\mathfrak{g}))^{\mathfrak{g}[[t]]}. \end{array}$$

It follows from the above discussion that the left vertical arrow is an isomorphism. Moreover, by Corollary 8.1.2 the lower horizontal arrow is surjective. Therefore the right vertical arrow is surjective. But it is also injective, according to Lemma 3.3.1. Therefore we obtain an isomorphism

$$\operatorname{gr}(V_{\kappa_c}(\mathfrak{g})^{\mathfrak{g}[[t]]}) \simeq (\operatorname{gr} V_{\kappa_c}(\mathfrak{g}))^{\mathfrak{g}[[t]]}.$$

In particular, this implies that the character of $\operatorname{gr} \mathfrak{z}(\widehat{\mathfrak{g}}) = \operatorname{gr}(V_{\kappa_c}(\mathfrak{g}))^{\mathfrak{g}[[t]]}$ is equal to that of $(\operatorname{gr} V_{\kappa_c}(\mathfrak{g})^{\mathfrak{g}[[t]]})$ given by formula (4.3.3). Since $\operatorname{ch} \mathfrak{z}(\widehat{\mathfrak{g}}) = \operatorname{ch}(\operatorname{gr} \mathfrak{z}(\widehat{\mathfrak{g}}))$, we find that

$$\operatorname{ch} \mathfrak{z}(\widehat{\mathfrak{g}}) = \prod_{i=1}^{\ell} \prod_{n_i \geq d_i + 1} (1 - q^{n_i})^{-1}. \tag{8.1.5}$$

Thus, we obtain the following (see Section 4.3.1):

Theorem 8.1.3 *The center $\mathfrak{z}(\widehat{\mathfrak{g}})$ is "as large as possible," i.e.,*

$$\operatorname{gr} \mathfrak{z}(\widehat{\mathfrak{g}}) = \operatorname{Inv} \mathfrak{g}^*[[t]].$$

Thus, there exist central elements $S_i \in \mathfrak{z}(\widehat{\mathfrak{g}}) \subset V_{\kappa_c}(\mathfrak{g})$ whose symbols are equal to $\overline{P}_{i,-1} \in \operatorname{Inv} \mathfrak{g}^*[[t]], i = 1, \ldots, \ell$, and such that

$$\mathfrak{z}(\widehat{\mathfrak{g}}) = \mathbb{C}[S_{i,(n)}]_{i=1,\ldots,\ell; n<0} |0\rangle,$$

where the $S_{i,(n)}$'s are the Fourier coefficients of the vertex operator $Y(S_i, z)$.

This is a non-trivial result which tells us a lot about the structure of the center. But we are not satisfied with it, because, as explained in Section 3.5.2, we would like to understand the geometric meaning of the center and in particular we want to know how the group $\operatorname{Aut}\mathcal{O}$ acts on $\mathfrak{z}(\widehat{\mathfrak{g}})$. This is expressed in Theorem 4.3.2, which identifies $\mathfrak{z}(\widehat{\mathfrak{g}})$ with the algebra of functions on $\operatorname{Op}_{^LG}(D)$ (note that Theorem 4.3.2 implies Theorem 8.1.3, see formula (4.3.2)). To prove this, we need to work a little harder and complete Steps 5 and 6 of our plan presented in Section 7.1.

8.1.2 The center and the classical $\mathcal{W}$-algebra

According to Proposition 7.3.6, $\mathfrak{z}(\widehat{\mathfrak{g}})$ is contained in the intersection of the kernels of the operators $\overline{V}_i[1], i = 1, \ldots, \ell$, on π_0. Now we compute the character of this intersection and compare it with the character formula (8.1.5) for $\mathfrak{z}(\widehat{\mathfrak{g}})$ to show that $\mathfrak{z}(\widehat{\mathfrak{g}})$ is equal to the intersection of the kernels of the operators $\overline{V}_i[1]$.

First, we identify this intersection with a classical limit of a one-parameter family of vertex algebras, called the $\mathcal{W}$-algebras. The $\mathcal{W}$-algebra $\mathcal{W}_\nu(\mathfrak{g})$ associated to a simple Lie algebra $\mathfrak{g}$ and an invariant inner product ν on $\mathfrak{g}$ was defined in [FF6] (see also [FB], Chapter 15) via the quantum Drinfeld–Sokolov reduction. For generic values of ν the vertex algebra $\mathcal{W}_\nu(\mathfrak{g})$ is equal to the intersection of the kernels of certain screening operators in a Heisenberg vertex algebra. Let us recall the definition of these operators.†

Consider another copy of the Heisenberg Lie algebra $\widehat{\mathfrak{h}}_\nu$ introduced in Section 6.2.1. To avoid confusion, we will denote the generators of this Heisenberg Lie algebra by

$$\mathbf{b}_{i,n}, \qquad i = 1, \ldots, \ell; n \in \mathbb{Z}.$$

We have the vertex operator

$$V^\nu_{-\alpha_i}(z) : \pi^\nu_0 \to \pi^\nu_{-\alpha_i}$$

defined by formula (7.2.13), and let $V^\nu_{-\alpha_i}[1] = \int V^\nu_{-\alpha_i}(z)dz$ be its residue. We call it a **$\mathcal{W}$-algebra screening operator** (to distinguish it from the Kac–Moody algebra screening operators defined above). Since

$$V^\nu_{-\alpha_i}(z) = Y_{\pi^\nu_0,\pi^\nu_{-\alpha_i}}(|-\alpha_i\rangle, z)$$

(in the notation of Section 7.2.1), we obtain that the intersection of the kernels of $V^\nu_{-\alpha_i}[1], i = 1, \ldots, \ell$, in π^ν_0 is a vertex subalgebra of π^ν_0. By Theorem 15.4.12 of [FB], for generic values of ν the $\mathcal{W}$-algebra $\mathcal{W}_\nu(\mathfrak{g})$ is equal to the intersection of the kernels of the operators $V^\nu_{-\alpha_i}[1], i = 1, \ldots, \ell$, in π^ν_0.

We are interested in the limit of $\mathcal{W}_\nu(\mathfrak{g})$ when $\nu \to \infty$. To define this limit, we fix an invariant inner product ν_0 on $\mathfrak{g}$ and denote by ϵ the ratio between

† In fact, we may use this property to *define* $\mathcal{W}_\nu(\mathfrak{g})$ for all ν, see [FF7].

ν_0 and ν. We have the following formula for the ith simple root $\alpha_i \in \mathfrak{h}^*$ as an element of $\mathfrak{h}$ using the identification between $\mathfrak{h}^*$ and $\mathfrak{h}$ induced by $\nu = \nu_0/\epsilon$:

$$\alpha_i = \epsilon \frac{2}{\nu_0(h_i, h_i)} h_i.$$

Let

$$\mathbf{b}'_{i,n} = \epsilon \frac{2}{\nu_0(h_i, h_i)} \mathbf{b}_{i,n}, \tag{8.1.6}$$

where the $\mathbf{b}_{i,n}$'s are the generators of $\widehat{\mathfrak{h}}_\nu$. Consider the $\mathbb{C}[\epsilon]$-lattice in $\pi_0^\nu \otimes_{\mathbb{C}} \mathbb{C}[\epsilon]$ spanned by all monomials in $\mathbf{b}'_{i,n}, i = 1, \ldots, \ell; n < 0$. We denote by $\pi_0^\vee$ the specialization of this lattice at $\epsilon = 0$; it is a commutative vertex algebra.

In the limit $\epsilon \to 0$, we obtain the following expansion of the operator $V^\nu_{-\alpha_i}[1]$:

$$V^\nu_{-\alpha_i}[1] = \epsilon \frac{2}{\nu_0(h_i, h_i)} \mathbf{V}_i[1] + \ldots,$$

where the dots denote terms of higher order in ϵ, and the operator $\mathbf{V}_i[1]$ acting on $\pi_0^\vee$ is given by the formula

$$\mathbf{V}_i[1] = \sum_{m \leq 0} \mathbf{V}_i[m] \, \mathbf{D}_{\mathbf{b}'_{i,m-1}}, \tag{8.1.7}$$

where

$$\mathbf{D}_{\mathbf{b}'_{i,m}} \cdot \mathbf{b}'_{j,n} = a_{ij} \delta_{n,m}, \tag{8.1.8}$$

(a_{ij}) is the Cartan matrix of $\mathfrak{g}$, and

$$\sum_{n \leq 0} \mathbf{V}_i[n] z^{-n} = \exp\left(- \sum_{m>0} \frac{\mathbf{b}'_{i,-m}}{m} z^m \right).$$

The intersection of the kernels of the operators $\mathbf{V}_i[1], i = 1, \ldots, \ell$, is a commutative vertex subalgebra of $\pi_0^\vee$, which we denote by $\mathbf{W}(\mathfrak{g})$ and call the **classical $\mathcal{W}$-algebra** associated to $\mathfrak{g}$.

Note that the structure of commutative vertex algebra on $\pi_0^\vee$ is independent of the choice of ν_0. The operators $\mathbf{V}_i[1]$ get rescaled if we change ν_0 and so their kernels are independent of ν_0. Therefore $\mathbf{W}(\mathfrak{g})$ is a commutative vertex subalgebra of $\pi_0^\vee$ that is independent of ν_0.†

8.1.3 The appearance of the Langlands dual Lie algebra

Comparing formulas (7.3.13) and (8.1.7), we find that after the substitution $b_{i,n} \mapsto -\mathbf{b}'_{i,n}$, the operators $\overline{V}_i[1]$ become the operators $\mathbf{V}_i[1]$, except that the matrix coefficient a_{ji} in formula (7.3.14) gets replaced by a_{ij} in formula

† However, as we will see below, both $\pi_0^\vee$ and $\mathbf{W}(\mathfrak{g})$ also carry vertex Poisson algebra structures, and those structures do depend on ν_0.

(8.1.8). This is not a typo! There is a serious reason for that: while the operator $\overline{V}_i[1]$ was obtained as the limit of a vertex operator corresponding to the ith *coroot* of $\mathfrak{g}$, the operator $\mathbf{V}_i[1]$ was obtained as the limit of a vertex operator corresponding to minus the ith *root* of $\mathfrak{g}$.

Under the exchange of roots and coroots, the Cartan matrix gets transposed. The transposed Cartan matrix of a simple Lie algebra $\mathfrak{g}$ is the Cartan matrix of another simple Lie algebra; namely, the **Langlands dual Lie algebra** of $\mathfrak{g}$, which is denoted by ${}^L\mathfrak{g}$. Because of this transposition, we may identify canonically the Cartan subalgebra ${}^L\mathfrak{h}$ of ${}^L\mathfrak{g}$ with the dual space $\mathfrak{h}^*$ to the Cartan subalgebra $\mathfrak{h}$ of $\mathfrak{g}$, so that the simple roots of $\mathfrak{g}$ (which are vectors in $\mathfrak{h}^*$) become the simple coroots of ${}^L\mathfrak{g}$ (which are vectors in ${}^L\mathfrak{h}$).

Let us identify π_0 corresponding to $\mathfrak{g}$ with $\pi_0^\vee$ corresponding to ${}^L\mathfrak{g}$ by sending $b_{i,n} \mapsto -\mathbf{b}'_{i,n}$. Then the operator $\overline{V}_i[1]$ attached to $\mathfrak{g}$ becomes the operator $\mathbf{V}_i[1]$ attached to ${}^L\mathfrak{g}$. Therefore the intersection of the kernels of the operators $\overline{V}_i[1], i = 1, \ldots, \ell$, on π_0, attached to a simple Lie algebra $\mathfrak{g}$, is isomorphic to the intersection of the kernels of the operators $\mathbf{V}_i[1], i = 1, \ldots, \ell$, on $\pi_0^\vee$, attached to ${}^L\mathfrak{g}$.

Using Proposition 7.3.6, we now find that $\mathfrak{z}(\widehat{\mathfrak{g}})$ is embedded into the intersection of the kernels of the operators $\mathbf{V}_i[1], i = 1, \ldots, \ell$, on $\pi_0^\vee$, i.e., into the classical $\mathcal{W}$-algebra $\mathbf{W}({}^L\mathfrak{g})$. Furthermore, we have the following result, whose proof will be postponed till Section 8.2.4 below.

Lemma 8.1.4 *The character of* $\mathbf{W}({}^L\mathfrak{g})$ *is equal to the character of* $\mathfrak{z}(\widehat{\mathfrak{g}})$ *given by formula* (8.1.5).

Therefore we obtain the following result.

Theorem 8.1.5 *The center* $\mathfrak{z}(\widehat{\mathfrak{g}})$ *is isomorphic, as a commutative vertex algebra, to the intersection of the kernels of the operators* $\overline{V}_i[1], i = 1, \ldots, \ell$, *on* π_0, *and hence to the classical* $\mathcal{W}$*-algebra* $\mathbf{W}({}^L\mathfrak{g})$.

This completes Step 5 of the plan presented in Section 7.1.

8.1.4 The vertex Poisson algebra structures

In addition to the structures of commutative vertex algebras, both $\mathfrak{z}(\widehat{\mathfrak{g}})$ and $\mathbf{W}({}^L\mathfrak{g})$ also carry the structures of **vertex Poisson algebra**, and we wish to show that the isomorphism of Theorem 8.1.5 is compatible with these structures.

We will not give a precise definition of vertex Poisson algebras here, referring the reader to [FB], Section 16.2. We recall that a vertex Poisson algebra P is in particular a vertex Lie algebra, and so we attach to it an ordinary Lie algebra

$$\mathrm{Lie}(P) = P \otimes \mathbb{C}((t)) / \operatorname{Im}(T \otimes 1 + 1 \otimes \partial_t)$$

(see [FB], Section 16.1.7).

According to Proposition 16.2.7 of [FB], if V_ϵ is a one-parameter family of vertex algebras, then the center $\mathcal{Z}(V_0)$ of V_0 acquires a natural vertex Poisson algebra structure. Namely, at $\epsilon = 0$ the polar part of the operation Y, restricted to $\mathcal{Z}(V_0)$, vanishes, so we define the operation Y_- on $\mathcal{Z}(V_0)$ as the ϵ-linear term of the polar part of Y.

Let us fix a non-zero inner product κ_0 on $\mathfrak{g}$, and let ϵ be the ratio between the inner products $\kappa - \kappa_c$ and κ_0. Consider the vertex algebras $V_\kappa(\mathfrak{g})$ as a one-parameter family using ϵ as a parameter. Then we obtain a vertex Poisson structure on $\mathfrak{z}(\widehat{\mathfrak{g}})$, the center of $V_{\kappa_c}(\mathfrak{g})$ (corresponding to $\epsilon = 0$). We will denote $\mathfrak{z}(\widehat{\mathfrak{g}})$, equipped with this vertex Poisson structure, by $\mathfrak{z}(\widehat{\mathfrak{g}})_{\kappa_0}$.

Next, consider the Heisenberg vertex algebra $\pi_0^{\kappa-\kappa_c}$ introduced in Section 6.2.1. From now on, to avoid confusion, we will write $\pi_0^{\kappa-\kappa_c}(\mathfrak{g})$ to indicate that it is associated to the Cartan subalgebra $\mathfrak{h}$ of $\mathfrak{g}$. Let ϵ be defined by the formula $\kappa - \kappa_c = \epsilon\kappa_0$ as above. We define the commutative vertex algebra $\pi_0(\mathfrak{g})$ as an $\epsilon \to 0$ limit of $\pi_0^{\kappa-\kappa_c}(\mathfrak{g})$ in the following way. Recall that $\pi_0^{\kappa-\kappa_c}(\mathfrak{g})$ is the Fock representation of the Heisenberg Lie algebra with generators $b_{i,n}, i = 1, \ldots, \ell; n \in \mathbb{Z}$. The commutation relations between them are as follows:

$$[b_{i,n}, b_{j,m}] = \epsilon n \kappa_0(h_i, h_j)\delta_{n,-m}. \tag{8.1.9}$$

Consider the $\mathbb{C}[\epsilon]$-lattice in $\pi_0^{\kappa-\kappa_c} \otimes_{\mathbb{C}} \mathbb{C}[\epsilon]$ spanned by all monomials in $b_{i,n}, i = 1, \ldots, \ell; n < 0$. Now $\pi_0(\mathfrak{g})$ is by definition the specialization of this lattice at $\epsilon = 0$.

This is a commutative vertex algebra, but since it is defined as the limit of a one-parameter family of vertex algebras, it acquires a vertex Poisson structure. This vertex Poisson structure is uniquely determined by the Poisson brackets

$$\{b_{i,n}, b_{j,m}\} = n\kappa_0(h_i, h_j)\delta_{n,-m}$$

in $\mathrm{Lie}(\pi_0(\mathfrak{g}))$, which immediately follow from formula (8.1.9). In particular, this vertex Poisson structure depends on κ_0, and so we will write $\pi_0(\mathfrak{g})_{\kappa_0}$ to indicate this dependence.

Recall that the homomorphism $w_{\kappa_c} : V_{\kappa_c}(\mathfrak{g}) \to W_{0,\kappa_c} = M_{\mathfrak{g}} \otimes \pi_0(\mathfrak{g})$ may be deformed to a homomorphism $w_\kappa : V_\kappa(\mathfrak{g}) \to W_{0,\kappa} = M_{\mathfrak{g}} \otimes \pi_0^{\kappa-\kappa_c}(\mathfrak{g})$. Therefore the ϵ-linear term of the polar part of the operation Y of $V_\kappa(\mathfrak{g})$, restricted to $\mathfrak{z}(\widehat{\mathfrak{g}})$, which is used in the definition of the vertex Poisson structure on $\mathfrak{z}(\widehat{\mathfrak{g}})$, may be computed by restricting to $\mathfrak{z}(\widehat{\mathfrak{g}}) \subset \pi_0(\mathfrak{g})$ the ϵ-linear term of the polar part of the operation Y of $\pi_0^{\kappa-\kappa_c}(\mathfrak{g}) \subset W_{0,\kappa}$. But the latter gives $\pi_0(\mathfrak{g})$ the structure of a vertex Poisson algebra, which we denote by $\pi_0(\mathfrak{g})_{\kappa_0}$. Therefore we obtain the following:

Lemma 8.1.6 *The embedding $\mathfrak{z}(\widehat{\mathfrak{g}})_{\kappa_0} \hookrightarrow \pi_0(\mathfrak{g})_{\kappa_0}$ is a homomorphism of vertex*

Poisson algebras. The corresponding map of local Lie algebras $\mathrm{Lie}(\mathfrak{z}(\widehat{\mathfrak{g}})_{\kappa_0}) \hookrightarrow \mathrm{Lie}(\pi_0(\mathfrak{g})_{\kappa_0})$ *is a Lie algebra homomorphism.*

On the other hand, consider the commutative vertex algebra $\pi_0^\vee(\mathfrak{g})$ and its subalgebra $\mathbf{W}(\mathfrak{g})$ defined in Section 8.1.2. The vertex algebra $\pi_0^\vee(\mathfrak{g})$ was defined as the limit of a one-parameter family of vertex algebras, namely, $\pi_0^\nu(\mathfrak{g})$, where $\nu = \nu_0/\epsilon$, as $\epsilon \to 0$. Therefore $\pi_0^\nu(\mathfrak{g})$ also carries a vertex Poisson algebra structure. We will denote the resulting vertex Poisson algebra by $\pi_0^\vee(\mathfrak{g})_{\nu_0}$. Its subalgebra $\mathbf{W}(\mathfrak{g})$ is obtained as the limit of a one-parameter family of vertex subalgebras of $\pi_0^\nu(\mathfrak{g})$; namely, of $\mathcal{W}_\nu(\mathfrak{g})$. Therefore $\mathbf{W}(\mathfrak{g})$ is a vertex Poisson subalgebra of $\pi_0^\vee(\mathfrak{g})_{\nu_0}$, which we will denote by $\mathbf{W}(\mathfrak{g})_{\nu_0}$.

Thus, we see that the objects appearing in the isomorphism of Theorem 8.1.5 carry natural vertex Poisson algebra structures. We claim that this isomorphism is in fact compatible with these structures.

Let us explain this more precisely. Note that the restriction of a non-zero invariant inner product κ_0 on $\mathfrak{g}$ to $\mathfrak{h}$ defines a non-zero inner product on $\mathfrak{h}^*$, which is the restriction of an invariant inner product $\kappa_0^\vee$ on ${}^L\mathfrak{g}$. To avoid confusion, let us denote by $\pi_0(\mathfrak{g})_{\kappa_0}$ and $\pi_0^\vee({}^L\mathfrak{g})_{\kappa_0^\vee}$ the classical limits of the Heisenberg vertex algebras defined above, with their vertex Poisson structures corresponding to κ_0 and $\kappa_0^\vee$, respectively. We have an isomorphism of vertex Poisson algebras

$$\begin{aligned} \imath : \pi_0(\mathfrak{g})_{\kappa_0} &\xrightarrow{\sim} \pi_0^\vee({}^L\mathfrak{g})_{\kappa_0^\vee}, \\ b_{i,n} &\mapsto -\mathbf{b}'_{i,n}, \end{aligned}$$

where $\mathbf{b}'_{i,n}$ is given by formula (8.1.6). The restriction of the isomorphism $\imath$ to the subspace $\bigcap_{1\le i\le \ell} \mathrm{Ker}\, \overline{V}_i[1]$ of $\pi_0(\mathfrak{g})_{\kappa_0}$ gives us an isomorphism of vertex Poisson algebras

$$\bigcap_{1\le i\le\ell} \mathrm{Ker}\, \overline{V}_i[1] \simeq \bigcap_{1\le i\le\ell} \mathrm{Ker}\, \mathbf{V}_i^\vee[1],$$

where by $\mathbf{V}_i^\vee[1]$ we denote the operator (8.1.7) attached to ${}^L\mathfrak{g}$. Recall that we have the following isomorphisms of vertex Poisson algebras:

$$\begin{aligned} \mathfrak{z}(\widehat{\mathfrak{g}})_{\kappa_0} &\simeq \bigcap_{1\le i\le\ell} \mathrm{Ker}\, \overline{V}_i[1], \\ \mathbf{W}({}^L\mathfrak{g})_{\kappa_0^\vee} &\simeq \bigcap_{1\le i\le\ell} \mathrm{Ker}\, \mathbf{V}_i^\vee[1]. \end{aligned}$$

Therefore we obtain the following stronger version of Theorem 8.1.5:

Theorem 8.1.7 *There is a commutative diagram of vertex Poisson algebras*

$$\begin{array}{ccc} \pi_0(\mathfrak{g})_{\kappa_0} & \xrightarrow{\sim} & \pi_0^\vee({}^L\mathfrak{g})_{\kappa_0^\vee} \\ \uparrow & & \uparrow \\ \mathfrak{z}(\widehat{\mathfrak{g}})_{\kappa_0} & \xrightarrow{\sim} & \mathbf{W}({}^L\mathfrak{g})_{\kappa_0^\vee} \end{array} \tag{8.1.10}$$

8.1.5 Aut $\mathcal{O}$*-module structures*

Both $\mathfrak{z}(\widehat{\mathfrak{g}})_{\kappa_0}$ and $\mathbf{W}({}^L\mathfrak{g})_{\kappa_0^\vee}$ carry actions of the group Aut $\mathcal{O}$, and we claim that the isomorphism of Theorem 8.1.7 intertwines these actions. To see tist, we describe the two actions as coming from the vertex Poisson algebra structures.

In both cases the action of the group Aut $\mathcal{O}$ is obtained by exponentiation of the action of the Lie algebra $\mathrm{Der}_0\,\mathcal{O} \subset \mathrm{Der}\,\mathcal{O}$. In the case of the center $\mathfrak{z}(\widehat{\mathfrak{g}})_{\kappa_0}$, the action of Der $\mathcal{O}$ is the restriction of the natural action on $V_{\kappa_c}(\mathfrak{g})$ which comes from its action on $\widehat{\mathfrak{g}}_{\kappa_c}$ (preserving its Lie subalgebra $\mathfrak{g}[[t]]$) by infinitesimal changes of variables. But away from the critical level, i.e., when $\kappa \neq \kappa_c$, the action of Der $\mathcal{O}$ is obtained through the action of the Virasoro algebra which comes from the conformal vector $\mathbf{S}_\kappa$, given by formula (6.2.5), which we rewrite as follows:

$$\mathbf{S}_\kappa = \frac{\kappa_0}{\kappa - \kappa_c} S_1,$$

where S_1 is given by formula (3.1.1) and κ_0 is the inner product used in that formula. Thus, the Fourier coefficients $L_n, n \geq -1$, of the vertex operator

$$Y(\mathbf{S}_\kappa, z) = \sum_{n \in \mathbb{Z}} L_n z^{-n-2}$$

generate the Der $\mathcal{O}$-action on $V_\kappa(\mathfrak{g})$ when $\kappa \neq \kappa_c$.

In the limit $\kappa \to \kappa_c$, we have $\mathbf{S}_\kappa = \epsilon^{-1} S_1$, where as before $\epsilon = \dfrac{\kappa - \kappa_c}{\kappa_0}$. Therefore the action of Der $\mathcal{O}$ is obtained through the vertex Poisson operation Y_- on $\mathfrak{z}(\widehat{\mathfrak{g}})_{\kappa_0}$, defined as the limit of ϵ^{-1} times the polar part of Y when $\epsilon \to 0$ (see [FB], Section 16.2), applied to $S_1 \in \mathfrak{z}(\widehat{\mathfrak{g}})_{\kappa_0}$.

In other words, the Der $\mathcal{O}$-action is generated by the Fourier coefficients $L_n, n \geq -1$, of the series

$$Y_-(S_1, z) = \sum_{n \geq -1} L_n z^{-n-2}.$$

Thus, we see that the natural Der $\mathcal{O}$-action (and hence Aut $\mathcal{O}$-action) on $\mathfrak{z}(\widehat{\mathfrak{g}})_{\kappa_0}$ is encoded in the vector $S_1 \in \mathfrak{z}(\widehat{\mathfrak{g}})_{\kappa_0}$ through the vertex Poisson algebra structure on $\mathfrak{z}(\widehat{\mathfrak{g}})_{\kappa_0}$. Note that this action endows $\mathfrak{z}(\widehat{\mathfrak{g}})_{\kappa_0}$ with a quasi-conformal structure (see Section 6.2.4).

Likewise, there is a Der $\mathcal{O}$-action on $\mathbf{W}({}^L\mathfrak{g})_{\kappa_0^\vee}$ coming from its vertex Poisson algebra structure. The vector generating this action is also equal to the limit of a conformal vector in the $\mathcal{W}$-algebra $\mathcal{W}_\nu({}^L\mathfrak{g})$ (which is a conformal vertex algebra) as $\nu \to \infty$. This conformal vector is unique because as we see from the character formula for $\mathcal{W}_\nu({}^L\mathfrak{g})$ given in the right hand side of (8.1.5) the homogeneous component of $\mathcal{W}_\nu({}^L\mathfrak{g})$ of degree two (where all conformal vectors live) is one-dimensional.† The limit of this vector as $\nu \to \infty$ gives rise to a vector in $\mathbf{W}({}^L\mathfrak{g})_{\kappa_0^\vee}$, which we denote by $\mathbf{t}$, such that the Fourier coefficients of $Y_-(\mathbf{t}, z)$ generate a Der $\mathcal{O}$-action on $\mathbf{W}({}^L\mathfrak{g})_{\kappa_0^\vee}$ (in the next section we will explain the geometric meaning of this action). Since such a vector is unique, it must be the image of $S_1 \in \mathfrak{z}(\widehat{\mathfrak{g}})_{\kappa_0}$ under the isomorphism $\mathfrak{z}(\widehat{\mathfrak{g}})_{\kappa_0} \simeq \mathbf{W}({}^L\mathfrak{g})_{\kappa_0^\vee}$.

Therefore we conclude that the Der $\mathcal{O}$-actions (and hence the corresponding Aut $\mathcal{O}$-actions) on both $\mathfrak{z}(\widehat{\mathfrak{g}})_{\kappa_0}$ and $\mathbf{W}({}^L\mathfrak{g})_{\kappa_0^\vee}$ are encoded, via the respective vertex Poisson structures, by certain vectors, which are in fact equal to the classical limits of conformal vectors. Under the isomorphism of Theorem 8.1.7 these vectors are mapped to each other. Thus, we obtain the following:

Proposition 8.1.8 *The commutative diagram* (8.1.10) *is compatible with the actions of* Der $\mathcal{O}$ *and* Aut $\mathcal{O}$.

We have already calculated in formula (6.2.5) the image of $\mathbf{S}_\kappa$ in $W_{0,\kappa}$ when $\kappa \neq \kappa_c$. By passing to the limit $\kappa \to \kappa_c$ we find that the image of $S_1 = \epsilon \mathbf{S}_\kappa$ belongs to $\pi_0(\mathfrak{g})_{\kappa_0} \subset W_{0,\kappa}$ and is equal to

$$\frac{1}{2}\sum_{i=1}^{\ell} b_{i,-1}b^i_{-1} - \rho_{-2},$$

where $\{b_i\}$ and $\{b^i\}$ are dual bases with respect to the inner product κ_0 used in the definition of S_1, restricted to $\mathfrak{h}$, and ρ is the element of $\mathfrak{h}$ corresponding to $\rho \in \mathfrak{h}^*$ under the isomorphism $\mathfrak{h}^* \simeq \mathfrak{h}$ induced by κ_0. Under the isomorphism of Theorem 8.1.7, this vector becomes the vector $S_1 \in \mathfrak{z}(\widehat{\mathfrak{g}})_{\kappa_0} \subset \pi_0(\mathfrak{g})_{\kappa_0}$, which is responsible for the Der $\mathcal{O}$ action on it.

The action of the corresponding operators $L_n \in \operatorname{Der}\mathcal{O}, n \geq -1$, on $\pi_0(\mathfrak{g})_{\kappa_0}$ is given by derivations of the algebra structure which are uniquely determined by formulas (6.2.13). Therefore the action of the L_n's on $\pi_0^\vee({}^L\mathfrak{g})_{\kappa_0^\vee}$ is as follows (recall that $b_{i,n} \mapsto -\mathbf{b}'_{i,n}$ under our isomorphism $\pi_0(\mathfrak{g})_{\kappa_0} \simeq \pi_0^\vee({}^L\mathfrak{g})_{\kappa_0^\vee}$):

$$\begin{aligned} L_n \cdot \mathbf{b}'_{i,m} &= -m\mathbf{b}'_{i,n+m}, \qquad -1 \leq n < -m, \\ L_n \cdot \mathbf{b}'_{i,-n} &= -n(n+1), \qquad n > 0, \\ L_n \cdot \mathbf{b}'_{i,m} &= 0, \qquad n > -m. \end{aligned} \tag{8.1.11}$$

These formulas determine the Der $\mathcal{O}$-action on $\pi_0^\vee({}^L\mathfrak{g})_{\kappa_0^\vee}$. By construction, $\mathbf{W}({}^L\mathfrak{g})_{\kappa_0^\vee} \subset \pi_0^\vee({}^L\mathfrak{g})_{\kappa_0^\vee}$ is preserved by this action.

† This determines this vector up to a scalar, which is fixed by the commutation relations.

Note in particular that the above actions of $\operatorname{Der}\mathcal{O}$ on $\pi_0(\mathfrak{g})_{\kappa_0}$ and $\pi_0^\vee({}^L\mathfrak{g})_{\kappa_0^\vee}$ (and hence on $\mathfrak{z}(\widehat{\mathfrak{g}})_{\kappa_0}$ and $\mathbf{W}({}^L\mathfrak{g})_{\kappa_0^\vee}$) are independent of the inner product κ_0.

8.2 Identification with the algebra of functions on opers

We have now identified the center $\mathfrak{z}(\widehat{\mathfrak{g}})_{\kappa_0}$ with the classical $\mathcal{W}$-algebra $\mathbf{W}({}^L\mathfrak{g})_{\kappa_0^\vee}$ in a way compatible with the vertex Poisson algebra structures and the $(\operatorname{Der}\mathcal{O}, \operatorname{Aut}\mathcal{O})$-actions. The last remaining step in our proof of Theorem 4.3.2 (Step 6 of our plan from Section 7.1) is the identification of $\mathbf{W}({}^L\mathfrak{g})_{\kappa_0^\vee}$ with the algebra $\operatorname{Op}_{^LG}(D)$ of functions on the space of ${}^LG$-opers on the disc. This is done in this section.

We start by introducing Miura opers and a natural map from generic Miura opers to opers called the Miura transformation. We then show that the algebras of functions on G-opers and generic Miura G-opers on the disc D are isomorphic to $\mathbf{W}(\mathfrak{g})_{\nu_0}$ and $\pi_0^\vee(\mathfrak{g})_{\nu_0}$, respectively. Furthermore, the homomorphism $\mathbf{W}(\mathfrak{g})_{\nu_0} \to \pi_0^\vee(\mathfrak{g})_{\nu_0}$ corresponding to the Miura transformation is precisely the embedding constructed in Section 8.1.2. This will enable us to identify the center $\mathfrak{z}(\widehat{\mathfrak{g}})$ with the algebra of functions on ${}^LG$-opers on D.

8.2.1 *Miura opers*

Let G be a simple Lie group of adjoint type.

A **Miura G-oper** on X, which is a smooth curve, or D, or $D^\times$, is by definition a quadruple $(\mathcal{F}, \nabla, \mathcal{F}_B, \mathcal{F}'_B)$, where $(\mathcal{F}, \nabla, \mathcal{F}_B)$ is a G-oper on X and $\mathcal{F}'_B$ is another B-reduction of $\mathcal{F}$ which is preserved by ∇.

Consider the space $\operatorname{MOp}_G(D)$ of Miura G-opers on the disc D. A B-reduction of $\mathcal{F}$ which is preserved by the connection ∇ is uniquely determined by a B-reduction of the fiber $\mathcal{F}_0$ of $\mathcal{F}$ at the origin $0 \in D$ (recall that the underlying G-bundles of all G-opers are isomorphic to each other). The set of such reductions is the $\mathcal{F}_0$-twist $(G/B)_{\mathcal{F}_0}$ of the flag manifold G/B. Let $\mathcal{F}_{\text{univ}}$ be the universal G-bundle on $\operatorname{Op}_G(D)$ whose fiber at the oper $(\mathcal{F}, \nabla, \mathcal{F}_B)$ is $\mathcal{F}_0$. Then we obtain that

$$\operatorname{MOp}_G(D) \simeq (G/B)_{\mathcal{F}_{\text{univ}}} = \mathcal{F}_{\text{univ}} \underset{G}{\times} G/B.$$

A Miura G-oper is called **generic** if the B-reductions $\mathcal{F}_B$ and $\mathcal{F}'_B$ are in generic relative position. We denote the space of generic Miura opers on D by $\operatorname{MOp}_G(D)_{\text{gen}}$. We have a natural forgetful morphism $\operatorname{MOp}_G(D)_{\text{gen}} \to \operatorname{Op}_G(D)$. The group

$$N_{\mathcal{F}_{B,0}} = \mathcal{F}_{B,0} \underset{B}{\times} N,$$

where $\mathcal{F}_{B,0}$ is the fiber of $\mathcal{F}_B$ at 0, acts on $(G/B)_{\mathcal{F}_0}$, and the subset of generic reductions is the open $N_{\mathcal{F}_{B,0}}$-orbit of $(G/B)_{\mathcal{F}_0}$. This orbit is in fact an $N_{\mathcal{F}_{B,0}}$-torsor. Therefore we obtain that the space of generic Miura opers with a fixed underlying G-oper $(\mathcal{F}, \nabla, \mathcal{F}_B)$ is a principal $N_{\mathcal{F}_{B,0}}$-bundle.

This may be rephrased as follows. Let $\mathcal{F}_{B,\text{univ}}$ be the B-reduction of the universal G-bundle $\mathcal{F}_{\text{univ}}$, whose fiber at the oper $(\mathcal{F}, \nabla, \mathcal{F}_B)$ is $\mathcal{F}_{B,0}$. Then

$$\mathrm{MOp}_G(D)_{\text{gen}} = \mathcal{F}_{B,\text{univ}} \underset{B}{\times} \mathcal{U}, \tag{8.2.1}$$

where $\mathcal{U} \simeq N$ is the open B-orbit in G/B.

Now we identify $\mathrm{MOp}_G(D)_{\text{gen}}$ with the space of H-connections. Consider the H-bundles $\mathcal{F}_H = \mathcal{F}_B/N$ and $\mathcal{F}'_H = \mathcal{F}'_B/N$ corresponding to a generic Miura oper $(\mathcal{F}, \nabla, \mathcal{F}_B, \mathcal{F}'_B)$ on X. If $\mathcal{P}$ is an H-bundle, then applying to it the automorphism w_0 of H, corresponding to the longest element of the Weyl group of G, we obtain a new H-bundle, which we denote by $w_0^*(\mathcal{P})$.

Lemma 8.2.1 *For a generic Miura oper $(\mathcal{F}, \nabla, \mathcal{F}_B, \mathcal{F}'_B)$ the H-bundle $\mathcal{F}'_H$ is isomorphic to $w_0^*(\mathcal{F}_H)$.*

Proof. Consider the vector bundles $\mathfrak{g}_{\mathcal{F}} = \mathcal{F} \underset{G}{\times} \mathfrak{g}$, $\mathfrak{b}_{\mathcal{F}_B} = \mathcal{F}_B \underset{B}{\times} \mathfrak{b}$ and $\mathfrak{b}_{\mathcal{F}'_B} = \mathcal{F}'_B \underset{B}{\times} \mathfrak{b}$. We have the inclusions $\mathfrak{b}_{\mathcal{F}_B}, \mathfrak{b}_{\mathcal{F}'_B} \subset \mathfrak{g}_{\mathcal{F}}$ which are in generic position. Therefore the intersection $\mathfrak{b}_{\mathcal{F}_B} \cap \mathfrak{b}_{\mathcal{F}'_B}$ is isomorphic to $\mathfrak{b}_{\mathcal{F}_B}/[\mathfrak{b}_{\mathcal{F}_B}, \mathfrak{b}_{\mathcal{F}_B}]$, which is the trivial vector bundle with fiber $\mathfrak{h}$. It naturally acts on the bundle $\mathfrak{g}_{\mathcal{F}}$ and under this action $\mathfrak{g}_{\mathcal{F}}$ decomposes into a direct sum of $\mathfrak{h}$ and the line subbundles $\mathfrak{g}_{F,\alpha}, \alpha \in \Delta$. Furthermore, $\mathfrak{b}_{\mathcal{F}_B} = \bigoplus_{\alpha \in \Delta_+} \mathfrak{g}_{F,\alpha}, \mathfrak{b}_{\mathcal{F}'_B} = \bigoplus_{\alpha \in \Delta_+} \mathfrak{g}_{F,w_0(\alpha)}$. Since the action of B on $\mathfrak{n}/[\mathfrak{n}, \mathfrak{n}]$ factors through $H = B/N$, we find that

$$\mathcal{F}_H \underset{H}{\times} \bigoplus_{i=1}^{\ell} \mathbb{C}_{\alpha_i} \simeq \bigoplus_{i=1}^{\ell} \mathfrak{g}_{\mathcal{F},\alpha_i}, \qquad \mathcal{F}'_H \underset{H}{\times} \bigoplus_{i=1}^{\ell} \mathbb{C}_{\alpha_i} \simeq \bigoplus_{i=1}^{\ell} \mathfrak{g}_{\mathcal{F},w_0(\alpha_i)}.$$

Therefore we obtain that

$$\mathcal{F}_H \underset{H}{\times} \mathbb{C}_{\alpha_i} \simeq \mathcal{F}'_H \underset{H}{\times} \mathbb{C}_{w_0(\alpha_i)}, \qquad i = 1, \ldots, \ell.$$

Since G is of adjoint type by our assumption, the above associated line bundles completely determine $\mathcal{F}_H$ and $\mathcal{F}'_H$, and the above isomorphisms imply that $\mathcal{F}'_H \simeq w_0^*(\mathcal{F}_H)$. □

Since the B-bundle $\mathcal{F}'_B$ is preserved by the oper connection ∇, we obtain a connection $\overline{\nabla}$ on $\mathcal{F}'_H$ and hence on $\mathcal{F}_H$. But according to Lemma 4.2.1, we have $\mathcal{F}_H \simeq \Omega^{\check{\rho}}$. Therefore we obtain a morphism β from the space $\mathrm{MOp}_G(D)_{\text{gen}}$ of generic Miura opers on D to the space $\mathrm{Conn}(\Omega^{\check{\rho}})_D$ of connections on the H-bundle $\Omega^{\check{\rho}}$ on D.

Explicitly, connections on the H-bundle $\Omega^{\check{\rho}}$ may be described by the operators

$$\overline{\nabla} = \partial_t + \mathbf{u}(t), \qquad \mathbf{u}(t) \in \mathfrak{h}[[t]],$$

where t is a coordinate on the disc D, which we use to trivialize $\Omega^{\check{\rho}}$. Let s be a new coordinate such that $t = \varphi(s)$. Then the same connection will appear as $\partial_s + \widetilde{\mathbf{u}}(s)$, where

$$\widetilde{\mathbf{u}}(s) = \varphi' \cdot \mathbf{u}(\varphi(s)) - \check{\rho}\frac{\varphi''}{\varphi'}. \tag{8.2.2}$$

This formula describes the action of the group $\operatorname{Aut}\mathcal{O}$ (and the Lie algebra $\operatorname{Der}\mathcal{O}$) on $\operatorname{Conn}(\Omega^{\check{\rho}})_D$.

Proposition 8.2.2 *The map* $\beta : \operatorname{MOp}_G(D)_{\text{gen}} \to \operatorname{Conn}(\Omega^{\check{\rho}})_D$ *is an isomorphism.*

Proof. We define a map τ in the opposite direction. Suppose we are given a connection $\overline{\nabla}$ on the H-bundle $\Omega^{\check{\rho}}$ on D. We associate to it a generic Miura oper as follows. Let us choose a splitting $H \to B$ of the homomorphism $B \to H$ and set $\mathcal{F} = \Omega^{\check{\rho}} \underset{H}{\times} G, \mathcal{F}_B = \Omega^{\check{\rho}} \underset{H}{\times} B$, where we consider the adjoint action of H on G and on B obtained through the above splitting. The choice of the splitting also gives us the opposite Borel subgroup B_-, which is the unique Borel subgroup in generic position with B containing H. Then we set $\mathcal{F}'_B = \Omega^{\check{\rho}} \underset{H}{\times} B_- w_0 B$.

Observe that the space of connections on $\mathcal{F}$ is isomorphic to the direct product

$$\operatorname{Conn}(\Omega^{\check{\rho}})_D \times \bigoplus_{\alpha \in \Delta} \omega^{\alpha(\check{\rho})+1}.$$

Its subspace corresponding to negative simple roots is isomorphic to the space $\left(\bigoplus_{i=1}^{\ell} \mathfrak{g}_{-\alpha_i}\right) \otimes \mathcal{O}$. Having chosen a basis element f_i of $\mathfrak{g}_{-\alpha_i}$ for each $i = 1, \ldots, \ell$, we now construct an element $p_{-1} = \sum_{i=1}^{\ell} f_i$ of this space. Now we set $\nabla = \overline{\nabla} + p_{-1}$. By construction, ∇ has the correct relative position with the B-reduction $\mathcal{F}_B$ and preserves the B-reduction $\mathcal{F}'_B$. Therefore the quadruple $(\mathcal{F}, \nabla, \mathcal{F}_B, \mathcal{F}'_B)$ is a generic Miura oper on D. We set $\tau(\overline{\nabla}) = (\mathcal{F}, \nabla, \mathcal{F}_B, \mathcal{F}'_B)$.

This map is independent of the choice of splitting $H \to B$ and of the generators $f_i, i = 1, \ldots, \ell$. Indeed, changing the splitting $H \to B$ amounts to a conjugation of the old splitting by an element of N. This is equivalent to applying to ∇ the gauge transformation by this element. Therefore it will not change the underlying Miura oper structure. Likewise, rescaling of the generators f_i may be achieved by a gauge transformation by a constant element of H, and this again does not change the Miura oper structure. It is clear from the construction that β and τ are mutually inverse isomorphisms. □

Under the isomorphism of Proposition 8.2.2, the natural forgetful morphism

$$\operatorname{MOp}_G(D)_{\text{gen}} \to \operatorname{Op}_G(D)$$

becomes a map

$$\mu : \mathrm{Conn}(\Omega^{\check{\rho}})_D \to \mathrm{Op}_G(D). \tag{8.2.3}$$

We call this map the **Miura transformation.**

The Miura transformation (8.2.3) gives rise to a homomorphism of the corresponding rings of functions $\widetilde{\mu} : \mathrm{Fun}\,\mathrm{Op}_G(D) \to \mathrm{Fun}\,\mathrm{Conn}(\Omega^{\check{\rho}})_D$. Each space has an action of $\mathrm{Der}\,\mathcal{O}$ and $\widetilde{\mu}$ is a $\mathrm{Der}\,\mathcal{O}$-equivariant homomorphism between these rings. We will now identify $\mathrm{Fun}\,\mathrm{Conn}(\Omega^{\check{\rho}})_D$ with $\pi_0^\vee(\mathfrak{g})$ and the image of $\widetilde{\mu}$ with the intersection of kernels of the $\mathcal{W}$-algebra screening operators. This will give us an identification of $\mathrm{Fun}\,\mathrm{Op}_G(D)$ with $\mathbf{W}(\mathfrak{g})$.

8.2.2 Explicit realization of the Miura transformation

As explained in Section 4.2.4, if we choose a coordinate t on the disc, we can represent each oper connection in the canonical form

$$\partial_t + p_{-1} + \sum_{i=1}^{\ell} v_i(t) \cdot \mathbf{c}_i, \qquad v_i(t) \in \mathbb{C}[[t]]$$

(see Section 4.2.4), where

$$v_i(t) = \sum_{n<0} v_{i,n} t^{-n-1}.$$

Thus,

$$\mathrm{Fun}\,\mathrm{Op}_G(D) = \mathbb{C}[v_{i,n_i}]_{i=1,\ldots,\ell; n_i<0}.$$

If we choose a Cartan subalgebra $\mathfrak{h}$ in $\mathfrak{b}$, then, according to Proposition 8.2.2, we can represent each generic Miura oper by a connection operator of the following type:

$$\partial_t + p_{-1} + \mathbf{u}(t), \qquad \mathbf{u}(t) \in \mathfrak{h}[[t]]. \tag{8.2.4}$$

Set $u_i(t) = \alpha_i(\mathbf{u}(t)), i = 1, \ldots, \ell$, and

$$u_i(t) = \sum_{n<0} u_{i,n} t^{-n-1}.$$

Then

$$\mathrm{Fun}\,\mathrm{MOp}_G(D)_{\mathrm{gen}} = \mathrm{Fun}\,\mathrm{Conn}(\Omega^{\check{\rho}})_D = \mathbb{C}[u_{i,n}]_{i=1,\ldots,\ell; n<0}.$$

Hence the Miura transformation gives rise to a homomorphism

$$\widetilde{\mu} : \mathbb{C}[v_{i,n_i}]_{i=1,\ldots,\ell; n_i<0} \to \mathbb{C}[u_{i,n}]_{i=1,\ldots,\ell; n<0}. \tag{8.2.5}$$

Example. We compute the Miura transformation μ in the case when $\mathfrak{g} = \mathfrak{sl}_2$. In this case an oper has the form

$$\partial_t + \begin{pmatrix} 0 & v(t) \\ 1 & 0 \end{pmatrix},$$

and a generic Miura oper has the form

$$\partial_t + \begin{pmatrix} \frac{1}{2}u(t) & 0 \\ 1 & -\frac{1}{2}u(t) \end{pmatrix}.$$

To compute μ, we need to find an element of $N[[t]]$ such that the corresponding gauge transformation brings the Miura oper into the oper form. We find that

$$\begin{pmatrix} 1 & -\frac{1}{2}u(t) \\ 0 & 1 \end{pmatrix} \left(\partial_t + \begin{pmatrix} \frac{1}{2}u(t) & 0 \\ 1 & -\frac{1}{2}u(t) \end{pmatrix} \right) \begin{pmatrix} 1 & \frac{1}{2}u(t) \\ 0 & 1 \end{pmatrix} = \partial_t + \begin{pmatrix} 0 & \frac{1}{4}u(t)^2 + \frac{1}{2}\partial_t u(t) \\ 1 & 0 \end{pmatrix}.$$

Therefore we obtain that

$$\mu(u(t)) = v(t) = \frac{1}{4}u(t)^2 + \frac{1}{2}\partial_t u(t),$$

which may also be written in the form

$$\partial_t^2 - v(t) = \left(\partial_t + \frac{1}{2}u(t)\right)\left(\partial_t - \frac{1}{2}u(t)\right).$$

It is this transformation that was originally introduced by R. Miura as the map intertwining the flows of the KdV hierarchy and the mKdV hierarchy.

In the case when $\mathfrak{g} = \mathfrak{sl}_2$ opers are projective connections and Miura opers are affine connections (see, e.g., [FB], Chapter 9). The Miura transformation is nothing but the natural map from affine connections to projective connections.

By construction, the Miura transformation (8.2.3) and hence the homomorphism (8.2.5) are $\operatorname{Der}\mathcal{O}$-equivariant. The action of $\operatorname{Der}\mathcal{O}$ on $\operatorname{Fun}\operatorname{Op}_G(D)$ is obtained from formulas (4.2.5), and the action on the algebra $\operatorname{Fun}\operatorname{Conn}(\Omega^{\check{\rho}})_D$ is obtained from formula (8.2.2). It translates into the following explicit formulas for the action of the generators $L_n = -t^{n+1}\partial_t, n \geq -1$, on the $u_{i,m}$'s:

$$\begin{aligned} L_n \cdot u_{i,m} &= -m u_{i,n+m}, \qquad -1 \leq n < -m, \\ L_n \cdot u_{i,-n} &= -n(n+1), \qquad n > 0, \\ L_n \cdot u_{i,m} &= 0, \qquad n > -m. \end{aligned} \tag{8.2.6}$$

8.2.3 Screening operators

Recall the realization (8.2.1) of $\operatorname{MOp}_G(D)_{\text{gen}}$ as an $\mathcal{F}_{B,\text{univ}}$-twist of the B-torsor $\mathcal{U} \subset G/B$. As explained in Section 4.2.4, the B-torsors $\mathcal{F}_{B,0}$ may be identified for all opers $(\mathcal{F}, \mathcal{F}_B, \nabla)$. Therefore we obtain that the group $N_{\mathcal{F}_{B,0}}$ acts transitively on the fibers of the map $\operatorname{MOp}_G(D)_{\text{gen}} \to \operatorname{Op}_G(D)$. According to Proposition 8.2.2, we have an isomorphism $\operatorname{MOp}_G(D)_{\text{gen}} \simeq \operatorname{Conn}(\Omega^{\check{\rho}})_D$. Therefore $N_{\mathcal{F}_{B,0}}$ acts transitively along the fibers of the Miura transformation μ. Therefore we obtain that the image of the homomorphism $\widetilde{\mu}$ is equal to the space of $N_{\mathcal{F}_{B,0}}$-invariants of $\operatorname{Fun}\operatorname{Conn}(\Omega^{\check{\rho}})_D$, and hence to the space of $\mathfrak{n}_{\mathcal{F}_{B,0}}$-invariants of $\operatorname{Fun}\operatorname{Conn}(\Omega^{\check{\rho}})_D$.

Let us fix a Cartan subalgebra $\mathfrak{h}$ in $\mathfrak{b}$ and a trivialization of $\mathcal{F}_{B,0}$. Using this trivialization, we identify the twist $\mathfrak{n}_{\mathcal{F}_{B,0}}$ with $\mathfrak{n}$. Now we choose the generators $e_i, i = 1, \ldots, \ell$, of $\mathfrak{n}$ with respect to the action of $\mathfrak{h}$ on $\mathfrak{n}$ in such a way that together with the previously chosen f_i they satisfy the standard relations of $\mathfrak{g}$. The $N_{\mathcal{F}_{B,0}}$-action on $\operatorname{Conn}(\Omega^{\check{\rho}})_D$ then gives rise to an infinitesimal action of e_i on $\operatorname{Conn}(\Omega^{\check{\rho}})_D$. We will now compute the corresponding derivation on $\operatorname{Fun}\operatorname{Conn}(\Omega^{\check{\rho}})_D$.

The action of e_i is given by the infinitesimal gauge transformation

$$\delta \mathbf{u}(t) = [x_i(t) \cdot e_i, \partial_t + p_{-1} + \mathbf{u}(t)], \tag{8.2.7}$$

where $x_i(t) \in \mathbb{C}[[t]]$ is such that $x_i(0) = 1$, and the right hand side of formula (8.2.7) belongs to $\mathfrak{h}[[t]]$. It turns out that these conditions determine $x_i(t)$ uniquely. Indeed, the right hand side of (8.2.7) reads

$$x_i(t) \cdot \check{\alpha}_i - u_i(t) x_i(t) \cdot e_i - \partial_t x_i(t) \cdot e_i.$$

Therefore it belongs to $\mathfrak{h}[[t]]$ if and only if

$$\partial_t x_i(t) = -u_i(t) x_i(t). \tag{8.2.8}$$

If we write

$$x_i(t) = \sum_{n \leq 0} x_{i,n} t^{-n},$$

and substitute it into formula (8.2.8), we obtain that the coefficients $x_{i,n}$ satisfy the following recurrence relation:

$$n x_{i,n} = \sum_{k+m=n; k<0; m \leq 0} u_{i,k} x_{i,m}, \qquad n < 0.$$

We find from this formula that

$$\sum_{n \leq 0} x_{i,n} t^{-n} = \exp\left(- \sum_{m>0} \frac{u_{i,-m}}{m} t^m \right). \tag{8.2.9}$$

Now we obtain that

$$\delta u_j(t) = \alpha_j(\delta \mathbf{u}(t)) = a_{ij} x_i(t),$$

where (a_{ij}) is the Cartan matrix of $\mathfrak{g}$. In other words, the operator e_i acts on the algebra $\operatorname{Fun}\operatorname{Conn}(\Omega^{\check{\rho}})_D = \mathbb{C}[u_{i,n}]$ by the derivation

$$\sum_{j=1}^{\ell} a_{ij} \sum_{n \geq 0} x_{i,n} \frac{\partial}{\partial u_{j,-n-1}}, \tag{8.2.10}$$

where $x_{i,n}$ are given by formula (8.2.9).

Now, the image of $\operatorname{Fun}\operatorname{Op}_G(D)$ under the Miura map $\widetilde{\mu}$ is the algebra of $\mathfrak{n}$-invariant functions on $\operatorname{Conn}(\Omega^{\check{\rho}})_D$. These are precisely the functions that are annihilated by the generators $e_i, i = 1, \ldots, \ell$, of $\mathfrak{n}$, which are given by formula (8.2.10). Therefore we obtain the following characterization of $\operatorname{Fun}\operatorname{Op}_G(D)$ as a subalgebra of $\operatorname{Fun}\operatorname{Conn}(\Omega^{\check{\rho}})_D$.

Proposition 8.2.3 *The image of* $\operatorname{Fun}\operatorname{Op}_G(D)$ *in* $\operatorname{Fun}\operatorname{Conn}(\Omega^{\check{\rho}})_D$ *under the Miura map* $\widetilde{\mu}$ *is equal to the intersection of the kernels of the operators given by formula* (8.2.10) *for* $i = 1, \ldots, \ell$.

8.2.4 Back to the $\mathcal{W}$-algebras

Comparing formula (8.2.10) with formula (8.1.7) we find that if we replace $\mathbf{b}'_{i,n}$ by $u_{i,n}$ in formula (8.1.7) for the $\mathcal{W}$-algebra screening operator $\mathbf{V}_i[1]$, then we obtain formula (8.2.10). Therefore the intersection of the kernels of the operators (8.2.10) is equal to the intersection of the kernels of the operators $\mathbf{V}_i[1], i = 1, \ldots, \ell$. But the latter is the classical $\mathcal{W}$-algebra $\mathbf{W}(\mathfrak{g})_{\nu_0}$. Hence we obtain the following commutative diagram:

$$
\begin{array}{ccc}
\pi_0^\vee(\mathfrak{g})_{\nu_0} & \xrightarrow{\sim} & \operatorname{Fun}\operatorname{Conn}(\Omega^{\check{\rho}})_D \\
\uparrow & & \uparrow \\
\mathbf{W}(\mathfrak{g})_{\nu_0} & \xrightarrow{\sim} & \operatorname{Fun}\operatorname{Op}_G(D)
\end{array}
\tag{8.2.11}
$$

where the top arrow is an isomorphism of algebras given on generators by the assignment $\mathbf{b}'_{i,n} \mapsto u_{i,n}$.

We can also compute the character of $\mathbf{W}(\mathfrak{g}) \simeq \operatorname{Fun}\operatorname{Op}_G(D)$ and prove Lemma 8.1.4. It follows from Proposition 8.2.2 that the action of the group N on $\operatorname{Conn}(\Omega^{\check{\rho}})_D$ is free. Therefore we obtain that

$$\operatorname{Fun}\operatorname{Conn}(\Omega^{\check{\rho}})_D \simeq \operatorname{Fun}\operatorname{Op}_G(D) \otimes \operatorname{Fun} N,$$

as vector spaces. The action of the generators e_i of the Lie algebra $\mathfrak{n}$ is given by the operators $\mathbf{V}_i[1]$, each having degree -1 with respect to the $\mathbb{Z}_+$-grading introduced above. Therefore a root generator $e_\alpha \in \mathfrak{n}, \alpha \in \Delta_+$, acts as an operator of degree $-\langle \alpha, \check{\rho} \rangle$. This means that the action of $\mathfrak{n}$ preserves the $\mathbb{Z}$-grading on $\operatorname{Fun}\operatorname{Conn}(\Omega^{\check{\rho}})_D$ if we equip $\mathfrak{n}$ with the negative of the principal gradation, for which $\deg e_\alpha = \langle \alpha, \check{\rho} \rangle$. With respect to this grading the character of $\operatorname{Fun} N$ is equal to

$$\prod_{\alpha \in \Delta_+} (1 - q^{\langle \alpha, \check{\rho} \rangle})^{-1} = \prod_{i=1}^{\ell} \prod_{n_i=1}^{d_i} (1 - q^{n_i})^{-1}.$$

On the other hand, the character of $\operatorname{Fun}\operatorname{Conn}(\Omega^{\check{\rho}})_D$ is equal to

$$\prod_{n>0} (1 - q^n)^{-\ell}.$$

Therefore the character of $\operatorname{Fun}\operatorname{Op}_G(D)$ is equal to

$$\prod_{n>0}(1-q^n)^{-\ell}\prod_{i=1}^{\ell}\prod_{n_i=1}^{d_i}(1-q^{n_i})^{-1}.$$

Thus, we obtain that the character of $\operatorname{Fun}\operatorname{Op}_G(D) \simeq \mathbf{W}(\mathfrak{g})_{\nu_0}$ is given by the right hand side of formula (8.1.5). This proves Lemma 8.1.4.

We also obtain from the above isomorphism a vertex Poisson algebra structure on $\operatorname{Fun}\operatorname{Op}_G(D)$ (the latter structure may alternatively be defined by means of the Drinfeld–Sokolov reduction, as we show below). As before, these structures depend on the choice of inner product ν_0, which we will sometimes use as a subscript to indicate which Poisson structure we consider.

The above diagram gives us a geometric interpretation of the $\mathcal{W}$-algebra screening operators $\mathbf{V}_i[1]$: they correspond to the action of the generators e_i of $\mathfrak{n}$ on $\operatorname{Fun}\operatorname{Conn}(\Omega^{\check{\rho}})_D$.

Comparing formulas (8.2.6) and (8.1.11), we find that the top isomorphism in (8.2.11) is $\operatorname{Der}\mathcal{O}$-equivariant. Since the subspaces $\mathbf{W}(\mathfrak{g})_{\nu_0}$ and $\operatorname{Fun}\operatorname{Op}_G(D)$ are stable under the $\operatorname{Der}\mathcal{O}$-action, we find that the bottom isomorphism is also $\operatorname{Der}\mathcal{O}$-equivariant.†

Furthermore, the action of $\operatorname{Der}\mathcal{O}$ on all of the above algebras is independent of the inner product ν_0. Thus we obtain the following:

Theorem 8.2.4 *The diagram* (8.2.11) *is compatible with the action of* $\operatorname{Der}\mathcal{O}$.

8.2.5 Completion of the proof

Now let us replace $\mathfrak{g}$ by its Langlands dual Lie algebra ${}^L\mathfrak{g}$. Then we obtain the following commutative diagram

$$\begin{array}{ccc} \pi_0^\vee({}^L\mathfrak{g})_{\nu_0} & \xrightarrow{\sim} & \operatorname{Fun}\operatorname{Conn}(\Omega^{\rho})_{D,\nu_0} \\ \uparrow & & \uparrow \\ \mathbf{W}({}^L\mathfrak{g})_{\nu_0} & \xrightarrow{\sim} & \operatorname{Fun}\operatorname{Op}_{{}^LG}(D)_{\nu_0} \end{array} \qquad (8.2.12)$$

of Poisson vertex algebras, which is compatible with the action of $\operatorname{Der}\mathcal{O}$ and $\operatorname{Aut}\mathcal{O}$.

This completes the sixth, and last, step of our plan from Section 7.1. We have now assembled all the pieces needed to describe the center $\mathfrak{z}(\widehat{\mathfrak{g}})$ of $V_{\kappa_c}(\mathfrak{g})$.

Combining the commutative diagrams (8.2.12) and (8.1.10) and taking into account Theorem 8.2.4, Theorem 8.1.7 and Proposition 8.1.8 we come to the

† We remark that the screening operators, understood as derivations of $\pi_0^\vee(\mathfrak{g})$ or $\operatorname{Fun}\operatorname{Conn}(\Omega^{\check{\rho}})_D$, do not commute with the action of $\operatorname{Der}\mathcal{O}$. However, we can make them commute with $\operatorname{Der}\mathcal{O}$ if we consider them as operators acting from $\pi_0^\vee(\mathfrak{g})$ to another module, isomorphic to $\pi_0^\vee(\mathfrak{g})$ as a vector space, but with a modified action of $\operatorname{Der}\mathcal{O}$. Since we will not use this fact here, we refer a curious reader to [FF7] for more details.

following result (here by ${}^LG$ we understand the group of inner automorphisms of ${}^L\mathfrak{g}$).

Theorem 8.2.5 *There is an isomorphism* $\mathfrak{z}(\widehat{\mathfrak{g}})_{\kappa_0} \simeq \operatorname{Fun}\operatorname{Op}_{{}^LG}(D)_{\kappa_0^\vee}$ *which preserves the vertex Poisson structures and the* $\operatorname{Der}\mathcal{O}$*-module structures on both sides. Moreover, it fits into a commutative diagram of vertex Poisson algebras equipped with* $\operatorname{Der}\mathcal{O}$*-action:*

$$\begin{array}{ccc} \pi_0(\mathfrak{g})_{\kappa_0} & \xrightarrow{\sim} & \operatorname{Fun}\operatorname{Conn}(\Omega^\rho)_{D,\kappa_0^\vee} \\ \uparrow & & \uparrow \\ \mathfrak{z}(\widehat{\mathfrak{g}})_{\kappa_0} & \xrightarrow{\sim} & \operatorname{Fun}\operatorname{Op}_{{}^LG}(D)_{\kappa_0^\vee} \end{array} \tag{8.2.13}$$

where the upper arrow is given on the generators by the assignment $b_{i,n} \mapsto -u_{i,n}$.

In particular, we have now proved Theorem 4.3.2.

Note that Theorem 8.2.5 is consistent with the description of $\pi_0(\mathfrak{g})$ given previously in Section 6.2.5. According to this description, the $b_i(t)$'s transform under the action of $\operatorname{Der}\mathcal{O}$ as components of a connection on the ${}^LH$-bundle $\Omega^{-\rho}$ on the disc $D = \operatorname{Spec}\mathbb{C}[[t]]$. Therefore $\pi_0(\mathfrak{g})$ is identified, in a $\operatorname{Der}\mathcal{O}$-equivariant way, with the algebra $\operatorname{Fun}\operatorname{Conn}(\Omega^\rho)_D$ of functions on the space of connections on $\Omega^{-\rho}$. Thus, the top isomorphism in the diagram (8.2.13) may be expressed as an isomorphism

$$\operatorname{Fun}\operatorname{Conn}(\Omega^{-\rho})_D \simeq \operatorname{Fun}\operatorname{Conn}(\Omega^\rho)_D. \tag{8.2.14}$$

Under this isomorphism a connection $\partial_t + \mathbf{u}(t)$ on Ω^ρ is mapped to the dual connection $\partial_t - \mathbf{u}(t)$ on the dual ${}^LH$-bundle $\Omega^{-\rho}$. This precisely corresponds to the map $b_{i,n} \mapsto -u_{i,n}$ on the generators of the two algebras, which appears in Theorem 8.2.5.

Given any disc D_x, we may consider the twists of our algebras by the $\operatorname{Aut}\mathcal{O}$-torsor $\mathcal{A}ut_x$, defined as in Section 6.2.6. We will mark them by the subscript x. Then we obtain that

$$\mathbf{W}(\mathfrak{g})_{\nu_0,x} \simeq \operatorname{Fun}\operatorname{Op}(D_x)_{\nu_0},$$
$$\mathfrak{z}(\widehat{\mathfrak{g}})_{\kappa_0,x} \simeq \operatorname{Fun}\operatorname{Op}_{{}^LG}(D_x)_{\kappa_0^\vee}.$$

8.2.6 The associated graded algebras

In this section we describe the isomorphism $\mathfrak{z}(\widehat{\mathfrak{g}}) \simeq \operatorname{Fun}\operatorname{Op}_{{}^LG}(D)$ of Theorem 8.2.5 at the level of associated graded spaces. We start by describing the filtrations on $\mathfrak{z}(\widehat{\mathfrak{g}})$ and $\operatorname{Fun}\operatorname{Op}_{{}^LG}(D)$.

The filtration on $\mathfrak{z}(\widehat{\mathfrak{g}})$ is induced by the Poincaré–Birkhoff–Witt filtration on the universal enveloping algebra $U(\widehat{\mathfrak{g}}_{\kappa_c})$, see Section 3.3.3. By Theorem 8.1.5, $\operatorname{gr} \mathfrak{z}(\widehat{\mathfrak{g}}) = (\operatorname{gr} V_{\kappa_c}(\mathfrak{g}))^{\mathfrak{g}[[t]]}$. But

$$\operatorname{gr} V_{\kappa_c}(\mathfrak{g}) = \operatorname{Sym} \mathfrak{g}((t))/\mathfrak{g}[[t]] \simeq \operatorname{Fun} \mathfrak{g}^*[[t]]dt,$$

independently of the choice of coordinate t and inner product on $\mathfrak{g}$. In Proposition 3.4.2 we gave a description of $(\operatorname{gr} V_{\kappa_c}(\mathfrak{g}))^{\mathfrak{g}[[t]]}$. The coordinate-independent version of this description is as follows. Let $C_{\mathfrak{g}} = \operatorname{Spec} (\operatorname{Fun} \mathfrak{g}^*)^G$ and

$$C_{\mathfrak{g},\Omega} = \Omega \underset{\mathbb{C}^\times}{\times} C_{\mathfrak{g}},$$

where $\Omega = \mathbb{C}[[t]]dt$ is the topological module of differentials on $D = \operatorname{Spec} \mathbb{C}[[t]]$. By Proposition 3.4.2 we have a canonical and coordinate-independent isomorphism

$$\operatorname{gr} \mathfrak{z}(\widehat{\mathfrak{g}}) \simeq \operatorname{Fun} C_{\mathfrak{g},\Omega}. \tag{8.2.15}$$

Note that a choice of homogeneous generators $\overline{P}_i, i = 1, \ldots, \ell$, of $(\operatorname{Fun} \mathfrak{g}^*)^G$ gives us an identification

$$C_{\mathfrak{g},\Omega} \simeq \bigoplus_{i=1}^{\ell} \Omega^{\otimes(d_i+1)},$$

but we prefer not to use it as there is no natural choice for such generators.

Next we consider the map

$$a : \mathfrak{z}(\widehat{\mathfrak{g}}) \to \pi_0(\mathfrak{g})$$

which is equal to the restriction of the embedding $V_{\kappa_c}(\mathfrak{g}) \to W_{0,\kappa_c}$ to $\mathfrak{z}(\widehat{\mathfrak{g}})$. In the proof of Proposition 7.1.1 we described a filtration on W_{0,κ_c} compatible with the PBW filtration on $V_{\kappa_c}(\mathfrak{g})$. This implies that the map $\mathfrak{z}(\widehat{\mathfrak{g}}) \to \pi_0(\mathfrak{g})$ is also compatible with filtrations. According to the results of Section 6.2.5, we have a canonical identification

$$\pi_0(\mathfrak{g}) = \operatorname{Fun} \operatorname{Conn}(\Omega^{-\rho})_D.$$

The space $\operatorname{Conn}(\Omega^{-\rho})_D$ of connections on the ${}^L H$-bundle $\Omega^{-\rho}$ is an affine space over the vector space ${}^L\mathfrak{h} \otimes \Omega = \mathfrak{h}^* \otimes \Omega$. Therefore we find that

$$\operatorname{gr} \pi_0(\mathfrak{g}) = \operatorname{Fun} \mathfrak{h}^* \otimes \Omega. \tag{8.2.16}$$

We have the Harish-Chandra isomorphism $(\operatorname{Fun} \mathfrak{g}^*)^G \simeq (\operatorname{Fun} \mathfrak{h}^*)^W$, where W is the Weyl group of $\mathfrak{g}$, and hence an embedding $(\operatorname{Fun} \mathfrak{g}^*)^G \to \operatorname{Fun} \mathfrak{h}^*$. This embedding gives rise to an embedding

$$\operatorname{Fun} C_{\mathfrak{g},\Omega} \to \operatorname{Fun} \mathfrak{h}^* \otimes \Omega.$$

It follows from the proof of Proposition 7.1.1 that this is precisely the map

$$\operatorname{gr} a : \operatorname{gr} \mathfrak{z}(\widehat{\mathfrak{g}}) \to \operatorname{gr} \pi_0(\mathfrak{g})$$

under the identifications (8.2.15) and (8.2.16). Thus, we have now described the associated graded of the left vertical map in the commutative diagram (8.2.13).

Next, we describe the associated graded of the right vertical map in the commutative diagram (8.2.13). Let

$$C^{\vee}_{\mathfrak{g}} = \mathfrak{g}/G = \operatorname{Spec}(\operatorname{Fun}\mathfrak{g})^G$$

and

$$C^{\vee}_{\mathfrak{g},\Omega} = \Omega \underset{\mathbb{C}^\times}{\times} C^{\vee}_{\mathfrak{g}}.$$

Following [BD1], Section 3.1.14, we identify $\operatorname{Op}_G(D)$ with an affine space modeled on $C^{\vee}_{\mathfrak{g},\Omega}$. In other words, we have a natural filtration on $\operatorname{Fun}\operatorname{Op}_G(D)$ such that

$$\operatorname{gr}\operatorname{Op}_G(D) = \operatorname{Fun} C^{\vee}_{\mathfrak{g},\Omega}. \tag{8.2.17}$$

Constructing such a filtration is the same as constructing a flat $\mathbb{C}[h]$-algebra such that its specialization at $h = 1$ is $\operatorname{Fun}\operatorname{Op}_G(D)$, and the specialization at $h = 0$ is $\operatorname{Fun} C^{\vee}_{\mathfrak{g},\Omega}$. Consider the algebra of functions on the space $\operatorname{Op}_{G,h}(X)$ of h-opers on D. The definition of an h-oper is the same as that of an oper, except that we consider instead of a connection of the form (4.2.1) an h-connection

$$h\partial_t + \sum_{i=1}^{\ell} \psi_i(t) f_i + \mathbf{v}(t).$$

One shows that this algebra is flat over $\mathbb{C}[h]$ by proving that each h-oper has a canonical form

$$h\partial_t + p_{-1} + \mathbf{v}(t), \qquad \mathbf{v}(t) \in V_{\text{can}}((t)),$$

in exactly the same way as in the proof of Lemma 4.2.2.

It is clear that $\operatorname{Op}_{G,1}(D) = \operatorname{Op}_G(D)$. In order to see that $\operatorname{Op}_{G,0}(D) = C^{\vee}_{\mathfrak{g},\Omega}$, observe that it follows from the definition that $\operatorname{Op}_{G,0}(D) = \Gamma(D, \Omega \underset{\mathbb{C}^\times}{\times} V_{\text{can}})$, where the action of $\mathbb{C}^\times$ on V_{can} is given by $a \mapsto a \operatorname{Ad}_{\rho(a)}$. But it follows from [Ko] that we have a canonical isomorphism of $\mathbb{C}^\times$-spaces $V_{\text{can}} \simeq \mathfrak{g}/G$. This implies that $\operatorname{Op}_{G,0}(D) = C^{\vee}_{\mathfrak{g},\Omega}$ and hence (8.2.17).

Now consider the space of connections $\operatorname{Conn}(\Omega^\rho)_D$ on the ${}^L H$-bundle Ω^ρ. It is an affine space over the vector space ${}^L\mathfrak{h} \otimes \Omega$. Therefore

$$\operatorname{gr}\operatorname{Fun}\operatorname{Conn}(\Omega^\rho)_D = \operatorname{Fun}{}^L\mathfrak{h} \otimes \Omega. \tag{8.2.18}$$

We have the Harish-Chandra isomorphism $(\operatorname{Fun}\mathfrak{g})^G \simeq (\operatorname{Fun}\mathfrak{h})^W$, where W is the Weyl group of $\mathfrak{g}$, and hence an embedding $(\operatorname{Fun}\mathfrak{g})^G \to \operatorname{Fun}\mathfrak{h}$. This embedding gives rise to an embedding

$$\operatorname{Fun} C^{\vee}_{\mathfrak{g},\Omega} \to \operatorname{Fun}\mathfrak{h} \otimes \Omega.$$

The explicit construction of the Miura transformation $\operatorname{Conn}(\Omega^\rho)_D \to \operatorname{Op}_{^L G}(D)$ given in the proof of Proposition 8.2.2 implies that the map

$$\operatorname{Fun} \operatorname{Op}_{^L G}(D) \to \operatorname{Fun} \operatorname{Conn}(\Omega^\rho)_D$$

preserves filtrations. Furthermore, with respect to the identifications (8.2.17) and (8.2.18), its associated graded map is nothing but the homomorphism

$$\operatorname{Fun} C^\vee_{^L\mathfrak{g},\Omega} \to \operatorname{Fun} {}^L\mathfrak{h} \otimes \Omega.$$

This describes the associated graded of the right vertical map in the commutative diagram (8.2.13).

Next, the associated graded of the upper horizontal isomorphism is given by the natural composition

$$\operatorname{gr} \pi_0(\mathfrak{g}) \simeq \operatorname{Fun} \mathfrak{h}^* \otimes \Omega = \operatorname{Fun} {}^L\mathfrak{h} \otimes \Omega \simeq \operatorname{gr} \operatorname{Fun} \operatorname{Conn}(\Omega^\rho)_D,$$

multiplied by the operator $(-1)^{\deg}$ which takes the value $(-1)^n$ on elements of degree n (this is due to the map $b_{i,n} \mapsto -u_{i,n}$ in Theorem 8.2.5).

Note that we have canonical isomorphisms

$$(\operatorname{Fun} \mathfrak{g}^*)^G = (\operatorname{Fun} \mathfrak{h}^*)^W = (\operatorname{Fun} {}^L\mathfrak{h})^W = (\operatorname{Fun} {}^L\mathfrak{g})^{^L G},$$

which give rise to a canonical identification

$$\operatorname{Fun} C_{\mathfrak{g},\Omega} = \operatorname{Fun} C^\vee_{^L\mathfrak{g},\Omega}.$$

Now, all maps in the diagram (8.2.13) preserve the filtrations. The commutativity of the diagram implies that the associated graded of the lower horizontal isomorphism

$$\mathfrak{z}(\widehat{\mathfrak{g}}) \simeq \operatorname{Fun} \operatorname{Op}_{^L G}(D) \tag{8.2.19}$$

is equal to the composition

$$\operatorname{gr} \mathfrak{z}(\widehat{\mathfrak{g}}) \simeq \operatorname{Fun} C_{\mathfrak{g},\Omega} = \operatorname{Fun} C^\vee_{^L\mathfrak{g},\Omega} \simeq \operatorname{gr} \operatorname{Fun} \operatorname{Op}_{^L G}(D)$$

multiplied by the operator $(-1)^{\deg}$ which takes the value $(-1)^n$ on elements of degree n. Thus, we obtain the following:

Theorem 8.2.6 *The isomorphism* (8.2.19) *preserves filtrations. The corresponding associated graded algebras are both isomorphic to* $\operatorname{Fun} C_{\mathfrak{g},\Omega}$. *The corresponding isomorphism of the associated graded algebras is equal to* $(-1)^{\deg}$.

8.3 The center of the completed universal enveloping algebra

Recall the completion $\widetilde{U}_\kappa(\widehat{\mathfrak{g}})$ of the universal enveloping algebra of $\widehat{\mathfrak{g}}_\kappa$ defined in Section 2.1.2. Let $Z(\widehat{\mathfrak{g}})$ be its center. In Theorem 4.3.6 we have derived from

Theorem 4.3.2 (which we have now proved, see Theorem 8.2.5) the following $\operatorname{Aut}\mathcal{O}$-equivariant isomorphism:

$$Z(\widehat{\mathfrak{g}}) \simeq \operatorname{Fun}\operatorname{Op}_{{}^L G}(D^\times). \tag{8.3.1}$$

In this section we discuss various properties of this isomorphism.

8.3.1 Isomorphism between $Z(\widehat{\mathfrak{g}})$ and $\operatorname{Fun}\operatorname{Op}_{{}^L G}(D^\times)$

Recall the $\operatorname{Aut}\mathcal{O}$-equivariant isomorphism of vertex Poisson algebras

$$\mathfrak{z}(\widehat{\mathfrak{g}})_{\kappa_0} \simeq \operatorname{Fun}\operatorname{Op}_{{}^L G}(D)_{\kappa_0^\vee} \tag{8.3.2}$$

established in Theorem 8.2.5. For a vertex Poisson algebra P, the Lie algebra structure on $U(P) = \operatorname{Lie}(P)$ gives rise to a Poisson algebra structure on the commutative algebra $\widetilde{U}(P)$. In particular, $\widetilde{U}(\mathfrak{z}(\widehat{\mathfrak{g}})_{\kappa_0}) = Z(\widehat{\mathfrak{g}})$ (see Proposition 4.3.4) is a Poisson algebra.

The corresponding Poisson structure on $Z(\widehat{\mathfrak{g}})_{\kappa_0}$ may be described as follows. Consider $\widetilde{U}_\kappa(\widehat{\mathfrak{g}})$ as the one-parameter family A_ϵ of associative algebras depending on the parameter $\epsilon = (\kappa - \kappa_c)/\kappa_0$. Then $Z(\widehat{\mathfrak{g}})$ is the center of A_0. Given $x, y \in Z(\widehat{\mathfrak{g}})$, let $\widetilde{x}, \widetilde{y}$ be their liftings to A_ϵ. Then the Poisson bracket $\{x, y\}$ is defined as the ϵ-linear term in the commutator $[\widetilde{x}, \widetilde{y}]$ considered as a function of ϵ (it is independent of the choice of the liftings).† We denote the center $Z(\widehat{\mathfrak{g}})$ equipped with this Poisson structure by $Z(\widehat{\mathfrak{g}})_{\kappa_0}$.

Likewise, the vertex Poisson algebra $\operatorname{Fun}\operatorname{Op}_{{}^L G}(D)_{\kappa_0^\vee}$ gives rise to a topological Poisson algebra $\widetilde{U}(\operatorname{Fun}\operatorname{Op}_{{}^L G}(D)_{\kappa_0^\vee})$. According to Lemma 4.3.5, the latter is isomorphic to $\operatorname{Fun}\operatorname{Op}_{{}^L G}(D^\times)$ in an $\operatorname{Aut}\mathcal{O}$-equivariant way. Therefore the isomorphism (8.3.2) gives rise to an isomorphism of topological Poisson algebras

$$Z(\widehat{\mathfrak{g}})_{\kappa_0} \simeq \operatorname{Fun}\operatorname{Op}_{{}^L G}(D^\times)_{\kappa_0^\vee}.$$

Here we use the subscript $\kappa_0^\vee$ to indicate the dependence of the Poisson structure on the inner product $\kappa_0^\vee$ on ${}^L\mathfrak{g}$. We will give another definition of this Poisson structure in the next section. Thus, we obtain the following result, which was originally conjectured by V. Drinfeld.

Theorem 8.3.1 *The center $Z(\widehat{\mathfrak{g}})_{\kappa_0}$ is isomorphic, as a Poisson algebra, to the Poisson algebra* $\operatorname{Fun}\operatorname{Op}_{{}^L G}(D^\times)_{\kappa_0^\vee}$. *Moreover, this isomorphism is* $\operatorname{Aut}\mathcal{O}$-*equivariant.*

† The fact that this is indeed a Poisson structure was first observed by V. Drinfeld following the work [Ha] of T. Hayashi.

8.3.2 The Poisson structure on $\operatorname{Fun}\operatorname{Op}_G(D^\times)_{\nu_0}$

We have obtained above a Poisson structure on the algebra $\operatorname{Fun}\operatorname{Op}_G(D^\times)$ from the vertex Poisson structure on $\operatorname{Fun}\operatorname{Op}_G(D)$, which was in turn obtained by realizing $\operatorname{Fun}\operatorname{Op}_G(D)$ as a vertex Poisson subalgebra of $\pi^\vee(\mathfrak{g})_{\nu_0}$ (namely, as the classical $\mathcal{W}$-algebra $\mathbf{W}(\mathfrak{g})_{\nu_0}$). Now we explain how to obtain this Poisson structure by using the Hamiltonian reduction called the **Drinfeld–Sokolov reduction** [DS].

We start with the Poisson manifold $\operatorname{Conn}_{\mathfrak{g}}$ of connections on the trivial G-bundle on $D^\times$, i.e., operators of the form $\nabla = \partial_t + A(t)$, where $A(t) \in \mathfrak{g}((t))$. The Poisson structure on this manifold comes from its identification with a hyperplane in the dual space to the affine Kac–Moody algebra $\widehat{\mathfrak{g}}_{\nu_0}$, where ν_0 is a non-zero invariant inner product on $\mathfrak{g}$. Indeed, the topological dual space to $\widehat{\mathfrak{g}}_{\nu_0}$ may be identified with the space of all λ-connections on the trivial bundle on $D^\times$, see [FB], Section 16.4. Namely, we split $\widehat{\mathfrak{g}}_{\nu_0} = \mathfrak{g}((t)) \oplus \mathbb{C}\mathbf{1}$ as a vector space. Then a λ-connection $\partial_t + A(t)$ gives rise to a linear functional on $\widehat{\mathfrak{g}}_{\nu_0}$ which takes the value

$$\lambda b + \operatorname{Res} \nu_0(A(t), B(t))dt$$

on $(B(t) + b\mathbf{1}) \in \mathfrak{g}((t)) \oplus \mathbb{C}\mathbf{1} = \widehat{\mathfrak{g}}_{\nu_0}$. Note that under this identification the action of $\operatorname{Aut}\mathcal{O}$ by changes of coordinates and the coadjoint (resp., gauge) action of $G((t))$ on the space of λ-connections (resp., the dual space to $\widehat{\mathfrak{g}}_{\nu_0}$) agree. The space $\operatorname{Conn}_{\mathfrak{g}}$ is now identified with the hyperplane in $\widehat{\mathfrak{g}}^*_{\nu_0}$ which consists of those functionals which take the value 1 on the central element $\mathbf{1}$.

The dual space $\widehat{\mathfrak{g}}^*_{\nu_0}$ carries a canonical Poisson structure called the Kirillov-Kostant structure (see, e.g., [FB], Section 16.4.1 for more details). Because $\mathbf{1}$ is a central element, this Poisson structure restricts to the hyperplane which we have identified with $\operatorname{Conn}_{\mathfrak{g}}$. Therefore we obtain a Poisson structure on $\operatorname{Conn}_{\mathfrak{g}}$.

The group $N((t))$ acts on $\operatorname{Conn}_{\mathfrak{g}}$ by gauge transformations. This action corresponds to the coadjoint action of $N((t))$ on $\widehat{\mathfrak{g}}^*_{\nu_0}$ and is Hamiltonian, the moment map being the surjection $m : \operatorname{Conn}_{\mathfrak{g}} \to \mathfrak{n}((t))^*$ dual to the embedding $\mathfrak{n}((t)) \to \widehat{\mathfrak{g}}_{\nu_0}$. We pick a one-point coadjoint $N((t))$-orbit in $\mathfrak{n}((t))^*$ represented by the linear functional ψ which is equal to the composition $\mathfrak{n}((t)) \to \mathfrak{n}/[\mathfrak{n},\mathfrak{n}]((t)) = \bigoplus_{i=1}^{\ell} \mathbb{C}((t)) \cdot e_i$ and the functional

$$(x_i(t))_{i=1}^{\ell} \mapsto \sum_{i=1}^{\ell} \operatorname{Res}_{t=0} x_i(t)dt.$$

One shows in the same way as in the proof of Lemma 4.2.2 that the action of $N((t))$ on $m^{-1}(\psi)$ is free. Moreover, the quotient $m^{-1}(\psi)/N((t))$, which is the Poisson reduced manifold, is canonically identified with the space of G-opers on $D^\times$. Therefore we obtain a Poisson structure on the topological algebra of functions $\operatorname{Fun}\operatorname{Op}_G(D^\times)$.

Thus, we now have two Poisson structures on $\operatorname{Fun}\operatorname{Op}_G(D^\times)$ associated to a non-zero invariant inner product ν_0 on $\mathfrak{g}$.

Lemma 8.3.2 *The two Poisson structures coincide.*

Proof. In [FB], Sections 15.4 and 16.8, we defined a complex $C^\bullet_\infty(\mathfrak{g})$ and showed that its zeroth cohomology is canonically isomorphic to $\mathbf{W}(\mathfrak{g})_{\nu_0}$, equipped with the vertex Poisson algebra structure introduced in Section 8.1.4 (and all other cohomologies vanish). This complex is a vertex Poisson algebra version of the BRST complex computing the result of the Drinfeld–Sokolov reduction described above. In particular, we identify $\operatorname{Fun}\operatorname{Op}_G(D^\times)$, equipped with the Poisson structure obtained via the Drinfeld–Sokolov reduction, with $\widetilde{U}(\mathbf{W}(\mathfrak{g})_{\nu_0})$, with the Poisson structure corresponding to the above vertex Poisson structure on $\mathbf{W}(\mathfrak{g})_{\nu_0}$. $\square$

8.3.3 The Miura transformation as the Harish-Chandra homomorphism

The Harish-Chandra homomorphism, which we have already discussed in Section 5.1, is a homomorphism from the center $Z(\mathfrak{g})$ of $U(\mathfrak{g})$, where $\mathfrak{g}$ is a simple Lie algebra, to the algebra $\operatorname{Fun}\mathfrak{h}^*$ of polynomials on $\mathfrak{h}^*$. It identifies $Z(\mathfrak{g})$ with the algebra $(\operatorname{Fun}\mathfrak{h}^*)^W$ of W-invariant polynomials on $\mathfrak{h}^*$. To construct this homomorphism, one needs to assign a central character to each $\lambda \in \mathfrak{h}^*$. This central character is just the character with which the center acts on the Verma module $M_{\lambda-\rho}$.

In the affine case, we construct a similar homomorphism from the center $Z(\widehat{\mathfrak{g}})$ of $\widetilde{U}_{\kappa_c}(\widehat{\mathfrak{g}})$ to the topological algebra $\operatorname{Fun}\operatorname{Conn}(\Omega^{-\rho})_{D^\times}$ of functions on the space of connections on the ${}^L H$-bundle $\Omega^{-\rho}$ on $D^\times$. According to Corollary 6.1.5, points of $\operatorname{Conn}(\Omega^{-\rho})_{D^\times}$ parameterize Wakimoto modules of critical level. Thus, for each $\overline{\nabla} \in \operatorname{Conn}(\Omega^{-\rho})_{D^\times}$ we have the Wakimoto module $W_{\overline{\nabla}}$ of critical level. The following theorem describes the affine analogue of the Harish-Chandra homomorphism.

Note that $\operatorname{Fun}\operatorname{Conn}(\Omega^{-\rho})_{D^\times}$ is the completion of the polynomial algebra in $b_{i,n}, i = 1, \ldots, \ell; n \in \mathbb{Z}$, with respect to the topology in which the base of open neighborhoods of 0 is formed by the ideals generated by $b_{i,n}, n < N$. In the same way as in the proof of Lemma 4.3.5 we show that it is isomorphic to $\widetilde{U}(\operatorname{Fun}\operatorname{Conn}(\Omega^{-\rho})_D)$. We also define the topological algebra $\operatorname{Fun}\operatorname{Conn}(\Omega^{\rho})_{D^\times}$ as $\widetilde{U}(\operatorname{Fun}\operatorname{Conn}(\Omega^{\rho})_D)$. It is the completion of the polynomial algebra in $u_{i,n}, i = 1, \ldots, \ell; n \in \mathbb{Z}$, with respect to the topology in which the base of open neighborhoods of 0 is formed by the ideals generated by $u_{i,n}, n < N$. The isomorphism (8.2.14) gives rise to an isomorphism of the topological algebras

$$\operatorname{Fun}\operatorname{Conn}(\Omega^{-\rho})_{D^\times} \to \operatorname{Fun}\operatorname{Conn}(\Omega^{\rho})_{D^\times},$$

under which $b_{i,n} \mapsto -u_{i,n}$.

The natural forgetful morphism $\mathrm{MOp}_{{}^L G}(D^\times)_{\mathrm{gen}} \to \mathrm{Op}_{{}^L G}(D^\times)$ and an identification

$$\mathrm{MOp}_{{}^L G}(D^\times)_{\mathrm{gen}} \simeq \mathrm{Conn}(\Omega^\rho)_{D^\times}$$

constructed in the same way as in Proposition 8.2.2 give rise to a map

$$\mu : \mathrm{Conn}(\Omega^\rho)_{D^\times} \to \mathrm{Op}_{{}^L G}(D^\times). \tag{8.3.3}$$

This is the Miura transformation on the punctured disc $D^\times$, which is analogous to the Miura transformation (8.2.3) on the disc D.

We recall the general construction of semi-infinite parabolic induction described in Corollary 6.3.2. We apply it in the case of the Borel subalgebra $\mathfrak{b} \subset \mathfrak{g}$. Let M be a module over the vertex algebra $M_{\mathfrak{g}}$, or, equivalently, a module over the Weyl algebra $\mathcal{A}^{\mathfrak{g}}$. Let R be a module over the commutative vertex algebra $\pi_0 = \mathrm{Fun}\,\mathrm{Conn}(\Omega^{-\rho})_D$, or, equivalently, a smooth module over the commutative algebra $\mathrm{Fun}\,\mathrm{Conn}(\Omega^{-\rho})_{D^\times}$. Then the homomorphism

$$w_{\kappa_c} : V_{\kappa_c}(\mathfrak{g}) \to M_{\mathfrak{g}} \otimes \pi_0$$

of Theorem 6.1.6 gives rise to the structure of a $V_{\kappa_c}(\mathfrak{g})$-module, or, equivalently, a smooth $\widehat{\mathfrak{g}}_{\kappa_c}$–module on the tensor product $M_{\mathfrak{g}} \otimes R$.

Now we obtain from Theorem 8.2.5 the following result.

Theorem 8.3.3 *The action of $Z(\widehat{\mathfrak{g}})$ on the $\widehat{\mathfrak{g}}_{\kappa_c}$-module $M \otimes R$ does not depend on M and factors through a homomorphism*

$$Z(\widehat{\mathfrak{g}}) \to \mathrm{Fun}\,\mathrm{Conn}(\Omega^{-\rho})_{D^\times}$$

and the action of $\mathrm{Fun}\,\mathrm{Conn}(\Omega^{-\rho})_{D^\times}$ on R. The corresponding morphism $\mathrm{Conn}(\Omega^{-\rho})_{D^\times} \to \mathrm{Spec}\,Z(\widehat{\mathfrak{g}})$ fits into a commutative diagram

$$\begin{array}{ccc} \mathrm{Conn}(\Omega^{-\rho})_{D^\times} & \xrightarrow{\sim} & \mathrm{Conn}(\Omega^{\rho})_{D^\times} \\ \downarrow & & \downarrow \\ \mathrm{Spec}\,Z(\widehat{\mathfrak{g}}) & \xrightarrow{\sim} & \mathrm{Op}_{{}^L G}(D^\times) \end{array} \tag{8.3.4}$$

where the right vertical arrow is the Miura transformation (8.3.3) *on $D^\times$.*

In particular, if $R = \mathbb{C}_{\overline{\nabla}}$ is the one-dimensional module corresponding to a point $\overline{\nabla} \in \mathrm{Conn}(\Omega^{-\rho})_{D^\times}$, then the center $Z(\widehat{\mathfrak{g}})$ acts on the corresponding Wakimoto module $W_{\overline{\nabla}} = M_{\mathfrak{g}} \otimes \mathbb{C}_{\overline{\nabla}}$ via a central character corresponding to the Miura transformation of $\overline{\nabla}$.

Proof. The action of $U(V_{\kappa_c}(\mathfrak{g}))$, and hence of $\widetilde{U}_{\kappa_c}(\widehat{\mathfrak{g}})$, on $W_{\overline{\nabla}}$ is obtained through the homomorphism of vertex algebras $V_{\kappa_c}(\mathfrak{g}) \to M_{\mathfrak{g}} \otimes \pi_0(\mathfrak{g})$. In particular, the action of $Z(\widehat{\mathfrak{g}})$ on $W_{\overline{\nabla}}$ is obtained through the homomorphism of commutative vertex algebras $\mathfrak{z}(\widehat{\mathfrak{g}}) \to \pi_0(\mathfrak{g})$. But $\pi_0(\mathfrak{g}) = \mathrm{Fun}\,\mathrm{Conn}(\Omega^{-\rho})_D$, according to Section 6.2.5. Therefore the statement of the theorem follows from Theorem 8.2.5 by applying the functor of enveloping algebras $V \mapsto \widetilde{U}(V)$ introduced in Section 3.2.3. □

Thus, we see that the affine analogue of the Harish-Chandra homomorphism is the right vertical arrow of the diagram (8.3.4), which is nothing but the Miura transformation for the Langlands dual group! In particular, its image (which in the finite-dimensional case consists of W-invariant polynomials on $\mathfrak{h}^*$) is described as the intersection of the kernels of the $\mathcal{W}$-algebra screening operators.

9

Structure of $\widehat{\mathfrak{g}}$-modules of critical level

In the previous chapters we have described the center $\mathfrak{z}(\widehat{\mathfrak{g}})$ of the vertex algebra $V_{\kappa_c}(\mathfrak{g})$ at the critical level. As we have shown in Section 3.3.2, $\mathfrak{z}(\widehat{\mathfrak{g}})$ is isomorphic to the algebra of endomorphisms of the $\widehat{\mathfrak{g}}_{\kappa_c}$-module $V_{\kappa_c}(\mathfrak{g})$, which commute with the action of $\widehat{\mathfrak{g}}_{\kappa_c}$. We have identified this algebra with the algebra of functions on the space $\mathrm{Op}_{^LG}(D)$ of ^LG-opers on the disc D.

In this chapter we obtain similar results about the algebras of endomorphisms of the Verma modules and the Weyl modules of critical level. We show that both are quotients of the center $Z(\widehat{\mathfrak{g}})$ of the completed enveloping algebra $\widetilde{U}_{\kappa_c}(\widehat{\mathfrak{g}})$, which is isomorphic to the algebra of functions on the space $\mathrm{Op}_{^LG}(D)$ of ^LG-opers on the punctured disc $D^\times$. In the case of Verma modules, the algebra of endomorphisms is identified with the algebra of functions on the space of opers with regular singularities and fixed residue, and in the case of Weyl modules it is the algebra of functions of opers with regular singularities, fixed residue and trivial monodromy. Thus, the geometry of opers is reflected in the representation theory of affine algebras of critical level. Understanding this connection between representation theory and geometry is important for the development of the local Langlands correspondence that we will discuss in the next chapter.

We begin this chapter by introducing the relevant subspaces of the space $\mathrm{Op}_{^LG}(D^\times)$: opers with regular singularities (in Section 9.1) and nilpotent opers (in Section 9.2), and explain the interrelations between them. We then consider in Section 9.3 the restriction of the Miura transformation defined in Section 8.2.1 to opers with regular singularities and nilpotent opers. We relate the fibers of the Miura transformation over the nilpotent opers to the Springer fibers. This will allow us to obtain families of Wakimoto modules parameterized by the Springer fibers. Next, we discuss in Section 9.4 some categories of representations of $\widehat{\mathfrak{g}}_{\kappa_c}$ that "live" over the spaces of opers with regular singularities. Finally, in Sections 9.5 and 9.6 we describe the algebras of endomorphisms of the Verma modules and the Weyl modules of critical level.

The results of Section 9.1 are due to A. Beilinson and V. Drinfeld [BD1], and most of the results of the remaining sections of this chapter were obtained by D. Gaitsgory and myself in [FG2, FG6].

9.1 Opers with regular singularity

In this section we introduce, following Beilinson and Drinfeld [BD1], the space of opers on the disc with regular singularities at the origin.

9.1.1 Definition

Recall that the space $\mathrm{Op}_G(D)$ (resp., $\mathrm{Op}_G(D^\times)$) of G-opers on D (resp., $D^\times$) is the quotient of the space of operators of the form (4.2.1) where $\psi_i(t)$ and $\mathbf{v}(t)$ take values in $\mathbb{C}[[t]]$ (resp., in $\mathbb{C}((t))$) by the action of $B[[t]]$ (resp., $B((t))$).

A G**-oper on** D **with regular singularity** is by definition (see [BD1], Section 3.8.8) a $B[[t]]$-conjugacy class of operators of the form

$$\nabla = \partial_t + t^{-1}\left(\sum_{i=1}^{\ell} \psi_i(t) f_i + \mathbf{v}(t)\right), \tag{9.1.1}$$

where $\psi_i(t) \in \mathbb{C}[[t]], \psi_i(0) \neq 0$, and $\mathbf{v}(t) \in \mathfrak{b}[[t]]$.

Equivalently, it is an $N[[t]]$-equivalence class of operators

$$\nabla = \partial_t + \frac{1}{t}\left(p_{-1} + \mathbf{v}(t)\right), \qquad \mathbf{v}(t) \in \mathfrak{b}[[t]]. \tag{9.1.2}$$

Denote by $\mathrm{Op}_G^{\mathrm{RS}}(D)$ the space of opers on D with regular singularity.

More generally, we consider, following [BD1], Section 3.8.8, the space $\mathrm{Op}_G^{\mathrm{ord}_k}(D)$ of opers with singularity of order less than or equal to $k > 0$ as the space of $N[[t]]$-equivalence classes of operators

$$\nabla = \partial_t + \frac{1}{t^k}\left(p_{-1} + \mathbf{v}(t)\right), \qquad \mathbf{v}(t) \in \mathfrak{b}[[t]].$$

9.1.2 Residue

Following [BD1], we associate to an oper with regular singularity its **residue**. For an operator (9.1.2) the residue is by definition equal to $p_{-1}+\mathbf{v}(0)$. Clearly, under gauge transformations by an element $x(t)$ of $N[[t]]$ the residue gets conjugated by $x(0) \in N$. Therefore its projection onto

$$\mathfrak{g}/G = \mathrm{Spec}(\mathrm{Fun}\,\mathfrak{g})^G = \mathrm{Spec}(\mathrm{Fun}\,\mathfrak{h})^W = \mathfrak{h}/W$$

is well-defined. Thus, we obtain a morphism

$$\mathrm{res} : \mathrm{Op}_G^{\mathrm{RS}}(D) \to \mathfrak{h}/W.$$

Given $\varpi \in \mathfrak{h}/W$, we denote by $\mathrm{Op}_G^{\mathrm{RS}}(D)_\varpi$ the space of opers with regular singularity and residue ϖ.

9.1.3 Canonical representatives

Suppose that we are given an oper with regular singularity represented by a $B[[t]]$-conjugacy class of a connection ∇ of the form (9.1.1). Recall that the set $E = \{d_1, \dots, d_\ell\}$ is the set of exponents of $\mathfrak{g}$ counted with multiplicity and the subspace $V^{\mathrm{can}} = \bigoplus_{i \in E} V_i^{\mathrm{can}}$ of $\mathfrak{g}$ with a basis $p_1, \dots, p_\ell$ introduced in Section 4.2.4.

In the same way as in the proof of Lemma 4.2.2 we use the gauge action of $B[[t]]$ to bring ∇ to the form

$$\partial_t + t^{-1}\left(p_{-1} + \sum_{j \in E} v_j(t) p_j\right), \qquad v_j(t) \in \mathbb{C}[[t]]. \tag{9.1.3}$$

The residue of this operator is equal to

$$p_{-1} + \sum_{j \in E} v_j(0) p_j. \tag{9.1.4}$$

Denote by $\mathfrak{g}_{\mathrm{can}}$ the affine subspace of $\mathfrak{g}$ consisting of all elements of the form $p_{-1} + \sum_{j \in E} y_j p_j$. Recall from [Ko] that the adjoint orbit of any regular elements in the Lie algebra $\mathfrak{g}$ contains a unique element which belongs to $\mathfrak{g}_{\mathrm{can}}$ and the corresponding morphism $\mathfrak{g}_{\mathrm{can}} \to \mathfrak{h}/W$ is an isomorphism. Thus, we obtain that when we bring an oper with regular singularity to the canonical form (9.1.3), its residue is realized as an element of $\mathfrak{g}_{\mathrm{can}}$.

Consider the natural morphism $\mathrm{Op}_G^{\mathrm{RS}}(D) \to \mathrm{Op}_G(D^\times)$ taking the $B[[t]]$-equivalence class of operators of the form (9.1.1) to its $B((t))$-equivalence class.

Proposition 9.1.1 ([DS], Prop. 3.8.9) *The map $\mathrm{Op}_G^{\mathrm{RS}}(D) \to \mathrm{Op}_G(D^\times)$ is injective. Its image consists of those G-opers on $D^\times$ whose canonical representatives have the form*

$$\partial_t + p_{-1} + \sum_{j \in E} t^{-j-1} c_j(t) p_j, \qquad c_j(t) \in \mathbb{C}[[t]]. \tag{9.1.5}$$

Moreover, the residue of this oper is equal to

$$p_{-1} + \left(c_1(0) + \frac{1}{4}\right) p_1 + \sum_{j \in E, j > 1} c_j(0) p_j. \tag{9.1.6}$$

Proof. First, we bring an oper with regular singularity to the form (9.1.3). Next, we apply the gauge transformation by $\check{\rho}(t)^{-1}$ and obtain the following

$$\partial_t + p_{-1} + \check{\rho} t^{-1} + \sum_{j \in E} t^{-j-1} v_j(t) p_j, \qquad v_j(t) \in \mathbb{C}[[t]].$$

Finally, applying the gauge transformation by $\exp(-p_1/2t)$ we obtain the operator

$$\partial_t + p_{-1} + t^{-2}\left(v_1(t) - \frac{1}{4}\right)p_1 + \sum_{j \in E, j>1} t^{-j-1} v_j(t) p_j, \qquad v_j(t) \in \mathbb{C}[[t]]. \tag{9.1.7}$$

Thus, we obtain an isomorphism between the space of opers with regular singularity and the space of opers on $D^\times$ of the form (9.1.5), and, in particular, we find that the map $\mathrm{Op}_G^{\mathrm{RS}}(D) \to \mathrm{Op}_G(D^\times)$ is injective. Moreover, comparing formula (9.1.7) with formula (9.1.5), we find that $c_1(t) = v_1(t) - \frac{1}{4}$ and $c_j(t) = v_j(t)$ for $j > 1$. By (9.1.4), the residue of the oper given by formula (9.1.5) is equal to (9.1.6). □

As a corollary, we obtain that for any point ϖ in $\mathfrak{h}/W \simeq \mathfrak{g}_{\mathrm{can}}$ the opers with regular singularity and residue ϖ form an affine subspace of $\mathrm{Op}_G^{\mathrm{RS}}(D)$.

Using the isomorphism (4.2.9), we can phrase Proposition 9.1.1 in a coordinate-independent way. Denote by $\mathcal{P}roj_{\geq -2}$ the space of projective connections on $D^\times$ having pole of order at most two at the origin and by $\omega_{\geq -(d_j+1)}^{\otimes (d_j+1)}$ the space of (d_j+1)-differentials on $D^\times$ having pole of order at most (d_j+1) at the origin.

Corollary 9.1.2 *There is an isomorphism between* $\mathrm{Op}_G^{\mathrm{RS}}(D)$ *and the space*

$$\mathcal{P}roj_{\geq -2} \times \bigoplus_{j>1} \omega_{\geq -(d_j+1)}^{\otimes (d_j+1)}.$$

Under this isomorphism, $\mathrm{Op}_G^{\mathrm{RS}}(D)_\varpi$ *corresponds to the affine subspace consisting of the* ℓ*-tuples* $(\eta_1, \ldots, \eta_\ell)$ *satisfying the following condition. Denote by* $\eta_{i,-d_i-1}$ *the* t^{-d_i-1}*-coefficient of the expansion of* η_i *in powers of* t *(it is independent of the choice of* t*). Then the conjugacy class in* $\mathfrak{g}/G = \mathfrak{h}/W$ *of*

$$p_{-1} + \left(\eta_{1,-2} + \frac{1}{4}\right)p_1 + \sum_{j>1} \eta_{j,-j-1} p_j$$

is equal to ϖ.

Suppose that we are given a regular oper

$$\partial_t + p_{-1} + \mathbf{v}(t), \qquad \mathbf{v}(t) \in \mathfrak{b}[[t]].$$

Using the gauge transformation with $\check{\rho}(t) \in B((t))$, we bring it to the form

$$\partial_t + \frac{1}{t}\left(p_{-1} - \check{\rho} + t \cdot \check{\rho}(t)(\mathbf{v}(t))\check{\rho}(t)^{-1}\right).$$

If $\mathbf{v}(t)$ is regular, then so is $\check{\rho}(t)\mathbf{v}(t)\check{\rho}(t)^{-1}$. Therefore this oper has the form (9.1.2), and its residue is equal to $\varpi(-\check{\rho})$, where ϖ is the projection $\mathfrak{h} \to \mathfrak{h}/W$. Since the map $\mathrm{Op}_G^{\mathrm{RS}}(D) \to \mathrm{Op}_G(D^\times)$ is injective (by Proposition 9.1.1), we find that the (affine) space $\mathrm{Op}_{^LG}(D)$ is naturally realized as a subspace of the space of opers with regular singularity and residue $\varpi(-\check{\rho})$.

9.2 Nilpotent opers

Now we introduce another realization of opers with regular singularity whose residue has the form $\varpi(-\check{\lambda}-\check{\rho})$, where $\check{\lambda}$ is the dominant integral weight, i.e., such that $\langle \alpha_i, \check{\lambda}\rangle \in \mathbb{Z}_+$, and $\varpi(-\check{\lambda}-\check{\rho})$ is the projection of $(-\check{\lambda}-\check{\rho}) \in \mathfrak{h}$ onto $\mathfrak{h}/W$. This realization will be convenient for us because it reveals a "secondary" residue invariant of an oper, which is an element of $\mathfrak{n}/B$ (as opposed to the "primary" residue, defined above, which is an element of $\mathfrak{h}/W$).

To illustrate the difference between the two realizations, consider the case of $\mathfrak{g} = \mathfrak{sl}_2$. As before, we identify $\mathfrak{h}$ with $\mathbb{C}$ by sending $\check{\rho} \mapsto 1$. An $\mathfrak{sl}_2$-oper with residue $-1-\check{\lambda}$, where $\check{\lambda} \in \mathbb{Z}_+$, is the $N[[t]]$-gauge equivalence class of an operator of the form

$$\partial_t + \begin{pmatrix} -\frac{1+\check{\lambda}}{2t} + a(t) & b(t) \\ \frac{1}{t} & \frac{1+\check{\lambda}}{2t} - a(t) \end{pmatrix},$$

where $a(t), b(t) \in \mathbb{C}[[t]]$. This operator may be brought to the canonical form by applying gauge transformations as in the proof of Proposition 9.1.1. Namely, we apply the gauge transformation with

$$\begin{pmatrix} 1 & -\frac{1}{2t} \\ 0 & 1 \end{pmatrix} \begin{pmatrix} t^{-1/2} & 0 \\ 0 & t^{1/2} \end{pmatrix} \begin{pmatrix} 1 & \frac{1+\check{\lambda}}{2} - ta(t) \\ 0 & 1 \end{pmatrix}$$

to obtain the operator in the canonical form

$$\partial_t + \begin{pmatrix} 0 & \frac{1}{t^2}\left(\frac{1+\check{\lambda}}{2} - ta(t)\right)^2 - \frac{1}{4t^2} + \frac{1}{t}(b(t)+a(t)) + a'(t) \\ 1 & 0 \end{pmatrix}.$$

On the other hand, if we apply the gauge transformation with the matrix

$$(\check{\rho}+\check{\lambda})(t)^{-1} = \begin{pmatrix} t^{(-1-\check{\lambda})/2} & 0 \\ 0 & t^{(1+\check{\lambda})/2} \end{pmatrix},$$

then we obtain the following operator:

$$\partial_t + \begin{pmatrix} a(t) & \frac{b(t)}{t^{\check{\lambda}+1}} \\ t^{\check{\lambda}} & -a(t) \end{pmatrix}.$$

Let $b(t) = \sum_{n \geq 0} b_n t^n$. Then, applying the gauge transformation with

$$\begin{pmatrix} 1 & \sum_{n=0}^{\check{\lambda}-1} \frac{b_n}{-\check{\lambda}-n} t^{-\check{\lambda}-n} \\ 0 & 1 \end{pmatrix},$$

we obtain an operator of the form

$$\partial_t + \begin{pmatrix} \widetilde{a}(t) & \frac{\widetilde{b}(t)}{t} \\ t^{\check{\lambda}} & -\widetilde{a}(t) \end{pmatrix},$$

where $\widetilde{a}(t), \widetilde{b}(t) \in \mathbb{C}[[t]]$.

This is an example of a nilpotent oper. Its residue, as a connection (not to be confused with the residue as an oper!), is the nilpotent matrix

$$\begin{pmatrix} 0 & \widetilde{b}_0 \\ 0 & 0 \end{pmatrix},$$

which is well-defined up to B-conjugation. From this we find that the monodromy of this oper is conjugate to the matrix

$$\begin{pmatrix} 1 & 2\pi i \widetilde{b}_0 \\ 0 & 1 \end{pmatrix}.$$

Thus, in addition to the canonical form, we obtain another realization of opers with integral residue. The advantage of this realization is that we have a natural residue map and so we can see more clearly what the monodromic properties of these opers are.

In this section we explain how to generalize this construction to other simple Lie algebras, following [FG2].

9.2.1 Definition

Let $\check{\lambda} \in \mathfrak{h}$ be a dominant integral coweight. Consider the space of operators of the form

$$\nabla = \partial_t + \sum_{i=1}^{\ell} t^{\langle \alpha_i, \check{\lambda} \rangle} \psi_i(t) f_i + \mathbf{v}(t) + \frac{v}{t}, \tag{9.2.1}$$

where $\psi_i(t) \in \mathbb{C}[[t]], \psi_i(0) \neq 0$, $\mathbf{v}(t) \in \mathfrak{b}[[t]]$ and $v \in \mathfrak{n}$. The group $B[[t]]$ acts on this space by gauge transformations. Since $B[[t]]$ is a subgroup of the group $B((t))$ which acts freely on the space of all operators of them form (4.2.1) (see Section 4.2.4), it follows that this action of $B[[t]]$ is free. We define the space $\mathrm{Op}_G^{\mathrm{nilp},\check{\lambda}}$ of **nilpotent G-opers of coweight $\check{\lambda}$** as the quotient of the space of all operators of the form (9.2.1) by $B[[t]]$.

We have a map

$$\beta_{\check{\lambda}} : \mathrm{Op}_G^{\mathrm{nilp},\check{\lambda}} \to \mathrm{Op}_G(D^\times)$$

taking the $B[[t]]$-equivalence class of operator ∇ of the form (9.2.1) to its $B((t))$-equivalence class. This map factors through a map

$$\delta_{\check{\lambda}} : \mathrm{Op}_G^{\mathrm{nilp},\check{\lambda}} \to \mathrm{Op}_G^{\mathrm{RS}}(D)_{\varpi(-\check{\lambda}-\check{\rho})},$$

where $\varpi(-\check{\lambda} - \check{\rho})$ is the projection of $(-\check{\lambda} - \check{\rho}) \in \mathfrak{h}$ onto $\mathfrak{h}/W$, which is constructed as follows.

Given an operator ∇ of the form (9.2.1), consider the operator

$$(\check{\lambda}+\check{\rho})(t)\nabla(\check{\lambda}+\check{\rho})(t)^{-1} = \partial_t + t^{-1}\left(\sum_{i=1}^{\ell}\psi_i(t)f_i + \mathbf{w}(t)\right),$$

$$\mathbf{w}(t) \in (\check{\lambda}+\check{\rho})(t)\mathfrak{b}[[t]](\check{\lambda}+\check{\rho})(t)^{-1}, \quad \mathbf{w}(0) = -\check{\lambda}-\check{\rho}. \quad (9.2.2)$$

Moreover, under this conjugation the $B[[t]]$-equivalence class of ∇ maps to the equivalence class of $(\check{\lambda}+\check{\rho})(t)\nabla(\check{\lambda}+\check{\rho})(t)^{-1}$ with respect to the subgroup

$$(\check{\lambda}+\check{\rho})(t)B[[t]](\check{\lambda}+\check{\rho})(t)^{-1} \subset B[[t]].$$

Now, the desired map $\delta_{\check{\lambda}}$ assigns to the $B[[t]]$-equivalence class of ∇ the $B[[t]]$-equivalence class of $(\check{\lambda}+\check{\rho})(t)\nabla(\check{\lambda}+\check{\rho})(t)^{-1}$. By construction, the latter is an oper with regular singularity and residue $\varpi(-\check{\lambda}-\check{\rho})$.

9.2.2 *Nilpotent opers and opers with regular singularity*

Clearly, the map $\beta_{\check{\lambda}}$ is the composition of $\delta_{\check{\lambda}}$ and the inclusion

$$\mathrm{Op}_G^{\mathrm{RS}}(D)_{\varpi(-\check{\lambda}-\check{\rho})} \hookrightarrow \mathrm{Op}_G(D^\times).$$

Proposition 9.2.1 *The map $\delta_{\check{\lambda}}$ is an isomorphism for any dominant integral coweight $\check{\lambda}$.*

Proof. We will show that for any $\mathbb{C}$-algebra R the morphism $\delta_{\check{\lambda}}$ gives rise to a bijection at the level of R-points. This would imply that $\delta_{\check{\lambda}}$ is an isomorphism not only set-theoretically, but as a morphism of schemes.

Thus, we need to show that any operator of the form (9.1.1) with residue $\varpi(-\check{\lambda}-\check{\rho})$ may be brought to the form (9.2.2) over any R, and moreover, that if any two operators of the form (9.2.2) are conjugate under the gauge action of $g \in B(R[[t]])$, then $g \in (\check{\lambda}+\check{\rho})(t)B(R[[t]])(\check{\lambda}+\check{\rho})(t)^{-1}$.

We will use the notation

$$\check{\mu} = \check{\lambda}+\check{\rho} = \sum_i \check{\mu}_i\check{\omega}_i.$$

By our assumption on $\check{\mu}$, we have $\check{\mu}_i \geq 1$ for all i.

Note that

$$(\check{\lambda}+\check{\rho})(t)B(R[[t]])(\check{\lambda}+\check{\rho})(t)^{-1} = H(R[[t]])\cdot(\check{\lambda}+\check{\rho})(t)N(R[[t]])(\check{\lambda}+\check{\rho})(t)^{-1}.$$

There is a unique element of $H(R[[t]])$ that makes each function $\psi_i(t)$ in (9.1.1) equal to 1. Hence we need to show that any operator ∇ of the form (9.1.2) with residue $\varpi(-\check{\lambda}-\check{\rho})$ may be brought to the form

$$\nabla = \partial_t + t^{-1}(p_{-1}+\mathbf{v}(t)), \quad \mathbf{v}(t) \in \check{\mu}(t)\mathfrak{b}[[t]]\check{\mu}(t)^{-1}, \quad \mathbf{v}(0) = -\check{\mu}, \quad (9.2.3)$$

under the action of $N(R[[t]])$, and, moreover, that if any two operators of this form are conjugate under the gauge action of $g \in N(R[[t]])$, then $g \in$

$\check{\mu}(t)N(R[[t]])\check{\mu}(t)^{-1}$. In order to simplify notation, we will suppress R and write $N[[t]]$ for $N(R[[t]])$, etc.

Consider first the case when $\check{\lambda} = 0$, i.e., when $\check{\mu}_i = 1$ for all i. We will bring an operator (9.1.2) to the form (9.2.3) by induction. We have a decomposition

$$\check{\rho}(t)\mathfrak{b}[[t]]\check{\rho}(t)^{-1} = \mathfrak{h}[[t]] \oplus \bigoplus_{j=1}^{h} t^j \mathfrak{n}_j[[t]],$$

where $\mathfrak{n} = \bigoplus_{j>0} \mathfrak{n}_j$ is the decomposition of $\mathfrak{n}$ with respect to the principal gradation and h is the Coxeter number (see Section 4.2.4). We will assume that $h > 1$, for otherwise $\mathfrak{g} = \mathfrak{sl}_2$ and the operator is already in the desired form.

Recall that, by our assumption, the residue of ∇ is equal to $\varpi(-\check{\rho})$. Hence we can use the gauge action of the constant subgroup $N \subset N[[t]]$ to bring ∇ to the form

$$\partial_t + \frac{1}{t}(p_{-1} - \check{\rho}) + \mathbf{v}(t), \tag{9.2.4}$$

where $\mathbf{v}(t) \in \mathfrak{b}[[t]]$. Let is decompose $\mathbf{v}(t) = \sum_{j=0}^{h} \mathbf{v}_j(t)$ with respect to the principal gradation and apply the gauge transformation by

$$g = \exp\left(-\frac{1}{h-1} t\mathbf{v}_h(0)\right) \in N[[t]].$$

The constant coefficient of the principal degree h term in the new operator reads

$$\mathbf{v}_h(0) + \frac{1}{h-1}[-\mathbf{v}_h(0), -\check{\rho}] + \frac{1}{h-1}\mathbf{v}_h(0) = 0$$

(the second term comes from $-g\frac{\check{\rho}}{t}g^{-1}$ and the last term from $g\partial_t(g^{-1})$). Thus, we find that in the new operator $\mathbf{v}_{\text{new},h}(t) \in t\mathfrak{n}_h[[t]]$.

If $h = 2$, then we are done. Otherwise, we apply the gauge transformation by $\exp(-\frac{1}{h-2}t\mathbf{v}_{h-1}(0))$. (Here and below we denote by $\mathbf{v}_i(t)$ the components of the operator obtained after the latest gauge transformation.) We obtain in the same way that the new operator will have $\mathbf{v}_i(t) \in t\mathfrak{n}_i[[t]]$ for $i = h-1, h$. Continuing this way, we obtain an operator of the form (9.2.4) in which $\mathbf{v}_1(t) \in \mathfrak{n}_1[[t]]$ and $\mathbf{v}_j(t) \in t\mathfrak{n}_j[[t]]$ for $j = 2, \ldots, h$. (It is easy to see that each of these gauge transformations preserves the vanishing conditions that were obtained at the previous steps.)

Next, we wish to obtain an operator of the form (9.2.4), where $\mathbf{v}_1(t) \in \mathfrak{n}_1[[t]], \mathbf{v}_2(t) \in t\mathbb{C}[[t]]$ and $\mathbf{v}_j(t) \in t^2\mathfrak{n}_j[[t]]$ for $j = 3, \ldots, h$. This is done in the same way as above, by consecutively applying the gauge transformations with

$$\exp\left(-\frac{1}{j-2}t^2\mathbf{v}'_j(0)\right), \qquad j = h, h-1, \ldots, 3,$$

where $\mathbf{v}'_j(0)$ denotes the t-coefficient in the jth component of the operator obtained at the previous step. Clearly, this procedure works over an arbitrary $\mathbb{C}$-algebra R.

Proceeding like this, we bring our operator to the form

$$\partial_t + \frac{1}{t}\left(p_{-1} + \mathbf{v}(t)\right), \quad \mathbf{v}(t) \in \check{\rho}(t)\mathfrak{b}[[t]]\check{\rho}(t)^{-1}, \quad \mathbf{v}(0) = -\check{\rho}, \tag{9.2.5}$$

which is (9.2.3) with $\check{\mu} = \check{\rho}$. Now let us show that if ∇ and $g\nabla g^{-1}$ are of the form (9.2.5) and $g \in N[[t]]$, then g necessarily belongs to the subgroup $\check{\rho}(t)N[[t]]\check{\rho}(t)^{-1}$ (this is equivalent to saying that the intersection of the $N[[t]]$-equivalence class of the initial operator ∇ with the set of all operators of the form (9.2.5), where $\check{\mu} = \check{\rho}$, is a $\check{\rho}(t)N[[t]]\check{\rho}(t)^{-1}$-equivalence class).

Since the gauge action of any element of the constant subgroup $N \subset N[[t]]$ other than the identity will change the residue of the connection, we may assume without loss of generality that g belongs to the first congruence subgroup $N^{(1)}[[t]]$ of $N[[t]]$. Since the exponential map $\exp : t\mathfrak{n}[[t]] \to N^{(1)}[[t]]$ is an isomorphism, we may represent any element of $N^{(1)}[[t]]$ uniquely as the product of elements of the form $\exp(\mathbf{c}_{\alpha,n}t^n)$, where $\mathbf{c}_{\alpha,n} \in \mathfrak{n}_\alpha$, in any chosen order. In particular, we may write any element of $N^{(1)}[[t]]$ as the product of two elements, one of which belongs to the subgroup $\check{\rho}(t)N[[t]]\check{\rho}(t)^{-1}$, and the other is the product of elements of the form $\exp(\mathbf{c}_{\alpha,n}t^n)$, where $\alpha \in \Delta_+, 1 \leq n < \langle \alpha, \check{\rho} \rangle$. But we have used precisely the gauge transformation of an element of the second type to bring our operator to the form (9.2.5) and they were fixed uniquely by this process. Therefore applying any of them to the operator (9.2.5) will change its form. So the subgroup of $N^{(1)}[[t]]$ consisting of those transformations that preserve the form (9.2.5) is precisely the subgroup $\check{\mu}(t)N[[t]]\check{\mu}(t)^{-1}$.

This completes the proof in the case where $\check{\lambda} = 0, \check{\mu} = \check{\rho}$.

We will prove the statement of the proposition for a general $\check{\mu}$ by induction. Observe first that the same argument as above shows that any operator of the form (9.1.1) with residue $\varpi(\check{\mu})$ may be brought to the form

$$\partial_t + t^{-1}\left(p_{-1} + \mathbf{v}(t)\right), \quad \mathbf{v}(t) \in \check{\rho}(t)\mathfrak{b}[[t]]\check{\rho}(t)^{-1}, \quad \mathbf{v}(0) = -\check{\mu},$$

for any $\check{\mu} = \check{\lambda} + \check{\rho}$, where $\check{\lambda}$ is a dominant integral coweight. Let us show that this operator may be further brought to the form

$$\partial_t + t^{-1}\left(p_{-1} + \mathbf{v}(t)\right), \quad \mathbf{v}(t) \in \check{\mu}'(t)\mathfrak{b}[[t]]\check{\mu}'(t)^{-1}, \quad \mathbf{v}(0) = -\check{\mu}, \tag{9.2.6}$$

where $\check{\mu}'$ is any dominant integral coweight such that

$$1 \leq \langle \alpha_i, \check{\mu}' \rangle \leq \langle \alpha_i, \check{\mu} \rangle, \qquad i = 1, \ldots, \ell. \tag{9.2.7}$$

We already know that this is true for $\check{\mu}' = \check{\rho}$. Suppose we have proved it for some $\check{\mu}'$. Let us prove it for the coweight $\check{\mu} + \check{\omega}_i$, assuming that it still satisfies condition (9.2.7). So we have an operator of the form (9.2.6), and we wish to bring it to the same form where $\check{\mu}'$ is replaced by $\check{\mu}' + \check{\omega}_i$. Using the root decomposition $\mathfrak{n} = \bigoplus_{\alpha\in\Delta_+} \mathfrak{n}_\alpha$, we write $\mathbf{v}(t) = \sum_{\alpha\in\Delta_+} \mathbf{v}_\alpha(t)$, where $\mathbf{v}_\alpha(t) \in \mathfrak{n}_\alpha$. We have

$$\mathbf{v}_\alpha(t) = \sum_{n\geq\langle\alpha,\check{\mu}'\rangle} \mathbf{v}_{\alpha,n} t^n. \tag{9.2.8}$$

We will proceed by induction. At the first step we consider the components $\mathbf{v}_\alpha(t)$, where α is a simple root (i.e., those which belong to $\mathfrak{n}_1$). We have $\langle\alpha, \check{\mu}' + \check{\omega}_i\rangle = \langle\alpha, \check{\mu}'\rangle$ for all such α's, except for $\alpha = \alpha_i$, for which we have $\langle\alpha, \check{\mu}' + \check{\omega}_i\rangle = \langle\alpha, \check{\mu}'\rangle + 1$.

Let us write $\check{\mu}_i = \langle\alpha_i, \check{\mu}\rangle, \check{\mu}_i' = \langle\alpha_i, \check{\mu}'\rangle$. Then condition (9.2.7) for $\check{\mu}' + \check{\omega}_i$ will become

$$2 \leq \check{\mu}_i' + 1 \leq \check{\mu}_i. \tag{9.2.9}$$

Apply the gauge transformation by

$$\exp\left(\frac{1}{\check{\mu}_i' - \check{\mu}_i} t^{\check{\mu}_i'} \mathbf{v}_{\alpha_i,\check{\mu}_i'}\right)$$

(note that the denominator is non-zero due condition (9.2.9)). As the result, we obtain an operator of the form (9.2.6) where $\mathbf{v}_{\alpha_i}(t) \in t^{\check{\mu}_i'+1}\mathbb{C}[[t]]$. At the same time, we do not spoil condition (9.2.8) for any other $\alpha \in \Delta_+$, because $\langle\alpha, \check{\mu}'\rangle \geq \langle\alpha, \check{\rho}\rangle$ by our assumption on $\check{\mu}'$.

Next, we move on to roots of height two (i.e., those which belong to $\mathfrak{n}_2$). Let us apply consecutively the gauge transformations by

$$\exp\left(\frac{1}{\langle\alpha, \check{\mu}'\rangle - \langle\alpha, \check{\mu}\rangle} t^{\langle\alpha,\check{\mu}'\rangle} \mathbf{v}_{\alpha,\langle\alpha,\check{\mu}'\rangle}\right)$$

for those α's for which $\langle\alpha, \check{\mu}' + \check{\omega}_i\rangle > \langle\alpha, \check{\mu}'\rangle$. Continuing this process across all values of the principal gradation, we obtain an operator which satisfies condition (9.2.8) for all α's for which $\langle\alpha, \check{\mu}'\rangle = \langle\alpha, \check{\mu}' + \check{\omega}_i\rangle$ and the condition

$$\mathbf{v}_\alpha(t) = \sum_{n\geq\langle\alpha,\check{\mu}'\rangle+1} \mathbf{v}_{\alpha,n} t^n$$

for all α's for which $\langle\alpha, \check{\mu}' + \check{\omega}_i\rangle > \langle\alpha, \check{\mu}'\rangle$.

At the next step we obtain in a similar way an operator which satisfies a stronger condition, namely, that in addition to the above we have

$$\mathbf{v}_\alpha(t) = \sum_{n\geq\langle\alpha,\check{\mu}'\rangle+2} \mathbf{v}_{\alpha,n} t^n$$

for those α's for which $\langle\alpha, \check{\mu}' + \check{\omega}_i\rangle > \langle\alpha, \check{\mu}'\rangle + 1$. Condinuing this way, we finally arrive at an operator for which

$$\mathbf{v}_\alpha(t) = \sum_{n\geq\langle\alpha,\check{\mu}'+\check{\omega}_i\rangle} \mathbf{v}_{\alpha,n} t^n,$$

for all $\alpha \in \Delta_+$. In other words, it has the form (9.2.3) where $\check{\mu}'$ is replaced by $\check{\mu}' + \check{\omega}_i$. This completes the inductive step.

Therefore we obtain that any operator of the form (9.1.2) with residue $\varpi(-\check{\mu})$ may be brought to the form (9.2.3) (clearly, this construction works over any $\mathbb{C}$-algebra R). Finally, we prove in the same way as in the case when $\check{\mu} = \check{\rho}$ that if any two operators of this form are conjugate under the gauge action of $g \in N[[t]]$, then $g \in \check{\mu}(t)N[[t]]\check{\mu}(t)^{-1}$. $\square$

By Proposition 9.1.1 and Proposition 9.2.1, the map $\beta_{\check{\lambda}} : \mathrm{Op}_G^{\mathrm{nilp},\check{\lambda}} \to \mathrm{Op}_G(D^\times)$ is an injection. In particular, using Proposition 9.1.1 we obtain canonical representatives of nilpotent opers. In what follows we will not distinguish between $\mathrm{Op}_G^{\mathrm{nilp},\check{\lambda}}$ and its image in $\mathrm{Op}_G(D^\times)$.

9.2.3 Residue

Let $\mathcal{F}_B$ be the B-bundle on the disc D underlying an oper $\chi \in \mathrm{Op}_G^{\mathrm{nilp},\check{\lambda}}$, and let $\mathcal{F}_{B,0}$ be its fiber at $0 \in D$. One can show in the same way as in Section 4.2.4 that the B-bundles $\mathcal{F}_B$ underlying all $\check{\lambda}$-nilpotent opers on D are canonically identified. Denote by $\mathfrak{n}_{\mathcal{F}_{B,0}}$ the $\mathcal{F}_{B,0}$-twist of $\mathfrak{n}$. Then for each operator of the form (9.2.1), v is a well-defined vector in $\mathfrak{n}_{\mathcal{F}_{B,0}}$. Therefore we obtain a morphism

$$\mathrm{Res}_{\mathcal{F},\check{\lambda}} : \mathrm{Op}_G^{\mathrm{nilp},\check{\lambda}} \to \mathfrak{n}_{\mathcal{F}_{B,0}}, \qquad \chi \mapsto v.$$

For each choice of trivialization of $\mathcal{F}_{B,0}$ we obtain an identification $\mathfrak{n}_{\mathcal{F}_{B,0}} \simeq \mathfrak{n}$. Changes of the trivialization lead to the adjoint action of B on $\mathfrak{n}$. Therefore we have a canonical map $\mathfrak{n}_{\mathcal{F}_{B,0}} \to \mathfrak{n}/B$. Its composition with $\mathrm{Res}_{\mathcal{F},\check{\lambda}}$ is a morphism

$$\mathrm{Res}_{\check{\lambda}} : \mathrm{Op}_G^{\mathrm{nilp},\check{\lambda}} \to \mathfrak{n}/B.$$

It maps an operator of the form (9.2.1) to the projection of v onto $\mathfrak{n}/B$. We call it the **residue morphism**. Note that any nilpotent G-oper, viewed as a connection on $D^\times$, has a monodromy operator, which is a conjugacy class in G. The following statement is clear.

Lemma 9.2.2 *Any nilpotent G-oper χ of coweight $\check{\lambda}$ has unipotent monodromy. Its monodromy is trivial if and only if* $\mathrm{Res}_{\check{\lambda}}(\chi) = 0$.

Let $\mathrm{Op}_G^{\check{\lambda}}$ be the preimage of $0 \in \mathfrak{n}/B$ in $\mathrm{Op}_G^{\mathrm{nilp},\check{\lambda}}$ under the map $\mathrm{Res}_{\check{\lambda}}$. Equivalently, this is the space of $B[[t]]$-equivalence classes of operators of the form

$$\nabla = \partial_t + \sum_{i=1}^{\ell} t^{\langle \alpha_i, \check{\lambda} \rangle} \psi_i(t) f_i + \mathbf{v}(t), \qquad (9.2.10)$$

where $\psi_i(t) \in \mathbb{C}[[t]], \psi_i(0) \neq 0$, $\mathbf{v}(t) \in \mathfrak{b}[[t]]$. This space was originally introduced by V. Drinfeld (unpublished).

The above lemma and Proposition 9.2.1 imply that $\mathrm{Op}_G^{\check{\lambda}}$ is the submanifold in $\mathrm{Op}_G^{\mathrm{RS}}(D)_{\varpi(-\check{\lambda}-\check{\rho})}$ corresponding to those opers with regular singularity that have trivial monodromy.

Consider the case when $\check{\lambda} = 0$ in more detail. We will denote $\mathrm{Op}_G^{\mathrm{nilp},0}$ simply by $\mathrm{Op}_G^{\mathrm{nilp}}$ and call its points **nilpotent G-opers**. These are the $B[[t]]$-equivalence classes of operators of the form

$$\partial_t + \sum_{i=1}^{\ell} \psi_i(t) f_i + \mathbf{v}(t) + \frac{v}{t}, \tag{9.2.11}$$

where $\psi_i(t) \in \mathbb{C}[[t]], \psi_i(0) \neq 0$, $\mathbf{v}(t) \in \mathfrak{b}[[t]]$ and $v \in \mathfrak{n}$, or, equivalently, the $N[[t]]$-equivalence classes of operators of the form

$$\partial_t + p_{-1} + \mathbf{v}(t) + \frac{v}{t}, \tag{9.2.12}$$

where $\mathbf{v}(t) \in \mathfrak{b}[[t]]$ and $v \in \mathfrak{n}$. The residue morphism, which we denote in this case simply by Res, maps such an operator to the image of v in $\mathfrak{n}/B$. Note that the fiber over $0 \in \mathfrak{n}/B$ is just the locus $\mathrm{Op}_G^0 = \mathrm{Op}_G(D)$ of regular opers on the disc D.

Proposition 9.1.1 identifies $\mathrm{Op}_G^{\mathrm{nilp}}$ with the preimage of $\varpi(-\check{\rho})$ under the map $\mathrm{res} : \mathrm{Op}_G^{\mathrm{RS}}(D) \to \mathfrak{h}/W$ defined in Section 9.1.1. Using Proposition 9.2.1 and Proposition 9.1.1, we obtain the following description of $\mathrm{Op}_G^{\mathrm{nilp}}$.

Corollary 9.2.3 *There is an isomorphism*

$$\mathrm{Op}_G^{\mathrm{nilp}} \simeq \mathcal{P}roj_{\geq -1} \times \bigoplus_{j \geq 2} \omega_{\geq -d_j}^{\otimes(d_j+1)}. \tag{9.2.13}$$

9.2.4 More general forms for opers with regular singularity

The argument we used in the proof of Proposition 9.2.1 allows us to construct more general representatives for opers with regular singularities.

Consider the space $\mathrm{Op}^{\mathrm{RS}}(D)_{\varpi(-\check{\mu})}$, where $\check{\mu}$ is an arbitrary element of $\mathfrak{h}$. An element of $\mathrm{Op}^{\mathrm{RS}}(D)_{\varpi(-\check{\mu})}$ is the same as an $N^{(1)}[[t]]$-equivalence class of operators of the form

$$\partial_t + \frac{1}{t}(p_{-1} - \check{\mu}) + \mathbf{v}(t), \qquad \mathbf{v}(t) \in \mathfrak{b}[[t]]. \tag{9.2.14}$$

Let $\check{\lambda}'$ be a dominant integral coweight such that

$$\langle \alpha, \check{\lambda}' \rangle < \langle \alpha, \check{\mu} \rangle, \qquad \text{if} \quad \langle \alpha, \check{\mu} \rangle \in \mathbb{Z}_{>0}, \alpha \in \Delta_+, \tag{9.2.15}$$

where $\mathbb{Z}_{>0}$ is the set of positive integers.

Set $\check{\mu}' = \check{\lambda}' + \check{\rho}$. Let $\mathrm{Op}(D)_{\check{\mu}}^{\check{\mu}'}$ be the space of gauge equivalence classes of the operators of the form (9.2.6) by the gauge action of the group $\check{\mu}'(t)N[[t]]\check{\mu}'(t)^{-1}$. Then we have a natural map

$$\delta_{\check{\mu}}^{\check{\mu}'} : \mathrm{Op}(D)_{\check{\mu}}^{\check{\mu}'} \to \mathrm{Op}^{\mathrm{RS}}(D)_{\varpi(-\check{\mu})}$$

taking these equivalence classes to the $N[[t]]$-equivalence classes. Applying verbatim the argument used in the proof of Proposition 9.2.1, we obtain the following result.

Proposition 9.2.4 *If $\check{\lambda}'$ satisfies condition* (9.2.15), *and $\check{\mu}' = \check{\lambda}' + \check{\rho}$, then the map $\delta_{\check{\mu}}^{\check{\mu}'}$ is an isomorphism.*

In particular, suppose that $\langle \alpha, \check{\mu} \rangle \notin \mathbb{Z}_{>0}$ for all $\alpha \in \Delta_+$. Note that any point in $\mathfrak{h}/W$ may be represented as $\varpi(-\check{\mu})$ where $\check{\mu}$ satisfies this condition. Then the statement of Proposition 9.2.4 is valid for any dominant integral weight $\check{\lambda}'$. Let us set $\check{\lambda}' = n\check{\rho}, n \in \mathbb{Z}_+$. Then we have $\check{\mu}' = (n+1)\check{\rho}$ and $\check{\mu}'(t)N[[t]]\check{\mu}'(t)^{-1}$ is the $(n+1)$st congruence subgroup of $N[[t]]$. Taking the limit $n \to \infty$, we obtain the following:

Corollary 9.2.5 *Suppose that $\check{\mu}$ satisfies the property that $\langle \alpha, \check{\mu} \rangle \notin \mathbb{Z}_{>0}$ for all $\alpha \in \Delta_+$. Then any oper with regular singularity and residue $\varpi(-\check{\mu})$ may be written uniquely in the form*

$$\partial_t + \frac{1}{t}(p_{-1} - \check{\mu}) + \mathbf{v}(t), \qquad \mathbf{v}(t) \in \mathfrak{h}[[t]],$$

or, equivalently, in the form

$$\partial_t + p_{-1} - \frac{\check{\mu} + \check{\rho}}{t} + \mathbf{v}(t), \qquad \mathbf{v}(t) \in \mathfrak{h}[[t]]. \tag{9.2.16}$$

Thus, there is a bijection between $\mathrm{Op}^{\mathrm{RS}}(D)_{\varpi(-\check{\mu})}$ and the space of operators of the form (9.2.16).

9.2.5 A characterization of nilpotent opers using the Deligne extension

Recall that given a G-bundle $\mathcal{F}$ with a connection ∇ on the punctured disc $D^\times$ which has unipotent monodromy, there is a canonical extension of $\mathcal{F}$, defined by P. Deligne [Del], to a G-bundle $\overline{\mathcal{F}}$ on the disc D such that the connection has regular singularity and its residue is nilpotent. Consider a G-oper $(\mathcal{F}, \mathcal{F}_B, \nabla)$ on $D^\times$. Suppose that it has unipotent monodromy. Any G-bundle on $D^\times$ with a connection whose monodromy is unipotent has a canonical extension to a bundle on D with a connection with regular singularity and nilpotent residue, called the Deligne extension. Thus, our bundle $\mathcal{F}$ on $D^\times$ has the Deligne extension to D, which we denote by $\overline{\mathcal{F}}$.

Since the flag variety is proper, the B-reduction $\mathcal{F}_B$ of $\mathcal{F}$ on $D^\times$ extends to a B-reduction of $\overline{\mathcal{F}}$ on D, which we denote by $\overline{\mathcal{F}}_B$. We refer to $(\overline{\mathcal{F}}, \overline{\mathcal{F}}_B, \nabla)$ as the Deligne extension of the oper $(\mathcal{F}, \mathcal{F}_B, \nabla)$ to D. By construction, the connection ∇ has regular singularity and its residue is an element of

$$\mathfrak{g}_{\overline{\mathcal{F}}_{B,0}} = \overline{\mathcal{F}}_B \underset{\overline{\mathcal{F}}_B}{\times} \mathfrak{g},$$

where $\overline{\mathcal{F}}_{B,0}$ is the fiber of $\overline{\mathcal{F}}_B$ at the origin.

Recall the H-torsor $\mathbf{O}$ which is an open dense subset in $\mathfrak{n}^\perp/\mathfrak{b} \subset \mathfrak{g}/\mathfrak{b}$. Suppose we are given a function f on D with values in $\mathfrak{n}^\perp/\mathfrak{b}$ which takes values in $\mathbf{O}$ over $D^\times$. Given an integral coweight $\check{\lambda}$, we will say that f vanishes to order $\check{\lambda}$ at the origin if the corresponding map $D^\times \to \mathbf{O}$ may be obtained as the restriction of the cocharacter $\check{\lambda} : \mathbb{C}^\times \to \mathbf{O}$ under an embedding $D^\times \to \mathbb{C}^\times$ and an identification $\mathbf{O} \simeq H$. Note that this definition still makes sense if instead of a function with values in $\mathfrak{n}^\perp/\mathfrak{b}$ we take a section of a twist of $\mathfrak{n}^\perp/\mathfrak{b}$ by an H-bundle on D.

We will say that $(\overline{\mathcal{F}}, \overline{\mathcal{F}}_B, \nabla)$ satisfies the $\check{\lambda}$-oper condition at the origin if the one-form $\nabla/\overline{\mathcal{F}}_B$ on D taking values in $(\mathfrak{n}^\perp/\mathfrak{b})_{\overline{\mathcal{F}}_B}$ (and in $\mathbf{O}_{\mathcal{F}_B} \subset (\mathfrak{g}/\mathfrak{b})_{\mathcal{F}_B}$ over $D^\times$) vanishes to order $\check{\lambda}$ at the origin. The following statement follows directly from the definitions.

Proposition 9.2.6 *Let $\check{\lambda}$ be an integral dominant coweight. A G-oper on $D^\times$ is a nilpotent G-oper of coweight $\check{\lambda}$ if and only if it has unipotent monodromy and its Deligne extension satisfies the $\check{\lambda}$-oper condition at the origin.*

9.3 Miura opers with regular singularities

In Section 8.2.1 we introduced the spaces of Miura opers and Cartan connections and the Miura transformation to the space of opers. In this section we consider Miura opers and Cartan connections with regular singularities and the restriction of the Miura transformation to the space of Cartan connections with regular singularities. We show that the fiber of the Miura transformation over a nilpotent oper χ is isomorphic to the Springer fiber of the residue of χ. The results of this section were obtained in [FG2].

9.3.1 Connections and opers with regular singularities

In Section 8.2.1 we defined the space $\operatorname{Conn}(\Omega^{\check{\rho}})_D$ of connections on the H-bundle $\Omega^{\check{\rho}}$ on the disc $D = \operatorname{Spec} \mathbb{C}[[t]]$. Now we consider the space $\operatorname{Conn}(\Omega^{\check{\rho}})_{D^\times}$ of connections on this bundle on the punctured disc $D^\times = \operatorname{Spec} \mathbb{C}((t))$. Elements of $\operatorname{Conn}(\Omega^{\check{\rho}})_{D^\times}$ are represented by operators of the form

$$\overline{\nabla} = \partial_t + \mathbf{u}(t), \qquad \mathbf{u}(t) \in \mathfrak{h}((t)), \tag{9.3.1}$$

and under changes of coordinates they transform according to formula (8.2.2).

Now let $\operatorname{Conn}(\Omega^{\check{\rho}})_D^{\mathrm{RS}}$ be the space of all connections on the H-bundle $\Omega^{\check{\rho}}$ on D with *regular singularity*, that is, of the form

$$\overline{\nabla} = \partial_t + \frac{1}{t}\mathbf{u}(t), \qquad \mathbf{u}(t) \in \mathfrak{h}[[t]]. \tag{9.3.2}$$

We have a map

$$\operatorname{res}_{\mathfrak{h}} : \operatorname{Conn}(\Omega^{\check{\rho}})_D^{\mathrm{RS}} \to \mathfrak{h}$$

assigning to a connection its residue $\mathbf{u}(0) \in \mathfrak{h}$. For $\check{\lambda} \in \mathfrak{h}$ let $\operatorname{Conn}(\Omega^{\check{\rho}})_D^{\check{\lambda}}$ be the subspace of connections with residue $\check{\lambda}$.

Recall that we have the Miura transformation (8.3.3)

$$\mu : \operatorname{Conn}(\Omega^{\check{\rho}})_{D^\times} \to \operatorname{Op}_G(D^\times) \tag{9.3.3}$$

defined as follows: given an operator $\overline{\nabla}$ of the form (9.3.1), we define the corresponding G-oper as the $N((t))$-gauge equivalence class of the operator $\nabla + p_{-1}$.

Suppose that $\overline{\nabla}$ is an element of $\operatorname{Conn}(\Omega^{\check{\rho}})_D^{\mathrm{RS}}$ given by formula (9.3.2). Then $\mu(\overline{\nabla})$ is by definition the $N((t))$-equivalence class of the operator

$$\nabla = \partial_t + p_{-1} + t^{-1}\mathbf{u}(t).$$

Applying gauge transformation with $\check{\rho}(t)$, we identify it with the $N((t))$-equivalence class of the operator

$$\partial_t + t^{-1}(p_{-1} - \check{\rho} + \mathbf{u}(t)),$$

which is an oper with regular singularity and residue $\varpi(-\check{\rho} + \mathbf{u}(0))$.

Thus, we obtain a morphism

$$\mu^{\mathrm{RS}} : \operatorname{Conn}(\Omega^{\check{\rho}})_D^{\mathrm{RS}} \to \operatorname{Op}_G^{\mathrm{RS}}(D). \tag{9.3.4}$$

and a commutative diagram

$$\begin{array}{ccc} \operatorname{Conn}(\Omega^{\check{\rho}})_D^{\mathrm{RS}} & \xrightarrow{\mu^{\mathrm{RS}}} & \operatorname{Op}_G^{\mathrm{RS}}(D) \\ {\scriptstyle \operatorname{res}_{\mathfrak{h}}}\downarrow & & \downarrow{\scriptstyle \operatorname{res}} \\ \mathfrak{h} & \longrightarrow & \mathfrak{h}/W \end{array} \tag{9.3.5}$$

where the lower horizontal map is the composition of the map $\check{\lambda} \mapsto \check{\lambda} - \check{\rho}$ and the projection $\varpi : \mathfrak{h} \to \mathfrak{h}/W$.

It is easy to see that the preimage of any $\chi \in \operatorname{Op}_G^{\mathrm{RS}}(D)$ under the Miura transformation (9.3.3) belongs to $\operatorname{Conn}(\Omega^{\check{\rho}})_D^{\mathrm{RS}} \subset \operatorname{Conn}(\Omega^{\check{\rho}})_{D^\times}$. In other words, the oper corresponding to a Cartan connection with irregular singularity necessarily has irregular singularity. Thus, μ^{RS} is the restriction of μ to $\operatorname{Op}_G^{\mathrm{RS}}(D)$. The above commutative diagram then implies that for any $\check{\lambda} \in \mathfrak{h}$

we have an isomorphism†

$$\mu^{-1}(\mathrm{Op}_G^{\mathrm{RS}}(D)_{\varpi(-\check{\lambda}-\check{\rho})}) = \bigsqcup_{w \in W} \mathrm{Conn}(\Omega^{\check{\rho}})_D^{\check{\rho}-w(\check{\lambda}+\check{\rho})}. \tag{9.3.6}$$

In addition, Corollary 9.2.5 implies the following:

Proposition 9.3.1 *If $\check{\lambda} \in \mathfrak{h}$ is such that $\langle \alpha, \check{\lambda} \rangle \notin \mathbb{Z}_+$ for all $\alpha \in \Delta_+$, then the restriction of the map μ^{RS} to $\mathrm{Conn}(\Omega^{\check{\rho}})_D^{-\check{\lambda}}$ is an isomorphism*

$$\mathrm{Conn}(\Omega^{\check{\rho}})_D^{-\check{\lambda}} \simeq \mathrm{Op}_G^{\mathrm{RS}}(D)_{\varpi(-\check{\lambda}-\check{\rho})}.$$

Thus, if $\check{\lambda}$ satisfies the conditions of the Proposition 9.3.1, then the stratum corresponding to $w = 1$ in the right hand side of (9.3.6) gives us a section of the restriction of the Miura transformation μ to $\mathrm{Op}_G^{\mathrm{RS}}(D)_{\varpi(-\check{\lambda}-\check{\rho})})$.

The following construction will be useful in what follows. For an H-bundle $\mathcal{F}_H$ on D and an integral coweight $\check{\lambda}$, let $\mathcal{F}_H(\check{\lambda}) = \mathcal{F}_H(\check{\lambda} \cdot 0)$, where 0 denotes the origin in D (the closed point of D), be the H-bundle on D such that for any character $\mu : H \to \mathbb{C}^\times$ we have

$$\mathcal{F}_H(\check{\lambda}) \underset{\mathbb{C}^\times}{\times} \mathbb{C}_\mu = (\mathcal{F}_H \times \mathbb{C}_\mu)(\langle \mu, \check{\lambda} \rangle \cdot 0).$$

Note that the restrictions of the bundles $\mathcal{F}_H(\check{\lambda})$ and $\mathcal{F}_H$ to $D^\times$ are canonically identified. Under this identification, a connection ∇ on $\mathcal{F}_H(\check{\lambda})$ becomes the connection $\check{\lambda}(t)^{-1}\nabla\check{\lambda}(t)$ on $\mathcal{F}_H$.

Denote by $\mathrm{Conn}(\Omega^{\check{\rho}}(\check{\lambda}))_D$ the space of connections on the H-bundle $\Omega^{\check{\rho}}(\check{\lambda})$ on D. We have a natural embedding

$$\mathrm{Conn}(\Omega^{\check{\rho}}(\check{\lambda}))_D \subset \mathrm{Conn}(\Omega^{\check{\rho}})_{D^\times}$$

obtained by restricting a connection to $D^\times$. The image of $\mathrm{Conn}(\Omega^{\check{\rho}}(\check{\lambda}))_D$ in $\mathrm{Conn}(\Omega^{\check{\rho}})_{D^\times}$ coincides with the space $\mathrm{Conn}(\Omega^{\check{\rho}})_D^{\check{\lambda}} \subset \mathrm{Conn}(\Omega^{\check{\rho}})_{D^\times}$. Indeed, given a connection $\nabla \in \mathrm{Conn}(\Omega^{\check{\rho}}(\check{\lambda})_D$, the connection $\check{\lambda}(t)^{-1}\nabla\check{\lambda}(t)$ is in $\mathrm{Conn}(\Omega^{\check{\rho}})_D^{\check{\lambda}}$.

In what follows we will use the notation $\mathrm{Conn}(\Omega^{\check{\rho}})_D^{\check{\lambda}}$ for both of these spaces keeping in mind the above two distinct realizations.

9.3.2 The case of integral dominant coweights

Let $\check{\lambda}$ be an integral dominant coweight of G. A **nilpotent Miura G-oper of coweight** $\check{\lambda}$ on the disc D is by definition a quadruple $(\mathcal{F}, \nabla, \mathcal{F}_B, \mathcal{F}'_B)$, where $(\mathcal{F}, \nabla, \mathcal{F}_B)$ is a nilpotent G-oper of coweight $\check{\lambda}$ on D and $\mathcal{F}'_B$ is another B-reduction of $\mathcal{F}$ stable under $\nabla_{t\partial_t} = t\nabla$. Let $\mathrm{MOp}_G^{\check{\lambda}}$ be the variety of nilpotent Miura G-opers of coweight $\check{\lambda}$.

† Here and throughout this chapter we consider the set-theoretic preimage of μ. The corresponding scheme-theoretic preimage may be non-reduced (see examples in Section 10.4.6).

The restriction of a nilpotent Miura G-oper of coweight $\check{\lambda}$ from D to $D^\times$ is a Miura oper on $D^\times$, as defined in Section 8.2.1. Let $\mathrm{MOp}_G(D^\times)$ be the set of Miura opers on $D^\times$. We have a natural forgetful map $\mathrm{MOp}_G^{\check{\lambda}} \to \mathrm{Op}_G^{\mathrm{nilp},\check{\lambda}}$ which fits into the following Cartesian diagram:

$$\begin{array}{ccc} \mathrm{MOp}_G^{\check{\lambda}} & \longrightarrow & \mathrm{MOp}_G(D^\times) \\ \downarrow & & \downarrow \\ \mathrm{Op}_G^{\mathrm{nilp},\check{\lambda}} & \longrightarrow & \mathrm{Op}_G(D^\times) \end{array}$$

The proof of the next lemma is due to V. Drinfeld (see [FG2], Lemma 3.2.1). Recall from Section 8.2.1 that a Miura oper $(\mathcal{F}, \nabla, \mathcal{F}_B, \mathcal{F}'_B)$ is called generic if the B-reductions $\mathcal{F}_B$ and $\mathcal{F}'_B$ are in generic relative position.

Lemma 9.3.2 *Any Miura oper on the punctured disc $D^\times$ is generic.*

Proof. Let us choose a coordinate t on the disc D and trivialize the B-bundle $\mathcal{F}_B$. Then we identify the sections of $\mathcal{F}$ over $D^\times$ with $G((t))$, the reduction $\mathcal{F}'_B$ with a right coset $g \cdot B((t)) \subset G((t))$ and the connection ∇ with the operator

$$\partial_t + A(t), \qquad A(t) = \sum_{i=1}^{\ell} \psi_i(t) f_i + \mathbf{v}(t), \tag{9.3.7}$$

where $\psi_i(t) \in \mathbb{C}((t))^\times$ and $\mathbf{v}(t) \in \mathfrak{b}[[t]]$.

We have the Bruhat decomposition $G((t)) = \bigsqcup_{w \in W} B((t)) w B((t))$. Therefore we may write $g = g_1 w g_2$ for some $w \in W$ and $g_1, g_2 \in B((t))$. Hence the space of sections over $D^\times$ of the Borel subalgebra $\mathfrak{b}_{\mathcal{F}'_B} \subset \mathfrak{g}_{\mathcal{F}'_B} = \mathfrak{g}_{\mathcal{F}_B}$ is equal to

$$\mathfrak{b}' = g\mathfrak{b}((t))g^{-1} = g_1(w\mathfrak{b}((t))w^{-1})g_1^{-1}.$$

The statement of the lemma is equivalent to the statement that the element w appearing in this formula is the longest element w_0 of the Weyl group W.

To see that, consider the Cartan subalgebra $\mathfrak{h}' = g_1 \mathfrak{h}((t)) g_1^{-1}$. Then, since $g_1 \in B((t))$, we have $\mathfrak{h}' \subset \mathfrak{b}' \cap \mathfrak{b}((t))$. We have the following Cartan decomposition of the Lie algebra $\mathfrak{g}((t))$ with respect to $\mathfrak{h}'$:

$$\mathfrak{g}((t)) = \mathfrak{h}' \oplus \bigoplus_{\alpha \in \Delta} \mathfrak{g}'_\alpha, \qquad \mathfrak{g}'_\alpha = g_1 \mathfrak{g}_\alpha g_1^{-1},$$

where $\mathfrak{g}_\alpha$ is the root subspace of $\mathfrak{g}$ corresponding to $\alpha \in \Delta$. Let us decompose $A(t)$ appearing in formula (9.3.7) with respect to this decomposition:

$$A(t) = A_0 + \sum_{\alpha \in \Delta} A_\alpha.$$

Then because of the oper condition we have $A_{-\alpha_i} \neq 0$. On the other hand, since $\mathcal{F}'_B$ is preserved by ∇, we have $A(t) \in \mathfrak{b}'$. This means that the roots $-\alpha_i$ are positive roots with respect to $\mathfrak{b}'$. Therefore all negative roots with respect to $\mathfrak{b}((t))$ are positive roots with respect to $\mathfrak{b}'$. This is possible if and only if $\mathfrak{b}' = g_1 \mathfrak{b}_- g_1^{-1}$, where $\mathfrak{b}_- = w_0 \mathfrak{b} w_0^{-1}$ and $g_1 \in B((t))$. □

In the same way as in Lemma 8.2.2 we show that there is a bijection between the set of generic Miura opers on $D^\times$ and the set $\mathrm{Conn}(\Omega^{\check{\rho}})_{D^\times}$ of connections on the H-bundle $\Omega^{\check{\rho}}$ on $D^\times$. Then Lemma 9.3.2 implies:

Lemma 9.3.3 *There is a bijection*

$$\mathrm{MOp}_G(D^\times) \simeq \mathrm{Conn}(\Omega^{\check{\rho}})_{D^\times}.$$

Thus, any nilpotent Miura oper becomes generic, when restricted to the punctured disc, and hence defines a connection on $\Omega^{\check{\rho}}$. Our goal now is to understand the relationship between the residue of this connection and the relative position between the B-reductions $\mathcal{F}_B$ and $\mathcal{F}'_B$ at $0 \in D$.

Let $\mathcal{N} \subset \mathfrak{g}$ be the cone of nilpotent elements in $\mathfrak{g}$ and $\widetilde{\mathcal{N}}$ the **Springer variety** of pairs $(x, \mathfrak{b}')$, where $x \in \mathcal{N}$ and $\mathfrak{b}'$ is a Borel subalgebra of $\mathfrak{g}$ that contains x. Let $\widetilde{\mathfrak{n}}$ be the restriction of $\widetilde{\mathcal{N}}$ to $\mathfrak{n} \subset \mathcal{N}$. Thus,

$$\widetilde{\mathfrak{n}} = \{(x, \mathfrak{b}') \,|\, x \in \mathfrak{n}, x \in \mathfrak{b}'\}.$$

Recall that in Section 9.2.3 we defined the residue $\mathrm{Res}_{\check{\lambda}}(\chi)$ of a nilpotent oper χ of coweight $\check{\lambda}$.

Lemma 9.3.4 *The reduction $\mathcal{F}'_B$ of a nilpotent oper χ of coweight $\check{\lambda}$ is uniquely determined by its fiber at the origin. It must be stable under the residue of the underlying nilpotent oper,* $\mathrm{Res}_{\check{\lambda}}(\chi)$.

To see that, recall that $\mathcal{F}'_B$ may be completely described by choosing line subbundles $\mathcal{L}^\nu$ in the vector bundle $V^\nu_{\mathcal{F}} = \mathcal{F} \underset{G}{\times} V^\nu$, where V^ν is an irreducible representation of G of highest weight ν, satisfying the Plücker relations (see, e.g., [FGV1], Section 2.1.2). Then the above statement is a corollary of the following:

Lemma 9.3.5 *Suppose we are given a first order differential equation with regular singularity*

$$(\partial_t + A(t))\Psi(t) = 0, \qquad A(t) = \sum_{i \geq -1} A_i t^i,$$

on a finite-dimensional vector space V. Suppose that A_{-1} is a nilpotent operator on V. Then this equation has a solution $\Psi(t) \in V[[t]]$ with the initial condition $\Psi(0) = v \in V$ if and only if $A_{-1}(v) = 0$.

Proof. We expand $\Psi(t) = \sum_{j \geq 0} \Psi_j t^j$. Then the above equation is equivalent to the system of linear equations

$$n\Psi_n + \sum_{i+j=n-1} A_i(\Psi_j) = 0. \tag{9.3.8}$$

The first of these equations reads $A_{-1}(\Psi_0) = 0$, so this is indeed a necessary condition. If this condition is satisfied, then the Ψ_n's with $n > 0$ may be found recursively, because the nilpotency of A_{-1} implies that the operator $(n \operatorname{Id} + A_{-1}), n > 0$, is invertible. □

This implies that we have a natural isomorphism

$$\operatorname{MOp}_G^{\check\lambda} \simeq \operatorname{Op}_G^{\text{nilp},\check\lambda} \underset{\mathfrak{n}/B}{\times} \widetilde{\mathfrak{n}}/B, \tag{9.3.9}$$

where the map $\operatorname{Op}_G^{\text{nilp},\check\lambda} \to \mathfrak{n}/B$ is the residue map $\operatorname{Res}_{\check\lambda}$. Thus, the fiber of $\operatorname{MOp}_G^{\check\lambda}$ over $\chi \in \operatorname{Op}_G^{\text{nilp},\check\lambda}$ is the (reduced) **Springer fiber** of $\operatorname{Res}_{\check\lambda}(\chi)$:

$$\operatorname{Sp}_{\operatorname{Res}_{\check\lambda}(\chi)} = \{\mathfrak{b}' \in G/B \mid \operatorname{Res}_{\check\lambda}(\chi) \in \mathfrak{b}'\}. \tag{9.3.10}$$

Given $w \in W$, we will say that a Borel subgroup B' (or the corresponding Lie algebra $\mathfrak{b}'$) is in relative position w with B (resp., $\mathfrak{b}$) if the point of the flag variety G/B corresponding to B' (resp., $\mathfrak{b}'$) lies in the B-orbit $Bw^{-1}w_0B$. Denote by $\widetilde{\mathcal{N}}_w$ the subvariety of $\widetilde{\mathcal{N}}$ which consists of those pairs $(x, \mathfrak{b}')$ for which $\mathfrak{b}'$ is in relative position w with our fixed Borel subalgebra $\mathfrak{b}$. Let $\widetilde{\mathfrak{n}}_w$ be the restriction of $\widetilde{\mathcal{N}}_w$ to $\mathfrak{n}$, respectively. The group B naturally acts on $\widetilde{\mathfrak{n}}$ preserving $\widetilde{\mathfrak{n}}_w$.

Let $\operatorname{MOp}_G^{\check\lambda,w}$ be the subvariety of $\operatorname{MOp}_G^{\check\lambda}$ consisting of those Miura opers for which the fiber of $\mathcal{F}'_B$ at the origin is in relative position w with B. Then we have an identification

$$\operatorname{MOp}_G^{\check\lambda,w} = \operatorname{Op}_G^{\text{nilp},\check\lambda} \underset{\mathfrak{n}/B}{\times} \widetilde{\mathfrak{n}}_w/B. \tag{9.3.11}$$

9.3.3 Connections and Miura opers

According to Proposition 9.2.1, for any integral dominant coweight $\check\lambda$ we have an isomorphism

$$\operatorname{Op}_G^{\text{RS}}(D)_{-\varpi(\check\lambda+\rho)} \simeq \operatorname{Op}_G^{\text{nilp},\check\lambda}.$$

Hence the restriction of the morphism μ^{RS} (see formula (9.3.4)) to the subspace $\operatorname{Conn}(\Omega^{\check\rho})_D^{-w(\check\lambda+\check\rho)+\check\rho} \subset \operatorname{Conn}(\Omega^{\check\rho})_D^{\text{RS}}$ gives us a morphism

$$\mu_{-w(\check\lambda+\check\rho)+\check\rho} : \operatorname{Conn}(\Omega^{\check\rho})_D^{-w(\check\lambda+\check\rho)+\check\rho} \to \operatorname{Op}_G^{\text{nilp},\check\lambda}.$$

Now, it follows from the explicit construction of the Miura transformation μ (see Section 9.3.1) that for any nilpotent oper χ of coweight $\check\lambda$, which is in the image of the map $\mu_{-w(\check\lambda+\check\rho)+\check\rho}$, the Borel reduction

$$\mathcal{F}'_B = \Omega^{\check\rho} \underset{H}{\times} w_0 B$$

is stable under $t\nabla$. Therefore the map $\mu_{-w(\check{\lambda}+\check{\rho})+\check{\rho}}$ may be canonically lifted to a map

$$\widetilde{\mu}_{-w(\check{\lambda}+\check{\rho})+\check{\rho}} : \operatorname{Conn}(\Omega^{\check{\rho}})_D^{-w(\check{\lambda}+\check{\rho})+\check{\rho}} \to \operatorname{MOp}_G^{\check{\lambda}}. \tag{9.3.12}$$

This map fits into the Cartesian diagram

$$\begin{array}{ccc} \operatorname{Conn}(\Omega^{\check{\rho}})_D^{-w(\check{\lambda}+\check{\rho})+\check{\rho}} & \longrightarrow & \operatorname{Conn}(\Omega^{\check{\rho}})_{D^\times} \\ {\scriptstyle \widetilde{\mu}_{-w(\check{\lambda}+\check{\rho})+\check{\rho}}}\big\downarrow & & \big\downarrow \\ \operatorname{MOp}_G^{\check{\lambda}} & \longrightarrow & \operatorname{MOp}_G(D^\times) \end{array}$$

where the right vertical map is the bijection established in Lemma 9.3.3. The following result is proved in [FG2], Cor. 3.6.5 (for $\check{\lambda} = 0$).

Theorem 9.3.6 *The map $\widetilde{\mu}_{-w(\check{\lambda}+\check{\rho})+\check{\rho}}$ is an isomorphism*

$$\operatorname{Conn}(\Omega^{\check{\rho}})_D^{-w(\check{\lambda}+\check{\rho})+\check{\rho}} \simeq \operatorname{MOp}_G^{\check{\lambda},w} \subset \operatorname{MOp}_G^{\check{\lambda}}.$$

Proof. Suppose that we are given a point of $\operatorname{MOp}_G^{\check{\lambda}}$, i.e., a nilpotent oper of coweight $\check{\lambda}$ equipped with a Borel reduction $\mathcal{F}'_B$ that is stable under $t\nabla$. Since our nilpotent Miura oper is generic on $D^\times$, by Lemma 9.3.2, it follows from Lemma 8.2.1 that the H-bundle $\mathcal{F}'_B/N$ is isomorphic over $D^\times$ to $w_0^*(\mathcal{F}_B/N) = w_0^*(\Omega^{\check{\rho}})$. Thus, we obtain a connection on $w_0^*(\Omega^{\check{\rho}})$ and hence a connection on $\Omega^{\check{\rho}}$ over $D^\times$. Denote the latter by $\overline{\nabla}$. It is clear that $\overline{\nabla}$ has regular singularity, and so

$$\overline{\nabla} = \partial_t - \frac{1}{t}(\check{\mu} + t(\ldots)), \qquad \check{\mu} \in \mathfrak{h}.$$

Clearly, the Miura oper $(\mathcal{F}, \nabla, \mathcal{F}_B, \mathcal{F}'_B)$ on $D^\times$ obtained from $\overline{\nabla}$ by applying the bijection of Lemma 9.3.3, is the Miura oper that we have started with. Therefore the corresponding oper has the form

$$\partial_t + \frac{1}{t}\left(p_{-1} - \check{\rho} - \check{\mu} + t(\ldots)\right).$$

Hence its residue in $\mathfrak{h}/W$ is equal to $\varpi(-\check{\mu} - \check{\rho})$. But by our assumption the residue of this oper is equal to $\varpi(-\check{\lambda} - \check{\rho})$. Therefore we obtain that there exists $y \in W$ such that $-\check{\rho} - \check{\mu} = y(-\check{\lambda} - \check{\rho})$, i.e., $\check{\mu} = y(\check{\lambda} + \rho) - \check{\rho}$.

Thus, we obtain that the restriction of the bijection of Lemma 9.3.3 to $\operatorname{MOp}_G^{\check{\lambda}}$ is a bijection

$$\operatorname{MOp}_G^{\check{\lambda}} \simeq \bigsqcup_{y \in W} \operatorname{Conn}(\Omega^{\check{\rho}})_D^{-y(\check{\lambda}+\check{\rho})+\check{\rho}}.$$

We need to show that under this bijection $\operatorname{MOp}_G^{w,\check{\lambda}}$ corresponds precisely to $\operatorname{Conn}(\Omega^{\check{\rho}})_D^{-w(\check{\lambda}+\check{\rho})+\check{\rho}}$.

Recall that we have an isomorphism of the H-bundles $\mathcal{F}_H|_{D^\times} \simeq w_0^*(\mathcal{F}'_H)|_{D^\times}$

over $D^\times$, where $\mathcal{F}_H = \mathcal{F}_B/N, \mathcal{F}'_H = \mathcal{F}'_B/N$. Let us observe that if we have two H-bundles $\mathcal{F}_H$ and $\mathcal{F}''_H$ over D, then any isomorphism $\mathcal{F}_H|_{D^\times} \simeq \mathcal{F}''_H|_{D^\times}$ extends to an isomorphism $\mathcal{F}_H \simeq \mathcal{F}''_H(\check{\mu})$ over D, for some integral coweight $\check{\mu}$. By considering the transformation properties of a nilpotent oper of coweight $\check{\lambda}$ under changes of variables in the same way as as in the proof of Lemma 4.2.1, we find that

$$\mathcal{F}_H = \mathcal{F}_B/N \simeq \Omega^{\check{\rho}}(-\check{\lambda}).$$

The assertion of the theorem will follow if we show that our isomorphism $\mathcal{F}_H|_{D^\times} \simeq w_0^*(\mathcal{F}'_H)|_{D^\times}$ extends to an isomorphism

$$\mathcal{F}'_H \simeq w_0^*(\mathcal{F}_H((\check{\lambda}+\check{\rho}) - w(\check{\lambda}+\check{\rho})) \tag{9.3.13}$$

over D, because then

$$\mathcal{F}'_H \simeq w_0^*(\Omega^{\check{\rho}}(\check{\rho} - w(\check{\lambda}+\check{\rho})),$$

and so the induced connection on $\Omega^{\check{\rho}}$ on $D^\times$ is in the image of

$$\operatorname{Conn}(\Omega^{\check{\rho}}(\check{\rho} - w(\check{\lambda}+\check{\rho}))_D \simeq \operatorname{Conn}(\Omega^{\check{\rho}})_D^{\check{\rho}-w(\check{\lambda}+\check{\rho})},$$

as desired.

Let us trivialize the fiber of $\mathcal{F}$ at the origin in such a way that $\mathcal{F}_B$ gets identified with $B \subset G$. Without loss of generality, we may assume that under this trivialization $\mathcal{F}'_{B,0}$ becomes $w^{-1}w_0 B$. We need to show the following. For a dominant integral weight ν let V^ν be the finite-dimensional representation of $\mathfrak{g}$ of highest weight ν. For $w \in W$ we pick a non-zero vector $v_w \in V^\nu$ in the one-dimensional subspace of V^ν of weight $w(\nu)$. By our assumption, $\mathcal{F}'_B$ is preserved by $t\nabla$. Therefore by Lemma 9.3.4, the residue of ∇ stabilizes the vector $v_{w^{-1}w_0}$. Consider the associated vector bundle $V^\nu_{\mathcal{F}} = \mathcal{F} \underset{G}{\times} V^\nu$ with the connection induced by ∇. It follows from Lemma 9.3.5 that there is a unique single-valued horizontal section of this flat bundle whose value at $t=0$ is equal to $v_{w^{-1}w_0}$. We express $\Psi(t)$ as the sum of components of different weights. Let $\phi_w(t)v_{w_0}$, where $\phi_w(t) \in \mathbb{C}[[t]]$, be the component of $\Psi(t)$ of the lowest weight $w_0(\nu)$. By Lemma 9.3.2, $\mathcal{F}'_B$ and $\mathcal{F}_B$ are in generic position over $D^\times$. This implies that $\phi_w(t) \neq 0$. Let n_w be the order of vanishing of $\phi_w(t)$ at $t=0$.

We need to show that for each dominant integral weight ν we have

$$n_w = \langle w_0(\nu), w(\check{\lambda}+\check{\rho}) - (\check{\lambda}+\check{\rho})\rangle, \tag{9.3.14}$$

because this is equivalent to formula (9.3.13).

To compute the order of vanishing of $\phi_w(t)$, recall that the connection underlying a nilpotent oper of coweight $\check{\lambda}$ has the form

$$\nabla = \partial_t + \sum_{i=1}^{\ell} t^{\langle \alpha_i, \check{\lambda}\rangle} f_i + \mathbf{v}(t) + \frac{v}{t}, \tag{9.3.15}$$

where $\mathbf{v}(t) \in \mathfrak{b}[[t]], v \in \mathfrak{n}$. The sought-after horizontal section of $V^\nu_{\mathcal{F}}$ may be obtained by solving the equation $\nabla\Psi(t) = 0$ with values in $V^\nu[[t]]$. We solve this equation recursively, as in the proof of Lemma 9.3.5. Then we have to solve equations (9.3.8), in which on the right hand side we have a linear combination of operators in $\mathfrak{b}$, which preserve or increase the principal gradation, and the operators $f_i, i = 1, \ldots, \ell$, which lower it by 1. It is therefore clear that the leading term of the component $\phi_w(t) v_{w_0}$ of $\Psi(t)$ proportional to v_{w_0} is obtained by applying the operators $f_i, i = 1, \ldots, \ell$, to $v_{w^{-1}w_0}$. Hence it is the same as for the connection

$$\nabla = \partial_t + \sum_{i=1}^{\ell} t^{\langle \alpha_i, \check{\lambda}\rangle} \psi_i(t) f_i.$$

When we solve the recurrence relations (9.3.8) for this connection, each application of f_i leads to the increase the t-degree of $\Psi(t)$ by $\langle \alpha_i, \check{\lambda}\rangle + 1 = \langle \alpha_i, \check{\lambda} + \check{\rho}\rangle$. This is the "$(\check{\lambda} + \check{\rho})$-degree" of α_i. Therefore to "reach" v_{w_0} from $v_{w^{-1}w_0}$ (the initial condition for $\Psi(t)$) we need to increase the t-degree by the difference between the $(\check{\lambda} + \check{\rho})$-degrees of the vectors $v_{w^{-1}w_0}$ and v_{w_0}, i.e., by

$$\langle w^{-1}w_0(\nu), \check{\lambda} + \check{\rho}\rangle - \langle w_0(\nu), \check{\lambda} + \check{\rho}\rangle = \langle w_0(\nu), w(\check{\lambda} + \check{\rho}) - (\check{\lambda} + \check{\rho})\rangle,$$

which is n_w given by formula (9.3.14). Hence we find that $\phi_w(t) = c_w t^{n_w}$ for some $c_w \in \mathbb{C}$. If c_w were equal to 0, then the entire v_{w_0}-component of the solution $\Psi(t)$ would be equal to 0. But this would mean that the corresponding Borel reductions $\mathcal{F}'_B$ and $\mathcal{F}_B$ are not in generic position over $D^\times$. This contradicts Lemma 9.3.2. Hence $c_w \neq 0$, and so for any oper (9.3.15) the leading t-degree of $\phi_w(t)$ is indeed equal to n_w. This completes the proof. □

Combining the isomorphisms (9.3.12) and (9.3.11), we obtain an isomorphism

$$\operatorname{Conn}(\Omega^{\check{\rho}})_D^{\check{\rho} - w(\check{\lambda} + \check{\rho})} \simeq \operatorname{Op}_G^{\mathrm{nilp}, \check{\lambda}} \underset{\mathfrak{n}/B}{\times} \widetilde{\mathfrak{n}}_w / B. \tag{9.3.16}$$

Thus, we have a bijection of the sets of $\mathbb{C}$-points

$$\bigsqcup_{w \in W} \operatorname{Conn}(\Omega^{\check{\rho}})_D^{\check{\rho} - w(\check{\lambda} + \check{\rho})} \simeq \operatorname{Op}_G^{\mathrm{nilp}, \check{\lambda}} \underset{\mathfrak{n}/B}{\times} \widetilde{\mathfrak{n}} / B. \tag{9.3.17}$$

While (9.3.16) is an isomorphism of varieties, (9.3.17) is not, because the left hand side is a disjoint union of strata, which are "glued" together in the right hand side.

Recall that according to formula (9.3.6), the left hand side of (9.3.17) is precisely the preimage of $\operatorname{Op}_G^{\mathrm{nilp}, \check{\lambda}} \subset \operatorname{Op}_G(D^\times)$ under the Miura transformation μ. Thus, we see that under the isomorphism (9.3.17) the natural projection

$$\operatorname{Op}_G^{\mathrm{nilp}, \check{\lambda}} \underset{\mathfrak{n}/B}{\times} \widetilde{\mathfrak{n}} / B \to \operatorname{Op}_G^{\mathrm{nilp}, \check{\lambda}}$$

on the first factor coincides with the restriction of the Miura transformation μ (see (9.3.3)) to $\mathrm{Op}_G^{\mathrm{nilp},\check{\lambda}}$. This implies that the set of points in the fiber $\mu^{-1}(\chi)$ of the Miura transformation μ over $\chi \in \mathrm{Op}_G^{\mathrm{nilp},\check{\lambda}}$ is in bijection with the set of points of the Springer fiber $\mathrm{Sp}_{\mathrm{Res}_{\check{\lambda}}(\chi)}$ (see (9.3.10)).

As in the case of the isomorphism (9.3.17), this is only a bijection at the level of $\mathbb{C}$-points. The fiber $\mu^{-1}(\chi)$ is the union of strata

$$\mu^{-1}(\chi)_w = \mu^{-1}(\chi) \cap \mathrm{Conn}(\Omega^{\check{\rho}})_D^{\check{\rho}-w(\check{\lambda}+\check{\rho})}, \qquad w \in W,$$

and each of these strata is isomorphic, as a variety, to the subvariety

$$\mathrm{Sp}^w_{\mathrm{Res}_{\check{\lambda}}(\chi)} = \mathrm{Sp}_{\mathrm{Res}_{\check{\lambda}}(\chi)} \cap S_w,$$

consisting of those Borel subalgebras that are in relative position w with $\mathfrak{b}$; here S_w is the Schubert cell in G/B corresponding to w (see the proof of Lemma 9.3.2).

We summarize this in the following statement.

Theorem 9.3.7 *For any $\chi \in \mathrm{Op}_G^{\mathrm{nilp},\check{\lambda}}$, the fiber $\mu^{-1}(\chi)$ of the Miura transformation μ over χ is the union of strata*

$$\mu^{-1}(\chi)_w \subset \mathrm{Conn}(\Omega^{\check{\rho}})_D^{\check{\rho}-w(\check{\lambda}+\check{\rho})}, \qquad w \in W.$$

The stratum $\mu^{-1}(\chi)_w$ is isomorphic to

$$\mathrm{Sp}^w_{\mathrm{Res}_{\check{\lambda}}(\chi)} = \mathrm{Sp}_{\mathrm{Res}_{\check{\lambda}}(\chi)} \cap S_w,$$

where $\mathrm{Sp}_{\mathrm{Res}_{\check{\lambda}}(\chi)}$ is the Springer fiber of $\mathrm{Res}_{\check{\lambda}}(\chi)$. Hence the set of points of $\mu^{-1}(\chi)$ is in bijection with the set of points of $\mathrm{Sp}_{\mathrm{Res}_{\check{\lambda}}(\chi)}$.

For example, if $\mathrm{Res}_{\check{\lambda}}(\chi) = 0$, then $\mathrm{Sp}_{\mathrm{Res}_{\check{\lambda}}(\chi)}$ is the flag variety G/B and $\mathrm{Sp}^w_{\mathrm{Res}_{\check{\lambda}}(\chi)}$ is the Schubert cell S_w. Thus, we find that $\mu^{-1}(\chi)$ is in bijection with the set of points of G/B in this case.

Note that $\widetilde{\mathfrak{n}}_{w_0}$ is a single point (namely, $\mathfrak{b}' = \mathfrak{b}$), and so $\mu^{-1}(\chi)_{w_0} = \mathrm{Sp}^{w_0}_{\mathrm{Res}_{\check{\lambda}}(\chi)}$ consists of a single point. Therefore we have an isomorphism

$$\mathrm{Conn}(\Omega^{\check{\rho}})_D^{\check{\rho}-w_0(\check{\lambda}+\check{\rho})} \simeq \mathrm{Op}_G^{\mathrm{nilp},\check{\lambda}}. \tag{9.3.18}$$

This is actually a special case of the statement of Proposition 9.3.1 (see also Corollary 9.2.5), because if $\check{\lambda}$ is an integral dominant coweight then the coweight $\check{\mu} = w_0(\check{\lambda} + \check{\rho}) - \check{\rho}$ satisfies $\langle \alpha, \check{\mu} \rangle \notin \mathbb{Z}_+$ for all $\alpha \in \Delta_+$.

9.4 Categories of representations of $\widehat{\mathfrak{g}}$ at the critical level

In the previous sections we defined certain subvarieties of the space $\mathrm{Op}_G(D^\times)$ of G-opers on the punctured disc $D^\times$. Let us now switch from G to the Langlands dual group ${}^LG$ (which we will assume to be of adjoint type, as before). In particular, this means that we switch from the space $\mathrm{Conn}(\Omega^{\check{\rho}})_{D^\times}$

of connections on the H-bundle $\Omega^{\check{\rho}}$ and its subspaces defined above to the space $\mathrm{Conn}(\Omega^{\rho})$ of connections on the ${}^L H$-bundle Ω^{ρ} and the corresponding subspaces.

According to Theorem 8.3.1, the space $\mathrm{Op}_{^L G}(D^\times)$ is isomorphic to the spectrum of the center $Z(\widehat{\mathfrak{g}})$ of the completed enveloping algebra of $\widehat{\mathfrak{g}}$ of critical level. Thus, the category of all smooth $\widehat{\mathfrak{g}}_{\kappa_c}$-modules "lives" over $\mathrm{Op}_{^L G}(D^\times)$. In this section we identify certain subcategories of this category, which "live" over the subspaces $\mathrm{Op}^{\mathrm{RS}}_{^L G}(D)_{\varpi} \subset \mathrm{Op}_{^L G}(D^\times)$.

9.4.1 Compatibility with the finite-dimensional Harish-Chandra homomorphism

We start by describing a certain compatibility between the Miura transformation and the Harish-Chandra isomorphism $Z(\mathfrak{g}) \simeq (\mathrm{Fun}\,\mathfrak{h}^*)^W$, where $Z(\mathfrak{g})$ be the center of $U(\mathfrak{g})$

Consider the center $Z(\widehat{\mathfrak{g}})$ of $\widetilde{U}_{\kappa_c}(\widehat{\mathfrak{g}})$. We have a natural $\mathbb{Z}$-gradation on $Z(\widehat{\mathfrak{g}})$ by the operator $L_0 = -t\partial_t \in \mathrm{Der}\,\mathcal{O}$.

Let us denote by the subscript 0 the degree 0 part and by the subscript < 0 the span of all elements of negative degrees in any $\mathbb{Z}$-graded object.

Next, consider the Wakimoto modules $W_{\chi(t)}$, where

$$\chi(t) = \frac{\lambda - \rho}{t}, \qquad \lambda \in \mathfrak{h}^*$$

(see the proof of Theorem 6.4.1). These Wakimoto modules are $\mathbb{Z}$-graded, with the degree 0 part $(W_{\chi(t)})_0$ being the subspace $\mathbb{C}[a^*_{\alpha,0}|0\rangle]_{\alpha \in \Delta_+} \subset M_{\mathfrak{g}} \subset W_{\chi(t)}$. The Lie algebra $\mathfrak{g}$ preserves this subspace, and it follows immediately from our construction of Wakimoto modules that as a $\mathfrak{g}$-module, $(W_{\chi(t)})_0$ is isomorphic to the contragredient Verma module $M^*_{\lambda-\rho}$. The algebra

$$Z' = Z(\widehat{\mathfrak{g}})_0 / (Z(\widehat{\mathfrak{g}}) \cdot Z(\widehat{\mathfrak{g}})_{<0})_0$$

naturally acts on $(W_{\chi(t)})_0$ and commutes with $\mathfrak{g}$. Moreover, since its action necessarily factors through the action of $U(\mathfrak{g})$, we find that it factors through $Z(\mathfrak{g})$. Therefore varying $\lambda \in \mathfrak{h}^*$ we obtain a homomorphism

$$Z' \to (\mathrm{Fun}\,\mathfrak{h}^*)^W, \tag{9.4.1}$$

which factors through the Harish-Chandra homomorphism $Z(\mathfrak{g}) \to (\mathrm{Fun}\,\mathfrak{h}^*)^W$.

On the other hand, recall from Section 9.1.1 the space

$$\mathrm{Op}^{\mathrm{RS}}_{^L G}(D) \subset \mathrm{Op}_{^L G}(D^\times)$$

of ${}^L G$-opers with regular singularity and its subspace $\mathrm{Op}^{\mathrm{RS}}_{^L G}(D^\times)_{\varpi}$ of opers with residue $\varpi \in {}^L\mathfrak{h}/W = \mathfrak{h}^*/W$ (recall that $\mathfrak{h}^* = {}^L\mathfrak{h}$).

We have a $\mathbb{Z}$-gradation on $\mathrm{Fun}\,\mathrm{Op}_{^L G}(D^\times)$ induced by the operator $L_0 =$

$-t\partial_t \in \operatorname{Der} \mathcal{O}$. If we write the oper connection in the canonical form of Lemma 4.2.2,

$$\nabla = \partial_t + p_{-1} + \sum_{j=1}^{\ell} v_j(t) \cdot p_j, \qquad v_j(t) = \sum_{n \in \mathbb{Z}} v_{j,n} t^{-n-1},$$

then we obtain from the transformation formulas of Section 4.2.4 that

$$L_0 \cdot v_{j,n} = (-n + d_j) v_{j,n}.$$

Therefore the generators $v_{j,n}$ are homogeneous, and those of strictly negative degrees are $v_{j,n}, n > d_j$, which by Corollary 9.1.2 generate the ideal of $\operatorname{Op}_G^{\mathrm{RS}}(D)$ in $\operatorname{Fun} \operatorname{Op}_G(D^\times)$. Therefore the ideal of $\operatorname{Op}_G^{\mathrm{RS}}(D)$ is generated by $\operatorname{Fun} \operatorname{Op}_{^L G}(D^\times)_{<0}$. Clearly, the residue map

$$\operatorname{res} : \operatorname{Op}_G^{\mathrm{RS}}(D) \to \mathfrak{h}^*/W$$

gives rise to an isomorphism

$$(\operatorname{Fun} \mathfrak{h}^*)^W \simeq (\operatorname{Fun} \operatorname{Op}_{^L G}^{\mathrm{RS}}(D))_0.$$

We also have the space

$$\operatorname{Conn}(\Omega^\rho)_D^{\mathrm{RS}} \subset \operatorname{Conn}(\Omega^\rho)_{D^\times}$$

of connections with regular singularity on the $^L H$-bundle Ω^ρ on $D^\times$ defined in Section 9.3.1. The Miura transformation restricts to a map (see (9.3.4))

$$\mu^{\mathrm{RS}} : \operatorname{Conn}(\Omega^\rho)_D^{\mathrm{RS}} \to \operatorname{Op}_{^L G}^{\mathrm{RS}}(D).$$

We have a $\mathbb{Z}$-gradation on $\operatorname{Fun} \operatorname{Conn}(\Omega^\rho)_{D^\times}$ induced by the operator $L_0 = -t\partial_t \in \operatorname{Der} \mathcal{O}$. If we write $\overline{\nabla} \in \operatorname{Conn}(\Omega^\rho)_{D^\times}$ in the form $\overline{\nabla} = \partial_t + \mathbf{u}(t)$, where

$$u_i(t) = \alpha_i(\mathbf{u}(t)) = \sum_{n<0} u_{i,n} t^{-n-1},$$

then we have

$$L_0 \cdot u_{i,n} = -n u_{i,n}.$$

Therefore the generators $u_{i,n}$ are homogeneous, and those of strictly negative degrees are $v_{i,n}, n > 0$. Those generate the ideal of $\operatorname{Conn}(\Omega^\rho)_D^{\mathrm{RS}}$ in $\operatorname{Fun} \operatorname{Conn}(\Omega^\rho)_{D^\times}$. The residue map

$$\operatorname{res}_{^L\mathfrak{h}} : \operatorname{Op}_G^{\mathrm{RS}}(D) \to \mathfrak{h}^*$$

defined in Section 9.3.1 gives rise to an isomorphism

$$\operatorname{Fun} \mathfrak{h}^* \to (\operatorname{Fun} \operatorname{Conn}(\Omega^\rho)_D^{\mathrm{RS}})_0.$$

We have a commutative diagram (9.3.5)

$$\begin{array}{ccc} \mathrm{Conn}(\Omega^\rho)_D^{\mathrm{RS}} & \xrightarrow{\mu^{\mathrm{RS}}} & \mathrm{Op}_{^LG}^{\mathrm{RS}}(D) \\ {\scriptstyle \mathrm{res}_{^L\mathfrak{h}}}\downarrow & & \downarrow{\scriptstyle \mathrm{res}} \\ \mathfrak{h}^* & \longrightarrow & \mathfrak{h}^*/W \end{array}$$

It gives rise to a commutative diagram of algebra homomorphisms

$$\begin{array}{ccc} \mathrm{Fun}\,\mathfrak{h}^* & \xrightarrow{\sim} & (\mathrm{Fun}\,\mathrm{Conn}(\Omega^\rho)_D^{\mathrm{RS}})_0 \\ \uparrow & & \uparrow \\ (\mathrm{Fun}\,\mathfrak{h}^*)^W & \xrightarrow{\sim} & (\mathrm{Fun}\,\mathrm{Op}_{^LG}^{\mathrm{RS}}(D))_0 \end{array} \tag{9.4.2}$$

Now, since the ideal of $\mathrm{Op}_{^LG}^{\mathrm{RS}}(D)$ in $\mathrm{Fun}\,\mathrm{Op}_{^LG}(D^\times)$ is generated by the subalgebra $\mathrm{Fun}\,\mathrm{Op}_{^LG}(D^\times)_{<0}$ we obtain that the isomorphism of Theorem 8.3.1 gives rise to an isomorphism of algebras

$$Z' \simeq (\mathrm{Fun}\,\mathrm{Op}_{^LG}(D^\times)/(\mathrm{Fun}\,\mathrm{Op}_{^LG}(D^\times)_{<0}))_0 = (\mathrm{Fun}\,\mathrm{Op}_{^LG}^{\mathrm{RS}}(D))_0. \tag{9.4.3}$$

Therefore commutative diagram (9.4.2) implies that the map (9.4.1) is an isomorphism.

Recall that in the isomorphism of Theorem 8.3.3 we have $b_{i,n} \mapsto -u_{i,n}$. Therefore we obtain the following:

Proposition 9.4.1 ([F4],Prop. 12.8) *There is a commutative diagram of isomorphisms*

$$\begin{array}{ccc} Z(\mathfrak{g}) & \xrightarrow{\sim} & (\mathrm{Fun}\,\mathrm{Op}_{^LG}^{\mathrm{RS}}(D))_0 \\ \downarrow & & \uparrow \\ (\mathrm{Fun}\,\mathfrak{h}^*)^W & \xrightarrow{\sim} & (\mathrm{Fun}\,\mathfrak{h}^*)^W \end{array}$$

where the lower horizontal arrow is given by $f \mapsto f^-$, $f^-(\lambda) = f(-\lambda)$.

9.4.2 Induction functor

We define a functor from the category of $\mathfrak{g}$-modules to the category of $\widehat{\mathfrak{g}}_{\kappa_c}$-modules on which $\mathbf{1}$ acts as the identity. Let M be a $\mathfrak{g}$-module. Define the induced $\widehat{\mathfrak{g}}_{\kappa_c}$-module $\mathrm{Ind}(M)$ as

$$\mathrm{Ind}(M) = \mathrm{Ind}_{\mathfrak{g}[[t]]\oplus\mathbf{1}}^{\widehat{\mathfrak{g}}_{\kappa_c}} M,$$

where $\mathfrak{g}[[t]]$ acts on M through the evaluation homomorphism $\mathfrak{g}[[t]] \to \mathfrak{g}$ and $\mathbf{1}$ acts as the identity.

In particular, if $M = M_\lambda$, the Verma module of highest weight $\lambda \in \mathfrak{h}^*$, then the induced module $\mathrm{Ind}(M_\lambda)$ is the Verma module over $\widehat{\mathfrak{g}}_{\kappa_c}$ of critical level, denoted by $\mathbb{M}_{\lambda,\kappa_c}$, which we have encountered previously.

Let again $Z(\mathfrak{g})$ be the center of $U(\mathfrak{g})$. The Harish-Chandra homomorphism $Z(\mathfrak{g}) \xrightarrow{\sim} \operatorname{Fun}(\mathfrak{h}^*)^W$ gives rise to an isomorphism $\mathfrak{h}^*/W \to \operatorname{Spec} Z(\mathfrak{g})$. Now each $\lambda \in \mathfrak{h}^*$ defines a point $\varpi(\lambda) \in \mathfrak{h}^*/W$ and hence a character of $Z(\mathfrak{g})$, which we also denote by $\varpi(\lambda)$. In particular, we find that $Z(\mathfrak{g})$ acts on the Verma module $M_{\lambda-\rho}$ through the central character $\varpi(\lambda)$.

Proposition 9.4.2 *Let M be any $\mathfrak{g}$-module. Consider* $\operatorname{Ind}(M)$ *as a module over the center $Z(\widehat{\mathfrak{g}})$ of $\widetilde{U}_{\kappa_c}(\widehat{\mathfrak{g}})$. Then* $\operatorname{Ind}(M)$ *is scheme theoretically supported on the subscheme $\operatorname{Op}^{\mathrm{RS}}_{^LG}(D) \subset \operatorname{Op}_{^LG}(D^\times)$, i.e., it is annihilated by the ideal of $\operatorname{Op}^{\mathrm{RS}}_{^LG}(D)$ in $Z(\widehat{\mathfrak{g}}) \simeq \operatorname{Fun}\operatorname{Op}_{^LG}(D^\times)$.*

Furthermore, if $Z(\mathfrak{g})$ acts on M through the character $\varpi(\lambda)$, then $\operatorname{Ind}(M)$ *is scheme theoretically supported on the subscheme $\operatorname{Op}^{\mathrm{RS}}_{^LG}(D)_{\varpi(-\lambda)}$.*

Proof. Define an action of $L_0 = -t\partial_t$ in such a way that it acts by 0 on the generating subspace $M \subset \operatorname{Ind}(M)$, and is compatible with the action of L_0 on $\widehat{\mathfrak{g}}_{\kappa_c}$. Then $\operatorname{Ind}(M)$ acquires a $\mathbb{Z}$-grading, which takes only non-negative values on $\operatorname{Ind}(M)$ with the degree 0 part being the subspace M. This implies that the subalgebra $Z(\widehat{\mathfrak{g}})_{<0}$ of $Z(\widehat{\mathfrak{g}})$ which is spanned by all elements of negative degrees acts by 0 on $\operatorname{Ind}(M)$. The operator L_0 also defines a $\mathbb{Z}$-grading on $\operatorname{Fun}\operatorname{Op}_{^LG}(D^\times)$, and the isomorphism $Z(\widehat{\mathfrak{g}}) \simeq \operatorname{Fun}\operatorname{Op}_{^LG}(D^\times)$ is compatible with this $\mathbb{Z}$-grading. Hence $Z(\widehat{\mathfrak{g}})_{<0}$ is mapped under this isomorphism to the subalgebra $\operatorname{Fun}\operatorname{Op}_{^LG}(D^\times)_{<0}$ spanned by all elements of negative degrees.

But the ideal of $\operatorname{Op}_{^LG}(D^\times)_{<0}$ in $\operatorname{Fun}\operatorname{Op}_{^LG}(D^\times)$ is precisely the ideal of the subspace $\operatorname{Op}^{\mathrm{RS}}_{^LG}(D^\times) \subset \operatorname{Op}_{^LG}(D^\times)$ of opers with regular singularities, as shown in Section 9.4.1. This implies the first statement of the proposition.

Next, consider the action of the degree 0 subalgebra $Z(\widehat{\mathfrak{g}})^0$ of $Z(\widehat{\mathfrak{g}})$ on $\operatorname{Ind}(M)$. It factors through the quotient

$$Z' = Z(\widehat{\mathfrak{g}})_0/(Z(\widehat{\mathfrak{g}}) \cdot Z(\widehat{\mathfrak{g}})_{<0})_0.$$

It follows from Proposition 9.4.1 that Z' is isomorphic to the center $Z(\mathfrak{g})$ of $U(\mathfrak{g})$ and its action on $M \subset \operatorname{Ind}(M)$, factors through this isomorphism. Furthermore, from Proposition 9.4.1 we have a commutative diagram

$$\begin{array}{ccc} Z' & \xrightarrow{\sim} & \operatorname{Fun}\operatorname{Op}^{\mathrm{RS}}_{^LG}(D) \\ \downarrow & & \uparrow \\ (\operatorname{Fun}\mathfrak{h}^*)^W & \xrightarrow{\sim} & (\operatorname{Fun}\mathfrak{h}^*)^W \end{array}$$

where the left vertical arrow is the composition of the isomorphism $Z' \simeq Z(\mathfrak{g})$ and the Harish-Chandra homomorphism, the right vertical arrow is induced by the residue map of Section 9.1.1

$$\operatorname{res} : \operatorname{Op}^{\mathrm{RS}}_{^LG}(D) \to \mathfrak{h}^*/W,$$

and the lower horizontal arrow is given by $f \mapsto f^-$, $f^-(\lambda) = f(-\lambda)$. This implies the second statement of the proposition. □

In particular, we obtain that the Verma module $\mathbb{M}_{\mu,\kappa_c}$ is scheme theoretically supported on the subscheme $\mathrm{Op}^{\mathrm{RS}}_{^LG}(D)_{\varpi(-\mu-\rho)}$.

9.4.3 Wakimoto modules and categories of $\widehat{\mathfrak{g}}$-modules

Consider the Lie subalgebra

$$\widehat{\mathfrak{n}}_+ = (\mathfrak{n}_+ \otimes 1) \oplus (g \otimes t\mathbb{C}[[t]])$$

of $\mathfrak{g}((t))$ and $\widehat{\mathfrak{g}}_{\kappa_c}$. The corresponding subgroup of $G((t))$ is denoted by I^0. It is the preimage of the Lie subgroup $N_+ \subset G$ corresponding to $\mathfrak{n}_+ \subset \mathfrak{g}$ under the evaluation map $G[[t]] \to G$. Note that $I^0 = [I, I]$, where I is the Iwahori subgroup defined in Section 10.2.2.

For a central character $\varpi(\nu) \in \operatorname{Spec} Z(\mathfrak{g}) \simeq \mathfrak{h}^*/W$, let $\mathcal{O}_{\varpi(\nu)}$ be the block of the category $\mathcal{O}$ of $\mathfrak{g}$-modules (with respect to a fixed Borel subalgebra $\mathfrak{b}$) whose objects are the $\mathfrak{g}$-modules on which $Z(\mathfrak{g})$ acts through the character $\varpi(\nu)$. Let $\widehat{\mathfrak{g}}_{\kappa_c}\text{-mod}^{I^0}_{\varpi(-\nu)}$ be the category of $\widehat{\mathfrak{g}}_{\kappa_c}$-modules M such that

- the action of the Lie subalgebra $\widehat{\mathfrak{n}}_+$ on M is locally nilpotent;
- the action of $Z(\widehat{\mathfrak{g}})$ on M factors through $\operatorname{Fun}\mathrm{Op}^{\mathrm{RS}}_{^LG}(D)_{\varpi(-\nu)}$.

On such modules the action of $\widehat{\mathfrak{n}}_+$ exponentiates to an action of I^0, and this explains the notation.

Then, according to Proposition 9.4.2, the induction functor Ind gives rise to a functor

$$\mathrm{Ind} : \mathcal{O}_{\varpi(\nu)} \to \widehat{\mathfrak{g}}_{\kappa_c}\text{-mod}^{I^0}_{\varpi(-\nu)} .$$

In particular, for any $\lambda \in \mathfrak{h}^*$ the Verma modules $\mathbb{M}_{w(\lambda+\rho)-\rho,\kappa_c}, w \in W$, are objects of the category $\widehat{\mathfrak{g}}_{\kappa_c}\text{-mod}^{I^0}_{\varpi(-\lambda-\rho)}$. Moreover, it is easy to see that any object of $\widehat{\mathfrak{g}}_{\kappa_c}\text{-mod}^{I^0}_{\varpi(-\lambda-\rho)}$ has a filtration such that the associated quotients are quotients of these Verma modules.

Now we want to show that some of the Wakimoto modules of critical level are also objects of the category $\widehat{\mathfrak{g}}_{\kappa_c}\text{-mod}^{I^0}_{\varpi(-\lambda-\rho)}$. The general construction of these modules and the corresponding action of the center $Z(\widehat{\mathfrak{g}})$ on them are described in Theorem 8.3.3.

Recall from Section 9.3.1 the subspace $\mathrm{Conn}(\Omega^\rho)^{-\lambda}_D \subset \mathrm{Conn}(\Omega^\rho)_{D^\times}$. Under the isomorphism $\mathrm{Conn}(\Omega^\rho)_{D^\times} \simeq \mathrm{Conn}(\Omega^{-\rho})_{D^\times}$ it becomes the subspace $\mathrm{Conn}(\Omega^{-\rho})^\lambda_D$ of connections of the form

$$\partial_t + \frac{\lambda}{t} + \mathbf{u}(t), \qquad \mathbf{u}(t) \in {}^L\mathfrak{h}[[t]] \tag{9.4.4}$$

on $\Omega^{-\rho}$. Note that $\operatorname{Fun}\mathrm{Conn}(\Omega^{-\rho})^\lambda_D$ is a smooth module over the algebra $\operatorname{Fun}\mathrm{Conn}(\Omega^{-\rho})_{D^\times}$. In fact, $\operatorname{Fun}\mathrm{Conn}(\Omega^{-\rho})^\lambda_D$ is naturally identified with the module π_λ^0 over the commutative vertex algebra π_0 (see Section 6.2.1). Therefore we obtain a $\widehat{\mathfrak{g}}_{\kappa_c}$-module structure on

$$W_\lambda \stackrel{\text{def}}{=} M_{\mathfrak{g}} \otimes \operatorname{Fun}\operatorname{Conn}(\Omega^{-\rho})_D^\lambda. \tag{9.4.5}$$

We claim that W_λ is an object of the category $\widehat{\mathfrak{g}}_{\kappa_c}\text{-mod}_{\varpi(-\lambda-\rho)}^{I^0}$.

Indeed, using the explicit formulas for the action of $\widehat{\mathfrak{g}}_{\kappa_c}$ given in Theorem 6.1.6 one checks that the Lie subalgebra $\widehat{\mathfrak{n}}_+$ acts on W_λ locally nilpotently. On the other hand, according to Section 9.3.1, the restriction of the Miura transformation to $\operatorname{Conn}(\Omega^{-\rho})_D^\lambda = \operatorname{Conn}(\Omega^\rho)_D^{-\lambda}$ takes values in

$$\operatorname{Op}_{^LG}^{\mathrm{RS}}(D)_{\varpi(-\lambda-\rho)} \subset \operatorname{Op}_{^LG}(D^\times).$$

Therefore we obtain from Theorem 8.3.3 that the action of $Z(\widehat{\mathfrak{g}})$ on W_λ factors through the quotient $\operatorname{Fun}\operatorname{Op}_{^LG}^{\mathrm{RS}}(D)_{\varpi(-\lambda-\rho)}$. Hence W_λ is an object of the category $\widehat{\mathfrak{g}}_{\kappa_c}\text{-mod}_{\varpi(-\lambda-\rho)}^{I^0}$.

This implies that any quotient of W_λ is also an object of $\widehat{\mathfrak{g}}_{\kappa_c}\text{-mod}_{\varpi(-\lambda-\rho)}^{I^0}$. The algebra $\operatorname{Fun}\operatorname{Conn}(\Omega^{-\rho})_D^\lambda$ acts on W_λ by endomorphisms which commute with the action of $\widehat{\mathfrak{g}}_{\kappa_c}$. Given $\overline{\nabla} \in \operatorname{Conn}(\Omega^{-\rho})_D^\lambda$, let $W_{\overline{\nabla}}$ be the quotient of W_λ by the maximal ideal of $\operatorname{Fun}\operatorname{Conn}(\Omega^{-\rho})_D^\lambda$ corresponding to the point $\overline{\nabla}$. This is just the Wakimoto module $M_{\mathfrak{g}} \otimes \mathbb{C}_{\overline{\nabla}}$ obtained from the one-dimensional module $\mathbb{C}_{\overline{\nabla}}$ over $\operatorname{Fun}\operatorname{Conn}(\Omega^{-\rho})_{D^\times}$ corresponding to $\overline{\nabla} \in \operatorname{Conn}(\Omega^{-\rho})_D^\lambda \subset \operatorname{Conn}(\Omega^{-\rho})_{D^\times}$. According to Theorem 8.3.3, the center $Z(\widehat{\mathfrak{g}})$ acts on $W_{\overline{\nabla}}$ via the central character $\mu(\overline{\nabla})$, where μ is the Miura transformation.

For $\chi \in \operatorname{Op}_{^LG}^{\mathrm{RS}}(D)$, let $\widehat{\mathfrak{g}}_{\kappa_c}\text{-mod}_\chi^{I^0}$ be the category of $\widehat{\mathfrak{g}}_{\kappa_c}$-modules M such that

- the action of the Lie subalgebra $\widehat{\mathfrak{n}}_+$ on M is locally nilpotent;
- the center $Z(\widehat{\mathfrak{g}})$ acts on M via the character $Z(\widehat{\mathfrak{g}}) \to \mathbb{C}$, corresponding to χ.

Suppose that the residue of χ is $\varpi(-\lambda-\rho) \in \mathfrak{h}^*/W$ for some $\lambda \in \mathfrak{h}^*$. Then the quotients $\mathbb{M}_{w(\lambda+\rho)-\rho,\kappa_c}(\chi)$ of the Verma modules $\mathbb{M}_{w(\lambda+\rho)-\rho,\kappa_c}$, $w \in W$, by the central character corresponding to χ are objects of this category. Moreover, it is easy to see that any object of $\widehat{\mathfrak{g}}_{\kappa_c}\text{-mod}_\chi^{I^0}$ has a filtration such that the associated quotients are quotients of $\mathbb{M}_{w(\lambda+\rho)-\rho,\kappa_c}(\chi)$.

The Wakimoto module $W_{\overline{\nabla}}$ is an object of $\widehat{\mathfrak{g}}_{\kappa_c}\text{-mod}_\chi^{I^0}$ if and only if $\overline{\nabla} \in \mu^{-1}(\chi)$. According to the isomorphism (9.3.6) (in which we replace G by ^LG and $\Omega^{\check{\rho}}$ by $\Omega^{-\rho}$, changing the sign of the residue of the connection), this means that

$$\overline{\nabla} \in \operatorname{Conn}(\Omega^{-\rho})_D^{w(\lambda+\rho)-\rho}, \qquad w \in W. \tag{9.4.6}$$

The Cartan subalgebra $\mathfrak{h} \otimes 1$ acts diagonally on $W_{\overline{\nabla}}$, but its action may be exponentiated to the action of the corresponding group H (and so $W_{\overline{\nabla}}$ is

I-equivariant) if and only if λ belongs to the set P^+ of dominant integral weights.

So let us assume that $\lambda \in P^+$. Then

$$\chi \in \mathrm{Op}^{\mathrm{RS}}_{^LG}(D)_{\varpi(-\lambda-\rho)} = \mathrm{Op}^{\mathrm{nilp},\lambda}_{^LG}$$

(see Section 9.2.1). We will discuss the corresponding categories $\widehat{\mathfrak{g}}_{\kappa_c}\text{-mod}_\chi^{I^0}$ in Section 10.4. They will appear as the categories associated by the local Langlands correspondence for loop groups to tamely ramified local systems on $D^\times$. The following fact will be very important for us in this context (see Section 10.4.5).

According to Theorem 9.3.7, the points of $\mu^{-1}(\chi)$ are in bijection with the points of the Springer fiber $\mathrm{Sp}_{\mathrm{Res}_{\check{\lambda}}(\chi)}$ of $\mathrm{Res}_{\check{\lambda}}(\chi)$. Therefore the Wakimoto modules $W_{\overline{\nabla}}$, where $\overline{\nabla} \in \mu^{-1}(\chi)$ give us a family of objects of the category $\widehat{\mathfrak{g}}_{\kappa_c}\text{-mod}_\chi^{I^0}$ parameterized by the points of $\mathrm{Sp}_{\mathrm{Res}_{\check{\lambda}}(\chi)}$.

In particular, if $\mathrm{Res}_{\check{\lambda}}(\chi) = 0$, and so $\chi \in \mathrm{Op}^{\lambda}_{^LG}$, then we have a family of Wakimoto modules in $\widehat{\mathfrak{g}}_{\kappa_c}\text{-mod}_\chi^{I^0}$ parameterized by the points of the flag variety $^LG/^LB$, which is the Springer fiber Sp_0.

Thus, we see that the category $\widehat{\mathfrak{g}}_{\kappa_c}\text{-mod}_\chi^{I^0}$ is quite complicated if $\chi \in \mathrm{Op}^{\mathrm{RS}}_{^LG}(D)_{\varpi(-\lambda-\rho)}$, where $\lambda \in P^+$. This represents one "extreme" example of the categories $\widehat{\mathfrak{g}}_{\kappa_c}\text{-mod}_\chi^{I^0}$. On the opposite "extreme" is the category $\widehat{\mathfrak{g}}_{\kappa_c}\text{-mod}_\chi^{I^0}$, where χ is as above and λ is a generic element of $\mathfrak{h}^*$ such that

$$\langle \alpha, w(\check{\lambda}+\rho)-\rho\rangle \notin \mathbb{Z}_+, \qquad \alpha \in \Delta_+, w \in W.$$

In this case the structure of this category is rather simple. Applying Proposition 9.3.1 (in which we again replace G by ^LG and $\Omega^{\check{\rho}}$ by $\Omega^{-\rho}$, changing the sign of the residue of the connection), we obtain that

$$\mu^{-1}(\chi) \cap \mathrm{Conn}(\Omega^{-\rho})_D^{w(\lambda+\rho)-\rho}$$

consists of a single point, which we denote by $\overline{\nabla}_w$. By (9.4.6), this implies that there are as many points in $\mu^{-1}(\chi)$ in this case as the number of elements in the Weyl group W, one in each of the subspaces $\mathrm{Conn}(\Omega^{-\rho})_D^{w(\lambda+\rho)-\rho}$ (being generic, λ is automatically regular). Therefore the category $\widehat{\mathfrak{g}}_{\kappa_c}\text{-mod}_\chi^{I^0}$ has $|W|$ inequivalent Wakimoto modules $W_{\overline{\nabla}_w}, w \in W$, in this case.

For each $w \in W$ we have a non-zero homomorphism $\mathbb{M}_{w(\lambda+\rho)-\rho,\kappa_c} \to W_{\overline{\nabla}_w}$. In the special case when these modules are $\mathbb{Z}$-graded (which means that χ is invariant under the vector field $L_0 = -t\partial_t$), we have shown in Section 6.4 that all of them are irreducible and this homomorphism is an isomorphism. Therefore it follows the same is true for generic χ (we expect that this is true for all χ with the residue $\varpi(-\lambda-\rho)$ with λ as above). We find that in this case the category $\widehat{\mathfrak{g}}_{\kappa_c}\text{-mod}_\chi^{I^0}$ has $|W|$ non-isomorphic irreducible objects $W_{\overline{\nabla}_w}, w \in W$. Analyzing the action of the Cartan subalgebra $\mathfrak{h} \otimes 1$ on them, we find that there can be no non-trivial extensions between them. Thus, any object of the category $\widehat{\mathfrak{g}}_{\kappa_c}\text{-mod}_\chi^{I^0}$ is a direct sum of copies of the irreducible Wakimoto modules $W_{\overline{\nabla}_w}, w \in W$.

For a general $\lambda \in \mathfrak{h}^*$ the structure of the category $\widehat{\mathfrak{g}}_{\kappa_c}\text{-mod}_\chi^{I^0}$ with $\chi \in \mathrm{Op}^{\mathrm{RS}}_{^LG}(D)_{\varpi(-\lambda-\rho)}$ is intermediate between the two extreme cases analyzed above.

9.5 Endomorphisms of the Verma modules

In this section and the next we identify the algebras of endomorphisms of some induced $\widehat{\mathfrak{g}}$-modules of critical level with algebras of functions on subvarieties of $\mathrm{Op}_{^LG}(D^\times)$ constructed earlier in this chapter.

Consider first the case of the vacuum module $V_{\kappa_c}(\mathfrak{g})$. According to formula (3.3.3), we have

$$\mathrm{End}_{\widehat{\mathfrak{g}}_{\kappa_c}} V_{\kappa_c}(\mathfrak{g}) \simeq \mathfrak{z}(\widehat{\mathfrak{g}}).$$

Therefore, by Theorem 4.3.2, we have the isomorphism

$$\mathrm{End}_{\widehat{\mathfrak{g}}_{\kappa_c}} V_{\kappa_c}(\mathfrak{g}) \simeq \mathrm{Fun}\,\mathrm{Op}_{^LG}(D). \tag{9.5.1}$$

Moreover, we find that the homomorphism $Z(\widehat{\mathfrak{g}}) \to \mathrm{End}_{\widehat{\mathfrak{g}}_{\kappa_c}} V_{\kappa_c}(\mathfrak{g})$ is surjective and corresponds to the homomorphism

$$\mathrm{Fun}\,\mathrm{Op}_{^LG}(D^\times) \to \mathrm{Fun}\,\mathrm{Op}_{^LG}(D).$$

Thus, $V_{\kappa_c}(\mathfrak{g})$ corresponds to the subvariety

$$\mathrm{Op}_{^LG}(D) \subset \mathrm{Op}_{^LG}(D^\times).$$

Here we generalize this result to the case of Verma modules, which turn out to correspond to the subvarieties $\mathrm{Op}^{\mathrm{RS}}_{^LG}(D)_\varpi$, and in the next section to the case of Weyl modules, which correspond to $\mathrm{Op}^{\check{\lambda}}_{^LG}$.

9.5.1 Families of Wakimoto modules

We consider first the Verma modules. Recall that a **Verma module** of level κ and highest weight $\lambda \in \mathfrak{h}^*$ is defined as the induced module (6.3.9). Since we will consider exclusively the critical level $\kappa = \kappa_c$, we will drop the subscript κ_c from the notation and write simply $\mathbb{M}_\lambda$.

Note that $\mathbb{M}_\lambda = \mathrm{Ind}(M_\lambda)$, where Ind is the induction functor introduced in Section 9.4.2.

We wish to describe the algebra of endomorphisms of $\mathbb{M}_\lambda$ for all highest weights $\lambda \in \mathfrak{h}^*$. In order to do this, we will identify each Verma module with a particular Wakimoto module, for which the algebra of endomorphisms is easy to compute. An example of such an identification is given by Proposition 6.3.3. This proposition suggests that we need to use more general

Wakimoto modules than the modules $W_\lambda = M_{\mathfrak{g}} \otimes \pi_\lambda$ considered above (see, e.g., Section 9.4.3). More precisely, we will use the Wakimoto obtained by the semi-infinite parabolic induction of Section 6.3 with respect to a Borel subalgebra of $\mathfrak{g}$ of the form $w\mathfrak{b}_- w^{-1}$, where $\mathfrak{b}_-$ is the standard Borel subalgebra that we have used before and w is an element of the Weyl group of $\mathfrak{g}$.

In this section we define a family of Wakimoto modules for each element w of the Weyl group of $\mathfrak{g}$. This generalizes the construction of the family W_λ, which corresponds to the case $w = 1$ (see Section 9.4.3), along the lines of the general construction presented in Section 6.3.

Recall the Weyl algebra $\mathcal{A}^{\mathfrak{g}}$ introduced in Section 5.3.3. We define the $\mathcal{A}^{\mathfrak{g}}$-module $M_{\mathfrak{g}}^w$ as the module generated by the vacuum vector $|0\rangle_w$ such that

$$a_{\alpha,n}|0\rangle_w = a^*_{\alpha,n}|0\rangle_w = 0, \qquad n > 0,$$

$$a_{\alpha,0}|0\rangle_w = 0, \quad \alpha \in \Delta_+ \cap w^{-1}(\Delta_+); \qquad a^*_{\alpha,0}|0\rangle_w = 0, \quad \alpha \in \Delta_+ \cap w^{-1}(\Delta_-).$$

According to Theorem 8.3.3, for each $\lambda \in \mathfrak{h}^*$ the tensor product $M_{\mathfrak{g}}^w \otimes \operatorname{Fun}\operatorname{Conn}(\Omega^{-\rho})^\lambda_D$ is a $\widehat{\mathfrak{g}}_{\kappa_c}$-module. Denote by ω the corresponding homomorphism

$$\widehat{\mathfrak{g}}_{\kappa_c} \to \operatorname{End}(M_{\mathfrak{g}}^w \otimes \operatorname{Fun}\operatorname{Conn}(\Omega^{-\rho})^\lambda_D).$$

We define a new $\widehat{\mathfrak{g}}_{\kappa_c}$-module, denoted by W_λ^w, as the w^{-1}-twist of $M_{\mathfrak{g}}^w \otimes \operatorname{Fun}\operatorname{Conn}(\Omega^{-\rho})^\lambda_D$. More precisely, we define a new action of $\widehat{\mathfrak{g}}_{\kappa_c}$ by the formula $x \mapsto \omega(\operatorname{Ad}\widetilde{w}^{-1} \cdot x)$, where $\widetilde{w}$ is the Tits lifting of w to the group G.

In particular, W_λ^1 is the module W_λ introduced in Section 9.4.3.

Recall that given a $\widehat{\mathfrak{g}}_{\kappa_c}$-module on which the operator $L_0 = -t\partial_t$ and the Lie subalgebra $\mathfrak{h} \otimes 1 \subset \widehat{\mathfrak{g}}_\kappa$ act diagonally with finite-dimensional common eigenspaces, we define the character of M as the formal series

$$\operatorname{ch} M = \sum_{\widehat{\lambda} \in (\mathfrak{h} \oplus \mathbb{C}L_0)^*} \dim M(\widehat{\lambda})\, e^{\widehat{\lambda}}, \tag{9.5.2}$$

where $M(\widehat{\lambda})$ is the eigenspace of L_0 and $\mathfrak{h} \otimes 1$ corresponding to the weight $\widehat{\lambda} : (\mathfrak{h} \oplus \mathbb{C}L_0) \to \mathbb{C}$.

Let us compute the character of the module W_λ^w. The operator L_0 on $M_{\mathfrak{g}}^w \otimes \operatorname{Fun}\operatorname{Conn}(\Omega^{-\rho})^\lambda_D$ is the sum of the corresponding operators on $M_{\mathfrak{g}}^w$ and $\operatorname{Fun}\operatorname{Conn}(\Omega^{-\rho})^\lambda_D$. The former is defined by the commutation relations

$$[L_0, a_{\alpha,n}] = -n a_{\alpha,n}, \qquad [L_0, a^*_{\alpha,n}] = -n a^*_{\alpha,n},$$

and the condition that $L_0|0\rangle_w = 0$. The latter is determined from the action of the vector field $-t\partial_t$ on the space $\operatorname{Conn}(\Omega^{-\rho})^\lambda_D$. If we write an element of $\operatorname{Conn}(\Omega^{-\rho})^\lambda_D$ in the form (9.4.4), and expand $u_i(t) = \langle \alpha_i, \mathbf{u}(t)\rangle$ in the series

$$u_i(t) = \sum_{n<0} u_{i,n} t^{-n-1},$$

we obtain an identification

$$\operatorname{Fun}\operatorname{Conn}(\Omega^{-\rho})_D^\lambda \simeq \mathbb{C}[u_{i,n}]_{i=1,\ldots,\ell;n<0}.$$

Then L_0 is just the derivation of this polynomial algebra completely determined by the formulas

$$L_0 \cdot u_{i,n} = -nu_{i,n}.$$

Next, we consider the action of $\mathfrak{h} = \mathfrak{h} \otimes 1 \subset \widehat{\mathfrak{g}}_{\kappa_c}$ on $M_{\mathfrak{g}}^w$. It follows from the original formulas given in Theorem 6.1.6 that for any $h \in \mathfrak{h}$ we have

$$[h, a_{\alpha,n}] = \langle \alpha, h\rangle a_{\alpha,n}, \qquad [h, a^*_{\alpha,n}] = -\langle \alpha, h\rangle a^*_{\alpha,n}.$$

After we twist the action of $\widehat{\mathfrak{g}}_{\kappa_c}$, and hence of $\mathfrak{h}$, by w we obtain new relations

$$[h, a_{\alpha,n}] = \langle w(\alpha), h\rangle a_{\alpha,n}, \qquad [h, a^*_{\alpha,n}] = -\langle w(\alpha), h\rangle a^*_{\alpha,n}. \tag{9.5.3}$$

Note that for any $\mu \in \mathfrak{h}^*$ we have $\langle \mu, w^{-1}(h)\rangle = \langle w(\mu), h\rangle$.

To compute the action of $\mathfrak{h}$ on $|0\rangle_w \otimes 1 \in M_{\mathfrak{g}}^w \otimes 1$, observe that according to the formula for the action of $h_{i,0}$ given in Theorem 6.1.6 we have

$$\begin{aligned}
h_{i,0} \cdot |0\rangle_w \otimes 1 &= \left(\langle w(\lambda), h_i\rangle - \sum_{\beta\in\Delta_+} \langle w(\beta), h_i\rangle a^*_{\beta,0} a_{\beta,0} \right) |0\rangle_w \otimes 1 \\
&= \left(\langle w(\lambda), h_i\rangle - \sum_{\beta\in\Delta_+\cap w^{-1}(\Delta_-)} \langle w(\beta), h_i\rangle \right) |0\rangle_w \otimes 1 \\
&= \langle w(\lambda) - w(w^{-1}(\rho) - \rho), h_i\rangle |0\rangle_w \otimes 1 \\
&= \langle w(\lambda+\rho) - \rho, h_i\rangle |0\rangle_w \otimes 1.
\end{aligned}$$

Therefore the $\mathfrak{h}$-weight of the vector $|0\rangle_w \otimes 1$ is equal to $w(\lambda+\rho) - \rho$.

Putting all of this together, we obtain the following formula for the character of the module W_λ^w:

$$\operatorname{ch} W_\lambda^w = e^{w(\lambda+\rho)-\rho} \prod_{\alpha\in\widehat{\Delta}_+} (1 - e^{-\alpha})^{-\operatorname{mult}(\alpha)}, \tag{9.5.4}$$

where $\widehat{\Delta}_+$ is the set of positive roots of $\widehat{\mathfrak{g}}$. Here the terms corresponding to the real roots come from the generators $a_{\alpha,n}$ and $a^*_{\alpha,n}$ of $\mathcal{A}^{\mathfrak{g}}$, and the terms corresponding to the imaginary roots come from the generators $u_{i,n}$ of $\operatorname{Fun}\operatorname{Conn}(\Omega^{-\rho})_D^\lambda$. Thus we obtain that the character of W_λ^w is equal to the character of the Verma module $\mathbb{M}_{w(\lambda+\rho)-\rho}$ with the highest weight $w(\lambda+\rho)-\rho$, whose character has been computed in the course of the proof of Proposition 6.3.3.

9.5.2 Verma modules and Wakimoto modules

Observe now that any weight ν of $\mathfrak{g}$ may be written (possibly, in different ways) in the form $\nu = w(\lambda+\rho)-\rho$, where λ has the property that $\langle\lambda,\check{\alpha}\rangle \notin \mathbb{Z}_+$ for any $\alpha \in \Delta_+$. Note that w is uniquely determined if $\nu+\rho$ is in the Weyl group orbit of a regular dominant integral weight. The following assertion establishes an isomorphism between the Verma module $\mathbb{M}_{w(\lambda+\rho)-\rho}$ and a Wakimoto module of a particular type.

Proposition 9.5.1 *For any weight λ such that $\langle\lambda,\check{\alpha}\rangle \notin \mathbb{Z}_+$ for all $\alpha \in \Delta_+$, and any element w of the Weyl group of $\mathfrak{g}$, the Verma module $\mathbb{M}_{w(\lambda+\rho)-\rho}$ is isomorphic to the Wakimoto module $W^w_{w(\lambda+\rho)-\rho}$.*

Proof. In the case when $w = w_0$ and $\lambda = -2\rho$ this is proved in Proposition 6.3.3 (note that $W^+_{0,\kappa_c} = W^{w_0}_{-2\rho}$). We will use a similar argument in general.

First, observe that the character formula implies that the vector $|0\rangle_w \otimes 1 \in W^w_{w(\lambda+\rho)-\rho}$ is a highest weight vector of highest weight $w(\lambda+\rho)-\rho$. Therefore there is a canonical homomorphism

$$\mathbb{M}_{w(\lambda+\rho)-\rho} \to W^w_{w(\lambda+\rho)-\rho}$$

taking the highest weight vector of $\mathbb{M}_{w(\lambda+\rho)-\rho}$ to $|0\rangle_w \otimes 1 \in W^w_{w(\lambda+\rho)-\rho}$. Since the characters of $\mathbb{M}_{w(\lambda+\rho)-\rho}$ and $W^w_{w(\lambda+\rho)-\rho}$ are equal, the proposition will follow if we show that this homomorphism is surjective, or, equivalently, that $W^w_{w(\lambda+\rho)-\rho}$ is generated by its highest weight vector $|0\rangle_w \otimes 1$.

Suppose that $W^w_{w(\lambda+\rho)-\rho}$ is not generated by the highest weight vector. Then there exists a homogeneous linear functional on $W^w_{w(\lambda+\rho)-\rho}$, whose weight is less than the highest weight $(\lambda, 0) \in \mathfrak{h}^* \oplus (\mathbb{C}L_0)^*$ and which is invariant under the Lie subalgebra

$$\widehat{\mathfrak{n}}_- = \mathfrak{n}_- \otimes \mathbb{C}[t^{-1}] \bigoplus \mathfrak{b}_+ \otimes \mathbb{C}[t^{-1}],$$

and in particular, under its Lie subalgebra

$$L_-\mathfrak{n}^w_+ = (w\mathfrak{n}_+w^{-1}) \cap \mathfrak{n}_- \otimes \mathbb{C}[t^{-1}] \bigoplus (w\mathfrak{n}_+w^{-1}) \cap \mathfrak{n}_+ \otimes t^{-1}\mathbb{C}[t^{-1}].$$

Therefore this functional factors through the space of $L_-\mathfrak{n}^w_+$-coinvariants of $W^w_{w(\lambda+\rho)-\rho}$. But it follows from the formulas for the action of $\widehat{\mathfrak{g}}_{\kappa_c}$ on $W^w_{w(\lambda+\rho)-\rho}$ that the action of the Lie subalgebra $L_-\mathfrak{n}^w_+$ is free, and the space of coinvariants is isomorphic to

$$\mathbb{C}[a^*_{\alpha,n}]_{\alpha\in\Delta_+, n\leq 0} \otimes \operatorname{Fun}\operatorname{Conn}(\Omega^{-\rho})_D^{w_0(\lambda+\rho)-\rho}.$$

Our functional has to take a non-zero value on a weight vector in this subspace of weight other than the highest weight $w(\lambda+\rho)-\rho$. Using formula (9.5.3), we obtain that the weight of $a^*_{\alpha,n}$ is $n - w(\alpha)$. Therefore the weights of vectors in this subspace are of the form

$$(w(\lambda+\rho)-\rho)+N\delta-\sum_j \gamma_j, \tag{9.5.5}$$

where $N \leq 0$ and each γ_j is a root in $w(\Delta_+)$. If such a functional exists, then $W^w_{w(\lambda+\rho)-\rho}$ has an irreducible subquotient of such highest weight. Since the characters of irreducible highest weight modules are linearly independent, and the characters of $W^w_{w(\lambda+\rho)-\rho}$ and $\mathbb{M}_{w(\lambda+\rho)-\rho}$ are equal, we obtain that then $\mathbb{M}_{w(\lambda+\rho)-\rho}$ must also have an irreducible subquotient whose highest weight is of the form (9.5.5).

Now recall the Kac–Kazhdan theorem [KK] describing the set of highest weights of irreducible subquotients of Verma modules (see Proposition 6.3.3). A weight $\widehat{\mu}=(\mu,n)$ appears in the decomposition of $\mathbb{M}_{\widehat{\nu}}$, where $\widehat{\nu}=(\nu,0)$ if and only $n \leq 0$ and either $\mu=0$ or there exists a finite sequence of weights $\mu_0,\ldots,\mu_m \in \mathfrak{h}^*$ such that $\mu_0=\mu, \mu_m=\nu$, $\mu_{i+1}=\mu_i \pm m_i\beta_i$ for some positive roots β_i and positive integers m_i which satisfy

$$\langle \mu_i+\rho,\check{\beta}_i\rangle = m_i. \tag{9.5.6}$$

Let us denote the set of such weights $\widehat{\mu}=(\mu,n)$ with a fixed weight $\widehat{\nu}=(\nu,0)$ by $S(\nu)$. It follows from the above description that if $(\mu,n)\in S(\nu)$, then $n\leq 0$, μ belongs the W-orbit of ν with respect to the ρ-shifted action, and moreover for any $\mu \in S(\nu)$, the difference $\mu-\nu$ is a linear combination of roots with integer coefficients.

Suppose now that a weight λ satisfies the property in the statement of the proposition that $\langle\lambda,\check{\alpha}\rangle \notin \mathbb{Z}_+$ for any $\alpha\in\Delta_+$. This implies that if $\mu = y(\lambda+\rho)-\rho$ and $\mu-\lambda=\sum_{i\in I} n_i\alpha_i$, where $n_i\in\mathbb{Z}$, then we necessarily have $n_i \geq 0$. Therefore for any weight $(\mu,0)\in S(\lambda)$ we have $\mu=\lambda+\sum_j \beta_j$, where $\beta_j\in\Delta_+$. This implies that if $\widehat{\mu}'=(\mu',n)\in S(w(\lambda+\rho)-\rho)$, then we have

$$\widehat{\mu}'=(w(\lambda+\rho)-\rho)-N\delta+\sum_j \beta_j, \qquad \beta_j\in w(\Delta_+), N\leq 0.$$

Such a weight cannot be of the form (9.5.5), unless it is $w(\lambda+\rho)-\rho$. Therefore $W^w_{w(\lambda+\rho)-\rho}$ is generated by its highest weight vector and hence $W^w_{w(\lambda+\rho)-\rho}$ is isomorphic to $\mathbb{M}_{w(\lambda+\rho)-\rho}$. □

9.5.3 Description of the endomorphisms of Verma modules

We now use Proposition 9.5.1 to describe the algebra $\mathrm{End}_{\widehat{\mathfrak{g}}_{\kappa_c}} \mathbb{M}_\nu$ of endomorphisms of the Verma module $\mathbb{M}_\nu$. Observe that since $\mathbb{M}_\nu$ is freely generated from its highest weight vector, such an endomorphism is uniquely determined by the image of the highest weight vector vector. The image of the highest weight vector can be an arbitrary $\widehat{\mathfrak{n}}_+$-invariant vector which has weight ν with respect to the Cartan subalgebra. We denote the space of such vectors by $(\mathbb{M}_\nu)_\nu^{\widehat{\mathfrak{n}}_+}$. Using the same argument as in the proof of the isomorphism (3.3.3), we obtain that

$$\mathrm{End}_{\widehat{\mathfrak{g}}_{\kappa_c}} \mathbb{M}_\nu \simeq (\mathbb{M}_\nu)_\nu^{\widehat{\mathfrak{n}}_+}. \tag{9.5.7}$$

Thus, for any λ that satisfies the conditions of Proposition 9.5.1 we obtain the following isomorphisms of vector spaces:

$$\begin{aligned}\mathrm{End}_{\widehat{\mathfrak{g}}_{\kappa_c}} \mathbb{M}_{w(\lambda+\rho)-\rho} &\simeq (\mathbb{M}_{w(\lambda+\rho)-\rho})^{\widehat{\mathfrak{n}}_+}_{w(\lambda+\rho)-\rho} \\ &\simeq (W^w_{w(\lambda+\rho)-\rho})^{\widehat{\mathfrak{n}}_+}_{w(\lambda+\rho)-\rho}.\end{aligned} \tag{9.5.8}$$

Lemma 9.5.2 *For any $\lambda \in \mathfrak{h}^*$ the space of $\widehat{\mathfrak{n}}_+$-invariant vectors of weight $w(\lambda+\rho)-\rho$ in $W^w_{w(\lambda+\rho)-\rho}$ is equal to its subspace $|0\rangle_w \otimes \mathrm{Fun}\,\mathrm{Conn}(\Omega^{-\rho})^\lambda_D$.*

Proof. It follows from the formulas of Theorem 6.1.6 for the action of $\widehat{\mathfrak{g}}_{\kappa_c}$ on the Wakimoto modules that all vectors in the subspace $|0\rangle_w \otimes \mathrm{Fun}\,\mathrm{Conn}(\Omega^{-\rho})^{w(\lambda+\rho)-\rho}_D$ are annihilated by $\widehat{\mathfrak{n}}_+$ and have weight $w(\lambda+\rho)-\rho$. Let us show that there are no other $\widehat{\mathfrak{n}}_+$-invariant vectors in $W^w_{w(\lambda+\rho)-\rho}$ of this weight.

An $\widehat{\mathfrak{n}}_+$-invariant vector is in particular annihilated by the Lie subalgebra

$$L_+\mathfrak{n}^w_+ = (w\mathfrak{n}_+w^{-1} \cap \mathfrak{n}_-) \otimes t\mathbb{C}[t] \bigoplus (w\mathfrak{n}_+w^{-1} \cap \mathfrak{n}_+) \otimes \mathbb{C}[t].$$

But it follows from the formulas of Theorem 6.1.6 that this Lie algebra acts cofreely on $M^w_{\mathfrak{g}} \otimes \mathrm{Fun}\,\mathrm{Conn}(\Omega^{-\rho})^\lambda_D$ and so the character of the space of $L_+\mathfrak{n}^w_+$-invariants is equal to the ratio between the character of $W^w_{w(\lambda+\rho)-\rho}$ and the character of the restricted dual of the universal enveloping algebra $U(L_+\mathfrak{n}^w_+)$. This ratio is equal to

$$e^{w(\lambda+\rho)-\rho} \prod_{n>0} (1-e^{-n\delta})^{-\ell} \prod_{\alpha \in w(\Delta_-)\cap\Delta_-, n\geq 0} (1-e^{-n\delta-\alpha})^{-1} \times \\ \times \prod_{\alpha\in w(\Delta_-)\cap\Delta_+, n>0} (1-e^{-n\delta+\alpha})^{-1}.$$

The character of the weight $w(\lambda+\rho)-\rho$ subspace of the space of $L_+\mathfrak{n}^w_+$-invariants in $M^w_{\mathfrak{g}} \otimes \mathrm{Fun}\,\mathrm{Conn}(\Omega^{-\rho})^\lambda_D$ is therefore equal to

$$e^{w(\lambda+\rho)-\rho} \prod_{n>0} (1-e^{-n\delta})^{-\ell},$$

which is precisely the character of the subspace $|0\rangle_w \otimes \mathrm{Fun}\,\mathrm{Conn}(\Omega^{-\rho})^\lambda_D$. This completes the proof. □

Recall from Theorem 8.3.3 that the action of the center $Z(\widehat{\mathfrak{g}})$ on $W^w_{w(\lambda+\rho)-\rho}$ factors through the homomorphism

$$Z(\widehat{\mathfrak{g}}) \simeq \mathrm{Fun}\,\mathrm{Op}_{{}^L G}(D^\times) \to \mathrm{Fun}\,\mathrm{Conn}(\Omega^{-\rho})_{D^\times}$$

induced by the Miura transformation. Therefore it factors through the homomorphism

$$\operatorname{Fun}\operatorname{Op}_{^LG}(D^\times) \to \operatorname{Fun}\operatorname{Conn}(\Omega^{-\rho})^\lambda_D.$$

But we know from the discussion of Section 9.3.1 that this homomorphism factors through the homomorphism

$$\operatorname{Fun}\operatorname{Op}^{\mathrm{RS}}_{^LG}(D)_{\varpi(-\lambda-\rho)} \to \operatorname{Fun}\operatorname{Conn}(\Omega^{-\rho})^\lambda_D.$$

Suppose that λ satisfies the condition of Proposition 9.5.1. Then this map is an isomorphism by Proposition 9.3.1.† Combining this with the isomorphism (9.5.8) and the statement of Lemma 9.5.2, we obtain the following result (see [FG2], Corollary 13.3.2).

Theorem 9.5.3 *The center* $Z(\widehat{\mathfrak{g}})$ *maps surjectively onto* $\operatorname{End}_{\widehat{\mathfrak{g}}_{\kappa_c}} \mathbb{M}_\nu$*, and we have a commutative diagram*

$$\begin{array}{ccc} Z(\widehat{\mathfrak{g}}) & \xrightarrow{\sim} & \operatorname{Fun}\operatorname{Op}_{^LG}(D^\times) \\ \downarrow & & \downarrow \\ \operatorname{End}_{\widehat{\mathfrak{g}}_{\kappa_c}} \mathbb{M}_\nu & \xrightarrow{\sim} & \operatorname{Fun}\operatorname{Op}^{\mathrm{RS}}_{^LG}(D)_{\varpi(-\nu-\rho)} \end{array}$$

for any weight ν.

Furthermore, the Verma module $\mathbb{M}_\nu$ *is free over the algebra*

$$\operatorname{End}_{\widehat{\mathfrak{g}}_{\kappa_c}} \mathbb{M}_\nu \simeq \operatorname{Fun}\operatorname{Op}^{\mathrm{RS}}_{^LG}(D)_{\varpi(-\nu-\rho)}.$$

9.6 Endomorphisms of the Weyl modules

Let V_λ be a finite-dimensional representation of the Lie algebra $\mathfrak{g}$ with highest weight λ, which is a dominant integral weight of $\mathfrak{g}$. We extend the action of $\mathfrak{g}$ to an action of $\mathfrak{g} \otimes \mathbb{C}[[t]] \oplus \mathbf{1}$ in such a way that $\mathfrak{g} \otimes t\mathbb{C}[[t]]$ acts trivially and $\mathbf{1}$ acts as the identity. Let $\mathbb{V}_{\lambda,\kappa}$ be the induced $\widehat{\mathfrak{g}}_\kappa$-module

$$\mathbb{V}_{\lambda,\kappa} = U(\widehat{\mathfrak{g}}_\kappa) \underset{U(\mathfrak{g}[[t]]\oplus\mathbb{C}\mathbf{1})}{\otimes} V_\lambda.$$

This is the **Weyl module** of level κ with highest weight λ.

In what follows we will only consider the Weyl modules of critical level $\kappa = \kappa_c$, so we will drop the subscript κ_c and denote them simply by $\mathbb{V}_\lambda$. In particular, $\mathbb{V}_0 = V_{\kappa_c}(\mathfrak{g})$, the vacuum module.

We note that $\mathbb{V}_\lambda = \operatorname{Ind}(V_\lambda)$, where Ind is the induction functor introduced in Section 9.4.2. Since V_λ is the finite-dimensional quotient of the Verma module M_λ over $\mathfrak{g}$, and we have $\operatorname{Ind}(M_\lambda) = \mathbb{M}_\lambda$, we obtain a homomorphism

$$\mathbb{M}_{\lambda,\kappa_c} \to \mathbb{V}_\lambda.$$

† Note that in order to apply the statement of Proposition 9.3.1 we need to replace G by ^LG and $\Omega^{\check{\rho}}$ by $\Omega^{-\rho}$, which changes the sign of the residue of the connection.

It is clear from the definitions that this homomorphism is surjective, and so the Weyl module $\mathbb{V}_\lambda$ is realized as a quotient of the Verma module $\mathbb{M}_\lambda$.

9.6.1 Statement of the result

Recall the space $\mathrm{Op}^{\lambda}_{^L G}$ of $^L G$-opers with regular singularity, residue $\varpi(-\rho-\lambda)$ and trivial monodromy, introduced in Section 9.2.3. In this section we will prove the following result, which generalizes the isomorphism (9.5.1) corresponding to $\lambda = 0$. For $\mathfrak{g} = \mathfrak{sl}_2$ it was proved in [F3], and for an arbitrary $\mathfrak{g}$ in [FG6] (the proof presented below is similar to that proof). It was also independently conjectured by Beilinson and Drinfeld (unpublished).

Theorem 9.6.1 *For any dominant integral weight λ the center $Z(\widehat{\mathfrak{g}})$ maps surjectively onto $\mathrm{End}_{\widehat{\mathfrak{g}}_{\kappa_c}} \mathbb{V}_\lambda$, and we have the following commutative diagram*

$$\begin{array}{ccc} Z(\widehat{\mathfrak{g}}) & \xrightarrow{\sim} & \mathrm{Fun}\,\mathrm{Op}_{^L G}(D^\times) \\ \downarrow & & \downarrow \\ \mathrm{End}_{\widehat{\mathfrak{g}}_{\kappa_c}} \mathbb{V}_\lambda & \xrightarrow{\sim} & \mathrm{Fun}\,\mathrm{Op}^{\lambda}_{^L G} \end{array}$$

Our strategy of the proof will be as follows: we will first construct a homomorphism $\mathbb{V}_\lambda \to W_\lambda$. Thus, we obtain the following maps:

$$\mathbb{M}_\lambda \twoheadrightarrow \mathbb{V}_\lambda \to W_\lambda, \tag{9.6.1}$$

and the corresponding maps of the algebras of endomorphisms

$$\mathrm{End}_{\widehat{\mathfrak{g}}_{\kappa_c}} \mathbb{M}_\lambda \to \mathrm{End}_{\widehat{\mathfrak{g}}_{\kappa_c}} \mathbb{V}_\lambda \to \mathrm{End}_{\widehat{\mathfrak{g}}_{\kappa_c}} W_\lambda. \tag{9.6.2}$$

According to Theorem 9.5.3 and the argument used in Lemma 6.3.4, the compsite map is the homomorphism of the algebras of functions corresponding to the morphism

$$\mu_\lambda : \mathrm{Conn}(\Omega^{-\rho})^{\lambda}_{D} \to \mathrm{Op}^{\mathrm{RS}}_{^L G}(D)_{\varpi(-\lambda-\rho)},$$

obtained by restriction from the Miura transformation. We therefore find that the image of this composite map is precisely $\mathrm{Fun}\,\mathrm{Op}^{\lambda}_{^L G} \subset \mathrm{Fun}\,\mathrm{Conn}(\Omega^{-\rho})^{\lambda}_{D} = \mathrm{End}_{\widehat{\mathfrak{g}}_{\kappa_c}} W_\lambda$.

On the other hand, we show that the second map in (9.6.2) is injective and its image is contained in $\mathrm{Fun}\,\mathrm{Op}^{\lambda}_{^L G}$. This implies that $\mathrm{End}_{\widehat{\mathfrak{g}}_{\kappa_c}} \mathbb{V}_\lambda$ is isomorphic to $\mathrm{Fun}\,\mathrm{Op}^{\lambda}_{^L G}$.

9.6.2 Weyl modules and Wakimoto modules

Let us now proceed with the proof and construct a homomorphism $\mathbb{V}_\lambda \to W_\lambda$. Recall that the Wakimoto module W_λ was defined by formula (9.4.5):

$$W_\lambda = M_{\mathfrak{g}} \otimes \mathrm{Fun}\,\mathrm{Conn}(\Omega^{-\rho})^{\lambda}_{D}. \tag{9.6.3}$$

In order to construct a map $\mathbb{V}_\lambda \to W_\lambda$, we observe (as we did previously, in Sections 6.2.3 and 9.4.1) that the action of the constant subalgebra $\mathfrak{g} \subset \widehat{\mathfrak{g}}_\kappa$ on the subspace

$$W_\lambda^0 = \mathbb{C}[a^*_{\alpha,0}]_{\alpha\in\Delta_+}|0\rangle \otimes 1 \subset M_{\mathfrak{g}} \otimes \operatorname{Fun}\operatorname{Conn}(\Omega^{-\rho})^\lambda_D$$

coincides with the natural action of $\mathfrak{g}$ on the contragredient Verma module M^*_λ realized as $\operatorname{Fun} N_+ = \mathbb{C}[y_\alpha]_{\alpha\in\Delta_+}$, where we substitute $a^*_{\alpha,0} \mapsto y_\alpha$ (see Section 5.2.4) for the definition of this action). In addition, the Lie subalgebra $\mathfrak{g}\otimes t\otimes\mathbb{C}[[t]] \subset \widehat{\mathfrak{g}}_{\kappa_c}$ acts by zero on the subspace W_λ^0, and $\mathbf{1}$ acts as the identity.

Therefore the injective $\mathfrak{g}$-homomorphism $V_\lambda \hookrightarrow M^*_\lambda$ gives rise to a non-zero $\widehat{\mathfrak{g}}_{\kappa_c}$-homomorphism

$$\imath_\lambda : \mathbb{V}_\lambda \to W_\lambda,$$

sending the generating subspace $V_\lambda \subset \mathbb{V}_\lambda$ to the image of V_λ in $W_\lambda^0 \simeq M^*_\lambda$.

Now we obtain a sequence of maps

$$\mathbb{M}_\lambda \twoheadrightarrow \mathbb{V}_\lambda \to W_\lambda. \tag{9.6.4}$$

According to formula (9.5.7), Theorem 9.5.3, and Proposition 9.2.1, we have

$$\operatorname{End}_{\widehat{\mathfrak{g}}_{\kappa_c}} \mathbb{M}_\lambda = (\mathbb{M}_\lambda)_\lambda^{\widehat{\mathfrak{n}}_+} \simeq \operatorname{Fun}\operatorname{Op}^{\mathrm{RS}}_{^LG}(D)_{\varpi(-\lambda-\rho)} \simeq \operatorname{Fun}\operatorname{Op}^{\mathrm{nilp},\lambda}_{^LG}.$$

Next, according to Lemma 9.5.2, we have

$$\operatorname{End}_{\widehat{\mathfrak{g}}_{\kappa_c}} W_\lambda = (W_\lambda)_\lambda^{\widehat{\mathfrak{n}}_+} = \operatorname{Fun}\operatorname{Conn}(\Omega^{-\rho})^\lambda_D.$$

Note that by formula (9.6.3), $\operatorname{Fun}\operatorname{Conn}(\Omega^{-\rho})^\lambda_D = |0\rangle \otimes \operatorname{Fun}\operatorname{Conn}(\Omega^{-\rho})^\lambda_D$ is naturally a subspace of W_λ.

Finally, note that

$$\operatorname{End}_{\widehat{\mathfrak{g}}_{\kappa_c}} \mathbb{V}_\lambda = (\mathbb{V}_\lambda \otimes V_\lambda^*)^{\mathfrak{g}[[t]]} = (\mathbb{V}_\lambda)_\lambda^{\widehat{\mathfrak{n}}_+}.$$

Indeed, any endomorphism of $\mathbb{V}_\lambda$ is uniquely determined by the image of the generating subspace V_λ. This subspace therefore defines a $\mathfrak{g}[[t]]$-invariant element of $(\mathbb{V}_\lambda \otimes V_\lambda^*)^{\mathfrak{g}[[t]]}$ and its highest weight vector gives rise to a vector in $(\mathbb{V}_\lambda)_\lambda^{\widehat{\mathfrak{n}}_+}$. In the same way as before, we obtain that the resulting maps are isomorphisms.

We obtain the following commutative diagram:

$$\begin{array}{ccccc}
(\mathbb{M}_\lambda)_\lambda^{\widehat{\mathfrak{n}}_+} & \longrightarrow & (\mathbb{V}_\lambda)_\lambda^{\widehat{\mathfrak{n}}_+} & \longrightarrow & (W_\lambda)_\lambda^{\widehat{\mathfrak{n}}_+} \\
\downarrow\wr & & \downarrow\wr & & \downarrow\wr \\
\operatorname{End}_{\widehat{\mathfrak{g}}_{\kappa_c}} \mathbb{M}_\lambda & \longrightarrow & \operatorname{End}_{\widehat{\mathfrak{g}}_{\kappa_c}} \mathbb{V}_\lambda & \longrightarrow & \operatorname{End}_{\widehat{\mathfrak{g}}_{\kappa_c}} W_\lambda \\
\downarrow\wr & & \downarrow\wr & & \downarrow\wr \\
\operatorname{Fun}\operatorname{Op}^{\mathrm{nilp},\lambda}_{^LG} & \longrightarrow & ? & \longrightarrow & \operatorname{Fun}\operatorname{Conn}(\Omega^{-\rho})^\lambda_D
\end{array} \tag{9.6.5}$$

This implies that we have a sequence of homomorphisms

$$\mathrm{End}_{\widehat{\mathfrak{g}}_{\kappa_c}} \mathbb{M}_\lambda \to \mathrm{End}_{\widehat{\mathfrak{g}}_{\kappa_c}} \mathbb{V}_\lambda \to \mathrm{End}_{\widehat{\mathfrak{g}}_{\kappa_c}} W_\lambda. \tag{9.6.6}$$

Moreover, it follows that $\mathrm{End}_{\widehat{\mathfrak{g}}_{\kappa_c}} \mathbb{V}_\lambda$ has a subquotient isomorphic to the image of the homomorphism

$$\mathrm{Fun}\,\mathrm{Op}_{^L G}^{\mathrm{nilp},\lambda} \to \mathrm{Fun}\,\mathrm{Conn}(\Omega^{-\rho})_D^\lambda \tag{9.6.7}$$

obtained from the above diagram.

By construction, the last homomorphism fits into a commutative diagram

$$\begin{array}{ccccc} Z(\widehat{\mathfrak{g}}) & \xrightarrow[\sim]{} & \mathrm{End}_{\widehat{\mathfrak{g}}_{\kappa_c}} \mathbb{M}_\lambda & \longrightarrow & \mathrm{End}_{\widehat{\mathfrak{g}}_{\kappa_c}} W_\lambda \\ \downarrow & & \downarrow\wr & & \downarrow\wr \\ \mathrm{Fun}\,\mathrm{Op}_{^L G}(D^\times) & \longrightarrow & \mathrm{Fun}\,\mathrm{Op}_{^L G}^{\mathrm{nilp},\lambda} & \longrightarrow & \mathrm{Fun}\,\mathrm{Conn}(\Omega^{-\rho})_D^\lambda \end{array}$$

and the composition

$$\mathrm{Fun}\,\mathrm{Op}_{^L G}(D^\times) \to \mathrm{Fun}\,\mathrm{Conn}(\Omega^{-\rho})_D^\lambda$$

of the maps on the bottom row corresponds to the natural action of $Z(\widehat{\mathfrak{g}})$ on W_λ. According to Theorem 8.3.3, this map is the Miura transformation map. Therefore the homomorphism (9.6.7) coincides with the homomorphism obtained by restriction of the Miura transformation. The fact that this homomorphism factors through $\mathrm{Fun}\,\mathrm{Op}_{^L G}^{\mathrm{nilp},\lambda}$ simply means that the image of $\mathrm{Conn}(\Omega^{-\rho})_D^\lambda \subset \mathrm{Conn}(\Omega^{-\rho})_{D^\times}$ under the Miura transformation belongs to $\mathrm{Op}_{^L G}^{\mathrm{nilp},\lambda}$. We already know that from formula (9.3.4) and the diagram (9.3.5). Moreover, we obtain from the diagram (9.6.5) that the image of the homomorphism

$$\mathrm{End}_{\widehat{\mathfrak{g}}_{\kappa_c}} \mathbb{V}_\lambda \to \mathrm{End}_{\widehat{\mathfrak{g}}_{\kappa_c}} W_\lambda$$

is nothing but the algebra of functions on the image of $\mathrm{Conn}(\Omega^{-\rho})_D^\lambda$ in $\mathrm{Op}_{^L G}^{\mathrm{nilp},\lambda}$ under the Miura map. Let us find what this image is.

9.6.3 Nilpotent opers and Miura transformation

Recall the space $\mathrm{MOp}_{^L G}^{\lambda,w} \subset \mathrm{MOp}_{^L G}^\lambda$ defined in formula (9.3.11). Consider the case when $w = 1$. The corresponding space $\mathrm{MOp}_{^L G}^{\lambda,1}$ parametrizes pairs $(\chi, \mathcal{F}'_{^L B,0})$, where $\chi = (\mathcal{F}, \nabla, \mathcal{F}_{^L B}) \in \mathrm{Op}_{^L G}^{\mathrm{nilp},\lambda}$ and $\mathcal{F}'_{^L B,0}$ is a $^L B$-reduction of the fiber at $0 \in D$ of the $^L G$-bundle $\mathcal{F}$ underlying χ, which is stable under the residue $\mathrm{Res}_\lambda(\chi)$ and is in generic relative position with the oper reduction $\mathcal{F}_{^L B,0}$. But since $\mathrm{Res}_\lambda(\chi) \in {}^L\mathfrak{n}_{\mathcal{F}_{^L G}}$, we find that such a reduction $\mathcal{F}'_{^L B,0}$ exists only if $\mathrm{Res}_\lambda(\chi) = 0$, so that $\chi \in \mathrm{Op}_{^L G}^\lambda$. The space of such reductions for an oper χ satisfying this property is a torsor over the group ${}^L N_{\mathcal{F}_{^L B},0}$ (it is isomorphic to the big cell of the flag variety ${}^L G/{}^L B$).

Thus, we obtain that the image of the forgetful morphism $\mathrm{MOp}_{^L G}^{\lambda,1} \to$

$\mathrm{Op}^{\mathrm{nilp},\lambda}_{^LG}$ is $\mathrm{Op}^{\lambda}_{^LG} \subset \mathrm{Op}^{\mathrm{nilp},\lambda}_{^LG}$, and $\mathrm{MOp}^{\lambda,1}_{^LG}$ is a principal $^LN_{\mathcal{F}_{^LB,0}}$-bundle over $\mathrm{Op}^{\lambda}_{^LG}$.

On the other hand, according to Theorem 9.3.6, we have an isomorphism

$$\mathrm{Conn}(\Omega^{-\rho})^{\lambda}_{D} \simeq \mathrm{MOp}^{\lambda,1}_{^LG}.$$

and the Miura transformation $\mathrm{Conn}(\Omega^{-\rho})^{\lambda}_{D} \to \mathrm{Op}^{\mathrm{nilp},\lambda}_{^LG}$ coincides with the forgetful map $\mathrm{MOp}^{\lambda,1}_{^LG} \to \mathrm{Op}^{\mathrm{nilp},\lambda}_{^LG}$. Thus, we obtain the following:

Lemma 9.6.2 *The image of* $\mathrm{Conn}(\Omega^{-\rho})^{\lambda}_{D}$ *in* $\mathrm{Op}^{\mathrm{nilp},\lambda}_{^LG}$ *under the Miura transformation is equal to* $\mathrm{Op}^{\lambda}_{^LG}$. *The map* $\mathrm{Conn}(\Omega^{-\rho})^{\lambda}_{D} \to \mathrm{Op}^{\lambda}_{^LG}$ *is a principal* $^LN_{\mathcal{F}_{^LB,0}}$*-bundle over* $\mathrm{Op}^{\lambda}_{^LG}$.

Let $(\mathbb{V}_\lambda)^{\widehat{\mathfrak{n}}_+,0}_{\lambda}$ be the subspace of $(\mathbb{V}_\lambda)^{\widehat{\mathfrak{n}}_+}_{\lambda} = \mathrm{End}_{\widehat{\mathfrak{g}}_{\kappa_c}} \mathbb{V}_\lambda$ which is the image of the top left horizontal homomorphism in (9.6.6). Then we obtain the following commutative diagram:

$$\begin{array}{ccc} (\mathbb{V}_\lambda)^{\widehat{\mathfrak{n}}_+}_{\lambda} & \longrightarrow & \mathrm{Fun}\,\mathrm{Conn}(\Omega^{-\rho})^{\lambda}_{D} \\ \uparrow & & \uparrow \\ (\mathbb{V}_\lambda)^{\widehat{\mathfrak{n}}_+,0}_{\lambda} & \longrightarrow & \mathrm{Fun}\,\mathrm{Op}^{\lambda}_{^LG} \end{array} \tag{9.6.8}$$

where the vertical maps are injective and the bottom horizontal map is surjective. We now wish to prove that the top horizontal map is injective and its image is contained in the image of the right vertical map. This would imply that the top horizontal map and the left vertical map are isomorphisms, and hence we will obtain the statement of Theorem 9.6.1.

As the first step, we will now obtain an explicit realization of

$$\mathrm{Fun}\,\mathrm{Op}^{\lambda}_{^LG} \subset \mathrm{Fun}\,\mathrm{Conn}(\Omega^{-\rho})^{\lambda}_{D}$$

as the intersection of kernels of some screening operators, generalizing the description in the case $\lambda = 0$ obtained in Section 8.2.3.

According to Lemma 9.6.2, $\mathrm{Fun}\,\mathrm{Op}^{\lambda}_{^LG}$ is realized as the subalgebra of the algebra $\mathrm{Fun}\,\mathrm{Conn}(\Omega^{-\rho})^{\lambda}_{D}$ which consists of the $^LN_{\mathcal{F}_{^LB,0}}$-invariant, or equivalently, $^L\mathfrak{n}_{^LB,0}$-invariant functions. An element of $\mathrm{Conn}(\Omega^{-\rho})^{\lambda}_{D}$ is represented by an operator of the form

$$\partial_t + \frac{\lambda}{t} + \mathbf{u}(t), \qquad \mathbf{u}(t) = \sum_{n \geq 0} \mathbf{u}_n t^{-n-1} \in {}^L\mathfrak{h}[[t]]. \tag{9.6.9}$$

Therefore

$$\mathrm{Fun}\,\mathrm{Conn}(\Omega^{-\rho})^{\lambda}_{D} = \mathbb{C}[u_{i,m}]_{i=1,\ldots,\ell; m<0}, \tag{9.6.10}$$

where $u_{i,m} = \langle \check{\alpha}_i, \mathbf{u}_m \rangle$. Here and below we are using a canonical identification $^L\mathfrak{h} = \mathfrak{h}^*$. We now compute explicitly the action of generators of $^L\mathfrak{n}_{^LB,0}$ on this space, by generalizing our computation in Section 8.2.3 (which corresponds to the case $\lambda = 0$).

For that we fix a Cartan subalgebra ${}^L\mathfrak{h}$ in ${}^L\mathfrak{b}$ and a trivialization of $\mathcal{F}_{^L B,0}$ such that ${}^L\mathfrak{n}_{\mathcal{F}_{^L B,0}}$ is identified with ${}^L\mathfrak{n}$, and ${}^L\mathfrak{b}_{\mathcal{F}'_{^L B,0}}$ is identified with ${}^L\mathfrak{b}_-$. Next, choose generators $e_i, i = 1, \ldots, \ell$, of ${}^L\mathfrak{n}$ with respect to the action of ${}^L\mathfrak{h}$ on ${}^L\mathfrak{n}$ in such a way that together with the previously chosen f_i they satisfy the standard relations of ${}^L\mathfrak{g}$. The ${}^L\mathfrak{n}_{\mathcal{F}_{^L B,0}}$-action on $\operatorname{Conn}(\Omega^{-\rho})_D^\lambda$ then gives rise to an infinitesimal action of the e_i's on $\operatorname{Conn}(\Omega^{-\rho})_D^\lambda$. We will now compute the corresponding derivation on $\operatorname{Fun}\operatorname{Conn}(\Omega^{-\rho})_D^\lambda$.

With respect to our trivialization of $\mathcal{F}_{^L B,0}$ the oper connection obtained by applying the Miura transformation to (9.6.9) reads as follows:†

$$\partial_t + \sum_{i=1}^{\ell} t^{\lambda_i} f_i - \mathbf{u}(t), \qquad \mathbf{u}(t) = \sum_{n \geq 0} \mathbf{u}_n t^{-n-1} \in {}^L\mathfrak{h}[[t]], \tag{9.6.11}$$

where $\lambda_i = \langle \check{\alpha}_i, \lambda \rangle$.

The infinitesimal gauge action of e_i on this connection operator is given by the formula

$$\delta \mathbf{u}(t) = -[x_i(t) \cdot e_i, \partial_t + \sum_{i=1}^{\ell} t^{\lambda_i} f_i - \mathbf{u}(t)], \tag{9.6.12}$$

where $x_i(t) \in \mathbb{C}[[t]]$ is such that $x_i(0) = 1$, and the right hand side of formula (9.6.12) belongs to ${}^L\mathfrak{h}[[t]]$. These conditions determine $x_i(t)$ uniquely. Indeed, the right hand side of (9.6.12) reads

$$-t^{\lambda_i} x_i(t) \cdot \alpha_i - u_i(t) x_i(t) \cdot e_i + \partial_t x_i(t) \cdot e_i,$$

where $u_i(t) = \langle \check{\alpha}_i, \mathbf{u}(t) \rangle$. Therefore it belongs to ${}^L\mathfrak{h}[[t]]$ if and only if

$$\partial_t x_i(t) = u_i(t) x_i(t). \tag{9.6.13}$$

If we write

$$x_i(t) = \sum_{n \leq 0} x_{i,n} t^{-n},$$

and substitute it into formula (9.6.13), we obtain that the coefficients $x_{i,n}$ satisfy the following recurrence relation:

$$n x_{i,n} = - \sum_{k+m=n; k<0; m \leq 0} u_{i,k} x_{i,m}, \qquad n < 0.$$

We find from this formula and the condition $x_i(0) = 1$ that

$$\sum_{n \leq 0} x_{i,n} t^{-n} = \exp \left(\sum_{m>0} \frac{u_{i,-m}}{m} t^m \right). \tag{9.6.14}$$

† The minus sign in front of $\mathbf{u}(t)$ is due to the fact that before applying the Miura transformation we switch from connections on $\Omega^{-\rho}$ to connections on Ω^{ρ}.

Now we obtain that

$$\delta \mathbf{u}(t) = -t^{\lambda_i} x_i(t) \cdot \alpha_i,$$

and so

$$\delta u_j(t) = -t^{\lambda_i} a_{ji} x_i(t),$$

where $a_{ji} = \langle \check{\alpha}_j, \alpha_i \rangle$ is the (j,i) entry of the Cartan matrix of $\mathfrak{g}$. In other words, the operator e_i acts on the algebra $\operatorname{Fun} \operatorname{Conn}(\Omega^{-\rho})^{\lambda}_D = \mathbb{C}[u_{i,n}]$ by the derivation

$$\widetilde{V}_i[\lambda_i + 1] = -\sum_{j=1}^{\ell} a_{ji} \sum_{n \geq \lambda_i} x_{i,n-\lambda_i} \frac{\partial}{\partial u_{j,-n-1}}, \tag{9.6.15}$$

where the $x_{i,n}$'s are given by formula (9.6.14).

Note that when $\lambda_i = 0, i = 1, \ldots, \ell$, this operator, which appeared in formula (8.2.10) (up to the substitution $u_{i,n} \mapsto -u_{i,n}$, $a_{ji} \mapsto a_{ij}$, because there we were considering G-opers and connections on $\Omega^{\check{\rho}}$), was identified with a limit of the (-1)st Fourier coefficient of a vertex operator between Fock representations of the Heisenberg vertex algebra (see Section 8.1.2). We will show in the proof of Proposition 9.6.7 that the operator (9.6.15) may be identified with the limit of the $-(\lambda_i + 1)$st Fourier coefficient of the same vertex operator.

Now recall that we have identified $\operatorname{Fun} \operatorname{Op}^{\lambda}_{^L G}$ with the algebra of ${}^L\mathfrak{n}$-invariant functions on $\operatorname{Conn}(\Omega^{-\rho})^{\lambda}_D$. These are precisely the functions that are annihilated by the generators $e_i, i = 1, \ldots, \ell$, of ${}^L\mathfrak{n}$, which are given by formula (9.6.15). Therefore we obtain the following characterization of $\operatorname{Fun} \operatorname{Op}^{\lambda}_{^L G}$ as a subalgebra of $\operatorname{Fun} \operatorname{Conn}(\Omega^{-\rho})^{\lambda}_D$.

Proposition 9.6.3 *The image of* $\operatorname{Fun} \operatorname{Op}^{\lambda}_{^L G}$ *in* $\operatorname{Fun} \operatorname{Conn}(\Omega^{-\rho})^{\lambda}_D$ *under the Miura map is equal to the intersection of the kernels of the operators* $\widetilde{V}_i[\lambda_i + 1], i = 1, \ldots, \ell$, *given by formula* (9.6.15).

As a corollary of this result, we obtain a character formula for $\operatorname{Fun} \operatorname{Op}^{\lambda}_{^L G}$ with respect to the $\mathbb{Z}_+$-grading defined by the operator $L_0 = -t\partial_t$. According to Lemma 9.6.2, the action of ${}^L N$ generated by the operators (9.6.15) on $\operatorname{Conn}(\Omega^{-\rho})^{\lambda}_D$ is free. Therefore we have an isomorphism of vector spaces

$$\operatorname{Fun} \operatorname{Conn}(\Omega^{-\rho})^{\lambda}_D \simeq \operatorname{Fun} \operatorname{Op}^{\lambda}_{^L G} \otimes \operatorname{Fun} {}^L N.$$

According to formula (9.6.15), the operator e_i is homogeneous of degree

$$-(\lambda_i + 1) = -\langle \check{\alpha}_i, \lambda + \rho \rangle$$

with respect to the $\mathbb{Z}_+$-grading introduced above. Therefore the degree of the root generator $e_{\check{\alpha}}, \check{\alpha} \in \Delta_+$, of ${}^L\mathfrak{n}$ is equal to $-\langle \check{\alpha}, \lambda + \rho \rangle$. Hence we find that the character of $\operatorname{Fun} {}^L N$ is equal to

$$\prod_{\check{\alpha}\in\check{\Delta}_+} (1 - q^{\langle\check{\alpha},\lambda+\rho\rangle})^{-1}.$$

On the other hand, since $\deg u_{i,n} = -n$, we find from formula (9.6.10) that the character of $\operatorname{Fun} \operatorname{Conn}(\Omega^{-\rho})_D^\lambda$ is equal to

$$\prod_{n>0} (1 - q^n)^{-\ell}.$$

Therefore we find that the character of $\operatorname{Fun} \operatorname{Op}_{{}^L G}^\lambda$ is the ratio of the two, which we rewrite in the form

$$\prod_{\check{\alpha}\in\check{\Delta}_+} \frac{1 - q^{\langle\check{\alpha},\lambda+\rho\rangle}}{1 - q^{\langle\check{\alpha},\rho\rangle}} \prod_{i=1}^{\ell} \prod_{n_i \geq d_i+1} (1 - q^{n_i})^{-1}, \tag{9.6.16}$$

using the identity

$$\prod_{\check{\alpha}\in\check{\Delta}_+} (1 - q^{\langle\check{\alpha},\rho\rangle}) = \prod_{i=1}^{\ell} \prod_{m_i=1}^{d_i} (1 - q^{m_i}).$$

9.6.4 Computations with jet schemes

In order to complete the proof of Theorem 9.6.1 it would suffice to show that the character of

$$\operatorname{End}_{\widehat{\mathfrak{g}}_{\kappa_c}} \mathbb{V}_\lambda = (\mathbb{V}_\lambda \otimes V_\lambda^*)^{\mathfrak{g}[[t]]}$$

is less than or equal to the character of $\operatorname{Fun} \operatorname{Op}_{{}^L G}^\lambda$ given by formula (9.6.16). In the case when $\lambda = 0$ we have shown in Theorem 8.1.3 that the natural map $\operatorname{gr}(\mathbb{V}_0)^{\mathfrak{g}[[t]]} \to (\operatorname{gr} \mathbb{V}_0)^{\mathfrak{g}[[t]]}$ is an isomorphism. This makes it plausible that the map

$$\operatorname{gr}(\mathbb{V}_\lambda \otimes V_\lambda^*)^{\mathfrak{g}[[t]]} \to (\operatorname{gr}(\mathbb{V}_\lambda \otimes V_\lambda^*))^{\mathfrak{g}[[t]]} \tag{9.6.17}$$

is an isomorphism for all λ. If we could show that the character of $(\operatorname{gr}(\mathbb{V}_\lambda \otimes V_\lambda^*))^{\mathfrak{g}[[t]]}$ is less than or equal to the character of $\operatorname{Fun} \operatorname{Op}_{{}^L G}^\lambda$, then this would imply that the map (9.6.17) is an isomorphism, as well as prove Theorem 9.6.1, so we would be done.

In this section and the next we will compute the character of $(\operatorname{gr}(\mathbb{V}_\lambda \otimes V_\lambda^*))^{\mathfrak{g}[[t]]}$. However, we will find that for general λ this character is greater than the character of $\operatorname{Fun} \operatorname{Op}_{{}^L G}^\lambda$. The equality is achieved if and only if λ is a minuscule weight. This means that the map (9.6.17) is an isomorphism if and only if λ is minuscule. Therefore this way we are able to prove Theorem 9.6.1 for minuscule λ only. Nevertheless, the results presented below may be of independent interest. In order to prove Theorem 9.6.1 for a general λ we will use another argument given in Section 9.6.6.

Thus, our task at hand is to compute the character of $(\mathrm{gr}(\mathbb{V}_\lambda \otimes V_\lambda^*))^{\mathfrak{g}[[t]]}$. The PBW filtration on $U_{\kappa_c}(\widehat{\mathfrak{g}})$ gives rise to a filtration on $\mathbb{V}_\lambda$ considered as the $\widehat{\mathfrak{g}}_{\kappa_c}$-module generated from the subspace V_λ. The corresponding associated graded space is

$$\mathrm{gr}\,\mathbb{V}_\lambda = V_\lambda \otimes \mathrm{Sym}(\mathfrak{g}((t))/\mathfrak{g}[[t]]) \simeq V_\lambda \otimes \mathrm{Fun}\,\mathfrak{g}^*[[t]].$$

Therefore we obtain an embedding

$$(\mathrm{gr}(\mathbb{V}_\lambda \otimes V_\lambda^*))^{\mathfrak{g}[[t]]} \hookrightarrow (V_\lambda \otimes V_\lambda^* \otimes \mathrm{Fun}\,\mathfrak{g}^*[[t]])^{\mathfrak{g}[[t]]}.$$

Thus, we can find an upper bound on the character of $\mathrm{End}_{\widehat{\mathfrak{g}}_{\kappa_c}} \mathbb{V}_\lambda$ by computing the character of $(V_\lambda \otimes V_\lambda^* \otimes \mathrm{Fun}\,\mathfrak{g}^*[[t]])^{\mathfrak{g}[[t]]}$.

The first step is the computation of the algebra of invariants in $\mathrm{Fun}\,\mathfrak{g}^*[[t]]$ under the action of the Lie algebra $t\mathfrak{g}[[t]] \subset \mathfrak{g}[[t]]$.

Recall the space

$$\mathcal{P} = \mathfrak{g}^*/G = \mathrm{Spec}(\mathrm{Fun}\,\mathfrak{g}^*)^G \simeq \mathrm{Spec}\,\mathbb{C}[\overline{P}_i]_{i=1,\ldots,\ell},$$

and the morphism

$$p : \mathfrak{g}^* \to \mathcal{P}.$$

Consider the corresponding jet schemes

$$J_N\mathcal{P} = \mathrm{Spec}\,\mathbb{C}[\overline{P}_{i,n}]_{i=1,\ldots,\ell;-N-1\leq n<0},$$

introduced in the proof of Theorem 3.4.2. Here and below we will allow N to be finite or infinite (the definition of jet schemes is given in Section 3.4.2.

According to Theorem 3.4.2, we have

$$(\mathrm{Fun}\,J_N\mathfrak{g}^*)^{J_NG} = \mathbb{C}[\overline{P}_{i,n}]_{i=1,\ldots,\ell;-N-1\leq n<0}.$$

In other words, the morphism

$$J_Np : J_N\mathfrak{g}^* \to J_N\mathcal{P}$$

factors through an isomorphism

$$J_N\mathfrak{g}^*/J_NG \simeq J_N\mathcal{P},$$

where, by definition,

$$J_N\mathfrak{g}^*/J_NG = \mathrm{Spec}(\mathrm{Fun}\,J_N\mathfrak{g}^*)^{J_NG}.$$

We have natural maps $\mathfrak{g}^* \to \mathcal{P}$ and $J_N\mathcal{P} \to \mathcal{P}$. Denote by $\widetilde{J_N\mathcal{P}}$ the fiber product

$$\widetilde{J_N\mathcal{P}} = J_N\mathcal{P} \underset{\mathcal{P}}{\times} \mathfrak{g}^*.$$

By definition, the algebra $\mathrm{Fun}\,\widetilde{J_N\mathcal{P}}$ of functions on $\widetilde{J_N\mathcal{P}}$ is generated by $\mathrm{Fun}\,J_N\mathcal{P} = (\mathrm{Fun}\,J_N\mathfrak{g}^*)^{J_NG}$ and $\mathrm{Fun}\,\mathfrak{g}^*$, with the only relation being that we identify elements of $(\operatorname{Fun} J_N\mathfrak{g}^*)^{J_NG}$ and $\operatorname{Fun}\mathfrak{g}^*$ obtained by pull-back of the same element of $(\operatorname{Fun}\mathfrak{g}^*)^G$.

Consider the morphism

$$\widetilde{J_Np} : J_N\mathfrak{g}^* \to \widetilde{J_N\mathcal{P}},$$
$$A(t) \mapsto (J_Np(A(t)), A(0))$$

and the corresponding homomorphism of algebras

$$\operatorname{Fun}\widetilde{J_N\mathcal{P}} \to \operatorname{Fun} J_N\mathfrak{g}^*. \tag{9.6.18}$$

Let $J_NG^{(1)}$ be the subgroup of J_NG of elements congruent to the identity modulo t. Note that the Lie algebra of J_NG is $J_N\mathfrak{g} = \mathfrak{g}\otimes(\mathbb{C}[[t]]/t^{N+1}\mathbb{C}[[t]])$ and the Lie algebra of $J_NG^{(1)}$ is $J_N\mathfrak{g}^{(1)} = \mathfrak{g}\otimes(t\mathbb{C}[[t]]/t^{N+1}\mathbb{C}[[t]])$. For all algebras that we consider in the course of this discussion the spaces of invariants under J_NG (resp., $J_NG^{(1)}$) are equal to the corresponding spaces of invariants under $J_N\mathfrak{g}$ (resp., $J_N\mathfrak{g}^{(1)}$).

It is clear that the image of the homomorphism (9.6.18) consists of elements of $\operatorname{Fun} J_N\mathfrak{g}^*$ that are invariant under the action of $J_N\mathfrak{g}^{(1)}$ (and hence $J_NG^{(1)}$).

Proposition 9.6.4 *The homomorphism* (9.6.18) *is injective and its image is equal to the algebra of $J_N\mathfrak{g}^{(1)}$-invariants in* $\operatorname{Fun} J_N\mathfrak{g}^*$ *(for both finite and infinite N).*

Proof. As in the proof of Theorem 3.4.2, consider the subset $\mathfrak{g}^*_{\text{reg}}$ of regular elements in $\mathfrak{g}^*$. Then we have a smooth and surjective morphism $p_{\text{reg}} : \mathfrak{g}^* \to \mathcal{P}$. The corresponding morphism $J_Np_{\text{reg}} : J_N\mathfrak{g}^*_{\text{reg}} \to J_N\mathcal{P}$ is also smooth and surjective. It follows from the definition of jet schemes that

$$J_N\mathfrak{g}^*_{\text{reg}} \simeq \mathfrak{g}^*_{\text{reg}} \times \left(\mathfrak{g}^*\otimes(t\mathbb{C}[[t]]/t^{N+1}\mathbb{C}[[t]])\right).$$

Therefore $J_N\mathfrak{g}^*_{\text{reg}}$ is a smooth subscheme of $J_N\mathfrak{g}^*$, which is open and dense, and its complement has codimension greater than 1.

As explained in the proof of Theorem 3.4.2, each fiber of J_Np_{reg} consists of a single orbit of the group J_NG. We used this fact to derive that

$$(\operatorname{Fun} J_N\mathfrak{g}^*)^{J_NG} = (\operatorname{Fun} J_N\mathfrak{g}^*_{\text{reg}})^{J_NG} \simeq J_N\mathcal{P}.$$

Now we wish to consider the invariants not with respect to the group J_NG, but with respect to the subgroup $J_NG^{(1)} \subset J_NG$. We again have

$$(\operatorname{Fun} J_N\mathfrak{g}^*)^{J_NG^{(1)}} = (\operatorname{Fun} J_N\mathfrak{g}^*_{\text{reg}})^{J_NG^{(1)}},$$

so we will focus on the latter algebra.

Let us assume first that N is finite.

We show first that the homomorphism (9.6.18) is injective. Let

$$\widetilde{J_N\mathcal{P}}_{\text{reg}} = J_N\mathcal{P}\underset{\mathcal{P}}{\times}\mathfrak{g}^*_{\text{reg}}.$$

Consider the morphism

$$\widetilde{J_N p_{\rm reg}} : J_N \mathfrak{g}^*_{\rm reg} \to \widetilde{J_N \mathcal{P}_{\rm reg}},$$
$$A(t) \mapsto (J_N p_{\rm reg}(A(t)), A(0)).$$

We claim that it is surjective. Indeed, suppose that we are given a point $(\phi, x) \in \widetilde{J_N \mathcal{P}_{\rm reg}}$. Since the map $J_N p_{\rm reg}$ is surjective, there exists $A(t) \in J_N \mathfrak{g}^*_{\rm reg}$ such that $J_N p_{\rm reg}(A(t)) = \phi$. But then $A(0)$ and x are elements of $\mathfrak{g}^*_{\rm reg}$ such that $p_{\rm reg}(A(0)) = p_{\rm reg}(x)$. We know that the group G acts transitively along the fibers of the map $p_{\rm reg}$. Therefore $x = g \cdot A(0)$ for some $g \in G$. Hence $\widetilde{J_N p_{\rm reg}}(g \cdot A(t)) = (\phi, x)$. Thus, $\widetilde{J_N p_{\rm reg}}$ is surjective, and therefore (9.6.18) is injective.

The group $J_N G^{(1)}$ acts along the fibers of the map $\widetilde{J_N p_{\rm reg}}$. We claim that each fiber is a single $J_N G^{(1)}$-orbit.

In order to prove this, we describe the stabilizers of the points of $J_N \mathfrak{g}^*_{\rm reg}$ under the action of the group $J_N G$. Any element of $J_N G$ may be written in the form

$$g(t) = g_0 \exp\left(\sum_{m=1}^{N} y_m t^m\right), \qquad g_0 \in G, y_m \in \mathfrak{g}.$$

The condition that $g(t)$ stabilizes a point

$$x(t) = \sum_{n=0}^{N} x_n t^n, \qquad x_0 \in \mathfrak{g}^*_{\rm reg}; x_n \in \mathfrak{g}^*, n > 0,$$

of $J_N \mathfrak{g}^*_{\rm reg}$ is equivalent to a system of recurrence relations on g_0 and $y_1, \ldots, y_m$. The first of these relations is $g_0 \cdot x_0 = x_0$, which means that g_0 belongs to the stabilizer of x_0. The next relation is

$$\mathrm{ad}^*_{x_0}(y_1) = x_1 - g_0^{-1} \cdot x_1.$$

It is easy to see that for a regular element x_0 and any $x_1 \in \mathfrak{g}^*$ the right hand side is always in the image of $\mathrm{ad}^*_{x_0}$. Therefore this relation may be resolved for y_1. But since the centralizer of x_0 has dimension $\ell = \mathrm{rank}(\mathfrak{g})$, we obtain an ℓ-dimensional family of solutions for y_1.

We continue solving these relations by induction. The mth recurrence relation has the form: $\mathrm{ad}^*_{x_0}(y_m)$ equals an element of $\mathfrak{g}^*$ that is completely determined by the solutions of the previous relations. If this element of $\mathfrak{g}^*$ is in the image of $\mathrm{ad}^*_{x_0}$, then the mth relation may be resolved for y_m, and we obtain an ℓ-dimensional family of solutions. Otherwise, there are no solutions, and the recurrence procedure stops. This means that there is no $g(t) \in J_N G$ with $g(0) = g_0$ in the stabilizer of $x(t)$.

It follows that the maximal possible dimension of the centralizer of any $x(t) \in J_N \mathfrak{g}^*_{\rm reg}$ in $J_N G$ is equal to $(N+1)\ell$. This bound is realized if and only if for generic $g_0 \in \mathrm{Stab}_{x_0}(G)$ all recurrence relations may be resolved (i.e., the right hand side of the mth relation is in the image of $\mathrm{ad}^*_{x_0}$ for all $m = 1, \ldots, N$). But if these relations may be resolved for generic g_0, then they may be resolved for *all* $g_0 \in \mathrm{Stab}_{x_0}(G)$. Thus, we obtain that the centralizer of $x(t) \in J_N\mathfrak{g}^*_{\mathrm{reg}}$ in J_NG has dimension $(N+1)\ell$ if and only if for any $g_0 \in \mathrm{Stab}_{x_0}(G)$ all recurrence relations may be resolved.

However, observe that the dimension of $J_N\mathfrak{g}^*_{\mathrm{reg}}$ is equal to $(N+1)\dim\mathfrak{g}$, and the dimension of $J_N\mathcal{P}$ is equal to $(N+1)\ell$. Therefore the dimension of a generic fiber of the map J_Np_{reg} is equal to $(N+1)(\dim\mathfrak{g} - \ell)$. Since the group J_NG acts transitively along the fibers, we find that the dimension of the stabilizer of a generic point of $J_Ng^*_{\mathrm{reg}}$ is equal to $(N+1)\ell$. Therefore for a generic $x(t) \in J_N\mathfrak{g}^*_{\mathrm{reg}}$ all of the above recurrence relations may be resolved, for any $g_0 \in \mathrm{Stab}_{x_0}$. But then they may be resolved for all $x(t)$ and $g_0 \in \mathrm{Stab}_{x_0}$. In particular, we find that for any $x(t) \in J_N\mathfrak{g}^*_{\mathrm{reg}}$ and $g_0 \in \mathrm{Stab}_{x_0}(G)$ there exists $g(t) \in J_NG$ such that $g(0) = g_0$ and $g(t) \cdot x(t) = x(t)$.

Now we are ready to prove that each fiber of $\widetilde{J_Np_{\mathrm{reg}}}$ is a single $J_NG^{(1)}$-orbit. Consider two points $x_1(t), x_2(t) \in J_N\mathfrak{g}^*_{\mathrm{reg}}$ that belong to the same fiber of the map $\widetilde{J_Np_{\mathrm{reg}}}$. In particular, we have $x_1(0) = x_2(0) = x_0$. These two points also belong to the same fiber of the map J_Np_{reg}. Therefore there exists $g(t) \in J_NG$ such that $x_2(t) = g(t) \cdot x_1(t)$. This means, in particular, that $g(0) \cdot x_0 = x_0$. According to the previous discussion, there exists $\widetilde{g}(t) \in J_NG$ such that $\widetilde{g}(0) = g(0)$ and $\widetilde{g}(t) \cdot x_2(t) = x_2(t)$. Therefore $\widetilde{g}(t)^{-1}g(t) \in J_NG^{(1)}$ has the desired property: it transforms $x_1(t)$ to $x_2(t)$.

Thus, we obtain that each fiber of $\widetilde{J_Np_{\mathrm{reg}}}$ is a single $J_NG^{(1)}$-orbit. In the same way as in the proof of Theorem 3.4.2 this implies that the algebra of $J_NG^{(1)}$-invariants in $\mathrm{Fun}\, J_N\mathfrak{g}^*_{\mathrm{reg}}$, and hence in $\mathrm{Fun}\, J_N\mathfrak{g}^*$, is equal to the image of $\mathrm{Fun}\,\widetilde{J_N\mathcal{P}}$ under the injective homomorphism (9.6.18). This completes the proof for finite values of N.

Finally, consider the case of infinite jet schemes. As before, we will denote $J_\infty X$ by JX. By definition, any element A of $\mathrm{Fun}\,\widetilde{J\mathcal{P}}$ comes by pull-back from an element A_N of $\mathrm{Fun}\,\widetilde{J\mathcal{P}}$ for sufficiently large N. The image of A in $\mathrm{Fun}\, J\mathfrak{g}^*$ is the pull-back of the image of A_N in $\mathrm{Fun}\, J_N\mathfrak{g}^*$. Therefore the injectivity of the homomorphism (9.6.18) for finite N implies its injectivity for $N = \infty$.

Any $JG^{(1)}$-invariant function on $J\mathfrak{g}^*_{\mathrm{reg}}$ comes by pull-back from a $J_NG^{(1)}$-invariant function on $J_N\mathfrak{g}^*_{\mathrm{reg}}$. Hence we obtain that the algebra of $JG^{(1)}$-invariant functions on $J\mathfrak{g}^*_{\mathrm{reg}}$ is equal to $\mathrm{Fun}\,\widetilde{J\mathcal{P}}$. $\square$

It is useful to note the following straightforward generalization of this proposition.

For N as above, let k be a non-negative integer less than N. Denote by $\widetilde{J_N^{(k)}\mathcal{P}}$ the fiber product

$$\widetilde{J_N^{(k)}\mathcal{P}} = J_N\mathcal{P} \underset{J_k\mathcal{P}}{\times} J_k\mathfrak{g}^*$$

with respect to the natural morphisms $J_N\mathcal{P} \to J_k\mathcal{P}$ and $J_k\mathfrak{g}^* \to J_k\mathcal{P}$.

Consider the morphism

$$\widetilde{J_N^{(k)}}p : J_N\mathfrak{g}^* \to \widetilde{J_N^{(k)}}\mathcal{P},$$
$$A(t) \mapsto (J_Np(A(t)), A(t) \mod (t^{k+1}))$$

and the corresponding homomorphism of algebras

$$\mathrm{Fun}\,\widetilde{J_N^{(k)}}\mathcal{P} \to \mathrm{Fun}\,J_N\mathfrak{g}^*. \tag{9.6.19}$$

Let $J_NG^{(k+1)}$ be the subgroup of J_NG of elements congruent to the identity modulo (t^{k+1}). Note that the Lie algebra of $J_NG^{(k+1)}$ is

$$J_N\mathfrak{g}^{(k+1)} = \mathfrak{g} \otimes (t^{k+1}\mathbb{C}[[t]]/t^{N+1}\mathbb{C}[[t]]).$$

By repeating the argument used in the proof of Proposition 9.6.4, we obtain the following generalization of Proposition 9.6.4 (which corresponds to the special case $k=0$):

Proposition 9.6.5 *The homomorphism* (9.6.19) *is injective and its image is equal to the algebra of $J_N\mathfrak{g}^{(k+1)}$-invariants in* $\mathrm{Fun}\,J_N\mathfrak{g}^*$ *(for both finite and infinite N).*

9.6.5 The invariant subspace in the associated graded space

The group G and the Lie algebra $\mathfrak{g}$ naturally act on the space of $JG^{(1)}$-invariant polynomial functions on $J\mathfrak{g}^* = \mathfrak{g}^*[[t]]$, preserving the natural $\mathbb{Z}_+$-grading on the algebra $\mathrm{Fun}\,\mathfrak{g}^*[[t]]$ defined by the degree of the polynomial. By Proposition 9.6.4, this space of $JG^{(1)}$-invariant functions is isomorphic to $\mathrm{Fun}\,\widetilde{J}\mathcal{P}$. As a $\mathbb{Z}_+$-graded $\mathfrak{g}$-module, the latter is isomorphic to

$$\mathrm{Fun}\,\mathfrak{g}^*/(\mathrm{Inv}\,\mathfrak{g}^*)_+ \underset{\mathbb{C}}{\otimes} \mathbb{C}[\overline{P}_{i,n}]_{i=1,\ldots,\ell;n<0}.$$

Here $(\mathrm{Inv}\,\mathfrak{g}^*)_+$ denotes the ideal in $\mathrm{Fun}\,\mathfrak{g}^*$ generated by the augmentation ideal of the subalgebra $\mathrm{Inv}\,\mathfrak{g}^*$ of $\mathfrak{g}$-invariants. The action of $\mathfrak{g}$ is non-trivial only on the first factor, and the $\mathbb{Z}_+$-grading is obtained by combining the gradings on both factors.

Therefore we have an isomorphism of $\mathbb{Z}_+$-graded vector spaces

$$\begin{aligned}(V_\lambda \otimes V_\lambda^* \otimes \mathrm{Fun}\,\mathfrak{g}^*[[t]])^{\mathfrak{g}[[t]]} \\ = \mathrm{Hom}_{\mathfrak{g}}(V_\lambda \otimes V_\lambda^*, \mathrm{Fun}\,\mathfrak{g}^*/(\mathrm{Inv}\,\mathfrak{g}^*)_+) \underset{\mathbb{C}}{\otimes} \mathbb{C}[\overline{P}_{i,n}]_{i=1,\ldots,\ell;n<0}.\end{aligned} \tag{9.6.20}$$

Let us compute the character of (9.6.20).

According to a theorem of B. Kostant [Ko], for any $\mathfrak{g}$-module V the $\mathbb{Z}_+$-graded space $\mathrm{Hom}_{\mathfrak{g}}(V, \mathrm{Fun}\,\mathfrak{g}^*/(\mathrm{Inv}\,\mathfrak{g}^*)_+)$ is isomorphic to the space of $\mathfrak{g}_e$-invariants of V, where e is a regular nilpotent element of $\mathfrak{g}$ and $\mathfrak{g}_e$ is its centralizer (which is an ℓ-dimensional commutative Lie subalgebra of $\mathfrak{g}$). Without loss of generality, we will choose e to be the element $\sum_\imath e_\imath$, where $\{e_\imath\}_{\imath=1,\dots,\ell}$ are generators of $\mathfrak{n}$. Then e is contained in the $\mathfrak{sl}_2$-triple $\{f, \check{\rho}, e\} \subset \mathfrak{g}$. The corresponding $\mathbb{Z}_+$-grading on the space $V^{\mathfrak{g}_e}$ is the $\mathbb{Z}_+$-grading defined by the action of $\check{\rho}$ with appropriate shift (see below).

Thus, we obtain that the character of the first factor

$$\mathrm{Hom}_{\mathfrak{g}}(V_\lambda \otimes V_\lambda^*, \mathrm{Fun}\,\mathfrak{g}^*/(\mathrm{Inv}\,\mathfrak{g}^*)_+) \tag{9.6.21}$$

in the tensor product appearing in the right hand side of (9.6.20) is equal to

$$(V_\lambda \otimes V_\lambda^*)^{\mathfrak{g}_e} = \mathrm{Hom}_{U(\mathfrak{g}_e)}(V_\lambda, V_\lambda).$$

If V_λ were a cyclic module over the polynomial algebra $U(\mathfrak{g}_e)$, generated by a lowest weight vector u_λ, then any homomorphism of $U(\mathfrak{g}_e)$-modules $V_\lambda \to V_\lambda$ would be uniquely determined by the image of u_λ. Hence we would obtain that

$$\mathrm{Hom}_{U(\mathfrak{g}_e)}(V_\lambda, V_\lambda) \simeq V_\lambda,$$

where the grading on the right hand side is the principal grading on V_λ minus $\langle\lambda, \check{\rho}\rangle$ (note that the weight of u_λ is equal to $-\langle\lambda, \check{\rho}\rangle$). Thus, we would find that the character of (9.6.21) is equal to the principal character of V_λ times $q^{-\langle\lambda,\check{\rho}\rangle}$. According to formula (10.9.4) of [K2], the latter is equal to

$$\mathrm{ch}\, V_\lambda = \prod_{\check{\alpha}\in\check{\Delta}_+} \frac{1-q^{\langle\check{\alpha},\lambda+\rho\rangle}}{1-q^{\langle\check{\alpha},\rho\rangle}}. \tag{9.6.22}$$

Therefore the sought-after character of $(V_\lambda \otimes V_\lambda^* \otimes \mathrm{Fun}\,\mathfrak{g}^*[[t]])^{\mathfrak{g}[[t]]}$, which is equal to the character of the right hand side of (9.6.20), would be given by formula (9.6.16).

However, V_λ is a cyclic module over $U(\mathfrak{g}_e)$ if and only if λ is a minuscule (i.e., each weight in V_λ has multiplicity one). The "only if" part follows from the fact that the dimension of $(V_\lambda \otimes V_\lambda^*)^{\mathfrak{g}_e}$ is equal to the dimension of the subspace of $V_\lambda \otimes V_\lambda^*$ of weight 0 (see [Ko]), which is equal to

$$\sum_\mu (\dim V_\lambda(\mu))^2 > \sum_\mu \dim V_\lambda(\mu) = \dim V_\lambda,$$

unless λ is minuscule (here $V_\lambda(\mu)$ denotes the weight μ component of V_λ). The "if" part is a result of V. Ginzburg, see [G1], Prop. 1.8.1 (see also [G2]).

Thus, for a minuscule weight λ the above calculation implies that the character of $\mathrm{End}_{\widehat{\mathfrak{g}}_{\kappa_c}} \mathbb{V}_\lambda = (\mathbb{V}_\lambda \otimes V_\lambda^*)^{\mathfrak{g}[[t]]}$, whose associated graded is contained in the space $(V_\lambda \otimes V_\lambda^* \otimes \mathrm{Fun}\,\mathfrak{g}^*[[t]])^{\mathfrak{g}[[t]]}$, is less than or equal to the character of $\mathrm{Fun}\,\mathrm{Op}^\lambda_{^LG}$ given by formula (9.6.16). Since we already know from the results of Section 9.6.2 that $\mathrm{Fun}\,\mathrm{Op}^\lambda_{^LG}$ is isomorphic to a subquotient of $\mathrm{End}_{\widehat{\mathfrak{g}}_{\kappa_c}} \mathbb{V}_\lambda$, this implies the statement of Theorem 9.6.1 for minuscule highest weights λ.

But for a non-minuscule highest weight λ we find that the character of

$(V_\lambda \otimes V_\lambda^* \otimes \operatorname{Fun} \mathfrak{g}^*[[t]])^{\mathfrak{g}[[t]]}$ is strictly greater than the character of $\operatorname{Fun} \operatorname{Op}^\lambda_{^LG}$, and hence the above computation cannot be used to prove Theorem 9.6.1. (Furthermore, this implies that the map (9.6.17) is an isomorphism if and only if λ is minuscule.) We will give another argument in the next section.

9.6.6 Completion of the proof

We will complete the proof of Theorem 9.6.1 by showing that the top horizontal homomorphism of the diagram (9.6.8),

$$(\mathbb{V}_\lambda \otimes V_\lambda^*)^{\mathfrak{g}[[t]]} = (\mathbb{V}_\lambda)_\lambda^{\widehat{\mathfrak{n}}_+} \to \operatorname{Fun} \operatorname{Conn}(\Omega^{-\rho})^\lambda_D, \tag{9.6.23}$$

is injective and its image is contained in the intersection of the kernels of the operators $\widetilde{V}_i[\lambda_i + 1], i = 1, \ldots, \ell$, given by formula (9.6.15).

The following statement (generalizing Proposition 7.1.1, corresponding to the case $\lambda = 0$) is proved in [FG6].

Theorem 9.6.6 *The map $\imath_\lambda : \mathbb{V}_\lambda \to W_\lambda$ is injective for any dominant integral weight λ.*

Therefore the map (9.6.23) is injective. By Lemma 9.5.2, $(W_\lambda)_\lambda^{\widehat{\mathfrak{n}}_+}$ is the subspace

$$\operatorname{Fun} \operatorname{Conn}(\Omega^{-\rho})^\lambda_D = \pi_\lambda \subset W_\lambda.$$

Therefore we obtain an injective map

$$(\mathbb{V})_\lambda^{\widehat{\mathfrak{n}}_+} \hookrightarrow \operatorname{Fun} \pi_\lambda. \tag{9.6.24}$$

Proposition 9.6.7 *The image of the map* (9.6.24) *is contained in the intersection of the kernels of the operators $\widetilde{V}_i[\lambda_i + 1], i = 1, \ldots, \ell$, given by formula* (9.6.15).

Proof. We use the same proof as in Proposition 7.3.6, which corresponds to the case $\lambda = 0$. In Section 7.3.5 we constructed, for any dominant integral weight λ, the screening operators

$$\overline{S}_i^{(\lambda_i)} : W_{\lambda,\kappa_c} \to \widetilde{W}_{0,\lambda,\kappa_c},$$

which commute with the action of $\widehat{\mathfrak{g}}_{\kappa_c}$. It is clear from formula (7.3.17) for these operators that they annihilate the highest weight vector in W_{λ,κ_c}. Since the image of $\mathbb{V}_\lambda$ in $W_\lambda = W_{\lambda,\kappa_c}$ is generated by this vector, we find that this image is contained in the intersection of the kernels of the operators $\overline{S}_i^{(\lambda_i)}$. But the image of $(\mathbb{V}_\lambda)_\lambda^{\widehat{\mathfrak{n}}_+}$ is also contained in the subspace $\pi_\lambda \subset W_\lambda$. Therefore it is contained in the intersection of the kernels of the operators obtained by resticting $\overline{S}_i^{(\lambda_i)}$ to $\pi_\lambda \subset W_\lambda$. As explained at the end of Section 7.3.5, this restriction is equal to the operator $\overline{V}_i[\lambda_i + 1]$ given by formula (7.3.18).

Comparing formula (7.3.18) to (9.6.15), we find that upon substitution $b_{i,n} \mapsto u_{i,n}$ the operator $\overline{V}_i[\lambda_i + 1]$ becoms the operator $\widetilde{V}_i[\lambda_i + 1]$ given by formula (9.6.15). This completes the proof. □

Now injectivity of the map (9.6.23), Proposition 9.6.7 and Proposition 9.6.3 imply that the map

$$(\mathbb{V}_\lambda)_\lambda^{\widehat{\mathfrak{n}}_+} \to \pi_\lambda = \operatorname{Fun}\operatorname{Conn}(\Omega^{-\rho})_D^\lambda$$

is injective and its image is contained in $\operatorname{Fun}\operatorname{Op}_{^L G}^\lambda \subset \operatorname{Fun}\operatorname{Conn}(\Omega^{-\rho})_D^\lambda$. As explained in Section 9.6.3 (see the discussion after the diagram (9.6.8)), this proves Theorem 9.6.1.

10
Constructing the Langlands correspondence

It is time to use the results obtained in the previous chapters in our quest for the local Langlands correspondence for loop groups. So we go back to the question posed at the end of Chapter 1: let

$$\mathrm{Loc}_{{}^L G}(D^\times) = \{\partial_t + A(t),\ A(t) \in {}^L\mathfrak{g}((t))\} \,/\, {}^L G((t)) \tag{10.0.1}$$

be the set of gauge equivalence classes of ${}^L G$-connections on the punctured disc $D^\times = \operatorname{Spec} \mathbb{C}((t))$. We had argued in Chapter 1 that $\mathrm{Loc}_{{}^L G}(D^\times)$ should be taken as the space of Langlands parameters for the loop group $G((t))$. Recall that the loop group $G((t))$ acts on the category $\widehat{\mathfrak{g}}_\kappa$-mod of (smooth) $\widehat{\mathfrak{g}}$-modules of level κ (see Section 1.3.6 for the definition of this category). We asked the following question:

> *Associate to each local Langlands parameter* $\sigma \in \mathrm{Loc}_{{}^L G}(D^\times)$ *a subcategory* $\widehat{\mathfrak{g}}_\kappa$-mod$_\sigma$ *of* $\widehat{\mathfrak{g}}_\kappa$-mod *which is stable under the action of the loop group* $G((t))$.

Even more ambitiously, we wish to represent the category $\widehat{\mathfrak{g}}_\kappa$-mod as "fibering" over the space of local Langlands parameters $\mathrm{Loc}_{{}^L G}(D^\times)$, with the categories $\widehat{\mathfrak{g}}_\kappa$-mod$_\sigma$ being the "fibers" and the group $G((t))$ acting along these fibers. If we could do that, then we would think of this fibration as a "spectral decomposition" of the category $\widehat{\mathfrak{g}}_\kappa$-mod over $\mathrm{Loc}_{{}^L G}(D^\times)$.

At the beginning of Chapter 2 we proposed a possible scenario for solving this problem. Namely, we observed that any abelian category may be thought of as "fibering" over the spectrum of its center. Hence our idea was to describe the center of the category $\widehat{\mathfrak{g}}_\kappa$-mod (for each value of κ) and see if its spectrum is related to the space $\mathrm{Loc}_{{}^L G}(D^\times)$ of Langlands parameters.

We have identified the center of the category $\widehat{\mathfrak{g}}_\kappa$-mod with the center $Z_\kappa(\widehat{\mathfrak{g}})$ of the associative algebra $\widetilde{U}_\kappa(\widehat{\mathfrak{g}})$, the completed enveloping algebra of $\widehat{\mathfrak{g}}$ of level κ, defined in Section 2.1.2. Next, we described the algebra $Z_\kappa(\widehat{\mathfrak{g}})$. According to Proposition 4.3.9, if $\kappa \neq \kappa_c$, the critical level, then $Z_\kappa(\widehat{\mathfrak{g}}) = \mathbb{C}$. Therefore our approach cannot work for $\kappa \neq \kappa_c$. However, we found that the center

$Z_{\kappa_c}(\widehat{\mathfrak{g}})$ at the critical level is highly non-trivial and indeed related to ${}^L G$-connections on the punctured disc.

In this chapter, following my joint works with D. Gaitsgory [FG1]–[FG6], we will use these results to formulate more precise conjectures on the local Langlands correspondence for loop groups and to provide some evidence for these conjectures. We will also discuss the implications of these conjectures for the global geometric Langlands correspondence.

We note that A. Beilinson has another proposal [Bei] for local geometric Langlands correspondence, using representations of affine Kac–Moody algebras of levels *less than critical.* It would be interesting to understand the connection between his proposal and ours.

Here is a more detailed plan of this chapter. In Section 10.1 we review the relation between local systems and opers. We expect that the true space of Langlands parameters for loop groups is the space of ${}^L G$-local systems on the punctured disc $D^\times$. On the other hand, the category of $\widehat{\mathfrak{g}}_{\kappa_c}$-modules fibers over the space of ${}^L G$-opers on $D^\times$. We will see that this has some non-trivial consequences. Next, in Section 10.2 we introduce the Harish-Chandra categories; these are the categorical analogues of the subspaces of invariant vectors under a compact subgroup $K \subset G(F)$. Our program is to describe the categories of Harish-Chandra modules over $\widehat{\mathfrak{g}}_{\kappa_c}$ in terms of the Langlands dual group ${}^L G$ and find connections and parallels between the corresponding objects in the classical Langlands correspondence. We implement this program in two important examples: the "unramified" case, where our local system on $D^\times$ is isomorphic to the trivial local system, and the "tamely ramified" case, where the corresponding connection has regular singularity and unipotent monodromy.

The unramified case is considered in detail in Section 10.3. We first recall that in the classical setting it is expressed by the Satake correspondence between isomorphism classes of unramified representations of the group $G(F)$ and semi-simple conjugacy classes in the Langlands dual group ${}^L G$. We then discuss the categorical version of this correspondence. We give two realizations of the categorical representations of the loop group associated to the trivial local system, one in terms of $\widehat{\mathfrak{g}}_{\kappa_c}$-modules and the other in terms of $\mathcal{D}$-modules on the affine Grassmannian, and the conjectural equivalence between them. In particular, we show that the corresponding category of Harish-Chandra modules is equivalent to the category of vector spaces, in agreement with the classical picture.

The tamely ramified case is discussed in Section 10.4. In the classical setting the corresponding irreducible representations of $G(F)$ are in one-to-one correspondence with irreducible modules over the affine Hecke algebra associated to $G(F)$. According to [KL, CG], these modules may be realized as subquotients of the algebraic K-theory of the Springer fibers associated to unipotent elements of ${}^L G$ (which play the role of the monodromy of the local system in the geometric setting). This suggests that in our categorical version the corresponding categories of Harish-Chandra modules over $\widehat{\mathfrak{g}}_{\kappa_c}$ should be equivalent to suitable categories of coherent sheaves on the Springer fibers (more precisely, we should consider here the derived categories). A precise conjecture to this effect was formulated in [FG2]. In Section 10.4 we discuss this conjecture and various supporting evidence for it, following [FG2]–[FG5]. We note that the appearance of the Springer fiber may be traced to the construction associating Wakimoto modules to Miura opers exposed in the previous chapter. We emphasize the parallels between the classical and the geometric pictures. We also present some explicit calculations in the case of $\mathfrak{g} = \mathfrak{sl}_2$, which should serve as an illustration of the general theory.

Finally, in Section 10.5 we explain what these results mean for the global geometric Langlands correspondence. We show how the localization functors link local categories of Harish-Chandra modules over $\widehat{\mathfrak{g}}_{\kappa_c}$ and global categories of Hecke eigensheaves on moduli stacks of G-bundles on curves with parabolic structures. We expect that, at least in the generic situation, the local and global categories are equivalent to each others. Therefore our results and conjectures on the local categories translate into explicit statements on the structure of the categories of Hecke eigensheaves. This way the representation theory of affine Kac–Moody algebras yields valuable insights into the global Langlands correspondence.

10.1 Opers vs. local systems

Over the course of several chapters we have given a detailed description of the center $Z_{\kappa_c}(\widehat{\mathfrak{g}})$ at the critical level. According to Theorem 4.3.6 and related results obtained in Chapter 8, $Z_{\kappa_c}(\widehat{\mathfrak{g}})$ is isomorphic to the algebra $\operatorname{Fun} \operatorname{Op}_{^LG}(D^\times)$ of functions on the space of ^LG-opers on the punctured disc $D^\times$, in a way that is compatible with various symmetries and structures on both algebras. Now observe that there is a one-to-one correspondence between points $\chi \in \operatorname{Op}_{^LG}(D^\times)$ and homomorphisms (equivalently, characters)

$$\operatorname{Fun} \operatorname{Op}_{^LG}(D^\times) \to \mathbb{C},$$

corresponding to evaluating a function at χ. Hence points of $\operatorname{Op}_{^LG}(D^\times)$ parameterize **central characters** $Z_{\kappa_c}(\widehat{\mathfrak{g}}) \to \mathbb{C}$.

Given a ^LG-oper $\chi \in \operatorname{Op}_{^LG}(D^\times)$, we define the category

$$\widehat{\mathfrak{g}}_{\kappa_c}\text{-mod}_\chi$$

as a full subcategory of $\widehat{\mathfrak{g}}_{\kappa_c}$-mod whose objects are $\widehat{\mathfrak{g}}$-modules of critical level (hence $\widetilde{U}_{\kappa_c}(\widehat{\mathfrak{g}})$-modules) on which the center $Z_{\kappa_c}(\widehat{\mathfrak{g}}) \subset \widetilde{U}_{\kappa_c}(\widehat{\mathfrak{g}})$ acts according to the central character corresponding to χ. Since the algebra $\operatorname{Op}_{^LG}(D^\times)$ acts on the category $\widehat{\mathfrak{g}}_{\kappa_c}$-mod, we may say that the category $\widehat{\mathfrak{g}}_{\kappa}$-mod "fibers" over the space $\mathrm{Op}_{^LG}(D^\times)$, in such a way that the fiber-category corresponding to $\chi \in \mathrm{Op}_{^LG}(D^\times)$ is the category $\widehat{\mathfrak{g}}_{\kappa_c}\text{-mod}_\chi$.

More generally, for any closed algebraic subvariety $Y \subset \mathrm{Op}_{^LG}(D^\times)$ (not necessarily a point), we have an ideal

$$I_Y \subset \mathrm{Fun}\,\mathrm{Op}_{^LG}(D^\times) \simeq Z_{\kappa_c}(\widehat{\mathfrak{g}})$$

of those functions that vanish on Y. We then have a full subcategory $\widehat{\mathfrak{g}}_{\kappa_c}\text{-mod}_Y$ of $\widehat{\mathfrak{g}}_{\kappa_c}\text{-mod}$ whose objects are $\widehat{\mathfrak{g}}$-modules of critical level on which I_Y acts by 0. This category is an example of a "base change" of the category $\widehat{\mathfrak{g}}_{\kappa_c}\text{-mod}$ with respect to the morphism $Y \to \mathrm{Op}_{^LG}(D^\times)$. It is easy to generalize this definition to an arbitrary affine scheme Y equipped with a morphism $Y \to \mathrm{Op}_{^LG}(D^\times)$. The corresponding base changed categories $\widehat{\mathfrak{g}}_{\kappa_c}\text{-mod}_Y$ may then be "glued" together, which allows us to define the base changed category $\widehat{\mathfrak{g}}_{\kappa_c}\text{-mod}_Y$ for any scheme Y mapping to $\mathrm{Op}_{^LG}(D^\times)$. It is not difficult to generalize this to a general notion of an abelian category fibering over an algebraic stack (see [Ga3]). A model example of a category fibering over an algebraic stack $\mathcal{Y}$ is the category of coherent sheaves on an algebraic stack $\mathcal{X}$ equipped with a morphism $\mathcal{X} \to \mathcal{Y}$.

Recall that the group $G((t))$ acts on $\widetilde{U}_{\kappa_c}(\widehat{\mathfrak{g}})$ and on the category $\widehat{\mathfrak{g}}_{\kappa_c}\text{-mod}$. According to Proposition 4.3.8, the action of $G((t))$ on $Z_{\kappa_c}(\widehat{\mathfrak{g}}) \subset \widetilde{U}_{\kappa_c}(\widehat{\mathfrak{g}})$ is trivial. Therefore the subcategories $\widehat{\mathfrak{g}}_{\kappa_c}\text{-mod}_\chi$ (and, more generally, $\widehat{\mathfrak{g}}_{\kappa_c}\text{-mod}_Y$) are stable under the action of $G((t))$. Thus, the group $G((t))$ acts "along the fibers" of the "fibration" $\widehat{\mathfrak{g}}_{\kappa_c}\text{-mod} \to \mathrm{Op}_{^LG}(D^\times)$ (see [FG2], Section 20, for more details).

The fibration $\widehat{\mathfrak{g}}_{\kappa_c}\text{-mod} \to \mathrm{Op}_{^LG}(D^\times)$ almost gives us the desired local Langlands correspondence for loop groups. But there is one important difference: we asked that the category $\widehat{\mathfrak{g}}_{\kappa_c}\text{-mod}$ fiber over the space $\mathrm{Loc}_{^LG}(D^\times)$ of local systems on $D^\times$. We have shown, however, that $\widehat{\mathfrak{g}}_{\kappa_c}\text{-mod}$ fibers over the space $\mathrm{Op}_{^LG}(D^\times)$ of ^LG-opers.

What is the difference between the two spaces? While a ^LG-local system is a pair $(\mathcal{F}, \nabla)$, where $\mathcal{F}$ is an ^LG-bundle and ∇ is a connection on $\mathcal{F}$, an ^LG-oper is a triple $(\mathcal{F}, \nabla, \mathcal{F}_{^LB})$, where $\mathcal{F}$ and ∇ are as before, and $\mathcal{F}_{^LB}$ is an additional piece of structure; namely, a reduction of $\mathcal{F}$ to a (fixed) Borel subgroup $^LB \subset {}^LG$ satisfying the transversality condition explained in Section 4.2.1. Thus, for any curve X we clearly have a forgetful map

$$\mathrm{Op}_{^LG}(X) \to \mathrm{Loc}_{^LG}(X).$$

The fiber of this map over $(\mathcal{F}, \nabla) \in \mathrm{Loc}_{^LG}(X)$ consists of all ^LB-reductions of $\mathcal{F}$ satisfying the transversality condition with respect to ∇.

It may well be that this map is not surjective, i.e., that the fiber of this map over a particular local system $(\mathcal{F}, \nabla)$ is empty. For example, if X is a projective curve and ^LG is a group of adjoint type, then there is a unique ^LG-bundle $\mathcal{F}_0$ such that the fiber over $(\mathcal{F}_0, \nabla)$ is non-empty. Furthermore, for this $\mathcal{F}_0$ the fiber over $(\mathcal{F}_0, \nabla)$ consists of one point for any connection ∇ (in other words, for each connection ∇ on $\mathcal{F}_0$ there is a unique ${}^L B$-reduction $\mathcal{F}_{{}^L B}$ of $\mathcal{F}_0$ satisfying the transversality condition with respect to ∇).

The situation is quite different when $X = D^\times$. In this case any ${}^L G$-bundle $\mathcal{F}$ may be trivialized. A connection ∇ therefore may be represented as a first order operator $\partial_t + A(t), A(t) \in {}^L\mathfrak{g}((t))$. However, the trivialization of $\mathcal{F}$ is not unique; two trivializations differ by an element of ${}^L G((t))$. Therefore the set of equivalence classes of pairs $(\mathcal{F}, \nabla)$ is identified with the quotient (10.0.1).

Suppose now that $(\mathcal{F}, \nabla)$ carries an oper reduction $\mathcal{F}_{{}^L B}$. Then we consider only those trivializations of $\mathcal{F}$ which come from trivializations of $\mathcal{F}_{{}^L B}$. There are fewer of these, since two trivializations now differ by an element of ${}^L B((t))$ rather than ${}^L G((t))$. Due to the oper transversality condition, the connection ∇ must have a special form with respect to any of these trivializations; namely,

$$\nabla = \partial_t + \sum_{i=1}^{\ell} \psi_i(t) f_i + \mathbf{v}(t),$$

where each $\psi_i(t) \neq 0$ and $\mathbf{v}(t) \in {}^L\mathfrak{b}((t))$ (see Section 4.2.2). Thus, we obtain a concrete realization of the space of opers as a space of gauge equivalence classes

$$\operatorname{Op}_{{}^L G}(D^\times) = \\ = \left\{ \partial_t + \sum_{i=1}^{\ell} \psi_i(t) f_i + \mathbf{v}(t), \ \psi_i \neq 0, \mathbf{v}(t) \in {}^L\mathfrak{b}((t)) \right\} \Big/ \ {}^L B((t)). \quad (10.1.1)$$

Now the map

$$\alpha : \operatorname{Op}_{{}^L G}(D^\times) \to \operatorname{Loc}_{{}^L G}(D^\times)$$

simply takes a ${}^L B((t))$-equivalence class of operators of the form (10.1.1) to its ${}^L G((t))$-equivalence class.

Unlike the case of projective curves X discussed above, we expect that the map α is **surjective** for any simple Lie group ${}^L G$. This has been proved in [De1] in the case of SL_n, and we conjecture it to be true in general.

Conjecture 10.1.1 *The map α is surjective for any simple Lie group ${}^L G$.*

Now we find ourselves in the following situation: we *expect* that there exists a category $\mathcal{C}$ fibering over the space $\operatorname{Loc}_{{}^L G}(D^\times)$ of "true" local Langlands parameters, equipped with a fiberwise action of the loop group $G((t))$. The fiber categories $\mathcal{C}_\sigma$ corresponding to various $\sigma \in \operatorname{Loc}_{{}^L G}(D^\times)$ should satisfy various, not yet specified, properties. Moreover, we expect $\mathcal{C}$ to be the *universal* category equipped with an action of $G((t))$. In other words, we expect that $\operatorname{Loc}_{{}^L G}(D^\times)$ is the universal parameter space for the categorical representations of $G((t))$ (at the moment we cannot formulate this property more precisely). The ultimate form of the local Langlands correspondence for loop groups should be, roughly, the following statement:

$$\boxed{\begin{array}{c}\text{categories fibering}\\ \text{over } \mathrm{Loc}_{^L G}(D^\times)\end{array}} \quad \Longleftrightarrow \quad \boxed{\begin{array}{c}\text{categories equipped}\\ \text{with action of } G((t))\end{array}} \tag{10.1.2}$$

The idea is that given a category $\mathcal{A}$ on the left hand side we construct a category on the right hand side by taking the "fiber product" of $\mathcal{A}$ and the universal category $\mathcal{C}$ over $\mathrm{Loc}_{^L G}(D^\times)$.

Now, we have *constructed* a category $\widehat{\mathfrak{g}}_{\kappa_c}$-mod, which fibers over a close cousin of the space $\mathrm{Loc}_{^L G}(D^\times)$ – namely, the space $\mathrm{Op}_{^L G}(D^\times)$ of $^L G$-opers – and is equipped with a fiberwise action of the loop group $G((t))$.

What should be the relationship between $\widehat{\mathfrak{g}}_{\kappa_c}$-mod and the conjectural universal category $\mathcal{C}$?

The idea of [FG2] is that the second fibration is a "base change" of the first one, that is, there is a Cartesian diagram

$$\begin{array}{ccc} \widehat{\mathfrak{g}}_{\kappa_c}\text{-mod} & \longrightarrow & \mathcal{C} \\ \downarrow & & \downarrow \\ \mathrm{Op}_{^L G}(D^\times) & \xrightarrow{\alpha} & \mathrm{Loc}_{^L G}(D^\times) \end{array} \tag{10.1.3}$$

that commutes with the action of $G((t))$ along the fibers of the two vertical maps. In other words,

$$\widehat{\mathfrak{g}}_{\kappa_c}\text{-mod} \simeq \mathcal{C} \underset{\mathrm{Loc}_{^L G}(D^\times)}{\times} \mathrm{Op}_{^L G}(D^\times).$$

Thus, $\widehat{\mathfrak{g}}_{\kappa_c}$-mod should arise as the category on the right hand side of the correspondence (10.1.2) attached to the category of quasicoherent sheaves on $\mathrm{Op}_{^L G}(D^\times)$ on the left hand side. The latter is a category fibering over $\mathrm{Loc}_{^L G}(D^\times)$, with respect to the map $\mathrm{Op}_{^L G}(D^\times) \to \mathrm{Loc}_{^L G}(D^\times)$.

At present, we do not have a definition of $\mathcal{C}$, and therefore we cannot make this isomorphism precise. But we will use it as our guiding principle. We will now discuss various corollaries of this conjecture and various pieces of evidence that make us believe that it is true.

In particular, let us fix a Langlands parameter $\sigma \in \mathrm{Loc}_{^L G}(D^\times)$ that is in the image of the map α (according to Conjecture 10.1.1, all Langlands parameters are). Let χ be an $^L G$-oper in the preimage of σ, $\alpha^{-1}(\sigma)$. Then, according to the above conjecture, the category $\widehat{\mathfrak{g}}_{\kappa_c}\text{-mod}_\chi$ is equivalent to the "would be" Langlands category $\mathcal{C}_\sigma$ attached to σ. Hence we may take $\widehat{\mathfrak{g}}_{\kappa_c}\text{-mod}_\chi$ as the **definition** of $\mathcal{C}_\sigma$.

The caveat is, of course, that we need to ensure that this definition is independent of the choice of χ in $\alpha^{-1}(\sigma)$. This means that for any two $^L G$-opers, χ and χ', in the preimage of σ, the corresponding categories, $\widehat{\mathfrak{g}}_{\kappa_c}\text{-mod}_\chi$ and $\widehat{\mathfrak{g}}_{\kappa_c}\text{-mod}_{\chi'}$, should be equivalent to each other, and this equivalence should commute with the action of the loop group $G((t))$. Moreover, we should expect that these equivalences are compatible with each other as we move along the fiber $\alpha^{-1}(\sigma)$. We will not try to make this condition more precise here (however, we will explain below in Conjecture 10.3.10 what this means for regular opers).

Even putting the questions of compatibility aside, we arrive at the following rather non-trivial conjecture.

Conjecture 10.1.2 *Suppose that $\chi, \chi' \in \operatorname{Op}_{{}^L G}(D^\times)$ are such that $\alpha(\chi) = \alpha(\chi')$, i.e., that the flat ${}^L G$-bundles on $D^\times$ underlying the ${}^L G$-opers χ and χ' are isomorphic to each other. Then there is an equivalence between the categories $\widehat{\mathfrak{g}}_{\kappa_c}\text{-mod}_\chi$ and $\widehat{\mathfrak{g}}_{\kappa_c}\text{-mod}_{\chi'}$ which commutes with the actions of the group $G((t))$ on the two categories.*

Thus, motivated by our quest for the local Langlands correspondence, we have found an unexpected symmetry in the structure of the category $\widehat{\mathfrak{g}}_{\kappa_c}$-mod of $\widehat{\mathfrak{g}}$-modules of critical level.

10.2 Harish–Chandra categories

As we explained in Chapter 1, the local Langlands correspondence for the loop group $G((t))$ should be viewed as a categorification of the local Langlands correspondence for the group $G(F)$, where F is a local non-archimedean field. This means that the categories $\mathcal{C}_\sigma$, equipped with an action of $G((t))$, that we are trying to attach to the Langlands parameters $\sigma \in \operatorname{Loc}_{{}^L G}(D^\times)$ should be viewed as categorifications of the smooth representations of $G(F)$ on complex vector spaces attached to the corresponding local Langlands parameters discussed in Section 1.1.5. Here we use the term "categorification" to indicate that we expect the Grothendieck groups of the categories $\mathcal{C}_\sigma$ to "look like" irreducible smooth representations of $G(F)$.

Our goal in this chapter is to describe the categories $\mathcal{C}_\sigma$ as categories of $\widehat{\mathfrak{g}}_{\kappa_c}$-modules. We begin by taking a closer look at the structure of irreducible smooth representations of $G(F)$.

10.2.1 Spaces of K-invariant vectors

It is known that an irreducible smooth representation (R, π) of $G(F)$ is automatically **admissible**, in the sense that for any open compact subgroup K, such as the Nth congruence subgroup K_N defined in Section 1.1.2, the space $R^{\pi(K)}$ of K-invariant vectors in R is finite-dimensional. Thus, while most of the irreducible smooth representations (R, π) of $G(F)$ are infinite-dimensional, they are filtered by the finite-dimensional subspaces $R^{\pi(K)}$ of K-invariant vectors, where K are smaller and smaller open compact subgroups. The space

$R^{\pi(K)}$ does not carry an action of $G(F)$, but it carries an action of the **Hecke algebra** $H(G(F),K)$.

By definition, $H(G(F),K)$ is the space of compactly supported K bi-invariant functions on $G(F)$. It is given an algebra structure with respect to the **convolution product**

$$(f_1 \star f_2)(g) = \int_{G(F)} f_1(gh^{-1}) f_2(h)\, dh, \tag{10.2.1}$$

where dh is the Haar measure on $G(F)$ normalized in such a way that the volume of the subgroup $K_0 = G(\mathcal{O})$ is equal to 1 (here $\mathcal{O}$ is the ring of integers of F; e.g., for $F = \mathbb{F}_q((t))$ we have $\mathcal{O} = \mathbb{F}_q[[t]]$). The algebra $H(G(F),K)$ acts on the space $R^{\pi(K)}$ by the formula

$$f \star v = \int_{G(F)} f_1(gh^{-1})(\pi(h)\cdot v)\, dh, \qquad v \in R^{\pi(K)}. \tag{10.2.2}$$

Studying the spaces of K-invariant vectors and their $\mathcal{H}(G(F),K)$-module structure gives us an effective tool for analyzing representations of the group $G(F)$, where $F = \mathbb{F}_q((t))$.

Can we find a similar structure in the categorical local Langlands correspondence for loop groups?

10.2.2 Equivariant modules

In the categorical setting a representation (R,π) of the group $G(F)$ is replaced by a category equipped with an action of $G((t))$, such as $\widehat{\mathfrak{g}}_{\kappa_c}\text{-mod}_\chi$. The open compact subgroups of $G(F)$ have obvious analogues for the loop group $G((t))$ (although they are, of course, not compact with respect to the usual topology on $G((t))$). For instance, we have the "maximal compact subgroup" $K_0 = G[[t]]$, or, more generally, the Nth congruence subgroup K_N, whose elements are congruent to 1 modulo $t^N\mathbb{C}[[t]]$. Another important example is the analogue of the **Iwahori subgroup**. This is the subgroup of $G[[t]]$, which we denote by I, whose elements $g(t)$ have the property that their value at 0, that is $g(0)$, belong to a fixed Borel subgroup $B \subset G$.

Now, in the categorical setting, an analogue of a vector in a representation of $G(F)$ is an object of our category, i.e., a smooth $\widehat{\mathfrak{g}}_{\kappa_c}$-module (M,ρ), where $\rho : \widehat{\mathfrak{g}}_{\kappa_c} \to \operatorname{End} M$. Hence for a subgroup $K \subset G((t))$ of the above type an analogue of a K-invariant vector in a representation of $G(F)$ is a smooth $\widehat{\mathfrak{g}}_{\kappa_c}$-module (M,ρ) which is stable under the action of K. Recall from Section 1.3.6 that for any $g \in G((t))$ we have a new $\widehat{\mathfrak{g}}_{\kappa_c}$-module (M,ρ_g), where $\rho_g(x) = \rho(\mathrm{Ad}_g(x))$. We say that (M,ρ) is stable under K, or that (M,ρ) is **weakly** K**-equivariant**, if there is a compatible system of isomorphisms between (M,ρ) and (M,ρ_k) for all $k \in K$. More precisely, this means that for each

$k \in K$ there exists a linear map $T_k^M : M \to M$ such that

$$T_k^M \rho(x)(T_k^M)^{-1} = \rho(\mathrm{Ad}_k(x))$$

for all $x \in \widehat{\mathfrak{g}}_{\kappa_c}$, and we have

$$T_1^M = \mathrm{Id}_M, \qquad T_{k_1}^M T_{k_2}^M = T_{k_1 k_2}^M.$$

Thus, M becomes a representation of the group K.† Consider the corresponding representation of the Lie algebra $\mathfrak{k} = \mathrm{Lie}\, K$ on M. Let us assume that the embedding $\mathfrak{k} \hookrightarrow \mathfrak{g}((t))$ lifts to $\mathfrak{k} \hookrightarrow \widehat{\mathfrak{g}}_{\kappa_c}$ (i.e., that the central extension cocycle is trivial on $\mathfrak{k}$). This is true, for instance, for any subgroup contained in $K_0 = G[[t]]$, or its conjugate. Then we also have a representation of $\mathfrak{k}$ on M obtained by restriction of ρ. In general, the two representations do not have to coincide. If they do coincide, then the module M is called **strongly K-equivariant**, or simply **K-equivariant**.

The pair $(\widehat{\mathfrak{g}}_{\kappa_c}, K)$ is an example of **Harish-Chandra pair**, that is, a pair $(\mathfrak{g}, H)$ consisting of a Lie algebra $\mathfrak{g}$ and a Lie group H whose Lie algebra is contained in $\mathfrak{g}$. The K-equivariant $\widehat{\mathfrak{g}}_{\kappa_c}$-modules are therefore called $(\widehat{\mathfrak{g}}_{\kappa_c}, K)$ **Harish-Chandra modules**. These are (smooth) $\widehat{\mathfrak{g}}_{\kappa_c}$-modules on which the action of the Lie algebra $\mathrm{Lie}\, K \subset \widehat{\mathfrak{g}}_{\kappa_c}$ may be exponentiated to an action of K (we will assume that K is connected). We denote by $\widehat{\mathfrak{g}}_{\kappa_c}\text{-mod}^K$ and $\widehat{\mathfrak{g}}_{\kappa_c}\text{-mod}_\chi^K$ the full subcategories of $\widehat{\mathfrak{g}}_{\kappa_c}\text{-mod}$ and $\widehat{\mathfrak{g}}_{\kappa_c}\text{-mod}_\chi$, respectively, whose objects are $(\widehat{\mathfrak{g}}_{\kappa_c}, K)$ Harish-Chandra modules.

We will stipulate that the analogues of K-invariant vectors in the category $\widehat{\mathfrak{g}}_{\kappa_c}\text{-mod}_\chi$ are $(\widehat{\mathfrak{g}}_{\kappa_c}, K)$ Harish-Chandra modules. Thus, while the categories $\widehat{\mathfrak{g}}_{\kappa_c}\text{-mod}_\chi$ should be viewed as analogues of smooth irreducible representations (R, π) of the group $G(F)$, the categories $\widehat{\mathfrak{g}}_{\kappa_c}\text{-mod}_\chi^K$ are analogues of the spaces of K-invariant vectors $R^{\pi(K)}$.

Next, we discuss the categorical analogue of the Hecke algebra $H(G(F), K)$.

10.2.3 Categorical Hecke algebras

We recall that $H(G(F), K)$ is the algebra of compactly supported K bi-invariant functions on $G(F)$. We realize it as the algebra of left K-invariant compactly supported functions on $G(F)/K$. In Section 1.3.3 we have already discussed the question of categorification of the algebra of functions on a homogeneous space like $G(F)/K$. Our conclusion was that the categorical analogue of this algebra, when $G(F)$ is replaced by the complex loop group $G((t))$, is the category of $\mathcal{D}$-modules on $G((t))/K$. More precisely, this quotient has the structure of an ind-scheme which is a direct limit of finite-dimensional

† In general, it is reasonable to modify the last condition to allow for a non-trivial two-cocycle and hence a non-trivial central extension of K; however, in the case of interest K does not have any non-trivial central extensions.

algebraic varieties with respect to closed embeddings. The appropriate notion of (right) $\mathcal{D}$-modules on such ind-schemes is formulated in [BD1] (see also [FG1, FG2]). As the categorical analogue of the algebra of left K-invariant functions on $G(F)/K$, we take the category $\mathcal{H}(G((t)), K)$ of K-equivariant $\mathcal{D}$-modules on the ind-scheme $G((t))/K$ (with respect to the left action of K on $G((t))/K$). We call it the **categorical Hecke algebra** associated to K.

It is easy to define the convolution of two objects of the category $\mathcal{H}(G((t)), K)$ by imitating formula (10.2.1). Namely, we interpret this formula as a composition of the operations of pulling back and integrating functions. Then we apply the same operations to $\mathcal{D}$-modules, thinking of the integral as push-forward. However, here one encounters two problems. The first problem is that for a general group K the morphisms involved will not be proper, and so we have to choose between the $*$- and !-push-forward. This problem does not arise, however, if K is such that $I \subset K \subset G[[t]]$, which will be our main case of interest. The second, and more serious, issue is that in general the push-forward is not an exact functor, and so the convolution of two $\mathcal{D}$-modules will not be a $\mathcal{D}$-module, but a complex, more precisely, an object of the corresponding K-equivariant (bounded) derived category $D^b(G((t))/K)^K$ of $\mathcal{D}$-modules on $G((t))/K$. We will not spell out the exact definition of this category here, referring the interested reader to [BD1] and [FG2]. The exception is the case of the subgroup $K_0 = G[[t]]$, when the convolution functor is exact and so we may restrict ourselves to the abelian category of K_0-equivariant $\mathcal{D}$-modules on $G((t))/K_0$.

The category $D^b(G((t))/K)^K$ has a monoidal structure, and as such it acts on the derived category of $(\widehat{\mathfrak{g}}_{\kappa_c}, K)$ Harish-Chandra modules (again, we refer the reader to [BD1, FG2] for the precise definition). In the special case when $K = K_0$, we may restrict ourselves to the corresponding abelian categories. This action should be viewed as the categorical analogue of the action of $H(G(F), K)$ on the space $R^{\pi(K)}$ of K-invariant vectors discussed above.

Our ultimate goal is understanding the "local Langlands categories" $\mathcal{C}_\sigma$ associated to the "local Langlands parameters" $\sigma \in \operatorname{Loc}_{{}^L G}(D^\times)$. We now have a candidate for the category $\mathcal{C}_\sigma$; namely, the category $\widehat{\mathfrak{g}}_{\kappa_c}\text{-mod}_\chi$, where $\sigma = \alpha(\chi)$. Therefore $\widehat{\mathfrak{g}}_{\kappa_c}\text{-mod}_\chi$ should be viewed as a categorification of a smooth representation (R, π) of $G(F)$. The corresponding category $\widehat{\mathfrak{g}}_{\kappa_c}\text{-mod}_\chi^K$ of $(\widehat{\mathfrak{g}}_{\kappa_c}, K)$ Harish-Chandra modules should therefore be viewed as a categorification of $R^{\pi(K)}$. This category is acted upon by the categorical Hecke algebra $\mathcal{H}(G((t)), K)$. Actually, this action does not preserve the abelian category $\widehat{\mathfrak{g}}_{\kappa_c}\text{-mod}_\chi^K$; rather $\mathcal{H}(G((t)), K)$ acts on the corresponding derived category.†

† This is the first indication that the local Langlands correspondence should assign to a local system $\sigma \in \operatorname{Loc}_{{}^L G}(D^\times)$ not an abelian, but a triangulated category (equipped with an action of $G((t))$). We will see in the examples considered below that these triangulated categories carry different t-structures. In other words, they may be interpreted as derived categories of different abelian categories, and we expect that one of them is $\widehat{\mathfrak{g}}_{\kappa_c}\text{-mod}_\chi$.

We summarize this analogy in the following table.

Classical theory	**Geometric theory**
Representation of $G(F)$ on a vector space R	Representation of $G((t))$ on a category $\widehat{\mathfrak{g}}_{\kappa_c}\text{-mod}_\chi$
A vector in R	An object of $\widehat{\mathfrak{g}}_{\kappa_c}\text{-mod}_\chi$
The subspace $R^{\pi(K)}$ of K-invariant vectors of R	The subcategory $\widehat{\mathfrak{g}}_{\kappa_c}\text{-mod}_\chi^K$ of $(\widehat{\mathfrak{g}}_{\kappa_c}, K)$ Harish-Chandra modules
Hecke algebra $H(G(F), K)$ acts on $R^{\pi(K)}$	Categorical Hecke algebra $\mathcal{H}(G((t)), K)$ acts on $\widehat{\mathfrak{g}}_{\kappa_c}\text{-mod}_\chi^K$

Now we may test our proposal for the local Langlands correspondence by studying the categories $\widehat{\mathfrak{g}}_{\kappa_c}\text{-mod}_\chi^K$ of Harish-Chandra modules and comparing their structure to the structure of the spaces $R^{\pi(K)}$ of K-invariant vectors of smooth representations of $G(F)$ in the known cases. Another possibility is to test Conjecture 10.1.2 when applied to the categories of Harish-Chandra modules.

In the next section we consider the case of the "maximal compact subgroup" $K_0 = G[[t]]$. We will show that the structure of the categories $\widehat{\mathfrak{g}}_{\kappa_c}\text{-mod}_\chi^{K_0}$ is compatible with the classical results about unramified representations of $G(F)$. We then take up the more complicated case of the Iwahori subgroup I in Section 10.4. Here we will also find the conjectures and results of [FG2] to be consistent with the known results about representations of $G(F)$ with Iwahori fixed vectors.

10.3 The unramified case

We first take up the case of the "maximal compact subgroup" $K_0 = G[[t]]$ of $G((t))$ and consider the categories $\widehat{\mathfrak{g}}_{\kappa_c}\text{-mod}_\chi$ which contain non-trivial K_0-equivariant objects.

10.3.1 *Unramified representations of* $G(F)$

These categories are the analogues of smooth representations of the group $G(F)$, where F is a local non-archimedian field (such as $\mathbb{F}_q((t))$), that contain non-zero K_0-invariant vectors. Such representations are called **unramified**. The classification of the irreducible unramified representations of $G(F)$ is the simplest case of the local Langlands correspondence discussed in Sections 1.1.4 and 1.1.5. Namely, we have a bijection between the sets of equivalence classes of the following objects:

$$\boxed{\begin{array}{c}\text{unramified admissible}\\ \text{homomorphisms } W'_F \to {}^L G\end{array}} \quad \Longleftrightarrow \quad \boxed{\begin{array}{c}\text{irreducible unramified}\\ \text{representations of } G(F)\end{array}} \tag{10.3.1}$$

where W'_F is the Weil–Deligne group introduced in Section 1.1.3.

By definition, unramified homomorphisms $W'_F \longrightarrow {}^L G$ are those which factor through the quotient

$$W'_F \to W_F \to \mathbb{Z}$$

(see Section 1.1.3 for the definitions of these groups and homomorphisms). Its admissibility means that its image in ${}^L G$ consists of semi-simple elements. Therefore the set on the left hand side of (10.3.1) is just the set of conjugacy classes of semi-simple elements of ${}^L G$. Thus, the above bijection may be reinterpreted as follows:

$$\boxed{\begin{array}{c}\text{semi-simple conjugacy}\\ \text{classes in } {}^L G\end{array}} \quad \Longleftrightarrow \quad \boxed{\begin{array}{c}\text{irreducible unramified}\\ \text{representations of } G(F)\end{array}} \tag{10.3.2}$$

To construct this bijection, we look at the **spherical Hecke algebra** $H(G(F), K_0)$. According to the Satake isomorphism [Sa], in the interpretation of Langlands [L], this algebra is commutative and isomorphic to the representation ring of the Langlands dual group ${}^L G$:

$$H(G(F), K_0) \simeq \operatorname{Rep} {}^L G. \tag{10.3.3}$$

We recall that $\operatorname{Rep} {}^L G$ consists of finite linear combinations $\sum_i a_i [V_i]$, where the V_i are finite-dimensional representations of ${}^L G$ (without loss of generality we may assume that they are irreducible) and $a_i \in \mathbb{C}$, with respect to the multiplication

$$[V] \cdot [W] = [V \otimes W].$$

Because $\operatorname{Rep} {}^L G$ is commutative, its irreducible modules are all one-dimensional. They correspond to characters $\operatorname{Rep} {}^L G \to \mathbb{C}$. We have a bijection

$$\boxed{\begin{array}{c}\text{semi-simple conjugacy}\\ \text{classes in } {}^L G\end{array}} \quad \Longleftrightarrow \quad \boxed{\begin{array}{c}\text{characters}\\ \text{of } \operatorname{Rep} {}^L G\end{array}} \tag{10.3.4}$$

where the character ϕ_γ corresponding to the conjugacy class γ is given by the formula†

$$\phi_\gamma : [V] \mapsto \operatorname{Tr}(\gamma, V).$$

Now, if (R, π) is a representation of $G(F)$, then the space $R^{\pi(K_0)}$ of K_0-invariant vectors in V is a module over $H(G(F), K_0)$. It is easy to show

† It is customary to multiply the right hand side of this formula, for irreducible representation V, by a scalar depending on q and the highest weight of V, but this is not essential for our discussion.

that this sets up a one-to-one correspondence between equivalence classes of irreducible unramified representations of $G(F)$ and irreducible $H(G(F), K_0)$-modules. Combining this with the bijection (10.3.4) and the isomorphism (10.3.3), we obtain the sought-after bijections (10.3.1) and (10.3.2).

In particular, we find that, because the Hecke algebra $H(G(F), K_0)$ is commutative, the space $R^{\pi(K_0)}$ of K_0-invariants of an irreducible representation, which is an irreducible $H(G(F), K_0)$-module, is either 0 or one-dimensional. If it is one-dimensional, then $H(G(F), K_0)$ acts on it by the character ϕ_γ for some γ:

$$H_V \star v = \phi_\gamma([V])v = \mathrm{Tr}(\gamma, V)v, \qquad v \in R^{\pi(K_0)}, [V] \in \mathrm{Rep}\, {}^L G, \qquad (10.3.5)$$

where H_V is the element of $H(G(F), K_0)$ corresponding to $[V]$ under the isomorphism (10.3.3) (see formula (10.2.2) for the definition of the convolution action). Thus, any $v \in R^{\pi(K_0)}$ is a **Hecke eigenvector**.

We now discuss the categorical analogues of these statements.

10.3.2 Unramified categories of $\widehat{\mathfrak{g}}_{\kappa_c}$-modules

In the categorical setting, the role of an irreducible representation (R, π) of $G(F)$ is played by the category $\widehat{\mathfrak{g}}_{\kappa_c}\text{-mod}_\chi$ for some $\chi \in \mathrm{Op}_{^L G}(D^\times)$. The analogue of an unramified representation is a category $\widehat{\mathfrak{g}}_{\kappa_c}\text{-mod}_\chi$ which contains non-zero $(\widehat{\mathfrak{g}}_{\kappa_c}, G[[t]])$ Harish-Chandra modules. This leads us to the following question: for what $\chi \in \mathrm{Op}_{^L G}(D^\times)$ does the category $\widehat{\mathfrak{g}}_{\kappa_c}\text{-mod}_\chi$ contain non-zero $(\widehat{\mathfrak{g}}_{\kappa_c}, G[[t]])$ Harish-Chandra modules?

We saw in the previous section that (R, π) is unramified if and only if it corresponds to an unramified Langlands parameter, which is a homomorphism $W'_F \to {}^L G$ that factors through $W'_F \to \mathbb{Z}$. Recall that in the geometric setting the Langlands parameters are ${}^L G$-local systems on $D^\times$. The analogues of unramified homomorphisms $W'_F \to {}^L G$ are those local systems on $D^\times$ which extend to the disc D, in other words, have no singularity at the origin $0 \in D$. Note that there is a unique, up to isomorphism local system on D. Indeed, suppose that we are given a regular connection on a ${}^L G$-bundle $\mathcal{F}$ on D. Let us trivialize the fiber $\mathcal{F}_0$ of $\mathcal{F}$ at $0 \in D$. Then, because D is contractible, the connection identifies $\mathcal{F}$ with the trivial bundle on D. Under this identification the connection itself becomes trivial, i.e., represented by the operator $\nabla = \partial_t$.

Therefore all regular ${}^L G$-local systems (i.e., those which extend to D) correspond to a single point of the set $\mathrm{Loc}_{^L G}(D^\times)$; namely, the equivalence class of the trivial local system σ_0.† From the point of view of the realization of $\mathrm{Loc}_{^L G}(D^\times)$ as the quotient (1.2.7) this simply means that there is a unique

† Note however that the trivial ${}^L G$-local system on D has a non-trivial group of automorphisms, namely, the group ${}^L G$ itself (it may be realized as the group of automorphisms of the fiber at $0 \in D$). Therefore if we think of $\mathrm{Loc}_{^L G}(D^\times)$ as a stack rather than as a set, then the trivial local system corresponds to a substack $\mathrm{pt}\,/{}^L G$.

${}^LG((t))$ gauge equivalence class containing all regular connections of the form $\partial_t + A(t)$, where $A(t) \in {}^L\mathfrak{g}[[t]]$.

The gauge equivalence class of regular connections is the unique local Langlands parameter that we may view as unramified in the geometric setting. Therefore, by analogy with the unramified Langlands correspondence for $G(F)$, we expect that the category $\widehat{\mathfrak{g}}_{\kappa_c}\text{-mod}_\chi$ contains non-zero $(\widehat{\mathfrak{g}}_{\kappa_c}, G[[t]])$ Harish-Chandra modules if and only if the ${}^LG$-oper $\chi \in \operatorname{Op}_{{}^LG}(D^\times)$ is ${}^LG((t))$ gauge equivalent to the trivial connection, or, in other words, χ belongs to the fiber $\alpha^{-1}(\sigma_0)$ over σ_0.

What does this fiber look like? Let P^+ be the set of dominant integral weights of G (equivalently, dominant integral coweights of ${}^LG$). In Section 9.2.3 we defined, for each $\lambda \in P^+$, the space $\operatorname{Op}^\lambda_{{}^LG}$ of ${}^LB[[t]]$-equivalence classes of operators of the form

$$\nabla = \partial_t + \sum_{i=1}^{\ell} t^{\langle \check{\alpha}_i, \lambda \rangle} \psi_i(t) f_i + \mathbf{v}(t), \tag{10.3.6}$$

where $\psi_i(t) \in \mathbb{C}[[t]], \psi_i(0) \neq 0$, $\mathbf{v}(t) \in {}^L\mathfrak{b}[[t]]$.

Lemma 10.3.1 *Suppose that the local system which underlies an oper $\chi \in \operatorname{Op}_{{}^LG}(D^\times)$ is trivial. Then χ belongs to the disjoint union of the subsets $\operatorname{Op}^\lambda_{{}^LG} \subset \operatorname{Op}_{{}^LG}(D^\times), \lambda \in P^+$.*

Proof. It is clear from the definition that any oper in $\operatorname{Op}^\lambda_{{}^LG}$ is regular on the disc D and is therefore ${}^LG((t))$ gauge equivalent to the trivial connection.

Now suppose that we have an oper $\chi = (\mathcal{F}, \nabla, \mathcal{F}_{{}^LB})$ such that the underlying ${}^LG$-local system is trivial. Then ∇ is ${}^LG((t))$ gauge equivalent to a regular connection, that is one of the form $\partial_t + A(t)$, where $A(t) \in {}^L\mathfrak{g}[[t]]$. We have the decomposition ${}^LG((t)) = {}^LG[[t]]{}^LB((t))$. The gauge action of ${}^LG[[t]]$ clearly preserves the space of regular connections. Therefore if an oper connection ∇ is ${}^LG((t))$ gauge equivalent to a regular connection, then its ${}^LB((t))$ gauge class already must contain a regular connection. The oper condition then implies that this gauge class contains a connection operator of the form (10.3.6) for some dominant integral weight λ of ${}^LG$. Therefore $\chi \in \operatorname{Op}^\lambda_{{}^LG}$. □

Thus, we see that the set of opers corresponding to the (unique) unramified Langlands parameter is the disjoint union $\bigsqcup_{\lambda \in P^+} \operatorname{Op}^\lambda_{{}^LG}$. We call such opers "unramified." The following result then confirms our expectation that the category $\widehat{\mathfrak{g}}_{\kappa_c}\text{-mod}_\chi$ corresponding to an unramified oper χ is also "unramified," that is, contains non-zero $G[[t]]$-equivariant objects, if and only if χ is unramified.

Lemma 10.3.2 *The category* $\widehat{\mathfrak{g}}_{\kappa_c}\text{-mod}_\chi$ *contains a non-zero*

$(\widehat{\mathfrak{g}}_{\kappa_c}, G[[t]])$ *Harish-Chandra module if and only if*

$$\chi \in \bigsqcup_{\lambda \in P^+} \mathrm{Op}^{\lambda}_{^L G}. \tag{10.3.7}$$

Proof. Let us write $G[[t]] = G \ltimes K_1$, where K_1 is the pro-unipotent pro-algebraic group, which is the first congruence subgroup of $G[[t]]$. A $\widehat{\mathfrak{g}}_{\kappa_c}$-module M is $G[[t]]$-equivariant if and only if the action of the Lie algebra $\mathfrak{g} \otimes t\mathbb{C}[[t]]$ of K_1 on M is locally nilpotent and exponentiates to an action of K_1 and in addition M decomposes into a direct sum of finite-dimensional representations under the action of the Lie subalgebra $\mathfrak{g} \otimes 1$.

Suppose now that the category $\widehat{\mathfrak{g}}_{\kappa_c}\text{-mod}_\chi$ contains a non-zero $(\widehat{\mathfrak{g}}_{\kappa_c}, G[[t]])$ Harish-Chandra module M. Since M is smooth, any vector $v \in M$ is annihilated by $\mathfrak{g} \otimes t^N\mathbb{C}[[t]]$ for some $N > 0$. Consider the $\mathfrak{g} \otimes t\mathbb{C}[[t]]$-submodule $M_v = U(\mathfrak{g} \otimes t\mathbb{C}[[t]])v$ of M. By construction, the action of $\mathfrak{g} \otimes t\mathbb{C}[[t]]$ on M_v factors through the quotient $(\mathfrak{g} \otimes t\mathbb{C}[[t]])/(\mathfrak{g} \otimes t^N\mathbb{C}[[t]])$. Since M is $G[[t]]$-equivariant, it follows that the action of this Lie algebra may be exponentiated to an algebraic action of the corresponding algebraic group K_1/K_N, where K_N is the Nth congruence subgroup of $G[[t]]$, which is a normal subgroup of K_1. The group K_1/K_N is a finite-dimensional unipotent algebraic group, and we obtain an algebraic representations of K_1/K_N on M_v. Any such representation contains an invariant vector.† Thus, we obtain that M_v, and hence M, contains a vector v' annihilated by $\mathfrak{g} \otimes t\mathbb{C}[[t]]$.

The $\mathfrak{g}[[t]]$-submodule (equivalently, $\mathfrak{g}$-submodule) of M generated by v' is a direct sum of irreducible finite-dimensional representations of $\mathfrak{g}$. Let us pick one of those irreducible representations and denote it by M'. By construction, M' is annihilated by $\mathfrak{g} \otimes t\mathbb{C}[[t]]$ and it is isomorphic to V_λ, the irreducible representation of $\mathfrak{g}$ with highest weight λ, as a $\mathfrak{g}$-module. Therefore there exists a non-zero homomorphism $\mathbb{V}_\lambda \to M$, where $\mathbb{V}_\lambda$ is the Weyl module introduced in Section 9.6, sending the generating subspace $V_\lambda \subset \mathbb{V}_\lambda$ to M'.

By our assumption, M is an object of the category $\widehat{\mathfrak{g}}_{\kappa_c}\text{-mod}_\chi$ on which the center $Z(\widehat{\mathfrak{g}})$ acts according to the central character corresponding to $\chi \in \mathrm{Op}_{^L G}(D^\times)$. The existence of a non-zero homomorphism $\mathbb{V}_\lambda \to M$ implies that χ belongs to the spectrum of the image of $Z(\widehat{\mathfrak{g}})$ in $\mathrm{End}_{\widehat{\mathfrak{g}}_{\kappa_c}} \mathbb{V}_\lambda$. According to Theorem 9.6.1, the latter spectrum is equal to $\mathrm{Op}^{\lambda}_{^L G}$. Therefore we obtain that $\chi \in \mathrm{Op}^{\lambda}_{^L G}$.

On the other hand, if $\chi \in \mathrm{Op}^{\lambda}_{^L G}$, then the quotient of $\mathbb{V}_\lambda$ by the central character corresponding to χ is non-zero, by Theorem 9.6.1. This quotient is clearly a $G[[t]]$-equivariant object of the category $\widehat{\mathfrak{g}}_{\kappa_c}\text{-mod}_\chi$. This completes the proof. □

The next question is to describe the category $\widehat{\mathfrak{g}}_{\kappa_c}\text{-mod}_\chi^{G[[t]]}$ of $(\widehat{\mathfrak{g}}_{\kappa_c}, G[[t]])$ Harish-Chandra modules for $\chi \in \mathrm{Op}^{\lambda}_{^L G}$.

† This follows by induction from the analogous statement for the additive group $\mathbb{G}_a$.

10.3.3 Categories of $G[[t]]$-equivariant modules

Let us recall from Section 10.3.1 that the space of K_0-invariant vectors in an unramified irreducible representation of $G(F)$ is always one-dimensional. According to our proposal, the category $\widehat{\mathfrak{g}}_{\kappa_c}\text{-mod}_\chi^{G[[t]]}$ should be viewed as a categorical analogue of this space. Therefore we expect it to be the simplest possible abelian category: the category of $\mathbb{C}$-vector spaces. Recall that here we assume that χ belongs to the union of the spaces $\mathrm{Op}^\lambda_{^LG}$, where $\lambda \in P^+$, for otherwise the category $\widehat{\mathfrak{g}}_{\kappa_c}\text{-mod}_\chi^{G[[t]]}$ would be trivial (i.e., the zero object would be the only object).

In this subsection we will prove, following [FG1] (see also [BD1]), that our expectation is in fact correct provided that $\lambda = 0$, in which case $\mathrm{Op}^0_{^LG} = \mathrm{Op}_{^LG}(D)$, and so

$$\chi \in \mathrm{Op}_{^LG}(D) \subset \mathrm{Op}_{^LG}(D^\times).$$

We will also conjecture that this is true if $\chi \in \mathrm{Op}^\lambda_{^LG}$ for all $\lambda \in P^+$.

Recall the vacuum module $\mathbb{V}_0 = V_{\kappa_c}(\mathfrak{g})$. According to the isomorphism (9.5.1), we have

$$\mathrm{End}_{\widehat{\mathfrak{g}}_{\kappa_c}} \mathbb{V}_0 \simeq \mathrm{Fun}\,\mathrm{Op}_{^LG}(D). \tag{10.3.8}$$

Let $\chi \in \mathrm{Op}_{^LG}(D) \subset \mathrm{Op}_{^LG}(D^\times)$. Then χ defines a character of the algebra $\mathrm{End}_{\widehat{\mathfrak{g}}_{\kappa_c}} \mathbb{V}_0$. Let $\mathbb{V}_0(\chi)$ be the quotient of $\mathbb{V}_0$ by the kernel of this character. Then we have the following result.

Theorem 10.3.3 *Let $\chi \in \mathrm{Op}_{^LG}(D) \subset \mathrm{Op}_{^LG}(D^\times)$. Then the category $\widehat{\mathfrak{g}}_{\kappa_c}\text{-mod}_\chi^{G[[t]]}$ is equivalent to the category of vector spaces: its unique, up to isomorphism, irreducible object is $\mathbb{V}_0(\chi)$ and any other object is isomorphic to a direct sum of copies of $\mathbb{V}_0(\chi)$.*

This theorem may be viewed as a categorical analogue of the local unramified Langlands correspondence discussed in Section 10.3.1. It also provides the first piece of evidence for Conjecture 10.1.2: we see that the categories $\widehat{\mathfrak{g}}_{\kappa_c}\text{-mod}_\chi^{G[[t]]}$ are equivalent to each other for all $\chi \in \mathrm{Op}_{^LG}(D)$.

It is more convenient to consider, instead of an individual regular ^LG-oper χ, the entire family $\mathrm{Op}^0_{^LG} = \mathrm{Op}_{^LG}(D)$ of regular opers on the disc D. Let $\widehat{\mathfrak{g}}_{\kappa_c}\text{-mod}_{\mathrm{reg}}$ be the full subcategory of the category $\widehat{\mathfrak{g}}_{\kappa_c}\text{-mod}$ whose objects are $\widehat{\mathfrak{g}}_{\kappa_c}$-modules on which the action of the center $Z(\widehat{\mathfrak{g}})$ factors through the homomorphism

$$Z(\widehat{\mathfrak{g}}) \simeq \mathrm{Fun}\,\mathrm{Op}_{^LG}(D^\times) \to \mathrm{Fun}\,\mathrm{Op}_{^LG}(D).$$

Note that the category $\widehat{\mathfrak{g}}_{\kappa_c}\text{-mod}_{\mathrm{reg}}$ is an example of a category $\widehat{\mathfrak{g}}_{\kappa_c}\text{-mod}_Y$ introduced in Section 10.1, in the case when $Y = \mathrm{Op}_{^LG}(D)$.

Let $\widehat{\mathfrak{g}}_{\kappa_c}\text{-mod}_{\mathrm{reg}}^{G[[t]]}$ be the corresponding $G[[t]]$-equivariant category. It is instructive to think of $\widehat{\mathfrak{g}}_{\kappa_c}\text{-mod}_{\mathrm{reg}}$ and $\widehat{\mathfrak{g}}_{\kappa_c}\text{-mod}_{\mathrm{reg}}^{G[[t]]}$ as categories fibered over $\mathrm{Op}_{^LG}(D)$, with the fibers over $\chi \in \mathrm{Op}_{^LG}(D)$ being $\widehat{\mathfrak{g}}_{\kappa_c}\text{-mod}_\chi$ and $\widehat{\mathfrak{g}}_{\kappa_c}\text{-mod}_\chi^{G[[t]]}$, respectively.

We will now describe the category $\widehat{\mathfrak{g}}_{\kappa_c}\text{-mod}_{\mathrm{reg}}^{G[[t]]}$. This description will in particular imply Theorem 10.3.3.

In order to simplify our formulas, in what follows we will use the following notation for $\mathrm{Fun}\,\mathrm{Op}_{^LG}(D)$:

$$\mathfrak{z} = \mathfrak{z}(\widehat{\mathfrak{g}}) = \mathrm{Fun}\,\mathrm{Op}_{^LG}(D).$$

Let $\mathfrak{z}$-mod be the category of modules over the commutative algebra $\mathfrak{z}$. Equivalently, this is the category of quasicoherent sheaves on $\mathrm{Op}_{^LG}(D)$.

By definition, any object of $\widehat{\mathfrak{g}}_{\kappa_c}\text{-mod}_{\mathrm{reg}}^{G[[t]]}$ is a $\mathfrak{z}$-module. Introduce the functors

$$\begin{aligned} \mathsf{F} : \widehat{\mathfrak{g}}_{\kappa_c}\text{-mod}_{\mathrm{reg}}^{G[[t]]} \to \mathfrak{z}\text{-mod}, \qquad & M \mapsto \mathrm{Hom}_{\widehat{\mathfrak{g}}_{\kappa_c}}(\mathbb{V}_0, M), \\ \mathsf{G} : \mathfrak{z}\text{-mod} \to \widehat{\mathfrak{g}}_{\kappa_c}\text{-mod}_{\mathrm{reg}}^{G[[t]]}, \qquad & \mathcal{F} \mapsto \mathbb{V}_0 \underset{\mathfrak{z}}{\otimes} \mathcal{F}. \end{aligned}$$

The following theorem has been proved in [FG1], Theorem 6.3 (important results in this direction were obtained earlier in [BD1]).

Theorem 10.3.4 *The functors* F *and* G *are mutually inverse equivalences of categories*

$$\widehat{\mathfrak{g}}_{\kappa_c}\text{-mod}_{\mathrm{reg}}^{G[[t]]} \simeq \mathfrak{z}\text{-mod}. \tag{10.3.9}$$

We will present the proof in the next section. Before doing this, let us note that it immediately implies Theorem 10.3.3. Indeed, for each $\chi \in \mathrm{Op}_{^LG}(D)$ the category $\widehat{\mathfrak{g}}_{\kappa_c}\text{-mod}_\chi^{G[[t]]}$ is the full subcategory of $\widehat{\mathfrak{g}}_{\kappa_c}\text{-mod}_{\mathrm{reg}}^{G[[t]]}$ whose objects are the $\widehat{\mathfrak{g}}_{\kappa_c}$-modules which are annihilated, as $\mathfrak{z}$-modules, by the maximal ideal I_χ of χ. By Theorem 10.3.4, this category is equivalent to the category of $\mathfrak{z}$-modules annihilated by I_χ. But this is the category of $\mathfrak{z}$-modules supported (scheme-theoretically) at the point χ, which is equivalent to the category of vector spaces.

10.3.4 Proof of Theorem 10.3.4

First of all, we note that the functor F is faithful, i.e., if $M \neq 0$, then $\mathsf{F}(M)$ is non-zero. Indeed, note that

$$\mathsf{F}(M) = \mathrm{Hom}_{\widehat{\mathfrak{g}}_{\kappa_c}}(\mathbb{V}_0, M) = M^{\mathfrak{g}[[t]]},$$

because any $\widehat{\mathfrak{g}}_{\kappa_c}$-homomorphism $\mathbb{V}_0 \to M$ is uniquely determined by the image in M of the generating vector of $\mathbb{V}_0$, which is $\mathfrak{g}[[t]]$-invariant. As shown in the proof of Lemma 10.3.2, any non-zero $\widehat{\mathfrak{g}}_{\kappa_c}$-module M in $\widehat{\mathfrak{g}}_{\kappa_c}\text{-mod}_{\mathrm{reg}}^{G[[t]]}$ necessarily contains a non-zero vector annihilated by $\mathfrak{g} \otimes t\mathbb{C}[[t]]$. The $\mathfrak{g}$-module generated by this vector has to be trivial, for otherwise the action of the center on it would not factor through $\mathfrak{z}$. Hence M contains a non-zero vector annihilated by $\mathfrak{g}[[t]]$, and so $M^{\mathfrak{g}[[t]]} \neq 0$.

Next, we observe that the functor G is left adjoint to the functor F, i.e., we have a compatible system of isomorphisms

$$\operatorname{Hom}(\mathcal{F}, \mathsf{F}(M)) \simeq \operatorname{Hom}(\mathsf{G}(\mathcal{F}), M). \tag{10.3.10}$$

In particular, taking $M = \mathsf{G}(\mathcal{F})$ in this formula, we obtain a compatible system of maps

$$\mathcal{F} \to \mathsf{F} \circ \mathsf{G}(\mathcal{F}),$$

i.e.,

$$\mathcal{F} \to \operatorname{Hom}_{\widehat{\mathfrak{g}}_{\kappa_c}}\left(\mathbb{V}_0, \mathbb{V}_0 \underset{\mathfrak{z}}{\otimes} \mathcal{F}\right). \tag{10.3.11}$$

We claim that this is in fact an isomorphism. For this we use the following two results.

Theorem 10.3.5 *The $\mathfrak{z}$-module $\mathbb{V}_0$ is free.*

Proof. Recall that we have the PBW filtration on $\mathbb{V}_0$, and the associated graded is isomorphic to $\operatorname{Fun}\mathfrak{g}^*[[t]]$ (see Section 3.3.3). The PBW filtration induces a filtration on $\mathfrak{z}(\widehat{\mathfrak{g}}) = (\mathbb{V}_0)^{\mathfrak{g}[[t]]}$, and the associated graded is isomorphic to $\operatorname{Inv}\mathfrak{g}^*[[t]] = (\operatorname{Fun}\mathfrak{g}^*[[t]])^{\mathfrak{g}[[t]]}$ (see Proposition 4.3.3). Now, according to [EF], $\operatorname{Fun}\mathfrak{g}^*[[t]]$ is a free module over $\operatorname{Inv}\mathfrak{g}^*[[t]]$. This implies in a straightforward way that the same holds for the original filtered objects, and so $\mathbb{V}_0$ is free over $\mathfrak{z} = \operatorname{Fun}\operatorname{Op}_{{}^LG}(D)$. □

This has the following immediate consequence.

Corollary 10.3.6 *The functor G is exact.*

Next, let $\widehat{\mathfrak{g}}_{\kappa_c}\text{-mod}^{G[[t]]}$ be the category of $(\widehat{\mathfrak{g}}_{\kappa_c}, G[[t]])$ Harish-Chandra modules (without any restrictions on the action of the center). Let

$$\operatorname{Ext}^i_{\widehat{\mathfrak{g}}, G[[t]]}(\mathbb{V}_0, M)$$

be the higher derived functors of the functor

$$M \mapsto \operatorname{Hom}_{\widehat{\mathfrak{g}}_{\kappa_c}}(\mathbb{V}_0, M)$$

from $\widehat{\mathfrak{g}}_{\kappa_c}\text{-mod}^{G[[t]]}$ to the category of vector spaces.

Theorem 10.3.7 ([FT]) *We have*

$$\operatorname{Ext}^i_{\widehat{\mathfrak{g}}, G[[t]]}(\mathbb{V}_0, \mathbb{V}_0) \simeq \Omega^i_{\mathfrak{z}}, \tag{10.3.12}$$

the space of differential forms of degree i on $\operatorname{Op}_{{}^LG}(D)$.

Note that this is a generalization of the isomorphism

$$\mathrm{Hom}_{\widehat{\mathfrak{g}}_{\kappa_c}}(\mathbb{V}_0, \mathbb{V}_0) \simeq \mathfrak{z},$$

corresponding to the case $i = 0$ in formula (10.3.12).

We will now prove the following isomorphisms:

$$\mathrm{Ext}^i_{\widehat{\mathfrak{g}}_{\kappa_c}, G[[t]]}(\mathbb{V}_0, \mathbb{V}_0) \underset{\mathfrak{z}}{\otimes} \mathcal{F} \simeq \mathrm{Ext}^i_{\widehat{\mathfrak{g}}_{\kappa_c}, G[[t]]}\left(\mathbb{V}_0, \mathbb{V}_0 \underset{\mathfrak{z}}{\otimes} \mathcal{F}\right), \qquad i \geq 0. \quad (10.3.13)$$

For $i = 0$ we will then obtain that the map (10.3.11) is an isomorphism, as needed.

In order to prove the isomorphism (10.3.13), we note that the functors

$$\mathcal{F} \mapsto \mathrm{Ext}^i_{\widehat{\mathfrak{g}}_{\kappa_c}, G[[t]]}\left(\mathbb{V}_0, \mathbb{V}_0 \underset{\mathfrak{z}}{\otimes} \mathcal{F}\right)$$

commute with taking the direct limits. This follows from their realization as the cohomology of the Chevalley complex computing the relative Lie algebra cohomology $H^i\left(\mathfrak{g}[[t]], \mathfrak{g}, \mathbb{V}_0 \underset{\mathfrak{z}}{\otimes} \mathcal{F}\right)$, as shown in [FT], Prop. 2.1 and Lemma 3.1. Therefore without loss of generality we may, and will, assume that $\mathcal{F}$ is a finitely presented $\mathfrak{z}$-module.

Since $\mathfrak{z}$ is isomorphic to a free polynomial algebra (see formula (4.3.1)), any such module admits a finite resolution

$$0 \to \mathcal{P}_n \to \ldots \to \mathcal{P}_1 \to \mathcal{P}_0 \to \mathcal{F} \to 0$$

by projective modules. According to standard results of homological algebra, the right hand side of (10.3.13) may be computed by the spectral sequence, whose term $E_1^{p,q}$ is isomorphic to

$$\mathrm{Ext}^p_{\widehat{\mathfrak{g}}_{\kappa_c}, G[[t]]}\left(\mathbb{V}_0, \mathbb{V}_0 \underset{\mathfrak{z}}{\otimes} \mathcal{P}_{-q}\right).$$

Since each $\mathcal{P}_j$ is projective, we have

$$\mathrm{Ext}^p_{\widehat{\mathfrak{g}}_{\kappa_c}, G[[t]]}\left(\mathbb{V}_0, \mathbb{V}_0 \underset{\mathfrak{z}}{\otimes} \mathcal{P}_{-q}\right) = \mathrm{Ext}^p_{\widehat{\mathfrak{g}}_{\kappa_c}, G[[t]]}(\mathbb{V}_0, \mathbb{V}_0) \underset{\mathfrak{z}}{\otimes} \mathcal{P}_{-q}.$$

Next, by Theorem 10.3.7, each $\mathrm{Ext}^p_{\widehat{\mathfrak{g}}_{\kappa_c}, G[[t]]}(\mathbb{V}_0, \mathbb{V}_0)$ is a free $\mathfrak{z}$-module. Therefore the second term of the spectral sequence is equal to

$$E_2^{p,0} = \mathrm{Ext}^p_{\widehat{\mathfrak{g}}_{\kappa_c}, G[[t]]}(\mathbb{V}_0, \mathbb{V}_0) \underset{\mathfrak{z}}{\otimes} \mathcal{F},$$

and $E^{p,q} = 0$ for $q \neq 0$. Therefore the spectral sequence degenerates in the second term and the result is the isomorphism (10.3.13).

Thus, we have proved that $\mathsf{F} \circ \mathsf{G} \simeq \mathrm{Id}$. Let us prove that $\mathsf{G} \circ \mathsf{F} \simeq \mathrm{Id}$.

Note that by adjunction (10.3.10) we have a map $\mathsf{G} \circ \mathsf{F} \to \mathrm{Id}$, i.e., a compatible system of maps

$$\mathsf{G} \circ \mathsf{F}(M) \to M, \qquad (10.3.14)$$

for M in $\widehat{\mathfrak{g}}_{\kappa_c}\text{-mod}_{\text{reg}}^{G[[t]]}$. We need to show that this map is an isomorphism.

Let us show that the map (10.3.14) is injective. If M' is the kernel of $\mathsf{G} \circ \mathsf{F}(M) \to M$, then by the left exactness of F, we would obtain that

$$\mathsf{F}(M') = \text{Ker}\big(\mathsf{F} \circ \mathsf{G} \circ \mathsf{F}(M) \to \mathsf{F}(M)\big) \simeq \text{Ker}\big(\mathsf{F}(M) \to \mathsf{F}(M)\big) = 0.$$

But we know that the functor F is faithful, so $M' = 0$.

Next, we show that the map (10.3.14) is surjective. Let M'' be the cokernel of $\mathsf{G} \circ \mathsf{F}(M) \to M$. We have the long exact sequence

$$0 \to \mathsf{F} \circ \mathsf{G} \circ \mathsf{F}(M) \to \mathsf{F}(M) \to \mathsf{F}(M'') \to R^1\mathsf{F}(\mathsf{G} \circ \mathsf{F}(M)) \to \dots, \qquad (10.3.15)$$

where $R^1\mathsf{F}$ is the first right derived functor of the functor F. Thus, we have

$$R^1\mathsf{F}(L) = \text{Ext}^1_{\widehat{\mathfrak{g}}_{\kappa_c}\text{-mod}_{\text{reg}}^{G[[t]]}}(\mathbb{V}_0, L).$$

To complete the proof, we need the following:

Lemma 10.3.8 *For any $\mathfrak{z}$-module $\mathcal{F}$ we have*

$$\text{Ext}^1_{\widehat{\mathfrak{g}}_{\kappa_c}\text{-mod}_{\text{reg}}^{G[[t]]}}\left(\mathbb{V}_0, \mathbb{V}_0 \underset{\mathfrak{z}}{\otimes} \mathcal{F}\right) = 0.$$

Proof. In the case when $\mathcal{F} = \mathfrak{z}$ this was proved in [BD1]. Here we follow the proof given in [FG1] in the general case.

Any element of

$$\text{Ext}^1_{\widehat{\mathfrak{g}}_{\kappa_c}\text{-mod}^{G[[t]]}}\left(\mathbb{V}_0, \mathbb{V}_0 \underset{\mathfrak{z}}{\otimes} \mathcal{F}\right) = \text{Ext}^1_{\widehat{\mathfrak{g}}_{\kappa_c}, G[[t]]}\left(\mathbb{V}_0, \mathbb{V}_0 \underset{\mathfrak{z}}{\otimes} \mathcal{F}\right)$$

defines an extension of the $\widehat{\mathfrak{g}}_{\kappa_c}$-modules

$$0 \to \mathbb{V}_0 \underset{\mathfrak{z}}{\otimes} \mathcal{F} \to \mathcal{M} \to \mathbb{V}_0 \to 0. \qquad (10.3.16)$$

Thus, $\mathcal{M}$ is an extension of two objects of the category $\widehat{\mathfrak{g}}_{\kappa_c}\text{-mod}_{\text{reg}}^{G[[t]]}$, on which the center $Z(\widehat{\mathfrak{g}})$ acts through the quotient $Z(\widehat{\mathfrak{g}}) \to \mathfrak{z}$. But the module $\mathcal{M}$ may not be an object of $\widehat{\mathfrak{g}}_{\kappa_c}\text{-mod}_{\text{reg}}^{G[[t]]}$. The statement of the proposition means that if $\mathcal{M}$ is an object of $\widehat{\mathfrak{g}}_{\kappa_c}\text{-mod}_{\text{reg}}^{G[[t]]}$, then it is necessarily split as a $\widehat{\mathfrak{g}}_{\kappa_c}$-module.

Let I be the ideal of $\mathfrak{z}(\widehat{\mathfrak{g}})$ in $Z(\widehat{\mathfrak{g}})$. Then $\mathcal{M}$ is an object of $\widehat{\mathfrak{g}}_{\kappa_c}\text{-mod}_{\text{reg}}^{G[[t]]}$ if and only if I acts on it by 0. Note that I necessarily vanishes on the submodule $\mathbb{V}_0 \underset{\mathfrak{z}}{\otimes} \mathcal{F}$ and the quotient $\mathbb{V}_0$. Therefore, choosing a linear splitting of (10.3.16) and applying I to the image of $\mathbb{V}_0$ in $\mathcal{M}$, we obtain a map

$$I \to \text{Hom}_{\widehat{\mathfrak{g}}_{\kappa_c}}\left(\mathbb{V}_0, \mathbb{V}_0 \underset{\mathfrak{z}}{\otimes} \mathcal{F}\right).$$

Moreover, any element of $I^2 \subset I$ maps to 0. Thus, we obtain a map

$$(I/I^2) \underset{\mathfrak{z}}{\otimes} \text{Ext}^1_{\widehat{\mathfrak{g}}_{\kappa_c}, G[[t]]}\left(\mathbb{V}_0, \mathbb{V}_0 \underset{\mathfrak{z}}{\otimes} \mathcal{F}\right) \to \text{Hom}_{\widehat{\mathfrak{g}}_{\kappa_c}}\left(\mathbb{V}_0, \mathbb{V}_0 \underset{\mathfrak{z}}{\otimes} \mathcal{F}\right).$$

We need to show that for any element of $\mathrm{Ext}^1_{\widehat{\mathfrak{g}}_{\kappa_c},G[[t]]}\left(\mathbb{V}_0,\mathbb{V}_0\underset{\mathfrak{z}}{\otimes}\mathcal{F}\right)$ the corresponding map

$$I/I^2 \to \mathrm{Hom}_{\widehat{\mathfrak{g}}_{\kappa_c}}\left(\mathbb{V}_0,\mathbb{V}_0\underset{\mathfrak{z}}{\otimes}\mathcal{F}\right)$$

is non-zero. Equivalently, we need to show that the kernel of the map

$$\mathrm{Ext}^1_{\widehat{\mathfrak{g}}_{\kappa_c},G[[t]]}\left(\mathbb{V}_0,\mathbb{V}_0\underset{\mathfrak{z}}{\otimes}\mathcal{F}\right)\to \mathrm{Hom}_{\mathfrak{z}}\left(I/I^2,\mathrm{Hom}_{\widehat{\mathfrak{g}}_{\kappa_c}}\left(\mathbb{V}_0,\mathbb{V}_0\underset{\mathfrak{z}}{\otimes}\mathcal{F}\right)\right)$$

is 0. In fact, this kernel is equal to $\mathrm{Ext}^1_{\widehat{\mathfrak{g}}_{\kappa_c}\text{-mod}^{G[[t]]}_{\mathrm{reg}}}\left(\mathbb{V}_0,\mathbb{V}_0\underset{\mathfrak{z}}{\otimes}\mathcal{F}\right)$, whose vanishing we wish to prove.

Now, according to Theorem 10.3.7 and formula (10.3.13), the last map may be rewritten as follows:

$$\Omega^1_{\mathfrak{z}}\underset{\mathfrak{z}}{\otimes}\mathcal{F}\to \mathrm{Hom}_{\mathfrak{z}}\left(I/I^2,\mathcal{F}\right). \tag{10.3.17}$$

In order to write down an explicit formula for this map, let us explain, following [FT], how to construct an isomorphism

$$\Omega^1_{\mathfrak{z}}\underset{\mathfrak{z}}{\otimes}\mathcal{F}\simeq \mathrm{Ext}^1_{\widehat{\mathfrak{g}}_{\kappa_c},G[[t]]}\left(\mathbb{V}_0,\mathbb{V}_0\underset{\mathfrak{z}}{\otimes}\mathcal{F}\right), \tag{10.3.18}$$

or, in other words, how to construct an extension $\mathcal{M}$ of the form (10.3.16) starting from an element of $\Omega^1_{\mathfrak{z}}\underset{\mathfrak{z}}{\otimes}\mathcal{F}$. For simplicity let us suppose that $\mathcal{F}=\mathfrak{z}$ (the general case is similar). Then as a vector space, the extension $\mathcal{M}$ is the direct sum of two copies of $\mathbb{V}_0$. By linearlity, we may assume that our element of $\Omega^1_{\mathfrak{z}}$ has the form BdA, where $A,B\in\mathfrak{z}$. Since $\mathbb{V}_0$ is generated by the vacuum vector $|0\rangle$, the extension $\mathcal{M}$ is uniquely determined by the action of $\mathfrak{g}[[t]]$ on the vector $|0\rangle$ in the summand corresponding to the quotient $\mathbb{V}_0$. We then set

$$X\cdot|0\rangle = Bcdot\lim_{\epsilon\to 0}\frac{1}{\epsilon}X\cdot A_\epsilon,\qquad X\in\mathfrak{g}[[t]].$$

Here A_ϵ is an arbitrary deformation of $A\in\mathfrak{z}(\widehat{\mathfrak{g}})\subset\mathbb{V}_0=V_{\kappa_c}(\mathfrak{g})$, considered as an element of the vacuum module $V_{\kappa_c+\epsilon\kappa_0}(\mathfrak{g})$ of level $\kappa_c+\epsilon\kappa_0$ (where κ_0 is a non-zero invariant inner product on $\mathfrak{g}$). Since $X\cdot A=0$ in $\mathbb{V}_0=V_{\kappa_c}(\mathfrak{g})$ for all $X\in\mathfrak{g}[[t]]$, the right hand side of this formula is well-defined.

Now recall that the center $Z(\widehat{\mathfrak{g}})$ is a Poisson algebra with the Poisson bracket defined in Section 8.3.1. Moreover, according to Theorem 8.3.1 and Lemma 8.3.2, this Poisson algebra is isomorphic to the Poisson algebra

$$\mathrm{Fun}\,\mathrm{Op}_G(D^\times)$$

with the Poisson algebra structure obtained via the Drinfeld–Sokolov reduction. It is easy to see that with respect to this Poisson structure the ideal I

is Poisson, i.e., $\{I, I\} \subset I$. In this situation the Poisson bracket map

$$\operatorname{Fun} \operatorname{Op}_G(D^\times) \to \Theta(\operatorname{Op}_G(D^\times)),$$

where $\Theta(\operatorname{Op}_G(D^\times))$ is the space of vector fields on $\operatorname{Op}_G(D^\times)$, induces a map

$$I/I^2 \to \Theta_{\mathfrak{z}}, \tag{10.3.19}$$

where $\Theta_{\mathfrak{z}}$ is the space of vector fields on $\operatorname{Op}_G(D)$.

It follows from the above description of the isomorphism (10.3.18) and the definition of the Poisson structure on $Z(\widehat{\mathfrak{g}})$ from Section 8.3.1 that the map (10.3.17) is the composition of the tautological isomorphism

$$\Omega^1_{\mathfrak{z}} \underset{\mathfrak{z}}{\otimes} \mathcal{F} \xrightarrow{\sim} \operatorname{Hom}_{\mathfrak{z}}(\Theta_{\mathfrak{z}}, \mathcal{F})$$

(note that $\Omega^1_{\mathfrak{z}}$ is a free $\mathfrak{z}$-module) and the map

$$\operatorname{Hom}_{\mathfrak{z}}(\Theta_{\mathfrak{z}}, \mathcal{F}) \to \operatorname{Hom}_{\mathfrak{z}}\left(I/I^2, \mathcal{F}\right) \tag{10.3.20}$$

induced by (10.3.19). But according to [BD1], Theorem 3.6.7, the map (10.3.19) is surjective (in fact, I/I^2 is isomorphic to the Lie algebroid on $\operatorname{Op}_{^LG}(D)$ corresponding to the universal ${}^LG$-bundle, and (10.3.19) is the corresponding anchor map, so its kernel is free as a $\mathfrak{z}$-module). Therefore we obtain that for any $\mathfrak{z}$-module $\mathcal{F}$ the map (10.3.20) is injective. Hence the map (10.3.17) is also injective. This completes the proof. $\square$

By Lemma 10.3.8, the sequence (10.3.15) gives us a short exact sequence

$$0 \to \mathsf{F} \circ \mathsf{G} \circ \mathsf{F}(M) \to \mathsf{F}(M) \to \mathsf{F}(M'') \to 0.$$

But the first arrow is an isomorphism because $\mathsf{F} \circ \mathsf{G} \simeq \operatorname{Id}$, which implies that $\mathsf{F}(M'') = 0$ and hence $M'' = 0$.

Theorem 10.3.4, and therefore Theorem 10.3.3, are now proved.

10.3.5 The action of the spherical Hecke algebra

In Section 10.3.1 we discussed irreducible unramified representations of the group $G(F)$, where F is a local non-archimedian field. We have seen that such representations are parameterized by conjugacy classes of the Langlands dual group ${}^LG$. Given such a conjugacy class γ, we have an irreducible unramified representation (R_γ, π_γ), which contains a one-dimensional subspace $(R_\gamma)^{\pi_\gamma(K_0)}$ of K_0-invariant vectors. The spherical Hecke algebra $H(G(F), K_0)$, which is isomorphic to $\operatorname{Rep} {}^LG$ via the Satake isomorphism, acts on this space by a character ϕ_γ, see formula (10.3.5).

In the geometric setting, we have argued that for any $\chi \in \operatorname{Op}_{^LG}(D)$ the category $\widehat{\mathfrak{g}}_{\kappa_c}\text{-mod}_\chi$, equipped with an action of the loop group $G((t))$, should be viewed as a categorification of (R_γ, π_γ). Furthermore, its subcategory $\widehat{\mathfrak{g}}_{\kappa_c}\text{-mod}_\chi^{G[[t]]}$ of $(\widehat{\mathfrak{g}}_{\kappa_c}, G[[t]])$ Harish-Chandra modules should be viewed as a "categorification" of the one-dimensional space $(R_\gamma)^{\pi_\gamma(K_0)}$. According to Theorem 10.3.3, the latter category is equivalent to the category of vector spaces, which is indeed the categorification of a one-dimensional vector space. So this result is consistent with the classical picture.

We now discuss the categorical analogue of the action of the spherical Hecke algebra on this one-dimensional space.

As explained in Section 10.2.3, the categorical analogue of the spherical Hecke algebra is the category of $G[[t]]$-equivariant $\mathcal{D}$-modules on the **affine Grassmannian** $\mathrm{Gr} = G((t))/G[[t]]$. We refer the reader to [BD1, FG2] for the precise definition of Gr and this category. There is an important property that is satisfied in the unramified case: the convolution functors with these $\mathcal{D}$-modules are exact, which means that we do not need to consider the derived category; the abelian category of such $\mathcal{D}$-modules will do. Let us denote this abelian category by $\mathcal{H}(G((t)), G[[t]])$. This is the categorical version of the spherical Hecke algebra $H(G((t)), G[[t]])$.

According to the results of [MV], the category $\mathcal{H}(G((t)), G[[t]])$ carries a natural structure of tensor category, which is equivalent to the tensor category $\mathcal{R}ep\,{}^L G$ of representations of ${}^L G$. This should be viewed as a categorical analogue of the Satake isomorphism. Thus, for each object V of $\mathcal{R}ep\,{}^L G$ we have an object of $\mathcal{H}(G((t)), G[[t]])$ which we denote by $\mathcal{H}_V$. What should be the analogue of the Hecke eigenvector property (10.3.5)?

As we explained in Section 10.2.3, the category $\mathcal{H}(G((t)), G[[t]])$ naturally acts on the category $\widehat{\mathfrak{g}}_{\kappa_c}\text{-mod}_\chi^{G[[t]]}$, and this action should be viewed as a categorical analogue of the action of $H(G(F), K_0)$ on $(R_\gamma)^{\pi_\gamma(K_0)}$.

Now, by Theorem 10.3.3, any object of $\widehat{\mathfrak{g}}_{\kappa_c}\text{-mod}_\chi^{G[[t]]}$ is a direct sum of copies of $\mathbb{V}_0(\chi)$. Therefore it is sufficient to describe the action of $\mathcal{H}(G((t)), G[[t]])$ on $\mathbb{V}_0(\chi)$. This action is described by the following statement, which follows from [BD1]: there exists a family of isomorphisms

$$\alpha_V : \mathcal{H}_V \star \mathbb{V}_0(\chi) \xrightarrow{\sim} \underline{V} \otimes \mathbb{V}_0(\chi), \qquad V \in \mathcal{R}ep\,{}^L G, \tag{10.3.21}$$

where $\underline{V}$ is the vector space underlying the representation V (see [FG2, FG4] for more details). Moreover, these isomorphisms are compatible with the tensor product structure on $\mathcal{H}_V$ (given by the convolution) and on $\underline{V}$ (given by tensor product of vector spaces).

In view of Theorem 10.3.3, this is not surprising. Indeed, it follows from the definition that $\mathcal{H}_V \star \mathbb{V}_0(\chi)$ is again an object of the category $\widehat{\mathfrak{g}}_{\kappa_c}\text{-mod}_\chi^{G[[t]]}$. Therefore it must be isomorphic to $U_V \otimes_{\mathbb{C}} \mathbb{V}_0(\chi)$, where U_V is a vector space. But then we obtain a functor

$$\mathcal{H}(G((t)), G[[t]]) \to \mathcal{V}ect, \qquad \mathcal{H}_V \mapsto U_V.$$

It follows from the construction that this is a tensor functor. Therefore the standard Tannakian formalism implies that U_V is isomorphic to $\underline{V}$.

The isomorphisms (10.3.21) should be viewed as the categorical analogues of the Hecke eigenvector conditions (10.3.5). The difference is that while in (10.3.5) the action of elements of the Hecke algebra on a K_0-invariant vector in R_γ amounts to multiplication by a scalar, the action of an object of the Hecke category $\mathcal{H}(G((t)), G[[t]])$ on the $G[[t]]$-equivariant object $\mathbb{V}_0(\chi)$ of $\widehat{\mathfrak{g}}_{\kappa_c}\text{-mod}_\chi$ amounts to multiplication by a *vector space*; namely, the vector space underlying the corresponding representation of ${}^L G$. It is natural to call a module satisfying this property a **Hecke eigenmodule**. Thus, we obtain that $\mathbb{V}_0(\chi)$ is a Hecke eigenmodule. This is in agreement with our expectation that the category $\widehat{\mathfrak{g}}_{\kappa_c}\text{-mod}_\chi^{G[[t]]}$ is a categorical version of the space of K_0-invariant vectors in R_γ.

One ingredient that is missing in the geometric case is the conjugacy class γ of ${}^L G$. We recall that in the classical Langlands correspondence this was the image of the Frobenius element of the Galois group $\text{Gal}(\overline{\mathbb{F}}_q/\mathbb{F}_q)$, which does not have an analogue in the geometric setting where our ground field is $\mathbb{C}$, which is algebraically closed. So while unramified local systems in the classical case are parameterized by the conjugacy classes γ, there is only one, up to an isomorphism, unramified local system in the geometric case. However, this local system has a large group of automorphisms; namely, ${}^L G$ itself. We will argue that what replaces γ in the geometric setting is the action of this group ${}^L G$ by automorphisms of the category $\widehat{\mathfrak{g}}_{\kappa_c}\text{-mod}_\chi$. We will discuss this in the next two sections.

10.3.6 Categories of representations and $\mathcal{D}$-modules

When we discussed the procedure of categorification of representations in Section 1.3.4, we saw that there are two possible scenarios for constructing categories equipped with an action of the loop group $G((t))$. In the first one we consider categories of $\mathcal{D}$-modules on the ind-schemes $G((t))/K$, where K is a "compact" subgroup of $G((t))$, such as $G[[t]]$ or the Iwahori subgroup. In the second one we consider categories of representations $\widehat{\mathfrak{g}}_{\kappa_c}\text{-mod}_\chi$. So far we have focused exclusively on the second scenario, but it is instructive to also discuss categories of the first type.

In the toy model considered in Section 1.3.3 we discussed the category of $\mathfrak{g}$-modules with fixed central character and the category of $\mathcal{D}$-modules on the flag variety G/B. We have argued that both could be viewed as categorifications of the representation of the group $G(\mathbb{F}_q)$ on the space of functions on $(G/B)(\mathbb{F}_q)$. These categories are equivalent, according to the Beilinson–Bernstein theory, with the functor of global sections connecting the two. Could something like this be true in the case of affine Kac–Moody algebras as well?

The affine Grassmannian $\text{Gr} = G((t))/G[[t]]$ may be viewed as the simplest possible analogue of the flag variety G/B for the loop group $G((t))$. Consider the category of $\mathcal{D}$-modules on $G((t))/G[[t]]$ (see [BD1, FG2] for the precise definition). We have a functor of global sections from this category to the category of $\mathfrak{g}((t))$-modules. In order to obtain $\widehat{\mathfrak{g}}_{\kappa_c}$-modules, we need to take instead the category $\mathcal{D}_{\kappa_c}$-mod of $\mathcal{D}$-modules twisted by a line bundle $\mathcal{L}_{\kappa_c}$. This is the unique line bundle $\mathcal{L}_{\kappa_c}$ on Gr which carries an action of $\widehat{\mathfrak{g}}_{\kappa_c}$ (such that the central element $\mathbf{1}$ is mapped to the identity) lifting the natural action of $\mathfrak{g}((t))$ on Gr. Then for any object $\mathcal{M}$ of $\mathcal{D}_{\kappa_c}$-mod, the space of global sections $\Gamma(\mathrm{Gr}, \mathcal{M})$ is a $\widehat{\mathfrak{g}}_{\kappa_c}$-module. Moreover, it is known (see [BD1, FG1]) that $\Gamma(\mathrm{Gr}, \mathcal{M})$ is in fact an object of $\widehat{\mathfrak{g}}_{\kappa_c}$-mod$_{\mathrm{reg}}$. Therefore we have a functor of global sections

$$\Gamma : \mathcal{D}_{\kappa_c}\text{-mod} \to \widehat{\mathfrak{g}}_{\kappa_c}\text{-mod}_{\mathrm{reg}} .$$

We note that the categories $\mathcal{D}$-mod and $\mathcal{D}_{\kappa_c}$-mod are equivalent under the functor $\mathcal{M} \mapsto \mathcal{M} \otimes \mathcal{L}_{\kappa_c}$. But the corresponding global sections functors are very different.

However, unlike in the Beilinson–Bernstein scenario, the functor Γ cannot possibly be an equivalence of categories. There are two reasons for this. First of all, the category $\widehat{\mathfrak{g}}_{\kappa_c}$-mod$_{\mathrm{reg}}$ has a large center, namely, the algebra $\mathfrak{z} = \mathrm{Fun}\,\mathrm{Op}_{^LG}(D)$, while the center of the category $\mathcal{D}_{\kappa_c}$-mod is trivial.† The second, and more serious, reason is that the category $\mathcal{D}_{\kappa_c}$-mod carries an additional symmetry, namely, an action of the tensor category $\mathcal{R}ep^L G$ of representations of the Langlands dual group ^LG, and this action trivializes under the functor Γ as we explain presently.

Over $\mathrm{Op}_{^LG}(D)$ there exists a canonical principal ^LG-bundle, which we will denote by $\mathcal{P}$. By definition, the fiber of $\mathcal{P}$ at $\chi = (\mathcal{F}, \nabla, \mathcal{F}_{^LB}) \in \mathrm{Op}_{^LG}(D)$ is $\mathcal{F}_0$, the fiber at $0 \in D$ of the ^LG-bundle $\mathcal{F}$ underlying χ. For an object $V \in \mathcal{R}ep\, ^LG$ let us denote by $\mathcal{V}$ the associated vector bundle over $\mathrm{Op}_{^LG}(D)$, i.e.,

$$\mathcal{V} = \mathcal{P} \underset{^LG}{\times} V.$$

Next, consider the category $\mathcal{D}_{\kappa_c}$-mod$^{G[[t]]}$ of $G[[t]]$-equivariant $\mathcal{D}_{\kappa_c}$-modules on Gr. It is equivalent to the category

$$\mathcal{D}\text{-mod}^{G[[t]]} = \mathcal{H}(G((t)), G[[t]])$$

considered above. This is a tensor category, with respect to the convolution functor, which is equivalent to the category $\mathcal{R}ep\, ^LG$. We will use the same notation $\mathcal{H}_V$ for the object of $\mathcal{D}_{\kappa_c}$-mod$^{G[[t]]}$ corresponding to $V \in \mathcal{R}ep\, ^LG$. The category $\mathcal{D}_{\kappa_c}$-mod$^{G[[t]]}$ acts on $\mathcal{D}_{\kappa_c}$-mod by convolution functors

$$\mathcal{M} \mapsto \mathcal{H}_V \star \mathcal{M}$$

which are exact. This amounts to a tensor action of the category $\mathcal{R}ep^L G$ on $\mathcal{D}_{\kappa_c}$-mod.

† Recall that we are under the assumption that G is a connected simply-connected algebraic group, and in this case Gr has one connected component. In general, the center of the category $\mathcal{D}_{\kappa_c}$-mod has a basis enumerated by the connected components of Gr and is isomorphic to the group algebra of the finite group $\pi_1(G)$.

Now, it follows from the results of [BD1] that there are functorial isomorphisms

$$\Gamma(\mathrm{Gr}, \mathcal{H}_V \star \mathcal{M}) \simeq \Gamma(\mathrm{Gr}, \mathcal{M}) \underset{\mathfrak{z}}{\otimes} \mathcal{V}, \qquad V \in \mathcal{R}ep\, {}^L G,$$

compatible with the tensor structure (see [FG2, FG4] for details). Thus, we see that there are non-isomorphic objects of $\mathcal{D}_{\kappa_c}$-mod, which the functor Γ sends to isomorphic objects of $\widehat{\mathfrak{g}}_{\kappa_c}$-mod$_{\mathrm{reg}}$. Therefore the category $\mathcal{D}_{\kappa_c}$-mod and the functor Γ need to be modified in order to have a chance to obtain a category equivalent to $\widehat{\mathfrak{g}}_{\kappa_c}$-mod$_{\mathrm{reg}}$.

In [FG2] it was shown how to modify the category $\mathcal{D}_{\kappa_c}$-mod, by simultaneously "adding" to it $\mathfrak{z}$ as a center, and "dividing" it by the above $\mathcal{R}ep\, {}^L G$-action. As the result, we obtain a candidate for a category that can be equivalent to $\widehat{\mathfrak{g}}_{\kappa_c}$-mod$_{\mathrm{reg}}$. This is the category of **Hecke eigenmodules** on Gr, denoted by $\mathcal{D}^{\mathrm{Hecke}}_{\kappa_c}$-mod$_{\mathrm{reg}}$.

By definition, an object of $\mathcal{D}^{\mathrm{Hecke}}_{\kappa_c}$-mod$_{\mathrm{reg}}$ is an object of $\mathcal{D}_{\kappa_c}$-mod, equipped with an action of the algebra $\mathfrak{z}$ by endomorphisms and a system of isomorphisms

$$\alpha_V : \mathcal{H}_V \star \mathcal{M} \xrightarrow{\sim} \mathcal{V} \underset{\mathfrak{z}}{\otimes} \mathcal{M}, \qquad V \in \mathcal{R}ep\, {}^L G,$$

compatible with the tensor structure.

The above functor Γ naturally gives rise to a functor

$$\Gamma^{\mathrm{Hecke}} : \mathcal{D}^{\mathrm{Hecke}}_{\kappa_c}\text{-mod}_{\mathrm{reg}} \to \widehat{\mathfrak{g}}_{\kappa_c}\text{-mod}_{\mathrm{reg}} . \tag{10.3.22}$$

This is in fact a general property. To explain this, suppose for simplicity that we have an abelian category $\mathcal{C}$ which is acted upon by the tensor category $\mathcal{R}ep\, H$, where H is an algebraic group; we denote this action by

$$\mathcal{M} \mapsto \mathcal{M} \star V, \qquad V \in \mathcal{R}ep\, H.$$

Let $\mathcal{C}^{\mathrm{Hecke}}$ be the category whose objects are collections $(\mathcal{M}, \{\alpha_V\}_{V \in \mathcal{R}ep\, H})$, where $\mathcal{M} \in \mathcal{C}$ and $\{\alpha_V\}$ is a compatible system of isomorphisms

$$\alpha_V : \mathcal{M} \star V \xrightarrow{\sim} \underline{V} \underset{\mathbb{C}}{\otimes} \mathcal{M}, \qquad V \in \mathcal{R}ep\, H,$$

where $\underline{V}$ is the vector space underlying V. The category $\mathcal{C}^{\mathrm{Hecke}}$ carries a natural action of the group H: for $h \in H$, we have

$$h \cdot (\mathcal{M}, \{\alpha_V\}_{V \in \mathcal{R}ep\, H}) = (\mathcal{M}, \{(h \otimes \mathrm{id}_{\mathcal{M}}) \circ \alpha_V\}_{V \in \mathcal{R}ep\, H}).$$

In other words, $\mathcal{M}$ remains unchanged, but the isomorphisms α_V get composed with h.

The category $\mathcal{C}$ may be reconstructed as the category of H-equivariant objects of $\mathcal{C}^{\mathrm{Hecke}}$ with respect to this action, see [AG]. Therefore one may think of $\mathcal{C}^{\mathrm{Hecke}}$ as the "de-equivariantization" of the category $\mathcal{C}$.

Suppose that we have a functor $\mathsf{G} : \mathcal{C} \to \mathcal{C}'$, such that we have functorial isomorphisms

$$\mathsf{G}(\mathcal{M} \star V) \simeq \mathsf{G}(\mathcal{M}) \underset{\mathbb{C}}{\otimes} \underline{V}, \qquad V \in \mathcal{R}ep\, H, \tag{10.3.23}$$

compatible with the tensor structure. Then, according to [AG], there exists a functor

$$\mathsf{G}^{\text{Hecke}} : \mathcal{C}^{\text{Hecke}} \to \mathcal{C}'$$

such that $\mathsf{G} \simeq \mathsf{G}^{\text{Hecke}} \circ \text{Ind}$, where the functor $\text{Ind} : \mathcal{C} \to \mathcal{C}^{\text{Hecke}}$ sends $\mathcal{M}$ to $\mathcal{M} \star \mathcal{O}_H$, where $\mathcal{O}_H$ is the regular representation of H. The functor $\mathsf{G}^{\text{Hecke}}$ may be explicitly described as follows: the isomorphisms α_V and (10.3.23) give rise to an action of the algebra $\mathcal{O}_H$ on $\mathsf{G}(\mathcal{M})$, and $\mathsf{G}^{\text{Hecke}}(\mathcal{M})$ is obtained by taking the fiber of $\mathsf{G}(\mathcal{M})$ at $1 \in H$.

In our case, we take $\mathcal{C} = \mathcal{D}_{\kappa_c}\text{-mod}$, $\mathcal{C}' = \widehat{\mathfrak{g}}_{\kappa_c}\text{-mod}_{\text{reg}}$, and $\mathsf{G} = \Gamma$. The only difference is that now we are working over the base $\text{Op}_{^L G}(D)$, which we have to take into account. Then the functor Γ factors as

$$\Gamma \simeq \Gamma^{\text{Hecke}} \circ \text{Ind},$$

where Γ^{Hecke} is the functor (10.3.22) (see [FG2, FG4] for more details). Moreover, the left action of the group $G((t))$ on Gr gives rise to its action on the category $\mathcal{D}^{\text{Hecke}}_{\kappa_c}\text{-mod}_{\text{reg}}$, and the functor Γ^{Hecke} intertwines this action with the action of $G((t))$ on $\widehat{\mathfrak{g}}_{\kappa_c}\text{-mod}_{\text{reg}}$.

The following was conjectured in [FG2]:

Conjecture 10.3.9 *The functor* Γ^{Hecke} *in formula* (10.3.22) *defines an equivalence of the categories* $\mathcal{D}^{\text{Hecke}}_{\kappa_c}\text{-mod}_{\text{reg}}$ *and* $\widehat{\mathfrak{g}}_{\kappa_c}\text{-mod}_{\text{reg}}$.

It was proved in [FG2] that the functor Γ^{Hecke}, when extended to the derived categories, is fully faithful. Furthermore, it was proved in [FG4] that it sets up an equivalence of the corresponding I^0-equivariant categories, where $I^0 = [I, I]$ is the radical of the Iwahori subgroup.

Let us specialize Conjecture 10.3.9 to a point $\chi = (\mathcal{F}, \nabla, \mathcal{F}_{^L B}) \in \text{Op}_{^L G}(D)$. Then on the right hand side we consider the category $\widehat{\mathfrak{g}}_{\kappa_c}\text{-mod}_\chi$, and on the left hand side we consider the category $\mathcal{D}^{\text{Hecke}}_{\kappa_c}\text{-mod}_\chi$. Its object consists of a $\mathcal{D}_{\kappa_c}$-module $\mathcal{M}$ and a collection of isomorphisms

$$\alpha_V : \mathcal{H}_V \star \mathcal{M} \xrightarrow{\sim} V_{\mathcal{F}_0} \otimes \mathcal{M}, \qquad V \in \mathcal{R}ep\, {}^L G. \tag{10.3.24}$$

Here $V_{\mathcal{F}_0}$ is the twist of the representation V by the ${}^L G$-torsor $\mathcal{F}_0$. These isomorphisms have to be compatible with the tensor structure on the category $\mathcal{H}(G((t)), G[[t]])$.

Conjecture 10.3.9 implies that there is a canonical equivalence of categories

$$\mathcal{D}^{\text{Hecke}}_{\kappa_c}\text{-mod}_\chi \simeq \widehat{\mathfrak{g}}_{\kappa_c}\text{-mod}_\chi\,. \tag{10.3.25}$$

It is this conjectural equivalence that should be viewed as an analogue of the Beilinson–Bernstein equivalence.

From this point of view, we may think of each of the categories $\mathcal{D}_{\kappa_c}^{\rm Hecke}\text{-mod}_\chi$ as the second incarnation of the sought-after Langlands category $\mathcal{C}_{\sigma_0}$ corresponding to the trivial ${}^L G$-local system.

Now we give another explanation why it is natural to view the category $\mathcal{D}_{\kappa_c}^{\rm Hecke}\text{-mod}_\chi$ as a categorification of an unramified representation of the group $G(F)$. First of all, observe that these categories are all equivalent to each other and to the category $\mathcal{D}_{\kappa_c}^{\rm Hecke}\text{-mod}$, whose objects are $\mathcal{D}_{\kappa_c}$-modules $\mathcal{M}$ together with a collection of isomorphisms

$$\alpha_V : \mathcal{H}_V \star \mathcal{M} \xrightarrow{\sim} \underline{V} \otimes \mathcal{M}, \qquad V \in \mathcal{R}ep\, {}^L G. \tag{10.3.26}$$

Comparing formulas (10.3.24) and (10.3.26), we see that there is an equivalence

$$\mathcal{D}_{\kappa_c}^{\rm Hecke}\text{-mod}_\chi \simeq \mathcal{D}_{\kappa_c}^{\rm Hecke}\text{-mod},$$

for each choice of trivialization of the ${}^L G$-torsor $\mathcal{F}_0$ (the fiber at $0 \in D$ of the principal ${}^L G$-bundle $\mathcal{F}$ on D underlying the oper χ).

Now recall from Section 10.3.1 that to each semi-simple conjugacy class γ in ${}^L G$ corresponds an irreducible unramified representation (R_γ, π_γ) of $G(F)$ via the Satake correspondence (10.3.2). It is known that there is a non-degenerate pairing

$$\langle , \rangle : R_\gamma \times R_{\gamma^{-1}} \to \mathbb{C},$$

in other words, $R_{\gamma^{-1}}$ is the representation of $G(F)$ which is contragredient to R_γ (it may be realized in the space of smooth vectors in the dual space to R_γ).

Let $v \in R_{\gamma^{-1}}$ be a non-zero vector such that $K_0 v = v$ (this vector is unique up to a scalar). It then satisfies the Hecke eigenvector property (10.3.5) (in which we need to replace γ by γ^{-1}). This allows us to embed R_γ into the space of smooth locally constant right K_0-invariant functions on $G(F)$ (equivalently, functions on $G(F)/K_0$), by using matrix coefficients, as follows:

$$u \in R_\gamma \mapsto f_u, \qquad f_u(g) = \langle u, gv \rangle.$$

The Hecke eigenvector property (10.3.5) implies that the functions f_u are right K_0-invariant and satisfy the condition

$$f \star H_V = \mathrm{Tr}(\gamma^{-1}, V) f, \tag{10.3.27}$$

where $\star$ denotes the convolution product (10.2.1). Let $C(G(F)/K_0)_\gamma$ be the space of smooth locally constant functions on $G(F)/K_0$ satisfying (10.3.27). It carries a representation of $G(F)$ induced by its left action on $G(F)/K_0$. We have constructed an injective map $R_\gamma \to C(G(R)/G(R))_\gamma$, and one can show that for generic γ it is an isomorphism.

Thus, we obtain a realization of an irreducible unramified representation of $G(F)$ in the space of functions on the quotient $G(F)/K_0$ satisfying the Hecke eigenfunction condition (10.3.27). The Hecke eigenmodule condition (10.3.26) may be viewed as a categorical analogue of (10.3.27). Therefore the category $\mathcal{D}_{\kappa_c}^{\mathrm{Hecke}}$-mod of twisted $\mathcal{D}$-modules on $\mathrm{Gr} = G((t))/K_0$ satisfying the Hecke eigenmodule condition (10.3.26), equipped with its $G((t))$-action, appears to be a natural categorification of the irreducible unramified representations of $G(F)$.

10.3.7 Equivalences between categories of modules

All opers in $\mathrm{Op}_{^LG}(D)$ correspond to one and the same ^LG-local system; namely, the trivial local system. Therefore, according to Conjecture 10.1.2, we expect that the categories $\widehat{\mathfrak{g}}_{\kappa_c}\text{-mod}_\chi$ are equivalent to each other. More precisely, for each isomorphism between the underlying local systems of any two opers in $\mathrm{Op}_{^LG}(D)$ we wish to have an equivalence of the corresponding categories, and these equivalences should be compatible with respect to the operation of composition of these isomorphisms.

Let us spell this out in detail. Let $\chi = (\mathcal{F}, \nabla, \mathcal{F}_{^LB})$ and $\chi' = (\mathcal{F}', \nabla', \mathcal{F}'_{^LB})$ be two opers in $\mathrm{Op}_{^LG}(D)$. Then an isomorphism between the underlying local systems $(\mathcal{F}, \nabla) \xrightarrow{\sim} (\mathcal{F}', \nabla')$ is the same as an isomorphism $\mathcal{F}_0 \xrightarrow{\sim} \mathcal{F}'_0$ between the ^LG-torsors $\mathcal{F}_0$ and $\mathcal{F}'_0$, which are the fibers of the ^LG-bundles $\mathcal{F}$ and $\mathcal{F}'$, respectively, at $0 \in D$. Let us denote this set of isomorphisms by $\mathrm{Isom}_{\chi,\chi'}$. Then we have

$$\mathrm{Isom}_{\chi,\chi'} = \mathcal{F}_0 \underset{^LG}{\times} {^LG} \underset{^LG}{\times} \mathcal{F}'_0,$$

where we twist ^LG by $\mathcal{F}_0$ with respect to the left action and by $\mathcal{F}'_0$ with respect to the right action. In particular,

$$\mathrm{Isom}_{\chi,\chi} = {^LG}_{\mathcal{F}_0} = \mathcal{F}_0 \underset{^LG}{\times} \mathrm{Ad}\, {^LG}$$

is just the group of automorphisms of $\mathcal{F}_0$.

It is instructive to combine the sets $\mathrm{Isom}_{\chi,\chi'}$ into a groupoid Isom over $\mathrm{Op}_{^LG}(D)$. Thus, by definition Isom consists of triples (χ, χ', ϕ), where $\chi, \chi' \in \mathrm{Op}_{^LG}(D)$ and $\phi \in \mathrm{Isom}_{\chi,\chi}$ is an isomorphism of the underlying local systems. The two morphisms $\mathrm{Isom} \to \mathrm{Op}_{^LG}(D)$ correspond to sending such a triple to χ and χ'. The identity morphism $\mathrm{Op}_{^LG}(D) \to \mathrm{Isom}$ sends χ to $(\chi, \chi, \mathrm{Id})$, and the composition morphism

$$\mathrm{Isom} \underset{\mathrm{Op}_{^LG}(D)}{\times} \mathrm{Isom} \to \mathrm{Isom}$$

corresponds to composing two isomorphisms.

Conjecture 10.1.2 has the following more precise formulation for regular opers.

Conjecture 10.3.10 *For each $\phi \in \mathrm{Isom}_{\chi,\chi'}$ there exists an equivalence*

$$E_\phi : \widehat{\mathfrak{g}}_{\kappa_c}\text{-mod}_\chi \to \widehat{\mathfrak{g}}_{\kappa_c}\text{-mod}_{\chi'},$$

which intertwines the actions of $G((t))$ on the two categories, such that $E_{\mathrm{Id}} = \mathrm{Id}$ and there exist isomorphisms $\beta_{\phi,\phi'} : E_{\phi\circ\phi'} \simeq E_\phi \circ E_{\phi'}$ satisfying

$$\beta_{\phi\circ\phi',\phi''}\beta_{\phi,\phi'} = \beta_{\phi,\phi'\circ\phi''}\beta_{\phi',\phi''}$$

for all isomorphisms ϕ, ϕ', ϕ'', whenever they may be composed in the appropriate order.

In other words, the groupoid Isom *over* $\mathrm{Op}_{^LG}(D)$ *acts on the category* $\widehat{\mathfrak{g}}_{\kappa_c}\text{-mod}_{\mathrm{reg}}$ *fibered over* $\mathrm{Op}_{^LG}(D)$*, preserving the action of $G((t))$ along the fibers.*

In particular, this conjecture implies that the group $^LG_{\mathcal{F}_0}$ acts on the category $\widehat{\mathfrak{g}}_{\kappa_c}\text{-mod}_\chi$ for any $\chi \in \mathrm{Op}_{^LG}(D)$.

Now we observe that Conjecture 10.3.9 implies Conjecture 10.3.10. Indeed, by Conjecture 10.3.9, there is a canonical equivalence of categories (10.3.25),

$$\mathcal{D}^{\mathrm{Hecke}}_{\kappa_c}\text{-mod}_\chi \simeq \widehat{\mathfrak{g}}_{\kappa_c}\text{-mod}_\chi .$$

It follows from the definition of the category $\mathcal{D}^{\mathrm{Hecke}}_{\kappa_c}\text{-mod}_\chi$ (namely, formula (10.3.24)) that for each isomorphism $\phi \in \mathrm{Isom}_{\chi,\chi'}$, i.e., an isomorphism of the ^LG-torsors $\mathcal{F}_0$ and $\mathcal{F}'_0$ underlying the opers χ and χ', there is a canonical equivalence

$$\mathcal{D}^{\mathrm{Hecke}}_{\kappa_c}\text{-mod}_\chi \simeq \mathcal{D}^{\mathrm{Hecke}}_{\kappa_c}\text{-mod}_{\chi'} .$$

Therefore we obtain the sought-after equivalence

$$E_\phi : \widehat{\mathfrak{g}}_{\kappa_c}\text{-mod}_\chi \to \widehat{\mathfrak{g}}_{\kappa_c}\text{-mod}_{\chi'} .$$

Furthermore, it is clear that these equivalences satisfy the conditions of Conjecture 10.3.10. In particular, they intertwine the actions of $G((t))$, because the action of $G((t))$ affects the $\mathcal{D}$-module $\mathcal{M}$ underlying an object of $\mathcal{D}^{\mathrm{Hecke}}_{\kappa_c}\text{-mod}_\chi$, but does not affect the isomorphisms α_V.

Equivalently, we can express this by saying that the groupoid Isom naturally acts on the category $\mathcal{D}^{\mathrm{Hecke}}_{\kappa_c}\text{-mod}_{\mathrm{reg}}$. By Conjecture 10.3.9, this gives rise to an action of Isom on $\widehat{\mathfrak{g}}_{\kappa_c}\text{-mod}_{\mathrm{reg}}$.

In particular, we construct an action of the group $(^LG)_{\mathcal{F}_0}$, the twist of ^LG by the ^LG-torsor $\mathcal{F}_0$ underlying a particular oper χ, on the category $\mathcal{D}^{\mathrm{Hecke}}_{\kappa_c}\text{-mod}_\chi$. Indeed, each element $g \in (^LG)_{\mathcal{F}_0}$ acts on the $\mathcal{F}_0$-twist $V_{\mathcal{F}_0}$ of any finite-dimensional representation V of ^LG. Given an object $(\mathcal{M}, (\alpha_V))$ of $\mathcal{D}^{\mathrm{Hecke}}_{\kappa_c}\text{-mod}_{\chi'}$, we construct a new object; namely, $(\mathcal{M}, ((g \otimes \mathrm{Id}_{\mathcal{M}}) \circ \alpha_V))$. Thus, we do not change the $\mathcal{D}$-module $\mathcal{M}$, but we change the isomorphisms α_V appearing in the Hecke eigenmodule condition (10.3.24) by composing them with the action of g on $V_{\mathcal{F}_0}$. According to Conjecture 10.3.9, the category

$\mathcal{D}_{\kappa_c}^{\text{Hecke}}$-mod$_\chi$ is equivalent to $\widehat{\mathfrak{g}}_{\kappa_c}$-mod$_\chi$. Therefore this gives rise to an action of the group $({}^LG)_{\mathcal{F}_0}$ on $\widehat{\mathfrak{g}}_{\kappa_c}$-mod$_\chi$. But this action is much more difficult to describe in terms of $\widehat{\mathfrak{g}}_{\kappa_c}$-modules. We will see examples of the action of these symmetries below.

10.3.8 Generalization to other dominant integral weights

We have extensively studied above the categories $\widehat{\mathfrak{g}}_{\kappa_c}$-mod$_\chi$ and $\widehat{\mathfrak{g}}_{\kappa_c}$-mod$_\chi^{G[[t]]}$ associated to regular opers $\chi \in \mathrm{Op}_{{}^LG}(D)$. However, by Lemma 10.3.1, the (set-theoretic) fiber of the map $\alpha : \mathrm{Op}_{{}^LG}(D^\times) \to \mathrm{Loc}_{{}^LG}(D^\times)$ over the trivial local system σ_0 is the disjoint union of the subsets $\mathrm{Op}_{{}^LG}^\lambda, \lambda \in P^+$. Here we discuss briefly the categories $\widehat{\mathfrak{g}}_{\kappa_c}$-mod$_\chi$ and $\widehat{\mathfrak{g}}_{\kappa_c}$-mod$_\chi^{G[[t]]}$ for $\chi \in \mathrm{Op}_{{}^LG}^\lambda$, where $\lambda \neq 0$.

Consider the Weyl module $\mathbb{V}_\lambda$ with highest weight λ defined in Section 9.6. According to Theorem 9.6.1, we have

$$\mathrm{End}_{\widehat{\mathfrak{g}}_{\kappa_c}} \mathbb{V}_\lambda \simeq \mathrm{Fun}\, \mathrm{Op}_{{}^LG}^\lambda . \tag{10.3.28}$$

Let $\chi \in \mathrm{Op}_{{}^LG}^\lambda \subset \mathrm{Op}_{{}^LG}(D^\times)$. Then χ defines a character of the algebra $\mathrm{End}_{\widehat{\mathfrak{g}}_{\kappa_c}} \mathbb{V}_\lambda$. Let $\mathbb{V}_\lambda(\chi)$ be the quotient of $\mathbb{V}_\lambda$ by the kernel of this character. The following conjecture of [FG6] is an analogue of Theorem 10.3.3:

Conjecture 10.3.11 *Let $\chi \in \mathrm{Op}_{{}^LG}^\lambda \subset \mathrm{Op}_{{}^LG}(D^\times)$. Then the category $\widehat{\mathfrak{g}}_{\kappa_c}$-mod$_\chi^{G[[t]]}$ is equivalent to the category of vector spaces: its unique, up to isomorphism, irreducible object is $\mathbb{V}_\lambda(\chi)$ and any other object is isomorphic to a direct sum of copies of $\mathbb{V}_\lambda(\chi)$.*

Note that this is consistent with Conjecture 10.1.2, which tells us that the categories $\widehat{\mathfrak{g}}_{\kappa_c}$-mod$_\chi^{G[[t]]}$ should be equivalent to each other for all opers which are gauge equivalent to the trivial local system on D.

As in the case $\lambda = 0$, it is useful to consider, instead of an individual ${}^LG$-oper χ, the entire family $\mathrm{Op}_{{}^LG}^\lambda$. Let $\widehat{\mathfrak{g}}_{\kappa_c}$-mod$_{\lambda,\text{reg}}$ be the full subcategory of the category $\widehat{\mathfrak{g}}_{\kappa_c}$-mod whose objects are $\widehat{\mathfrak{g}}_{\kappa_c}$-modules on which the action of the center $Z(\widehat{\mathfrak{g}})$ factors through the homomorphism

$$Z(\widehat{\mathfrak{g}}) \simeq \mathrm{Fun}\, \mathrm{Op}_{{}^LG}(D^\times) \to \mathfrak{z}_\lambda,$$

where we have set

$$\mathfrak{z}_\lambda = \mathrm{Fun}\, \mathrm{Op}_{{}^LG}^\lambda .$$

Let $\widehat{\mathfrak{g}}_{\kappa_c}$-mod$_{\lambda,\text{reg}}^{G[[t]]}$ be the corresponding $G[[t]]$-equivariant category. Thus, $\widehat{\mathfrak{g}}_{\kappa_c}$-mod$_{\lambda,\text{reg}}$ and $\widehat{\mathfrak{g}}_{\kappa_c}$-mod$_{\lambda,\text{reg}}^{G[[t]]}$ are categories fibered over $\mathrm{Op}_{{}^LG}^\lambda$, with the fibers over $\chi \in \mathrm{Op}_{{}^LG}^\lambda$ being $\widehat{\mathfrak{g}}_{\kappa_c}$-mod$_\chi$ and $\widehat{\mathfrak{g}}_{\kappa_c}$-mod$_\chi^{G[[t]]}$, respectively.

We have the following conjectural description of the category $\widehat{\mathfrak{g}}_{\kappa_c}$-mod$_{\lambda,\text{reg}}^{G[[t]]}$, which implies Conjecture 10.3.3 (see [FG6]).

Let $\mathfrak{z}_\lambda$-mod be the category of modules over the commutative algebra $\mathfrak{z}_\lambda$. Equivalently, this is the category of quasicoherent sheaves on the space $\mathrm{Op}^\lambda_{^L G}$.

By definition, any object of $\widehat{\mathfrak{g}}_{\kappa_c}\text{-mod}^{G[[t]]}_{\lambda,\mathrm{reg}}$ is a $\mathfrak{z}_\lambda$-module. Introduce the functors

$$\begin{aligned} \mathsf{F}_\lambda &: \widehat{\mathfrak{g}}_{\kappa_c}\text{-mod}^{G[[t]]}_{\lambda,\mathrm{reg}} \to \mathfrak{z}_\lambda\text{-mod}, && M \mapsto \mathrm{Hom}_{\widehat{\mathfrak{g}}_{\kappa_c}}(\mathbb{V}_\lambda, M), \\ \mathsf{G}_\lambda &: \mathfrak{z}_\lambda\text{-mod} \to \widehat{\mathfrak{g}}_{\kappa_c}\text{-mod}^{G[[t]]}_{\lambda,\mathrm{reg}}, && \mathcal{F} \mapsto \mathbb{V}_\lambda \underset{\mathfrak{z}_\lambda}{\otimes} \mathcal{F}. \end{aligned}$$

Conjecture 10.3.12 *The functors F_λ and G_λ are mutually inverse equivalences of categories.*

We expect that this conjecture may be proved along the lines of the proof of Theorem 10.3.4 presented in Section 10.3.4. More precisely, it should follow from the conjecture of [FG6], which generalizes the corresponding statements in the case $\lambda = 0$.

Conjecture 10.3.13
(1) *$\mathbb{V}_\lambda$ is free as a $\mathfrak{z}_\lambda$-module.*
(2)

$$\mathrm{Ext}^i_{\widehat{\mathfrak{g}},G[[t]]}(\mathbb{V}_\lambda, \mathbb{V}_\lambda) \simeq \Omega^i_{\mathfrak{z}_\lambda}, \tag{10.3.29}$$

the space of differential forms of degree i on $\mathrm{Op}^\lambda_{^L G}$.

10.4 The tamely ramified case

In the previous section we have considered categorical analogues of the irreducible unramified representations of a reductive group $G(F)$ over a local non-archimedian field F. We recall that these are the representations containing non-zero vectors fixed by the maximal compact subgroup $K_0 \subset G(F)$. The corresponding Langlands parameters are unramified admissible homomorphisms from the Weil–Deligne group W'_F to $^L G$, i.e., those which factor through the quotient

$$W'_F \to W_F \to \mathbb{Z},$$

and whose image in $^L G$ is semi-simple. Such homomorphisms are parameterized by semi-simple conjugacy classes in $^L G$.

We have seen that the categorical analogues of unramified representations of $G(F)$ are the categories $\widehat{\mathfrak{g}}_{\kappa_c}\text{-mod}_\chi$ (equipped with an action of the loop group $G((t))$), where χ is a $^L G$-oper on $D^\times$ whose underlying $^L G$-local system is trivial. These categories can be called unramified in the sense that they contain non-zero $G[[t]]$-equivariant objects. The corresponding Langlands parameter is the trivial $^L G$-local system σ_0 on $D^\times$, which should be viewed as an analogue of an unramified homomorphism $W'_F \to {^L G}$. However, the local system σ_0 is realized by many different opers, and this introduces an additional complication into our picture: at the end of the day we need to show that the categories $\widehat{\mathfrak{g}}_{\kappa_c}$-mod$_\chi$, where χ is of the above type, are equivalent to each other. In particular, Conjecture 10.3.10 describes what we expect to happen when $\chi \in \mathrm{Op}_{^LG}(D)$.

The next natural step is to consider categorical analogues of representations of $G(F)$ that contain vectors invariant under the Iwahori subgroup $I \subset G[[t]]$, the preimage of a fixed Borel subgroup $B \subset G$ under the evaluation homomorphism $G[[t]] \to G$. We begin this section by recalling a classification of these representations, due to D. Kazhdan and G. Lusztig [KL] and V. Ginzburg [CG]. We then discuss the categorical analogues of these representations following [FG2]–[FG5] and the intricate interplay between the classical and the geometric pictures. We close this section with some explicit calculations in the case of $\mathfrak{g} = \mathfrak{sl}_2$, which should serve as an illustration of the general theory.

10.4.1 Tamely ramified representations

The Langlands parameters corresponding to irreducible representations of $G(F)$ with I-invariant vectors are **tamely ramified** homomorphisms $W'_F \to {}^LG$. Recall from Section 1.1.3 that $W'_F = W_F \ltimes \mathbb{C}$. A homomorphism $W'_F \to {}^LG$ is called tamely ramified if it factors through the quotient

$$W'_F \to \mathbb{Z} \ltimes \mathbb{C}.$$

According to the relation (1.1.1), the group $\mathbb{Z} \ltimes \mathbb{C}$ is generated by two elements $F = 1 \in \mathbb{Z}$ (Frobenius) and $M = 1 \in \mathbb{C}$ (monodromy) satisfying the relation

$$FMF^{-1} = qM. \tag{10.4.1}$$

Under an admissible tamely ramified homomorphism the generator F goes to a semi-simple element $\gamma \in {}^LG$ and the generator M goes to a unipotent element $N \in {}^LG$. According to formula (10.4.1), they have to satisfy the relation

$$\gamma N \gamma^{-1} = N^q. \tag{10.4.2}$$

Alternatively, we may write $N = \exp(u)$, where u is a nilpotent element of ${}^L\mathfrak{g}$. Then this relation becomes

$$\gamma u \gamma^{-1} = qu.$$

Thus, we have the following bijection between the sets of equivalence classes

$$\boxed{\begin{array}{c}\text{tamely ramified admissible}\\ \text{homomorphisms } W'_F \to {}^LG\end{array}} \iff \boxed{\begin{array}{c}\text{pairs } \gamma \in {}^LG, \text{ semi-simple,}\\ u \in {}^L\mathfrak{g}, \text{ nilpotent}, \gamma u \gamma^{-1} = qu\end{array}} \tag{10.4.3}$$

In both cases equivalence relation amounts to conjugation by an element of ${}^LG$.

Now to each Langlands parameter of this type we wish to attach an irreducible smooth representation of $G(F)$, which contains non-zero I-invariant vectors. It turns out that if $G = GL_n$ there is indeed a bijection, proved in [BZ], between the sets of equivalence classes of the following objects:

tamely ramified admissible homomorphisms $W'_F \to GL_n$	$\Longleftrightarrow$	irreducible representations (R, π) of $GL_n(F)$, $R^{\pi(I)} \neq 0$

(10.4.4)

However, such a bijection is no longer true for other reductive groups: two new phenomena appear, which we discuss presently.

The first one is the appearance of L-**packets**. One no longer expects to be able to assign to a particular admissible homomorphism $W'_F \to {}^LG$ a single irreducible smooth representation of $G(F)$. Instead, a finite collection of such representations (more precisely, a collection of equivalence classes of representations) is assigned, called an L-packet. In order to distinguish representations in a given L-packet, one needs to introduce an additional parameter. We will see how this is done in the case at hand shortly. However, and this is the second subtlety alluded to above, it turns out that not all irreducible representations of $G(F)$ within the L-packet associated to a given tamely ramified homomorphism $W'_F \to {}^LG$ contain non-zero I-invariant vectors. Fortunately, there is a certain property of the extra parameter used to distinguish representations inside the L-packet that tells us whether the corresponding representation of $G(F)$ has I-invariant vectors.

In the case of tamely ramified homomorphisms $W'_F \to {}^LG$ this extra parameter is an irreducible representation ρ of the finite group $C(\gamma, u)$ of components of the simultaneous centralizer of γ and u in ${}^LG$, on which the center of ${}^LG$ acts trivially (see [Lu1]). In the case of $G = GL_n$ these centralizers are always connected, and so this parameter never appears. But for other reductive groups G this group of components is often non-trivial. The simplest example is when ${}^LG = G_2$ and u is a subprincipal nilpotent element of the Lie algebra ${}^L\mathfrak{g}$.† In this case for some γ satisfying $\gamma u \gamma^{-1} = qu$ the group of components $C(\gamma, u)$ is the symmetric group S_3, which has three irreducible representations (up to equivalence). Each of them corresponds to a particular member of the L-packet associated with the tamely ramified homomorphism $W'_F \to {}^LG$ defined by (γ, u). Thus, the L-packet consists of three (equivalence classes of) irreducible smooth representations of $G(F)$. However, not all of them contain non-zero I-invariant vectors.

The representations ρ of the finite group $C(\gamma, u)$ which correspond to representations of $G(F)$ with I-invariant vectors are distinguished by the following

† The term "subprincipal" means that the adjoint orbit of this element has codimension 2 in the nilpotent cone.

property. Consider the **Springer fiber** Sp_u. We recall (see formula (9.3.10)) that

$$\mathrm{Sp}_u = \{\mathfrak{b}' \in {}^L G/{}^L B \,|\, u \in \mathfrak{b}'\}. \tag{10.4.5}$$

The group $C(\gamma, u)$ acts on the homology of the variety Sp_u^γ of γ-fixed points of Sp_u. A representation ρ of $C(\gamma, u)$ corresponds to a representation of $G(F)$ with non-zero I-invariant vectors if and only if ρ occurs in the homology of Sp_u^γ, $H_\bullet(\mathrm{Sp}_u^\gamma)$.

In the case of G_2 the Springer fiber Sp_u of the subprincipal element u is a union of four projective lines connected with each other as in the Dynkin diagram of D_4. For some γ the set Sp_u^γ is the union of a projective line (corresponding to the central vertex in the Dynkin diagram of D_4) and three points (each in one of the remaining three projective lines). The corresponding group $C(\gamma, u) = S_3$ on Sp_u^γ acts trivially on the projective line and by permutation of the three points. Therefore the trivial and the two-dimensional representations of S_3 occur in $H_\bullet(\mathrm{Sp}_u^\gamma)$, but the sign representation does not. The irreducible representations of $G(F)$ corresponding to the first two contain non-zero I-invariant vectors, whereas the one corresponding to the sign representation of S_3 does not.

The ultimate form of the local Langlands correspondence for representations of $G(F)$ with I-invariant vectors is then as follows (here we assume, as in [KL, CG], that the group G is split and has connected center):

triples $(\gamma, u, \rho), \gamma u \gamma^{-1} = qu$, $\rho \in \mathcal{R}ep\, C(\gamma, u)$ occurs in $H_\bullet(\mathrm{Sp}_u^\gamma, \mathbb{C})$	$\Longleftrightarrow$	irreducible representations (R, π) of $G(F)$, $R^{\pi(I)} \neq 0$

(10.4.6)

Again, this should be understood as a bijection between two sets of equivalence classes of the objects listed. This bijection is due to [KL] (see also [CG]). It was conjectured by Deligne and Langlands, with a subsequent modification (addition of ρ) made by Lusztig.

How to set up this bijection? The idea is to replace irreducible representations of $G(F)$ appearing on the right hand side of (10.4.6) with irreducible modules over the corresponding Hecke algebra $H(G(F), I)$. Recall from Section 10.2.1 that this is the algebra of compactly supported I bi-invariant functions on $G(F)$, with respect to convolution. It naturally acts on the space of I-invariant vectors of any smooth representation of $G(F)$ (see formula (10.2.2)). Thus, we obtain a functor from the category of smooth representations of $G(F)$ to the category of $H(G(F), I)$. According to a theorem of A. Borel [Bor1], it induces a bijection between the set of equivalence classes of irreducible smooth representations of $G(F)$ with non-zero I-invariant vectors and the set of equivalence classes of irreducible $H(G(F), I)$-modules.

The algebra $H(G(F), I)$ is known as the **affine Hecke algebra** and has the standard description in terms of generators and relations. However, for our purposes we need another description, due to [KL, CG], which identifies it with the equivariant K-theory of the **Steinberg variety**

$$\mathrm{St} = \widetilde{\mathcal{N}} \underset{\mathcal{N}}{\times} \widetilde{\mathcal{N}},$$

where $\mathcal{N} \subset {}^L\mathfrak{g}$ is the nilpotent cone and $\widetilde{\mathcal{N}}$ is the **Springer resolution**

$$\widetilde{\mathcal{N}} = \{x \in \mathcal{N}, \mathfrak{b}' \in {}^LG/{}^LB \mid x \in \mathfrak{b}'\}.$$

Thus, a point of St is a triple consisting of a nilpotent element of ${}^L\mathfrak{g}$ and two Borel subalgebras containing it. The group ${}^LG \times \mathbb{C}^\times$ naturally acts on St, with ${}^LG$ conjugating members of the triple and $\mathbb{C}^\times$ acting by multiplication on the nilpotent elements,

$$a \cdot (x, \mathfrak{b}', \mathfrak{b}'') = (a^{-1}x, \mathfrak{b}', \mathfrak{b}''). \tag{10.4.7}$$

According to a theorem of [KL, CG], there is an isomorphism

$$H(G(F), I) \simeq K^{{}^LG \times \mathbb{C}^\times}(\mathrm{St}). \tag{10.4.8}$$

The right hand side is the ${}^LG \times \mathbb{C}^\times$-equivariant K-theory of St. It is an algebra with respect to a natural operation of convolution (see [CG] for details). It is also a free module over its center, isomorphic to

$$K^{{}^LG \times \mathbb{C}^\times}(\mathrm{pt}) = \mathrm{Rep}\, {}^LG \otimes \mathbb{C}[\mathbf{q}, \mathbf{q}^{-1}].$$

Under the isomorphism (10.4.8) the element $\mathbf{q}$ goes to the standard parameter $\mathbf{q}$ of the affine Hecke algebra $H(G(F), I)$ (here we consider $H(G(F), I)$ as a $\mathbb{C}[\mathbf{q}, \mathbf{q}^{-1}]$-module).

Now, the algebra $K^{{}^LG \times \mathbb{C}^\times}(\mathrm{St})$, and hence the algebra $H(G(F), I)$, has a natural family of modules, which are parameterized precisely by the conjugacy classes of pairs (γ, u) as above. On these modules $H(G(F), I)$ acts via a central character corresponding to a point in $\mathrm{Spec}(\mathrm{Rep}\, {}^LG \underset{\mathbb{C}}{\otimes} \mathbb{C}[\mathbf{q}, \mathbf{q}^{-1}])$, which is just a pair (γ, q), where γ is a semi-simple conjugacy class in ${}^LG$ and $q \in \mathbb{C}^\times$. In our situation q is the cardinality of the residue field of F (hence a power of a prime), but in what follows we will allow a larger range of possible values of q: all non-zero complex numbers except for the roots of unity. Consider the quotient of $H(G(F), I)$ by the central character defined by (γ, u). This is just the algebra $K^{{}^LG \times \mathbb{C}^\times}(\mathrm{St})$, specialized at (γ, q). We denote it by $K^{{}^LG \times \mathbb{C}^\times}(\mathrm{St})_{(\gamma,q)}$.

Now for a nilpotent element $u \in \mathcal{N}$ consider the Springer fiber Sp_u. The condition that $\gamma u \gamma^{-1} = qu$ means that u, and hence Sp_u, is stabilized by the action of $(\gamma, q) \in {}^LG \times \mathbb{C}^\times$ (see formula (10.4.7)). Let A be the smallest algebraic subgroup of ${}^LG \times \mathbb{C}^\times$ containing (γ, q). The algebra $K^{{}^LG \times \mathbb{C}^\times}(\mathrm{St})_{(\gamma,q)}$ naturally acts on the equivariant K-theory $K^A(\mathrm{Sp}_u)$ specialized at (γ, q),

$$K^A(\mathrm{Sp}_u)_{(\gamma,q)} = K^A(\mathrm{Sp}_u) \underset{\mathrm{Rep}\, A}{\otimes} \mathbb{C}_{(\gamma,q)}.$$

It is known that $K^A(\mathrm{Sp}_u)_{(\gamma,q)}$ is isomorphic to the homology $H_\bullet(\mathrm{Sp}_u^\gamma)$ of the γ-fixed subset of Sp_u (see [KL, CG]). Thus, we obtain that $K^A(\mathrm{Sp}_u)_{(\gamma,q)}$ is a module over $H(G(F), I)$.

Unfortunately, these $H(G(F), I)$-modules are not irreducible in general, and one needs to work harder to describe the irreducible modules over $H(G(F), I)$. For $G = GL_n$ one can show that each of these modules has a unique irreducible quotient, and this way one recovers the bijection (10.4.4). But for a general groups G the finite groups $C(\gamma, u)$ come into play. Namely, the group $C(\gamma, u)$ acts on $K^A(\mathrm{Sp}_u)_{(\gamma,q)}$, and this action commutes with the action of $K^{{}^LG\times\mathbb{C}^\times}(\mathrm{St})_{(\gamma,q)}$. Therefore we have a decomposition

$$K^A(\mathrm{Sp}_u)_{(\gamma,q)} = \bigoplus_{\rho\in\mathrm{Irrep}\, C(\gamma,u)} \rho\otimes K^A(\mathrm{Sp}_u)_{(\gamma,q,\rho)},$$

of $K^A(\mathrm{Sp}_u)_{(\gamma,q)}$ as a representation of $C(\gamma, u) \times H(G(F), I)$. One shows (see [KL, CG] for details) that each $H(G(F), I)$-module $K^A(\mathrm{Sp}_u)_{(\gamma,q,\rho)}$ has a unique irreducible quotient, and this way one obtains a parameterization of irreducible modules by the triples appearing in the left hand side of (10.4.6). Therefore we obtain that the same set is in bijection with the right hand side of (10.4.6). This is how the tame local Langlands correspondence (10.4.6), also known as the Deligne–Langlands conjecture, is proved.

10.4.2 Categories admitting $(\widehat{\mathfrak{g}}_{\kappa_c}, I)$ Harish-Chandra modules

We now wish to find categorical analogues of the above results in the framework of the categorical Langlands correspondence for loop groups.

As we explained in Section 10.2.2, in the categorical setting a representation of $G(F)$ is replaced by a category $\widehat{\mathfrak{g}}_{\kappa_c}$-mod$_\chi$ equipped with an action of $G((t))$, and the space of I-invariant vectors is replaced by the subcategory of $(\widehat{\mathfrak{g}}_{\kappa_c}, I)$ Harish-Chandra modules in $\widehat{\mathfrak{g}}_{\kappa_c}$-mod$_\chi$. Hence the analogue of the question as to which representations of $G(F)$ admit non-zero I-invariant vectors becomes the following question: for which χ does the category $\widehat{\mathfrak{g}}_{\kappa_c}$-mod$_\chi$ contain non-zero $(\widehat{\mathfrak{g}}_{\kappa_c}, I)$ Harish-Chandra modules? The answer is given by the following lemma.

Recall that in Section 9.1.2 we introduced the space $\mathrm{Op}^{\mathrm{RS}}_{{}^LG}(D)_\varpi$ of ${}^LG$-opers on D with regular singularity and residue $\varpi \in {}^L\mathfrak{h}/W = \mathfrak{h}^*/W$, where W is the Weyl group of ${}^LG$. Given $\mu \in \mathfrak{h}^*$, we write $\varpi(\mu)$ for the projection of μ onto $\mathfrak{h}^*/W$. Finally, let P be the set of integral (not necessarily dominant) weights of $\mathfrak{g}$, viewed as a subset of $\mathfrak{h}^*$.

Lemma 10.4.1 *The category $\widehat{\mathfrak{g}}_{\kappa_c}$-mod$_\chi$ contains a non-zero $(\widehat{\mathfrak{g}}_{\kappa_c}, I)$ Harish-*

Chandra module if and only if

$$\chi \in \bigsqcup_{\nu \in P/W} \operatorname{Op}^{\mathrm{RS}}_{{}^LG}(D)_{\varpi(\nu)}. \tag{10.4.9}$$

Proof. Let M be a non-zero $(\widehat{\mathfrak{g}}_{\kappa_c}, I)$ Harish-Chandra module in $\widehat{\mathfrak{g}}_{\kappa_c}\text{-mod}_\chi$. We show, in the same way as in the proof of Lemma 10.3.2, that it contains a vector v annihilated by $I^0 = [I, I]$ and such that $I/I^0 = H$ acts via a character $\nu \in P$. Therefore there is a non-zero homomorphism from the Verma module $\mathbb{M}_\nu$ to M sending the highest weight vector of $\mathbb{M}_\nu$ to v. According to Theorem 9.5.3, the action of the center $Z(\widehat{\mathfrak{g}}) \simeq \operatorname{Fun}\operatorname{Op}_{{}^LG}(D^\times)$ on $\mathbb{M}_\nu$ factors through $\operatorname{Fun}\operatorname{Op}^{\mathrm{RS}}_{{}^LG}(D)_{\varpi(-\nu-\rho)}$. Therefore $\chi \in \operatorname{Op}^{\mathrm{RS}}_{{}^LG}(D)_{\varpi(-\nu-\rho)}$.

On the other hand, suppose that χ belongs to $\operatorname{Op}^{\mathrm{RS}}_{{}^LG}(D)_{\varpi(\nu)}$ for some $\nu \in P$. Let $\mathbb{M}_\nu(\chi)$ be the quotient of the Verma module $\mathbb{M}_\nu$ by the central character corresponding to χ. It follows from Theorem 9.5.3 that $\mathbb{M}_\nu(\chi)$ is a non-zero object of $\widehat{\mathfrak{g}}_{\kappa_c}\text{-mod}_\chi$. On the other hand, it is clear that $\mathbb{M}_\nu$, and hence $\mathbb{M}_\nu(\chi)$, are I-equivariant $\widehat{\mathfrak{g}}_{\kappa_c}$-modules. □

Thus, the opers χ for which the corresponding category $\widehat{\mathfrak{g}}_{\kappa_c}\text{-mod}_\chi$ contain non-trivial I-equivariant objects are precisely the points of the subscheme (10.4.9) of $\operatorname{Op}_{{}^LG}(D^\times)$. The next question is what are the corresponding ${}^LG$-local systems.

Let $\operatorname{Loc}^{\mathrm{RS,tame}}_{{}^LG} \subset \operatorname{Loc}_{{}^LG}(D^\times)$ be the locus of ${}^LG$-local systems on $D^\times$ with regular singularity and unipotent monodromy. Such a local system is determined, up to an isomorphism, by the conjugacy class of its monodromy (see, e.g., [BV], Section 8). Therefore $\operatorname{Loc}^{\mathrm{RS,tame}}_{{}^LG}$ is an algebraic stack isomorphic to $\mathcal{N}/{}^LG$. The following result is proved in a way similar to the proof of Lemma 10.3.1.

Lemma 10.4.2 *If the local system underlying an oper $\chi \in \operatorname{Op}_{{}^LG}(D^\times)$ belongs to $\operatorname{Loc}^{\mathrm{RS,tame}}_{{}^LG}$, then χ belongs to the subset* (10.4.9) *of* $\operatorname{Op}_{{}^LG}(D^\times)$.

In other words, the subscheme (10.4.9) is precisely the (set-theoretic) preimage of $\operatorname{Loc}^{\mathrm{RS,tame}}_{{}^LG} \subset \operatorname{Loc}_{{}^LG}(D^\times)$ under the map $\alpha : \operatorname{Op}_{{}^LG}(D^\times) \to \operatorname{Loc}_{{}^LG}(D^\times)$.

This hardly comes as a surprise. Indeed, by analogy with the classical Langlands correspondence we expect that the categories $\widehat{\mathfrak{g}}_{\kappa_c}\text{-mod}_\chi$ containing non-trivial I-equivariant objects correspond to the Langlands parameters which are the geometric counterparts of tamely ramified homomorphisms $W'_F \to {}^LG$. The most obvious candidates for those are precisely the ${}^LG$-local systems on $D^\times$ with regular singularity and unipotent monodromy. For this reason we will call such local systems **tamely ramified** (this is reflected in the above notation).

Let us summarize: suppose that σ is a tamely ramified ${}^LG$-local system on $D^\times$, and let χ be a ${}^LG$-oper that is in the gauge equivalence class of σ. Then χ belongs to the subscheme (10.4.9), and the corresponding category $\widehat{\mathfrak{g}}_{\kappa_c}\text{-mod}_\chi$

contains non-zero I-equivariant objects, by Lemma 10.4.1. Let $\widehat{\mathfrak{g}}_{\kappa_c}\text{-mod}_\chi^I$ be the corresponding category of I-equivariant (or, equivalently, $(\widehat{\mathfrak{g}}_{\kappa_c}, I)$ Harish-Chandra) modules. This is our candidate for the categorification of the space of I-invariant vectors in an irreducible representation of $G(F)$ corresponding to a tamely ramified homomorphism $W'_F \to {}^L G$.

Note that according to Conjecture 10.1.2, the categories $\widehat{\mathfrak{g}}_{\kappa_c}\text{-mod}_\chi$ (resp., $\widehat{\mathfrak{g}}_{\kappa_c}\text{-mod}_\chi^I$) should be equivalent to each other for all χ which are gauge equivalent to each other as ${}^L G$-local systems.

In the next section, following [FG2], we will give a conjectural description of the categories $\widehat{\mathfrak{g}}_{\kappa_c}\text{-mod}_\chi^I$ for $\chi \in \mathrm{Op}_{^L G}^{\mathrm{RS}}(D)_{\varpi(-\rho)}$ in terms of the category of coherent sheaves on the Springer fiber corresponding to the residue of χ. This description in particular implies that at least the derived categories of these categories are equivalent to each other for the opers corresponding to the same local system. We have a similar conjecture for $\chi \in \mathrm{Op}_{^L G}^{\mathrm{RS}}(D)_{\varpi(\nu)}$ for other $\nu \in P$, which the reader may easily reconstruct from our discussion of the case $\nu = -\rho$.

10.4.3 Conjectural description of the categories of $(\widehat{\mathfrak{g}}_{\kappa_c}, I)$ Harish-Chandra modules

Let us consider one of the connected components of the subscheme (10.4.9), namely, $\mathrm{Op}_{^L G}^{\mathrm{RS}}(D)_{\varpi(-\rho)}$. Recall from Section 9.2.1 and Proposition 9.2.1 that this space is isomorphic to the space $\mathrm{Op}_{^L G}^{\mathrm{nilp}}$ of **nilpotent opers** on D (corresponding to the weight $\lambda = 0$). As explained in Section 9.2.3, it comes equipped with the residue maps

$$\mathrm{Res}_{\mathcal{F}} : \mathrm{Op}_{^L G}^{\mathrm{nilp}} \to {}^L\mathfrak{n}_{\mathcal{F}_{^L B,0}}, \qquad \mathrm{Res} : \mathrm{Op}_{^L G}^{\mathrm{nilp}} \to {}^L\mathfrak{n}/{}^L B = \widetilde{\mathcal{N}}/{}^L G.$$

For any $\chi \in \mathrm{Op}_{^L G}^{\mathrm{nilp}}$ the ${}^L G$-gauge equivalence class of the corresponding connection is a tamely ramified ${}^L G$-local system on $D^\times$. Moreover, its monodromy conjugacy class is equal to $\exp(2\pi i\, \mathrm{Res}(\chi))$.

We wish to describe the category $\widehat{\mathfrak{g}}_{\kappa_c}\text{-mod}_\chi^I$ of $(\widehat{\mathfrak{g}}_{\kappa_c}, I)$ Harish-Chandra modules with the central character $\chi \in \mathrm{Op}_{^L G}^{\mathrm{nilp}}$. However, here we face the first major complication as compared to the unramified case. While in the ramified case we worked with the abelian category $\widehat{\mathfrak{g}}_{\kappa_c}\text{-mod}_\chi^{G[[t]]}$, this does not seem to be possible in the tamely ramified case. So from now on we will work with the appropriate derived category $D^b(\widehat{\mathfrak{g}}_{\kappa_c}\text{-mod}_\chi)^I$. By definition, this is the full subcategory of the bounded derived category $D^b(\widehat{\mathfrak{g}}_{\kappa_c}\text{-mod}_\chi)$, whose objects are complexes with cohomologies in $\widehat{\mathfrak{g}}_{\kappa_c}\text{-mod}_\chi^I$.

Roughly speaking, the conjecture of [FG2] is that the category $D^b(\widehat{\mathfrak{g}}_{\kappa_c}\text{-mod}_\chi)^I$ is equivalent to the derived category $D^b(\mathrm{QCoh}(\mathrm{Sp}_{\mathrm{Res}_{\mathcal{F}}(\chi)}))$ of the category $\mathrm{QCoh}(\mathrm{Sp}_{\mathrm{Res}_{\mathcal{F}}(\chi)})$ of quasicoherent sheaves on the Springer fiber of $\mathrm{Res}_{\mathcal{F}}(\chi)$. However, we need to make some adjustments to this statement. These adjustments are needed to arrive at a "nice" statement, Conjecture 10.4.4 below. We now explain what these adjustments are and the reasons behind them.

The first adjustment is that we need to consider a slightly larger category of representations than $D^b(\widehat{\mathfrak{g}}_{\kappa_c}\text{-mod}_\chi)^I$. Namely, we wish to include extensions of I-equivariant $\widehat{\mathfrak{g}}_{\kappa_c}$-modules which are not necessarily I-equivariant, but only I^0-equivariant, where $I^0 = [I, I]$. To explain this more precisely, let us choose a Cartan subgroup $H \subset B \subset I$ and the corresponding Lie subalgebra $\mathfrak{h} \subset \mathfrak{b} \subset \operatorname{Lie} I$. We then have an isomorphism $I = H \ltimes I^0$. An I-equivariant $\widehat{\mathfrak{g}}_{\kappa_c}$-module is the same as a module on which $\mathfrak{h}$ acts diagonally with eigenvalues given by integral weights and the Lie algebra $\operatorname{Lie} I^0$ acts locally nilpotently. However, there may exist extensions between such modules on which the action of $\mathfrak{h}$ is no longer semi-simple. Such modules are called ***I*-monodromic**. More precisely, an I-monodromic $\widehat{\mathfrak{g}}_{\kappa_c}$-module is a module that admits an increasing filtration whose consecutive quotients are I-equivariant. It is natural to include such modules in our category. However, it is easy to show that an I-monodromic object of $\widehat{\mathfrak{g}}_{\kappa_c}\text{-mod}_\chi$ is the same as an I^0-equivariant object of $\widehat{\mathfrak{g}}_{\kappa_c}\text{-mod}_\chi$ for any $\chi \in \operatorname{Op}^{\mathrm{nilp}}_{{}^LG}$ (see [FG2]). Therefore instead of I-monodromic modules we will use I^0-equivariant modules. Denote by $D^b(\widehat{\mathfrak{g}}_{\kappa_c}\text{-mod}_\chi)^{I^0}$ the full subcategory of $D^b(\widehat{\mathfrak{g}}_{\kappa_c}\text{-mod}_\chi)$ whose objects are complexes with cohomologies in $\widehat{\mathfrak{g}}_{\kappa_c}\text{-mod}_\chi^{I^0}$.

The second adjustment has to do with the non-flatness of the Springer resolution $\widetilde{\mathcal{N}} \to \mathcal{N}$. By definition, the Springer fiber Sp_u is the fiber product $\widetilde{\mathcal{N}} \underset{\mathcal{N}}{\times} \mathrm{pt}$, where pt is the point $u \in \mathcal{N}$. This means that the structure sheaf of Sp_u is given by

$$\mathcal{O}_{\mathrm{Sp}_u} = \mathcal{O}_{\widetilde{\mathcal{N}}} \underset{\mathcal{O}_{\mathcal{N}}}{\otimes} \mathbb{C}. \tag{10.4.10}$$

However, because the morphism $\widetilde{\mathcal{N}} \to \mathcal{N}$ is not flat, this tensor product functor is not left exact, and there are non-trivial derived tensor products (the *Tor*'s). Our (conjectural) equivalence is not going to be an exact functor: it sends a general object of the category $\widehat{\mathfrak{g}}_{\kappa_c}\text{-mod}_\chi^{I^0}$ not to an object of the category of quasicoherent sheaves, but to a complex of sheaves, or, more precisely, an object of the corresponding derived category. Hence we are forced to work with derived categories, and so the higher derived tensor products need to be taken into account.

To understand better the consequences of this non-exactness, let us consider the following model example. Suppose that we have established an equivalence between the derived category $D^b(\mathrm{QCoh}(\widetilde{\mathcal{N}}))$ and another derived category $D^b(\mathcal{C})$. In particular, this means that both categories carry an action of the algebra $\operatorname{Fun}\mathcal{N}$ (recall that $\mathcal{N}$ is an affine algebraic variety). Let us suppose that the action of $\operatorname{Fun}\mathcal{N}$ on $D^b(\mathcal{C})$ comes from its action on the abelian category $\mathcal{C}$. Thus, $\mathcal{C}$ fibers over $\mathcal{N}$, and let $\mathcal{C}_u$ the fiber category corresponding to $u \in \mathcal{N}$. This is the full subcategory of $\mathcal{C}$ whose objects are objects of $\mathcal{C}$ on which the ideal of u in $\operatorname{Fun}\mathcal{N}$ acts by 0.† What is the category $D^b(\mathcal{C}_u)$ equivalent to?

It is tempting to say that it is equivalent to $D^b(\operatorname{QCoh}(\operatorname{Sp}_u))$. However, this does not follow from the equivalence of $D^b(\operatorname{QCoh}(\mathcal{N}))$ and $D^b(\mathcal{C})$ because of the tensor product (10.4.10) having non-trivial higher derived functors. Suppose that the category $\mathcal{C}$ is *flat over* $\mathcal{N}$; this means that all projective objects of $\mathcal{C}$ are flat as modules over $\operatorname{Fun}\mathcal{N}$. Then $D^b(\mathcal{C}_u)$ is equivalent to the category $D^b(\operatorname{QCoh}(\operatorname{Sp}_u^{\mathrm{DG}}))$, where $\operatorname{Sp}_u^{\mathrm{DG}}$ is the "DG fiber" of $\widetilde{\mathcal{N}} \to \mathcal{N}$ at u. By definition, a quasicoherent sheaf on $\operatorname{Sp}_u^{\mathrm{DG}}$ is a DG module over the DG algebra

$$\mathcal{O}_{\operatorname{Sp}_u^{\mathrm{DG}}} = \mathcal{O}_{\widetilde{\mathcal{N}}} \overset{L}{\underset{\mathcal{O}_{\mathcal{N}}}{\otimes}} \mathbb{C}_u, \tag{10.4.11}$$

where we now take the full derived functor of tensor product. Thus, the category $D^b(\operatorname{QCoh}(\operatorname{Sp}_u^{\mathrm{DG}}))$ may be thought of as the derived category of quasi-coherent sheaves on the "DG scheme" $\operatorname{Sp}_u^{\mathrm{DG}}$ (see [CK] for a precise definition of DG scheme).

Finally, the last adjustment is that we should consider the non-reduced Springer fibers. This means that instead of the Springer resolution $\widetilde{\mathcal{N}}$ we should consider the "thickened" Springer resolution

$$\widetilde{\widetilde{\mathcal{N}}} = {}^L\widetilde{\mathfrak{g}} \underset{{}^L\mathfrak{g}}{\times} \mathcal{N},$$

where ${}^L\widetilde{\mathfrak{g}}$ is the so-called **Grothendieck alteration**,

$${}^L\widetilde{\mathfrak{g}} = \{x \in {}^L\mathfrak{g}, \mathfrak{b}' \in {}^LG/{}^LB \mid x \in \mathfrak{b}'\},$$

which we have encountered previously in Section 7.1.1. The variety $\widetilde{\widetilde{\mathcal{N}}}$ is non-reduced, and the underlying reduced variety is the Springer resolution $\widetilde{\mathcal{N}}$. For instance, the fiber of $\widetilde{\mathcal{N}}$ over a regular element in $\mathcal{N}$ consists of a single point, but the corresponding fiber of $\widetilde{\widetilde{\mathcal{N}}}$ is the spectrum of the Artinian ring $h_0 = \operatorname{Fun}{}^L\mathfrak{h}/(\operatorname{Fun}{}^L\mathfrak{h})_+^W$. Here $(\operatorname{Fun}{}^L\mathfrak{h})_+^W$ is the ideal in $\operatorname{Fun}{}^L\mathfrak{h}$ generated by the augmentation ideal of the subalgebra of W-invariants. Thus, $\operatorname{Spec} h_0$ is the scheme-theoretic fiber of $\varpi : {}^L\mathfrak{h} \to {}^L\mathfrak{h}/W$ at 0. It turns out that in order to describe the category $D^b(\widehat{\mathfrak{g}}_{\kappa_c}\text{-mod}_\chi)^{I^0}$ we need to use the "thickened" Springer resolution.†

Let us summarize: in order to construct the sought-after equivalence of categories we take, instead of individual Springer fibers, the whole Springer resolution, and we further replace it by the "thickened" Springer resolution

† The relationship between $\mathcal{C}$ and $\mathcal{C}_u$ is similar to the relationship between $\widehat{\mathfrak{g}}_{\kappa_c}\text{-mod}$ and and $\widehat{\mathfrak{g}}_{\kappa_c}\text{-mod}_\chi$, where $\chi \in \operatorname{Op}_{{}^LG}(D^\times)$.

† This is explained in Section 10.4.6 for $\mathfrak{g} = \mathfrak{sl}_2$.

$\widetilde{\widetilde{\mathcal{N}}}$ defined above. In this version we will be able to formulate our equivalence so that we avoid DG schemes.

This means that instead of considering the categories $\widehat{\mathfrak{g}}_{\kappa_c}$-mod$_\chi$ for individual nilpotent opers χ, we should consider the "universal" category $\widehat{\mathfrak{g}}_{\kappa_c}$-mod$_{\text{nilp}}$, which is the "family version" of all of these categories. By definition, the category $\widehat{\mathfrak{g}}_{\kappa_c}$-mod$_{\text{nilp}}$ is the full subcategory of $\widehat{\mathfrak{g}}_{\kappa_c}$-mod whose objects have the property that the action of $Z(\widehat{\mathfrak{g}}) = \operatorname{Fun}\operatorname{Op}_{{}^LG}(D)$ on them factors through the quotient $\operatorname{Fun}\operatorname{Op}_{{}^LG}(D) \to \operatorname{Fun}\operatorname{Op}_{{}^LG}^{\text{nilp}}$. Thus, the category $\widehat{\mathfrak{g}}_{\kappa_c}$-mod$_{\text{nilp}}$ is similar to the category $\widehat{\mathfrak{g}}_{\kappa_c}$-mod$_{\text{reg}}$ that we have considered above. While the former fibers over $\operatorname{Op}_{{}^LG}^{\text{nilp}}$, the latter fibers over $\operatorname{Op}_{{}^LG}(D)$. The individual categories $\widehat{\mathfrak{g}}_{\kappa_c}$-mod$_\chi$ are now realized as fibers of these categories over particular opers χ.

Our naive idea was that for each $\chi \in \operatorname{Op}_{{}^LG}^{\text{nilp}}$ the category $D^b(\widehat{\mathfrak{g}}_{\kappa_c}\text{-mod}_\chi)^{I^0}$ is equivalent to $\operatorname{QCoh}(\operatorname{Sp}_{\operatorname{Res}_{\mathcal{F}}(\chi)})$. We would like to formulate now a "family version" of such an equivalence. To this end we form the fiber product

$$ {}^L\widetilde{\widetilde{\mathfrak{n}}} = {}^L\widetilde{\mathfrak{g}} \underset{{}^L\mathfrak{g}}{\times} {}^L\mathfrak{n}. $$

It turns out that this fiber product does not suffer from the problem of the individual Springer fibers, as the following lemma shows:

Lemma 10.4.3 ([FG2],Lemma 6.4) *The derived tensor product*

$$ \operatorname{Fun}{}^L\widetilde{\mathfrak{g}} \overset{L}{\underset{\operatorname{Fun}{}^L\mathfrak{g}}{\otimes}} \operatorname{Fun}{}^L\mathfrak{n} $$

is concentrated in cohomological dimension 0.

The variety ${}^L\widetilde{\widetilde{\mathfrak{n}}}$ may be thought of as the family of (non-reduced) Springer fibers parameterized by ${}^L\mathfrak{n} \subset {}^L\mathfrak{g}$. It is important to note that it is singular, reducible, and non-reduced. For example, if $\mathfrak{g} = \mathfrak{sl}_2$, it has two components, one of which is $\mathbb{P}^1$ (the Springer fiber at 0) and the other is the doubled affine line (i.e., $\operatorname{Spec}\mathbb{C}[x,y]/(y^2)$), see Section 10.4.6.

We note that the corresponding reduced scheme is ${}^L\widetilde{\mathfrak{n}}$ introduced in Section 9.3.2:

$$ {}^L\widetilde{\mathfrak{n}} = \widetilde{\mathcal{N}} \underset{\mathcal{N}}{\times} {}^L\mathfrak{n}. \tag{10.4.12} $$

However, the derived tensor product corresponding to (10.4.12) is not concentrated in cohomological dimension 0, and this is the reason why we prefer to use ${}^L\widetilde{\widetilde{\mathfrak{n}}}$ rather than ${}^L\widetilde{\mathfrak{n}}$.

Now we set

$$ \widetilde{\operatorname{MOp}}{}^0_{{}^LG} = \operatorname{Op}_{{}^LG}^{\text{nilp}} \underset{{}^L\mathfrak{n}/{}^LB}{\times} {}^L\widetilde{\widetilde{\mathfrak{n}}}/{}^LB, $$

where we use the residue morphism $\operatorname{Res} : \operatorname{Op}_{{}^LG}^{\text{nilp}} \to {}^L\mathfrak{n}/{}^LB$. Thus, informally

$\widetilde{\mathrm{MOp}}^0_{{}^L G}$ may be thought as the family over $\mathrm{Op}^{\mathrm{nilp}}_{{}^L G}$ whose fiber over $\chi \in \mathrm{Op}^{\mathrm{nilp}}_{{}^L G}$ is the Springer fiber of $\mathrm{Res}(\chi)$.

As the notation suggests, $\widetilde{\mathrm{MOp}}^0_{{}^L G}$ is closely related to the space of Miura opers whose underlying opers are nilpotent (see Section 10.4.5 below). More precisely, the reduced scheme of $\widetilde{\mathrm{MOp}}^0_{{}^L G}$ is the scheme $\mathrm{MOp}^0_{{}^L G}$ of nilpotent Miura ${}^L G$-opers of weight 0, introduced in Section 9.3.2:

$$\mathrm{MOp}^0_G \simeq \mathrm{Op}^{\mathrm{nilp}}_{{}^L G} \underset{{}^L\mathfrak{n}/{}^L B}{\times} {}^L\widetilde{\mathfrak{n}}/B$$

(see formulas (9.3.9) and (10.4.12)).

We also introduce the category $\widehat{\mathfrak{g}}_{\kappa_c}\text{-mod}^{I^0}_{\mathrm{nilp}}$ as the full subcategory of $\widehat{\mathfrak{g}}_{\kappa_c}\text{-mod}_{\mathrm{nilp}}$ whose objects are I^0-equivariant, and the corresponding derived category $D^b(\widehat{\mathfrak{g}}_{\kappa_c}\text{-mod}_{\mathrm{nilp}})^{I^0}$.

Now we can formulate the Main Conjecture of [FG2]:

Conjecture 10.4.4 *There is an equivalence of categories*

$$D^b(\widehat{\mathfrak{g}}_{\kappa_c}\text{-mod}_{\mathrm{nilp}})^{I^0} \simeq D^b(\mathrm{QCoh}(\widetilde{\mathrm{MOp}}^0_{{}^L G})), \tag{10.4.13}$$

which is compatible with the action of the algebra $\mathrm{Fun}\,\mathrm{Op}^{\mathrm{nilp}}_{{}^L G}$ *on both categories.*

Note that the action of $\mathrm{Fun}\,\mathrm{Op}^{\mathrm{nilp}}_{{}^L G}$ on the first category comes from the action of the center $Z(\widehat{\mathfrak{g}})$, and on the second category it comes from the fact that $\widetilde{\mathrm{MOp}}^0_{{}^L G}$ is a scheme over $\mathrm{Op}^{\mathrm{nilp}}_{{}^L G}$.

The conjectural equivalence (10.4.13) may be viewed as a categorical analogue of the local Langlands correspondence in the tamely ramified case, as we discuss in the next section.

An important feature of the equivalence (10.4.13) does not preserve the t-structure on the two categories. In other words, objects of the abelian category $\widehat{\mathfrak{g}}_{\kappa_c}\text{-mod}^{I^0}_{\mathrm{nilp}}$ are in general mapped under this equivalence to complexes in $D^b(\mathrm{QCoh}(\widetilde{\mathrm{MOp}}^0_{{}^L G}))$, and vice versa. We will see examples of this in the case of $\mathfrak{sl}_2$ in Section 10.4.6.

There are similar conjectures for the categories corresponding to the spaces $\mathrm{Op}^{\mathrm{nilp},\lambda}_{{}^L G}$ of nilpotent opers with dominant integral weights $\lambda \in P^+$.

In the next section we will discuss the connection between Conjecture 10.4.4 and the classical tamely ramified Langlands correspondence. We then present some evidence for this conjecture and consider in detail the example of $\mathfrak{g} = \mathfrak{sl}_2$.

10.4.4 Connection between the classical and the geometric settings

Let us discuss the connection between the equivalence (10.4.13) and the realization of representations of affine Hecke algebras in terms of K-theory of the Springer fibers. As we have explained, we would like to view the category $D^b(\widehat{\mathfrak{g}}_{\kappa_c}\text{-mod}_\chi)^{I^0}$ for $\chi \in \mathrm{Op}^{\mathrm{nilp}}_{{}^L G}$ as, roughly, a categorification of the space $R^{\pi(I)}$ of I-invariant vectors in an irreducible representation (R, π) of $G(F)$. Therefore, we expect that the Grothendieck group of the category $D^b(\widehat{\mathfrak{g}}_{\kappa_c}\text{-mod}_\chi)^{I^0}$ is somehow related to the space $R^{\pi(I)}$.

Let us try to specialize the statement of Conjecture 10.4.4 to a particular oper $\chi = (\mathcal{F}, \nabla, \mathcal{F}_{^LB}) \in \mathrm{Op}^{\mathrm{nilp}}_{^LG}$. Let $\widetilde{\mathrm{Sp}}^{\mathrm{DG}}_{\mathrm{Res}_{\mathcal{F}}(\chi)}$ be the DG fiber of $\widetilde{\mathrm{MOp}}^0_{^LG}$ over χ. By definition (see Section 9.2.3), the residue $\mathrm{Res}_{\mathcal{F}}(\chi)$ of χ is a vector in the twist of ${}^L\mathfrak{n}$ by the ${}^LB$-torsor $\mathcal{F}_{^LB,0}$. It follows that $\widetilde{\mathrm{Sp}}^{\mathrm{DG}}_{\mathrm{Res}_{\mathcal{F}}(\chi)}$ is the DG fiber over $\mathrm{Res}_{\mathcal{F}}(\chi)$ of the $\mathcal{F}_{^LB,0}$-twist of the Grothendieck alteration.

If we trivialize $\mathcal{F}_{^LB,0}$, then $u = \mathrm{Res}_{\mathcal{F}}(\chi)$ becomes an element of ${}^L\mathfrak{n}$. By definition, the (non-reduced) DG Springer fiber $\widetilde{\mathrm{Sp}}^{\mathrm{DG}}_u$ is the DG fiber of the Grothendieck alteration ${}^L\widetilde{\mathfrak{g}} \to {}^L\mathfrak{g}$ at u. In other words, the corresponding structure sheaf is the DG algebra

$$\mathcal{O}_{\widetilde{\mathrm{Sp}}^{\mathrm{DG}}_u} = \mathcal{O}_{^L\widetilde{\mathfrak{g}}} \overset{L}{\underset{\mathcal{O}_{^L\mathfrak{g}}}{\otimes}} \mathbb{C}_u$$

(compare with formula (10.4.11)).

To see what these DG fibers look like, let $u = 0$. Then the naive Springer fiber is just the flag variety ${}^LG/{}^LB$ (it is reduced in this case), and $\mathcal{O}_{\widetilde{\mathrm{Sp}}_0}$ is the structure sheaf of ${}^LG/{}^LB$. But the sheaf $\mathcal{O}_{\widetilde{\mathrm{Sp}}^{\mathrm{DG}}_0}$ is a sheaf of DG algebras, which is quasi-isomorphic to the complex of differential forms on ${}^LG/{}^LB$, with the zero differential. In other words, $\widetilde{\mathrm{Sp}}^{\mathrm{DG}}_0$ may be viewed as a "$\mathbb{Z}$-graded manifold" such that the corresponding supermanifold, obtained by replacing the $\mathbb{Z}$-grading by the corresponding $\mathbb{Z}/2\mathbb{Z}$-grading, is $\Pi T({}^LG/{}^LB)$, the tangent bundle to ${}^LG/{}^LB$ with the parity of the fibers changed from even to odd.

We expect that the category $\widehat{\mathfrak{g}}_{\kappa_c}\text{-mod}^{I^0}_{\mathrm{nilp}}$ is flat over $\mathrm{Op}^{\mathrm{nilp}}_{^LG}$. Therefore, specializing Conjecture 10.4.4 to a particular oper $\chi \in \mathrm{Op}^{\mathrm{nilp}}_{^LG}$, we obtain as a corollary an equivalence of categories

$$D^b(\widehat{\mathfrak{g}}_{\kappa_c}\text{-mod}_\chi)^{I^0} \simeq D^b(\mathrm{QCoh}(\widetilde{\mathrm{Sp}}^{\mathrm{DG}}_{\mathrm{Res}_{\mathcal{F}}(\chi)})). \tag{10.4.14}$$

This fits well with Conjecture 10.1.2 saying that the categories $\widehat{\mathfrak{g}}_{\kappa_c}\text{-mod}_{\chi_1}$ and $\widehat{\mathfrak{g}}_{\kappa_c}\text{-mod}_{\chi_2}$ (and hence $D^b(\widehat{\mathfrak{g}}_{\kappa_c}\text{-mod}_{\chi_1})^{I^0}$ and $D^b(\widehat{\mathfrak{g}}_{\kappa_c}\text{-mod}_{\chi_2})^{I^0}$) should be equivalent if the underlying local systems of the opers χ_1 and χ_2 are isomorphic. For nilpotent opers χ_1 and χ_2 this is so if and only if their monodromies are conjugate to each other. Since their monodromies are obtained by exponentiating their residues, this is equivalent to saying that the residues, $\mathrm{Res}_{\mathcal{F}}(\chi_1)$ and $\mathrm{Res}_{\mathcal{F}}(\chi_2)$, are conjugate with respect to the $\mathcal{F}_{^LB,0}$-twist of ${}^LG$. But in this case the DG Springer fibers corresponding to χ_1 and χ_2 are also isomorphic, and so $D^b(\widehat{\mathfrak{g}}_{\kappa_c}\text{-mod}_{\chi_1})^{I^0}$ and $D^b(\widehat{\mathfrak{g}}_{\kappa_c}\text{-mod}_{\chi_2})^{I^0}$ are equivalent to each other, by (10.4.14).

The Grothendieck group of the category $D^b(\mathrm{QCoh}(\widetilde{\mathrm{Sp}}^{\mathrm{DG}}_u))$, where u is any nilpotent element, is the same as the Grothendieck group of $\mathrm{QCoh}(\mathrm{Sp}_u)$. In other words, the Grothendieck group does not "know" about the DG or the non-reduced structure of $\widetilde{\mathrm{Sp}}_u^{\mathrm{DG}}$. Hence it is nothing but the algebraic K-theory $K(\mathrm{Sp}_u)$. As we explained at the end of Section 10.4.1, equivariant variants of this algebraic K-theory realize the "standard modules" over the affine Hecke algebra $H(G(F), I)$. Moreover, the spaces of I-invariant vectors $R^{\pi(I)}$ as above, which are naturally modules over the affine Hecke algebra, may be realized as subquotients of $K(\mathrm{Sp}_u)$. This indicates that the equivalences (10.4.14) and (10.4.13) are compatible with the classical results.

However, at first glance there are some important differences between the classical and the categorical pictures, which we now discuss in more detail.

In the construction of $H(G(F), I)$-modules outlined in Section 10.4.1 we had to pick a semi-simple element γ of ${}^L G$ such that $\gamma u \gamma^{-1} = qu$, where q is the number of elements in the residue field of F. Then we consider the specialized A-equivariant K-theory $K^A(\mathrm{Sp}_u)_{(\gamma,q)}$, where A is the the smallest algebraic subgroup of ${}^L G \times \mathbb{C}^\times$ containing (γ, q). This gives $K(\mathrm{Sp}_u)$ the structure of an $H(G(F), I)$-module. But this module carries a residual symmetry with respect to the group $C(\gamma, u)$ of components of the centralizer of γ and u in ${}^L G$, which commutes with the action of $H(G(F), I)$. Hence we consider the $H(G(F), I)$-module

$$K^A(\mathrm{Sp}_u)_{(\gamma,q,\rho)} = \mathrm{Hom}_{C(\gamma,u)}(\rho, K(\mathrm{Sp}_u)),$$

corresponding to an irreducible representation ρ of $C(\gamma, u)$. Finally, each of these components has a unique irreducible quotient, and this is an irreducible representation of $H(G(F), I)$, which is realized on the space $R^{\pi(I)}$, where (R, π) is an irreducible representation of $G(F)$ corresponding to (γ, u, ρ) under the bijection (10.4.6). How is this intricate structure reflected in the categorical setting?

Our category $D^b(\mathrm{QCoh}(\widetilde{\mathrm{Sp}}_u^{\mathrm{DG}}))$, where $u = \mathrm{Res}_{\mathcal{F}}(\chi)$, is a particular categorification of the (non-equivariant) K-theory $K(\mathrm{Sp}_u)$. Note that in the classical local Langlands correspondence (10.4.6), the element u of the triple (γ, u, ρ) is interpreted as the logarithm of the monodromy of the corresponding representation of the Weil–Deligne group W'_F. This is in agreement with the interpretation of $\mathrm{Res}_{\mathcal{F}}(\chi)$ as the logarithm of the monodromy of the ${}^L G$-local system on $D^\times$ corresponding to χ, which plays the role of the local Langlands parameter for the category $\widehat{\mathfrak{g}}_{\kappa_c}\text{-mod}_\chi$ (up to the inessential factor $2\pi i$).

But what about the other parameters, γ and ρ? And why does our category correspond to the non-equivariant K-theory of the Springer fiber, and not the equivariant K-theory, as in the classical setting?

The element γ corresponding to the Frobenius in W'_F does not seem to have an analogue in the geometric setting. We have already seen this above in the unramified case: while in the classical setting unramified local Langlands parameters are the semi-simple conjugacy classes γ in ${}^LG$, in the geometric setting we have only one unramified local Langlands parameter; namely, the trivial local system.

To understand better what is going on here, we revisit the unramified case. Recall that the spherical Hecke algebra $H(G(F), K_0)$ is isomorphic to the representation ring $\operatorname{Rep} {}^LG$. The one-dimensional space of K_0-invariants in an irreducible unramified representation (R, π) of $G(F)$ realizes a one-dimensional representation of $H(G(F), K_0)$, i.e., a homomorphism $\operatorname{Rep} {}^LG \to \mathbb{C}$. The unramified Langlands parameter γ of (R, π), which is a semi-simple conjugacy class in ${}^LG$, is the point in $\operatorname{Spec}(\operatorname{Rep} {}^LG)$ corresponding to this homomorphism. What is a categorical analogue of this homomorphism? The categorification of $\operatorname{Rep} {}^LG$ is the category $\mathcal{R}ep\, {}^LG$. The product structure on $\operatorname{Rep} {}^LG$ is reflected in the structure of tensor category on $\mathcal{R}ep\, {}^LG$. On the other hand, the categorification of the algebra $\mathbb{C}$ is the category $\mathcal{V}ect$ of vector spaces. Therefore a categorical analogue of a homomorphism $\operatorname{Rep} {}^LG \to \mathbb{C}$ is a functor $\mathcal{R}ep\, {}^LG \to \mathcal{V}ect$ respecting the tensor structures on both categories. Such functors are called the fiber functors. The fiber functors form a category of their own, which is equivalent to the category of ${}^LG$-torsors. Thus, any two fiber functors are isomorphic, but not canonically. In particular, the group of automorphisms of each fiber functor is isomorphic to ${}^LG$. (Incidentally, this is how ${}^LG$ is reconstructed from a fiber functor in the Tannakian formalism.) Thus, we see that, while in the categorical world we do not have analogues of semi-simple conjugacy classes γ (the points of $\operatorname{Spec}(\operatorname{Rep} {}^LG)$), their role is in some sense played by the group of automorphisms of a fiber functor.

This is reflected in the fact that, while in the categorical setting we have a unique unramified Langlands parameter – namely, the trivial ${}^LG$-local system σ_0 on $D^\times$ – this local system has a non-trivial group of automorphisms, namely, ${}^LG$. We therefore expect that the group ${}^LG$ should act by automorphisms of the Langlands category $\mathcal{C}_{\sigma_0}$ corresponding to σ_0, and this action should commute with the action of the loop group $G((t))$ on $\mathcal{C}_{\sigma_0}$. It is this action of ${}^LG$ that is meant to compensate for the lack of unramified Langlands parameters, as compared to the classical setting.

We have argued in Section 10.3 that the category $\widehat{\mathfrak{g}}_{\kappa_c}\text{-mod}_\chi$, where $\chi = (\mathcal{F}, \nabla, \mathcal{F}_{{}^LB}) \in \operatorname{Op}_{{}^LG}(D)$, is a candidate for the Langlands category $\mathcal{C}_{\sigma_0}$. Therefore we expect that the group ${}^LG$ (more precisely, its twist ${}^LG_{\mathcal{F}}$) acts on the category $\widehat{\mathfrak{g}}_{\kappa_c}\text{-mod}_\chi$. In Section 10.3.7 we showed how to obtain this action using the conjectural equivalence between $\widehat{\mathfrak{g}}_{\kappa_c}\text{-mod}_\chi$ and the category $\mathcal{D}^{\text{Hecke}}_{\kappa_c}\text{-mod}_\chi$ of Hecke eigenmodules on the affine Grassmannian Gr (see Conjecture 10.3.9). The category $\mathcal{D}^{\text{Hecke}}_{\kappa_c}\text{-mod}_\chi$ was defined in Section 10.3.6 as a "de-equivariantization" of the category $\mathcal{D}_{\kappa_c}\text{-mod}$ of twisted $\mathcal{D}$-modules on Gr with respect to the monoidal action of the category $\mathcal{R}ep\, {}^LG$.

Now comes a crucial observation that will be useful for understanding the way things work in the tamely ramified case: the category $\mathcal{R}ep\, {}^LG$ may be interpreted as the category of ${}^L G$-equivariant quasicoherent sheaves on the variety $\mathrm{pt} = \operatorname{Spec} \mathbb{C}$. In other words, $\mathcal{R}ep\, {}^L G$ may be interpreted as the category of quasicoherent sheaves on the stack $\mathrm{pt} / {}^L G$. The existence of monoidal action of the category $\mathcal{R}ep\, {}^L G$ on $\mathcal{D}_{\kappa_c}$-mod should be viewed as the statement that the category $\mathcal{D}_{\kappa_c}$-mod "fibers" over the stack $\mathrm{pt} / {}^L G$. The statement of Conjecture 10.3.9 may then be interpreted as saying that

$$\widehat{\mathfrak{g}}_{\kappa_c}\text{-mod}_\chi \simeq \mathcal{D}_{\kappa_c}\text{-mod} \underset{\mathrm{pt} / {}^L G}{\times} \mathrm{pt} .$$

In other words, if $\mathcal{C}$ is the conjectural Langlands category fibering over the stack $\operatorname{Loc}_{{}^L G}(D^\times)$ of all ${}^L G$-local systems on $D^\times$, then

$$\mathcal{D}_{\kappa_c}\text{-mod} \simeq \mathcal{C} \underset{\operatorname{Loc}_{{}^L G}(D^\times)}{\times} \mathrm{pt} / {}^L G,$$

whereas

$$\widehat{\mathfrak{g}}_{\kappa_c}\text{-mod}_\chi \simeq \mathcal{C} \underset{\operatorname{Loc}_{{}^L G}(D^\times)}{\times} \mathrm{pt},$$

where the morphism $\mathrm{pt} \to \operatorname{Loc}_{{}^L G}(D^\times)$ corresponds to the oper χ.

Thus, in the categorical setting there are two different ways to think about the trivial local system σ_0: as a point (defined by a particular ${}^L G$-bundle on D with connection, such as a regular oper χ), or as a stack $\mathrm{pt} / {}^L G$. The base change of the Langlands category in the first case gives us a category with an action of ${}^L G$, such as the categories $\widehat{\mathfrak{g}}_{\kappa_c}\text{-mod}_\chi$ or $\mathcal{D}^{\mathrm{Hecke}}_{\kappa_c}$-mod. The base change in the second case gives us a category with a monoidal action of $\mathcal{R}ep\, {}^L G$, such as the category $\mathcal{D}_{\kappa_c}$-mod. We can go back and forth between the two versions by applying the procedures of equivariantization and de-equivariantization with respect to ${}^L G$ and $\mathcal{R}ep\, {}^L G$, respectively.

Now we return to the tamely ramified case. The semi-simple element γ appearing in the triple (γ, u, ρ) plays the same role as the unramified Langlands parameter γ. However, now it must satisfy the identity $\gamma u \gamma^{-1} = qu$. Recall that the center Z of $H(G(F), I)$ is isomorphic to $\operatorname{Rep} {}^L G$, and so $\operatorname{Spec} Z$ is the set of all semi-simple elements in ${}^L G$. For a fixed nilpotent element u the equation $\gamma u \gamma^{-1} = qu$ cuts out a locus C_u in $\operatorname{Spec} Z$ corresponding to those central characters which may occur on irreducible $H(G(F), I)$-modules corresponding to u. In the categorical setting (where we set $q = 1$) the analogue of C_u is the centralizer $Z(u)$ of u in ${}^L G$, which is precisely the group $\operatorname{Aut}(\sigma)$ of automorphisms of a tame local system σ on $D^\times$ with monodromy $\exp(2\pi i u)$. On general grounds we expect that the group $\operatorname{Aut}(\sigma)$ acts on the Langlands category $\mathcal{C}_\sigma$, just as we expect the group ${}^L G$ of automorphisms of the trivial local system σ_0 to act on the category $\mathcal{C}_{\sigma_0}$. It is this action that replaces the parameter γ in the geometric setting.

In the classical setting we also have one more parameter, ρ. Let us recall that ρ is a representation of the group $C(\gamma, u)$ of connected components of the centralizer $Z(\gamma, u)$ of γ and u. But the group $Z(\gamma, u)$ is a subgroup of $Z(u)$, which becomes the group $\mathrm{Aut}(\sigma)$ in the geometric setting. Therefore one can argue that the parameter ρ is also absorbed into the action of $\mathrm{Aut}(\sigma)$ on the category $\mathcal{C}_\sigma$.

If we have an action of $\mathrm{Aut}(\sigma)$ on the category $\mathcal{C}_\sigma$, or on one of its many incarnations $\widehat{\mathfrak{g}}_{\kappa_c}\text{-mod}_\chi, \chi \in \mathrm{Op}^{\mathrm{nilp}}_{^LG}$, it means that these categories must be "de-equivariantized," just like the categories $\widehat{\mathfrak{g}}_{\kappa_c}\text{-mod}_\chi, \chi \in \mathrm{Op}_{^LG}(D)$, in the unramified case. This is the reason why in the equivalence (10.4.14) (and in Conjecture 10.4.4) we have the non-equivariant categories of quasicoherent sheaves (whose Grothendieck groups correspond to the non-equivariant K-theory of the Springer fibers).

However, there is also an equivariant version of these categories. Consider the substack $\mathrm{Loc}^{\mathrm{RS,tame}}_{^LG}$ of tamely ramified local systems in $\mathrm{Loc}_{^LG}(D^\times)$ introduced in Section 10.4.2. Since a tamely ramified local system is completely determined by the logarithm of its (unipotent) monodromy, this substack is isomorphic to $\mathcal{N}/{}^LG$. This substack plays the role of the substack $\mathrm{pt}/{}^LG$ corresponding to the trivial local system. Let us set

$$\mathcal{C}_{\mathrm{tame}} = \mathcal{C} \underset{\mathrm{Loc}_{^LG}(D^\times)}{\times} \mathcal{N}/{}^LG.$$

Then, according to our general conjecture expressed by the Cartesian diagram (10.1.3), we expect to have

$$\widehat{\mathfrak{g}}_{\kappa_c}\text{-mod}_{\mathrm{nilp}} \simeq \mathcal{C}_{\mathrm{tame}} \underset{\mathcal{N}/{}^LG}{\times} \mathrm{Op}^{\mathrm{nilp}}_{^LG}. \tag{10.4.15}$$

Let $D^b(\mathcal{C}_{\mathrm{tame}})^{I^0}$ be the I^0-equivariant derived category corresponding to $\mathcal{C}_{\mathrm{tame}}$. Combining (10.4.15) with Conjecture 10.4.4, and noting that

$$\widetilde{\mathrm{MOp}}^0_{^LG} \simeq \mathrm{Op}^{\mathrm{nilp}}_{^LG} \underset{\mathcal{N}/{}^LG}{\times} \widetilde{\widetilde{\mathcal{N}}}/{}^LG,$$

we obtain the following conjecture (see [FG2]):

$$D^b(\mathcal{C}_{\mathrm{tame}})^{I^0} \simeq D^b(\mathrm{QCoh}(\widetilde{\widetilde{\mathcal{N}}}/{}^LG)). \tag{10.4.16}$$

The category on the right hand side may be interpreted as the derived category of ${}^LG$-equivariant quasicoherent sheaves on the "thickened" Springer resolution $\widetilde{\widetilde{\mathcal{N}}}$.

Together, the conjectural equivalences (10.4.14) and (10.4.16) should be thought of as the categorical versions of the realizations of modules over the affine Hecke algebra in the K-theory of the Springer fibers.

One corollary of the equivalence (10.4.14) is the following: the classes of irreducible objects of the category $\widehat{\mathfrak{g}}_{\kappa_c}\text{-mod}^{I_0}_\chi$ in the Grothendieck group of $\widehat{\mathfrak{g}}_{\kappa_c}\text{-mod}^{I_0}_\chi$ give rise to a basis in the algebraic K-theory $K(\mathrm{Sp}_u)$, where $u = \mathrm{Res}_{\mathcal{F}}(\chi)$. Presumably, this basis is closely related to the bases in (the equivariant version of) this K-theory constructed by G. Lusztig in [Lu2] (from the perspective of unrestricted $\mathfrak{g}$-modules in positive characteristic).

10.4.5 Evidence for the conjecture

We now describe some evidence for Conjecture 10.4.4. It consists of the following four groups of results:

- Interpretation of the Wakimoto modules as $\widehat{\mathfrak{g}}_{\kappa_c}$-modules corresponding to the skyscraper sheaves on $\widetilde{\mathrm{MOp}}^0_{^LG}$.
- Proof of the equivalence of certain quotient categories of $D^b(\widehat{\mathfrak{g}}_{\kappa_c}\text{-mod}_{\mathrm{nilp}})^{I^0}$ and $D^b(\mathrm{QCoh}(\widetilde{\mathrm{MOp}}^0_{^LG}))$.
- Proof of the restriction of the equivalence (10.4.13) to regular opers.
- Connection to R. Bezrukavnikov's theory.

We start with the discussion of Wakimoto modules.

Suppose that we have proved the equivalence of categories (10.4.13). Then each quasicoherent sheaf on $\widetilde{\mathrm{MOp}}^0_{^LG}$ should correspond to an object of the derived category $D^b(\widehat{\mathfrak{g}}_{\kappa_c}\text{-mod}_{\mathrm{nilp}})^{I^0}$. The simplest quasicoherent sheaves on $\widetilde{\mathrm{MOp}}^0_{^LG}$ are the **skyscraper sheaves** supported at the $\mathbb{C}$-points of $\widetilde{\mathrm{MOp}}^0_{^LG}$. It follows from the definition that a $\mathbb{C}$-point of $\widetilde{\mathrm{MOp}}^0_{^LG}$, which is the same as a $\mathbb{C}$-point of the reduced scheme MOp^0_G, is a pair $(\chi, \mathfrak{b}')$, where $\chi = (\mathcal{F}, \nabla, \mathcal{F}_{^LB})$ is a nilpotent ^LG-oper in $\mathrm{Op}^{\mathrm{nilp}}_{^LG}$ and $\mathfrak{b}'$ is a point of the Springer fiber corresponding to $\mathrm{Res}_{\mathcal{F}}(\chi)$, which is the variety of Borel subalgebras in $^L\mathfrak{g}_{\mathcal{F}_0}$ that contain $\mathrm{Res}_{\mathcal{F}}(\chi)$. Thus, if Conjecture 10.4.4 is true, we should have a family of objects of the category $D^b(\widehat{\mathfrak{g}}_{\kappa_c}\text{-mod}_{\mathrm{nilp}})^{I^0}$ parameterized by these data. What are these objects?

The reader who has studied the previous chapter of this book will no doubt have already guessed the answer: these are the **Wakimoto modules**! Indeed, let us recall that Wakimoto modules of critical level are parameterized by the space $\mathrm{Conn}(\Omega^{-\rho})_{D^\times}$. The Wakimoto module corresponding to $\overline{\nabla} \in \mathrm{Conn}(\Omega^{-\rho})_{D^\times}$ is denoted by $W_{\overline{\nabla}}$ (see Section 9.4.3). According to Theorem 8.3.3, the center $Z(\widehat{\mathfrak{g}})$ acts on $W_{\overline{\nabla}}$ via the central character $\mu(\overline{\nabla})$, where μ is the Miura transformation. Moreover, as explained in Section 9.4.3, if $\chi \in \mathrm{Op}^{\mathrm{nilp}}_{^LG}$, then $W_{\overline{\nabla}}$ is an object of the category $\widehat{\mathfrak{g}}_{\kappa_c}\text{-mod}^I_\chi$ for any $\overline{\nabla} \in \mu^{-1}(\chi)$.

According to Theorem 9.3.7, the points of $\mu^{-1}(\chi)$ are in bijection with the points of the Springer fiber $\mathrm{Sp}_{\mathrm{Res}_{\mathcal{F}}(\chi)}$ corresponding to the nilpotent element $\mathrm{Res}_{\mathcal{F}}(\chi)$. Therefore to each point of $\mathrm{Sp}_{\mathrm{Res}_{\mathcal{F}}(\chi)}$ we have assigned a Wakimoto module, which is an object of the category $\widehat{\mathfrak{g}}_{\kappa_c}\text{-mod}^{I^0}_\chi$ (and hence of the corresponding derived category). In other words, Wakimoto modules are objects of the category $\widehat{\mathfrak{g}}_{\kappa_c}\text{-mod}^I_{\mathrm{nilp}}$ parameterized by the $\mathbb{C}$-points of $\widetilde{\mathrm{MOp}}^0_{^LG}$. It is natural to assume that they correspond to the skyscraper sheaves on $\widetilde{\mathrm{MOp}}^0_{^LG}$

under the equivalence (10.4.13). This was in fact one of our motivations for this conjecture.

Incidentally, this gives us a glimpse into how the group of automorphisms of the ${}^L G$-local system underlying the oper χ acts on the category $\widehat{\mathfrak{g}}_{\kappa_c}\text{-mod}_\chi$. This group is $Z(\mathrm{Res}_{\mathcal{F}}(\chi))$, the centralizer of the residue $\mathrm{Res}_{\mathcal{F}}(\chi)$, and it acts on the Springer fiber $\mathrm{Sp}_{\mathrm{Res}_{\mathcal{F}}(\chi)}$. Therefore $g \in Z(\mathrm{Res}_{\mathcal{F}}(\chi))$ sends the skyscraper sheaf supported at a point $p \in \mathrm{Sp}_{\mathrm{Res}_{\mathcal{F}}(\chi)}$ to the skyscraper sheaf supported at $g \cdot p$. Hence we expect that g sends the Wakimoto module corresponding to p to the Wakimoto module corresponding to $g \cdot p$.

If the Wakimoto modules indeed correspond to the skyscraper sheaves, then the equivalence (10.4.13) may be thought of as a kind of "spectral decomposition" of the category $D^b(\widehat{\mathfrak{g}}_{\kappa_c}\text{-mod}_{\mathrm{nilp}})^{I^0}$, with the basic objects being the Wakimoto modules $W_{\overline{\nabla}}$, where $\overline{\nabla}$ runs over the locus in $\mathrm{Conn}(\Omega^{-\rho})_{D^\times}$, which is isomorphic, pointwise, to $\widetilde{\mathrm{MOp}}^0_{{}^L G}$.

Note however that, even though the (set-theoretic) preimage of $\mathrm{Op}^{\mathrm{nilp}}_{{}^L G}$ in $\mathrm{Conn}(\Omega^{-\rho})_{D^\times}$ under the Miura transformation μ is equal to MOp^0_G, they are not isomorphic as algebraic varieties. Instead, according to Section 9.3.3, $\mu^{-1}(\mathrm{Op}^{\mathrm{nilp}}_{{}^L G})$ is the disjoint union of subvarieties

$$\mu^{-1}(\mathrm{Op}^{\mathrm{nilp}}_{{}^L G}) = \bigsqcup_{w \in W} \mathrm{Conn}(\Omega^{-\rho})_D^{w(\rho)-\rho},$$

where $\mathrm{Conn}(\Omega^{-\rho})_D^{w(\rho)-\rho}$ is the variety of connections with regular singularity and residue $w(\rho)-\rho$. On the other hand, $\mathrm{MOp}^0_{{}^L G}$ has a stratification (obtained from intersections with the Schubert cells on the flag variety ${}^L G/{}^L B$), with the strata $\mathrm{MOp}^{0,w}_{{}^L G}$ isomorphic to $\mathrm{Conn}(\Omega^{-\rho})_D^{w(\rho)-\rho}, w \in W$ (see Section 9.3.2 and Theorem 9.3.6). But these strata are "glued" together in $\mathrm{MOp}^0_{{}^L G}$ in a non-trivial way.

For each stratum, which is an affine algebraic variety (in fact, an infinite-dimensional affine space), we can associate a functor

$$\mathsf{F}_w : \mathrm{QCoh}(\mathrm{Conn}(\Omega^{-\rho})_D^{w(\rho)-\rho}) \to \widehat{\mathfrak{g}}_{\kappa_c}\text{-mod}^{I^0}_{\mathrm{nilp}},$$

using the Wakimoto module $W_{w(\rho)-\rho}$ defined by formula (9.4.3). The functor F_w maps an object of $\mathrm{QCoh}(\mathrm{Conn}(\Omega^{-\rho})_D^{w(\rho)-\rho})$, which is the same as a module $\mathcal{F}$ over the algebra $\mathrm{Fun}\,\mathrm{Conn}(\Omega^{-\rho})_D^{w(\rho)-\rho}$, to the $\widehat{\mathfrak{g}}_{\kappa_c}$-module

$$W_{w(\rho)-\rho} \underset{\mathrm{Fun}\,\mathrm{Conn}(\Omega^{-\rho})_D^{w(\rho)-\rho}}{\otimes} \mathcal{F}.$$

Informally, one can say that the category $D^b(\mathrm{QCoh}(\widetilde{\mathrm{MOp}}^0_{{}^L G}))$ is "glued" from the categories of quasi-coherent sheaves on the strata $\mathrm{Conn}(\Omega^{-\rho})_D^{w(\rho)-\rho}$. Conjecture 10.4.4 may therefore be interpreted as saying that the category $D^b(\widehat{\mathfrak{g}}_{\kappa_c}\text{-mod}_{\mathrm{nilp}})^{I^0}$ is "glued" from its subcategories obtained as the images of these functors. For more on this, see [FG5].

At this point it is instructive to compare the conjectural equivalence (10.4.13) with the equivalence (10.3.9) in the case of regular opers, which we rewrite in a more suggestive way as

$$\widehat{\mathfrak{g}}_{\kappa_c}\text{-mod}_{\text{reg}}^{G[[t]]} \simeq \text{QCoh}(\text{Op}_{^LG}(D)). \qquad (10.4.17)$$

There are two essential differences between the two statements. First of all, (10.4.17) is an equivalence between abelian categories, whereas in (10.4.13) we need to use the derived categories. The second difference is that while in (10.4.17) we have the category of quasicoherent sheaves on a smooth affine algebraic variety $\text{Op}_{^LG}(D)$ (actually, an infinite-dimensional affine space), in (10.4.13) we have the category of quasicoherent sheaves on a quasi-projective variety $\widetilde{\text{MOp}}^0_{^LG}$ which is highly singular and non-reduced. In some sense, most of the difficulties in proving (10.4.13) are caused by the non-affineness of $\widetilde{\text{MOp}}^0_{^LG}$.

However, there is a truncated version of the equivalence (10.4.13), which involves the category of quasicoherent sheaves on an affine subscheme of $\widetilde{\text{MOp}}^0_{^LG}$ and a certain quotient of the category $D^b(\widehat{\mathfrak{g}}_{\kappa_c}\text{-mod}_{\text{nilp}})^{I^0}$. Let us give a precise statement.

For each $i = 1, \ldots, \ell$, let $\mathfrak{sl}_2^{(i)}$ be the $\mathfrak{sl}_2$ subalgebra of $\mathfrak{g} = \mathfrak{g} \otimes 1 \subset \widehat{\mathfrak{g}}_{\kappa_c}$ generated by e_i, h_i, f_i. Let us call a $\widehat{\mathfrak{g}}_{\kappa_c}$-module M **partially integrable** if there exists $i = 1, \ldots, \ell$, such that the action of $\mathfrak{sl}_2^{(i)}$ on M may be exponentiated to an action of the corresponding group SL_2. It easy to see that partially integrable objects form a Serre subcategory in $\widehat{\mathfrak{g}}^{I^0}_{\text{nilp}}$. Let ${}^f\widehat{\mathfrak{g}}_{\kappa_c}\text{-mod}^{I^0}_{\text{nilp}}$ be the quotient category of $\widehat{\mathfrak{g}}_{\kappa_c}\text{-mod}^{I^0}_{\text{nilp}}$ by the subcategory of partially integrable objects. We will denote by ${}^fD^b(\widehat{\mathfrak{g}}_{\kappa_c}\text{-mod}_{\text{nilp}})^{I^0}$ the triangulated quotient category of $D^b(\widehat{\mathfrak{g}}_{\kappa_c}\text{-mod}_{\text{nilp}})^{I^0}$ by the subcategory whose objects have partially integrable cohomology.

On the other hand, consider the subscheme $\widetilde{\text{MOp}}^{0,w_0}_{^LG}$ of $\widetilde{\text{MOp}}^0_{^LG}$, which is the closure of the restriction of $\widetilde{\text{MOp}}^0_{^LG}$ to the locus of regular nilpotent elements in ${}^L\mathfrak{n}$ under the map $\widetilde{\text{MOp}}^0_{^LG} \to {}^L\mathfrak{n}$. As we have seen in Section 10.4.3, the non-reduced Springer fiber over regular nilpotent elements is isomorphic to $\text{Spec}\, h_0$, where $h_0 = \text{Fun}\, {}^L\mathfrak{h}/(\text{Fun}\, {}^L\mathfrak{h})^W_+$ is the Artinian algebra of dimension $|W|$. Therefore we have

$$\widetilde{\text{MOp}}^{0,w_0}_{^LG} \simeq \text{Op}^{\text{nilp}}_{^LG} \times \text{Spec}\, h_0.$$

Thus, we see that it is an affine subscheme of $\widetilde{\text{MOp}}^0_{^LG}$. The corresponding reduced scheme is in fact the stratum $\text{MOp}^{0,w_0}_{^LG} \subset \text{MOp}^0_{^LG}$ introduced in Section 9.3.2. The category of quasicoherent sheaves on $\widetilde{\text{MOp}}^{0,w_0}_{^LG}$ may be realized as a Serre quotient of the category of quasicoherent sheaves on $\widetilde{\text{MOp}}^0_{^LG}$.

The following result, proved in [FG2], shows that the two quotient categories are equivalent to each other (already at the level of abelian categories).

Theorem 10.4.5 *The categories* ${}^f\widehat{\mathfrak{g}}_{\kappa_c}\text{-mod}_{\text{nilp}}^{I^0}$ *and* $\text{QCoh}(\widetilde{\text{MOp}}_{{}^LG}^{0,w_0})$ *are equivalent to each other.*

This is the second piece of evidence for Conjecture 10.4.4 from the list given at the beginning of this section.

Let us now discuss the third piece of evidence. The categories on both sides of (10.4.13) fiber over $\text{Op}_{{}^LG}^{\text{nilp}}$; in other words, the algebra $\text{Fun}\,\text{Op}_{{}^LG}^{\text{nilp}}$ acts on objects of each of these categories. Consider the restrictions of these categories to the subscheme of regular opers $\text{Op}_{{}^LG}(D) \subset \text{Op}_{{}^LG}^{\text{nilp}}$. The corresponding abelian categories are defined as the full subcategories on which the action of $\text{Fun}\,\text{Op}_{{}^LG}^{\text{nilp}}$ factors through $\text{Fun}\,\text{Op}_{{}^LG}(D)$. The objects of the corresponding derived categories are those complexes whose cohomologies have this property.

The restriction of the category on the right hand side of (10.4.13) to the subscheme $\text{Op}_{{}^LG}(D)$ is the derived category of quasicoherent sheaves on the DG scheme

$$\widetilde{\text{MOp}}_{{}^LG}^{\text{reg}} = \widetilde{\text{MOp}}_{{}^LG}^{0} \underset{\text{Op}_{{}^LG}^{\text{nilp}}}{\overset{L}{\times}} \text{Op}_{{}^LG}(D).$$

The corresponding tensor product functor is not left exact, and so we are in the situation described in Section 10.4.3. Therefore we need to take the DG fiber product. Since the residue of any regular oper is equal to 0, we find that the fiber of $\widetilde{\text{MOp}}_{{}^LG}^{\text{reg}}$ over each oper $\chi \in \text{Op}_{{}^LG}(D)$ is isomorphic to the DG Springer fiber Sp_0^{DG}, which is the DG scheme $\Pi T({}^LG/{}^LB)$, as we discussed in Section 10.4.3.

On the other hand, the restriction of the category on the left hand side of (10.4.13) to $\text{Op}_{{}^LG}(D)$ is the category $D^b(\widehat{\mathfrak{g}}_{\kappa_c}\text{-mod}_{\text{reg}})^{I^0}$, the I^0-equivariant part of the derived category of $\widehat{\mathfrak{g}}_{\kappa_c}\text{-mod}_{\text{reg}}$. According to a theorem proved in [FG4] (which is the I^0-equivariant version of Conjecture 10.3.9), there is an equivalence of categories

$$\widehat{\mathfrak{g}}_{\kappa_c}\text{-mod}_{\text{reg}}^{I^0} \simeq \mathcal{D}_{\kappa_c}^{\text{Hecke}}\text{-mod}_{\text{reg}}^{I^0}$$

and hence the corresponding derived categories are also equivalent. On the other hand, it follows from the results of [ABG] that the derived category of $D^b(\mathcal{D}_{\kappa_c}^{\text{Hecke}}\text{-mod}_{\text{reg}})^{I^0}$ is equivalent to $D^b(\text{QCoh}(\widetilde{\text{MOp}}_{{}^LG}^{\text{reg}}))$. Therefore the restrictions of the categories appearing in (10.4.13) to $\text{Op}_{{}^LG}(D)$ are equivalent.

Finally, we discuss the fourth piece of evidence, connection with Bezrukavnikov's theory.

To motivate it, let us recall that in Section 10.3.5 we discussed the action of the categorical spherical Hecke algebra $\mathcal{H}(G((t)), G[[t]])$ on the category $\widehat{\mathfrak{g}}_{\kappa_c}\text{-mod}_{\chi}$, where χ is a regular oper. The affine Hecke algebra $H(G(F), I)$ also has a categorical analogue. Consider the **affine flag variety** $\text{Fl} = G((t))/I$. The categorical affine Hecke algebra is the category $\mathcal{H}(G((t)), I)$, which is the full subcategory of the derived category of $\mathcal{D}$-modules on $\text{Fl} = G((t))/I$

whose objects are complexes with I-equivariant cohomologies. This category naturally acts on the derived category $D^b(\widehat{\mathfrak{g}}_{\kappa_c}\text{-mod}_\chi)^I$. What does this action correspond to on the other side of the equivalence (10.4.13)?

The answer is given by a theorem of R. Bezrukavnikov [B], which may be viewed as a categorification of the isomorphism (10.4.8):

$$D^b(\mathcal{D}^{\mathrm{Fl}}_{\kappa_c}\text{-mod})^{I^0} \simeq D^b(\mathrm{QCoh}(\widetilde{\mathrm{St}})), \tag{10.4.18}$$

where $\mathcal{D}^{\mathrm{Fl}}_{\kappa_c}$-mod is the category of twisted $\mathcal{D}$-modules on Fl and $\widetilde{\mathrm{St}}$ is the "thickened" Steinberg variety

$$\widetilde{\mathrm{St}} = \widetilde{\mathcal{N}} \underset{\mathcal{N}}{\times} \widetilde{\widetilde{\mathcal{N}}} = \widetilde{\mathcal{N}} \underset{^L\mathfrak{g}}{\times} {}^L\widetilde{\mathfrak{g}}.$$

Morally, we expect that the two categories in (10.4.18) act on the two categories in (10.4.13) in a compatible way. However, strictly speaking, the left hand side of (10.4.18) acts like this:

$$D^b(\widehat{\mathfrak{g}}_{\kappa_c}\text{-mod}_{\mathrm{nilp}})^I \to D^b(\widehat{\mathfrak{g}}_{\kappa_c}\text{-mod}_{\mathrm{nilp}})^{I^0},$$

and the right hand side of (10.4.18) acts like this:

$$D^b(\mathrm{QCoh}(\mathrm{MOp}^0_{^LG})) \to D^b(\mathrm{QCoh}(\widetilde{\mathrm{MOp}^0_{^LG}})).$$

So one needs a more precise statement, which may be found in [B], Section 4.2. Alternatively, one can consider the corresponding actions of the affine braid group of ^LG, as in [B].

A special case of this compatibility concerns some special objects of the category $D^b(\mathcal{D}^{\mathrm{Fl}}_{\kappa_c}\text{-mod})^I$, the central sheaves introduced in [Ga1]. They correspond to the central elements of the affine Hecke algebra $H(G(F), I)$. These central elements act as scalars on irreducible $H(G(F), I)$-modules, as well as on the standard modules $K^A(\mathrm{Sp}_u)_{(\gamma,q,\rho)}$ discussed above. We have argued that the categories $\widehat{\mathfrak{g}}_{\kappa_c}\text{-mod}_\chi^{I^0}, \chi \in \mathrm{Op}^{\mathrm{nilp}}_{^LG}$, are categorical versions of these representations. Therefore it is natural to expect that its objects are "eigenmodules" with respect to the action of the central sheaves from $D^b(\mathcal{D}^{\mathrm{Fl}}_{\kappa_c}\text{-mod})^I$ (in the sense of Section 10.3.5). This has indeed been proved in [FG3].

This discussion indicates the intimate connection between the categories $D^b(\widehat{\mathfrak{g}}_{\kappa_c}\text{-mod}_{\mathrm{nilp}})$ and the category of twisted $\mathcal{D}$-modules on the affine flag variety, which is similar to the connection between $\widehat{\mathfrak{g}}_{\kappa_c}\text{-mod}_{\mathrm{reg}}$ and the category of twisted $\mathcal{D}$-modules on the affine Grassmannian, which we discussed in Section 10.3.6. A more precise conjecture relating $D^b(\widehat{\mathfrak{g}}_{\kappa_c}\text{-mod}_{\mathrm{nilp}})$ and $D^b(\mathcal{D}^{\mathrm{Fl}}_{\kappa_c}\text{-mod})$ was formulated in [FG2] (see the Introduction and Section 6), where we refer the reader for more details. This conjecture may be viewed as an analogue of Conjecture 10.3.9 for nilpotent opers. As explained in [FG2], this conjecture is supported by the results of [AB, ABG] (see also [B]). Together, these results and conjectures provide additional evidence for the equivalence (10.4.13).

10.4.6 The case of $\mathfrak{g} = \mathfrak{sl}_2$

In this section we will illustrate Conjecture 10.4.4 in the case when $\mathfrak{g} = \mathfrak{sl}_2$, thus explaining the key ingredients of this conjecture in more concrete terms.

First, let us discuss the structure of the scheme $\widetilde{\mathrm{MOp}}^0_{PGL_2}$. By definition,

$$\widetilde{\mathrm{MOp}}^0_{PGL_2} = \mathrm{Op}^{\mathrm{nilp}}_{PGL_2} \underset{\mathfrak{n}/PGL_2}{\times} \tilde{\tilde{\mathfrak{n}}}/PGL_2.$$

The variety $\tilde{\tilde{\mathfrak{n}}}$ maps to $\mathfrak{n} = \mathbb{C}$, and its fiber over $u \in \mathfrak{n}$ is the non-reduced Springer fiber of u, the variety of Borel subalgebras of $\mathfrak{sl}_2$ containing u.

There are two possibilities. If $u = 0$, then the fiber is $\mathbb{P}^1$, the flag variety of PGL_2. If $u \neq 0$, then it corresponds to a regular nilpotent element, and therefore there is a unique Borel subalgebra containing u. Therefore the reduced Springer fiber is a single point. However, the non-reduced Springer fiber is a double point $\operatorname{Spec} \mathbb{C}[\epsilon]/(\epsilon^2) = \operatorname{Spec} h_0$.

To see this, let us show that there are non-trivial $\mathbb{C}[\epsilon]/(\epsilon^2)$-points of the Grothendieck alteration $\widetilde{\mathfrak{sl}}_2$ which project onto a regular nilpotent element in $\mathfrak{sl}_2$. Without loss of generality we may choose this element to be represented by the matrix

$$e = \begin{pmatrix} 0 & 1 \\ 0 & 0 \end{pmatrix}.$$

Then a $\mathbb{C}[\epsilon]/(\epsilon^2)$-point of $\widetilde{\mathfrak{sl}}_2$ above e is a Borel subalgebra in $\mathfrak{sl}_2[\epsilon]/(\epsilon^2)$ whose reduction modulo ϵ contains e. Now observe that the Lie algebra $\mathfrak{b}[\epsilon]/(\epsilon^2)$ of upper triangular matrices in $\mathfrak{sl}_2[\epsilon]/(\epsilon^2)$ contains the element

$$\begin{pmatrix} 1 & 0 \\ -a\epsilon & 1 \end{pmatrix} e \begin{pmatrix} 1 & 0 \\ a\epsilon & 1 \end{pmatrix} = \begin{pmatrix} a\epsilon & 1 \\ 0 & -a\epsilon \end{pmatrix}$$

for any $a \in \mathbb{C}$. Therefore

$$e \in \begin{pmatrix} 1 & 0 \\ -a\epsilon & 1 \end{pmatrix} \mathfrak{b}[\epsilon]/(\epsilon^2) \begin{pmatrix} 1 & 0 \\ a\epsilon & 1 \end{pmatrix}, \qquad a \in \mathbb{C}.$$

Hence we find non-trivial $\mathbb{C}[\epsilon]/(\epsilon^2)$-points of the Springer fiber over e.

Thus, we obtain that the variety $\tilde{\tilde{\mathfrak{n}}}$ has two components: one of them is $\mathbb{P}^1$ and the other one is a double affine line $\widetilde{\mathbb{A}}^1$. These components correspond to two elements of the Weyl group of $\mathfrak{sl}_2$, which label possible relative positions of two Borel subalgebras in $\mathfrak{sl}_2$. The points on the line correspond to the case when the Borel subalgebra coincides with the fixed Borel subalgebra $\mathfrak{b}$ containing $\mathfrak{n}$, and generic points of $\mathbb{P}^1$ correspond to Borel subalgebras that are in generic relative position with $\mathfrak{b}$. These two components are "glued together" at one point. Thus, we find that $\tilde{\tilde{\mathfrak{n}}}$ is a singular reducible non-reduced quasi-projective variety.

This gives us a rough idea as to what $\widetilde{\mathrm{MOp}}^0_{PGL_2}$ looks like. It fibers over the space $\mathrm{Op}_{PGL_2}^{\mathrm{nilp}}$ of nilpotent PGL_2-opers. Let us choose a coordinate t on the disc and use it to identify $\mathrm{Op}_{PGL_2}^{\mathrm{nilp}}$ with the space of second order operators

$$\partial_t^2 - v(t), \qquad v(t) = \sum_{n \geq -1} v_n t^n.$$

It is easy to see that the residue map $\mathrm{Op}_{PGL_2}^{\mathrm{nilp}} \to \mathfrak{n} = \mathbb{C}$ is given by the formula $v(t) \mapsto v_{-1}$. Therefore, if we decompose $\mathrm{Op}_{PGL_2}^{\mathrm{nilp}}$ into the Cartesian product of a line with the coordinate v_{-1} and the infinite-dimensional affine space with the coordinates $v_n, n \geq 0$, we obtain that $\widetilde{\mathrm{MOp}}_{PGL_2}^0$ is isomorphic to the Cartesian product of $\widetilde{\widetilde{\mathfrak{n}}}$ and an infinite-dimensional affine space.

Thus, $\widetilde{\mathrm{MOp}}_{PGL_2}^0$ has two components, $\widetilde{\mathrm{MOp}}_{PGL_2}^{0,1}$ and $\widetilde{\mathrm{MOp}}_{PGL_2}^{0,w_0}$ corresponding to two elements, 1 and w_0, of the Weyl group of $\mathfrak{sl}_2$.

Conjecture 10.4.4 states that the derived category of quasicoherent sheaves on the scheme $\widetilde{\mathrm{MOp}}_{PGL_2}^0$ is equivalent to the equivariant derived category $D^b(\widehat{\mathfrak{sl}}_2\text{-mod}_{\mathrm{nilp}})^{I^0}$. We now consider examples of objects of these categories.

First, we consider the skyscraper sheaf at a point of $\widetilde{\mathrm{MOp}}_{PGL_2}^0$, which is the same as a point of $\mathrm{MOp}_{PGL_2}^0$. The variety $\mathrm{MOp}_{PGL_2}^0$ is a union of two strata

$$\mathrm{MOp}_{PGL_2}^0 = \mathrm{MOp}_{PGL_2}^{0,1} \bigsqcup \mathrm{MOp}_{PGL_2}^{0,w_0}.$$

By Theorem 9.3.6, we have

$$\mathrm{MOp}_{PGL_2}^{0,w} \simeq \mathrm{Conn}(\Omega^{-\rho})_D^{w(\rho)-\rho}$$

(here we switch from $\Omega^{\check{\rho}}$ to $\Omega^{-\rho}$, as before).

The stratum $\mathrm{MOp}_{PGL_2}^{0,1}$ for $w = 1$ fibers over the subspace of regular opers $\mathrm{Op}_{PGL_2}(D) \subset \mathrm{Op}_{PGL_2}^{\mathrm{nilp}}$ (those with zero residue, $v_{-1} = 0$). The corresponding space of connections $\mathrm{Conn}(\Omega^{-\rho})_D^{w(\rho)-\rho}$ is the space of regular connections on the line bundle $\Omega^{-1/2}$ on D. Using our coordinate t, we represent such a connection as an operator

$$\overline{\nabla} = \partial_t + u(t), \qquad u(t) = \sum_{n \geq 0} u_n t^n.$$

The map $\mathrm{MOp}_{PGL_2}^{0,1} \to \mathrm{Op}_{PGL_2}(D)$ is the Miura transformation

$$\partial_t + u(t) \mapsto (\partial_t - u(t))\,(\partial_t + u(t)) = \partial_t^2 - v(t),$$

where

$$v(t) = u(t)^2 - \partial_t u(t). \tag{10.4.19}$$

These formulas were previously obtained in Section 8.2.2, except that we have now rescaled $u(t)$ by a factor of 2 to simplify our calculations.

The fiber of the Miura transformation $\mathrm{Conn}(\Omega^{-1/2})_D \to \mathrm{Op}_{PGL_2}(D)$ at $v(t) \in \mathbb{C}[[t]] \simeq \mathrm{Op}_{PGL_2}(D)$ consists of all solutions $u(t) \in \mathbb{C}[[t]]$ of the Riccati equation (10.4.19). This equation amounts to an infinite system of algebraic equations on the coefficients u_n of $u(t)$, which may be solved recursively. These equations read

$$nu_n = -v_{n-1} + \sum_{i+j=n-1} u_i u_j, \qquad n \geq 1.$$

Thus, we see that all coefficients $u_n, n \geq 1$, are uniquely determined by these equations for each choice of u_0, which could be an arbitrary complex number. Therefore the fiber of this map is isomorphic to the affine line, as expected.

For each solution $u(t) \in \mathbb{C}[[t]]$ of these equations we have the Wakimoto module $W_{u(t)}$ as defined in Corollary 6.1.5. According to Theorem 8.3.3, the center $Z(\widehat{\mathfrak{sl}}_2)$ acts on $W_{u(t)}$ via the central character corresponding to the PGL_2-oper χ corresponding to the projective connection $\partial_t^2 - v(t)$. Thus, we obtain a family of Wakimoto modules in the category $\widehat{\mathfrak{sl}}_2\text{-mod}_\chi$ parameterized by the affine line. These are the objects corresponding to the skyscraper sheaves supported at the points of $\mathrm{MOp}^{0,1}_{PGL_2}$.

Now we consider the points of the other stratum

$$\mathrm{MOp}^{0,w_0}_{PGL_2} \simeq \mathrm{Conn}(\Omega^{-1/2})^{-1}_D.$$

The corresponding connections on $\Omega^{-1/2}$ have regular singularity at $0 \in D$ with the residue -1 (corresponding to -2ρ). With respect to our coordinate t they correspond to the operators

$$\overline{\nabla} = \partial_t + u(t), \qquad u(t) = \sum_{n \geq -1} u_n t^n, \quad u_{-1} = -1.$$

The Miura transformation $\mathrm{Conn}(\Omega^{-1/2})^{-1}_D \to \mathrm{Op}^{\mathrm{nilp}}_{PGL_2}$ is again given by formula (10.4.19). Let us determine the fiber of this map over a nilpotent oper corresponding to the projective connection $\partial_t^2 - v(t)$. We obtain the following system of equations on the coefficients u_n of $u(t)$:

$$(n+2)u_n = -v_{n-1} + \sum_{i+j=n-1; i,j \geq 0} u_i u_j, \qquad n \geq 0 \tag{10.4.20}$$

(here we use the fact that $u_{-1} = -1$). This system may be solved recursively, but now u_0 is determined by the first equation, and so there is no free parameter as in the previous case. Thus, we find that the map $\mathrm{Conn}(\Omega^{-1/2})^{-1}_D \to \mathrm{Op}^{\mathrm{nilp}}_{PGL_2}$ is an isomorphism, as expected.

If $v_{-1} = 0$, and so our oper $\chi = \partial^2 - v(t)$ is regular, then we obtain an additional point in the fiber of the Miura transformation over χ. Therefore we find that the set of points in the fiber over a regular oper χ is indeed in bijection with the set of points of the projective line. The Wakimoto module $W_{u(t)}$, where $u(t)$ is as above, is the object of the category $\widehat{\mathfrak{sl}}_2\text{-mod}^{I^0}_\chi$ corresponding to the extra point represented by $u(t)$. Thus, for each regular oper χ we obtain a family of Wakimoto modules in $\widehat{\mathfrak{sl}}_2\text{-mod}_\chi^{I^0}$ parameterized by $\mathbb{P}^1$, as expected.

Now suppose that $v_{-1} \neq 0$, so the residue of our oper χ is non-zero, and hence it has non-trivial unipotent monodromy. Then the set-theoretic fiber of the Miura transformation over χ consists of one point. We have one Wakimoto module $W_{u(t)}$ with $u(t) = -1/t + \ldots$ corresponding to this point. However, the fiber of $\widetilde{\text{MOp}}^0_{PGL_2}$ is a double point. Can we see this doubling effect from the point of view of the Miura transformation?

It turns out that we can. Let us consider the equation (10.4.19) with $v_{-1} \neq 0$ and $u(t) = \sum_{n\in\mathbb{Z}} u_n t^n$, where $u_n \in \mathbb{C}[\epsilon]/(\epsilon^2)$. It is easy to see that the general solution to this equation such that $u(t) \bmod \epsilon$ is the above solution $u_{\text{red}}(t) = -1/t + \sum_{n\geq 0} u_{\text{red},n} t^n$, has the form

$$u_a(t) = u_{\text{red}}(t) + \frac{a\epsilon}{t^2} + a\epsilon \sum_{n\geq -1} \delta_n t^n,$$

where a is an arbitrary complex number and $\delta_n \in \mathbb{C}$ are uniquely determined by (10.4.19). Thus, we obtain that the scheme-theoretic fiber of the Miura transformation $\text{Conn}(\Omega^{-\rho})_{D^\times} \to \text{Op}^{\text{nilp}}_{PGL_2}$ at χ of the above form (with $v_{-1} \neq 0$) is a double point, just like the fiber of $\widetilde{\text{MOp}}^0_{PGL_2}$.

This suggests that in general the scheme-theoretic fiber of the Miura transformation over a nilpotent oper $\chi \in \text{Op}^{\text{nilp}}_{^LG}$ is also isomorphic to the non-reduced Springer fiber $\text{Sp}_{\text{Res}_{\mathcal{F}}(\chi)}$. In [FG2] it was shown that this is indeed the case.

The existence of the above one-parameter family of solutions $u_a(t), a \in \mathbb{C}$, of (10.4.19) over $\mathbb{C}[\epsilon]/(\epsilon^2)$ means that there is a one-dimensional space of self-extensions of the Wakimoto module $W_{u_{\text{red}}(t)}$. Recall that this module is realized in the vector space $M_{\mathfrak{sl}_2}$, the Fock module of the Weyl algebra generated by $a_n, a_n^*, n \in \mathbb{Z}$. This extension is realized in $M_{\mathfrak{sl}_2} \otimes \mathbb{C}[\epsilon]/(\epsilon^2)$. The action of the Lie algebra $\widehat{\mathfrak{sl}}_2$ is given by the same formulas as before (see Section 6.1.2), except that now $b_n, n \in \mathbb{Z}$, acts as the linear operator on $\mathbb{C}[\epsilon]/(\epsilon^2)$ corresponding to the nth coefficient $u_{a,n}$ of $u_a(t)$.

In fact, it follows from Theorem 10.4.5 (or from an explicit calculation) that the category $\widehat{\mathfrak{sl}}_2\text{-mod}_\chi^{I^0}$ corresponding to $\chi \in \text{Op}^{\text{nilp}}_{PGL_2}$ with non-zero residue has a unique irreducible object; namely, $W_{u_{\text{red}}(t)}$. We note that this module is isomorphic to the quotient of the Verma module $\mathbb{M}_{-2}$ by the central character corresponding to χ. This module has a non-trivial self-extension described by the above formula, and the category $\widehat{\mathfrak{sl}}_2\text{-mod}_\chi^{I^0}$ is equivalent to the category $\text{QCoh}(\text{Spec}\, h_0)$, where $h_0 = \mathbb{C}[\epsilon]/(\epsilon^2)$.

Now let us consider the category $\widehat{\mathfrak{sl}}_2\text{-mod}_\chi^{I^0}$, where χ is a PGL_2-oper corresponding to the operator $\partial_t^2 - v(t)$, which is regular at $t = 0$ (and so its residue is $v_{-1} = 0$). We have constructed a family of objects $W_{u(t)}$ of this category, where $u(t)$ are the solutions of the Riccati equation (10.4.19), which are in one-to-one correspondence with points of $\mathbb{P}^1$ (for all but one of them $u(t)$ is regular at $t=0$ and the last one has the form $u(t) = -1/t+\ldots$). What is the structure of these Wakimoto modules? It is not difficult to see directly that each of them is an extension of the same two irreducible $\widehat{\mathfrak{sl}}_2$-modules, which are in fact the only irreducible objects of the category $\widehat{\mathfrak{sl}}_2\text{-mod}_\chi^{I^0}$. The first of them is the module $\mathbb{V}_0(\chi)$ introduced in Section 10.3.3, and the second module is obtained from $\mathbb{V}_0(\chi)$ by the twist with the involution of $\widehat{\mathfrak{sl}}_2$ defined on the Kac–Moody generators (see Appendix A.5) by the formula

$$e_0 \leftrightarrow e_1, \qquad h_0 \leftrightarrow h_1, \qquad f_0 \leftrightarrow f_1.$$

We denote this module by $\mathbb{V}_{-2}(\chi)$. While $\mathbb{V}_0(\chi)$ is generated by a vector annihilated by the "maximal compact" Lie subalgebra $\mathfrak{k}_0 = \mathfrak{sl}_2[[t]]$, $\mathbb{V}_{-2}(\chi)$ is generated by a vector annihilated by the Lie subalgebra $\mathfrak{k}_0'$, which is obtained by applying the above involution to $\mathfrak{sl}_2[[t]]$. These two Lie subalgebras, $\mathfrak{k}_0$ and $\mathfrak{k}_0'$, are the two "maximal compact" Lie subalgebras of $\widehat{\mathfrak{sl}}_2$ up two conjugation by $SL_2((t))$. In particular, the generating vector of $\mathbb{V}_{-2}(\chi)$ is annihilated by the Lie algebra $\widehat{\mathfrak{n}}_+$ and the Cartan subalgebra $\mathfrak{h}$ acts on it according to the weight -2 (corresponding to $-\alpha$). Thus, $\mathbb{V}_{-2}(\chi)$ is a quotient of the Verma module $\mathbb{M}_{-2}$.

We have a two-dimensional space of extensions $\mathrm{Ext}^1(\mathbb{V}_{-2}(\chi), \mathbb{V}_0(\chi))$. Since the extensions corresponding to vectors in $\mathrm{Ext}^1(\mathbb{V}_{-2}(\chi), \mathbb{V}_0(\chi))$ that are proportional are isomorphic as $\widehat{\mathfrak{sl}}_2$-modules, we obtain that the isomorphism classes of extensions are parameterized by points of $\mathbb{P}^1$. They are represented precisely by our Wakimoto modules $W_{u(t)}$. Thus, our $\mathbb{P}^1$, the fiber of the Miura transformation at a regular oper, acquires a useful interpretation as the projectivization of the space of extensions of the two irreducible objects of the category $\widehat{\mathfrak{sl}}_2\text{-mod}_\chi^{I^0}$.

According to our conjectures, the derived category $D^b(\widehat{\mathfrak{sl}}_2\text{-mod}_\chi)^{I^0}$ is equivalent to the derived category of quasicoherent sheaves on the DG Springer fiber at 0, which is $\Pi T\mathbb{P}^1$. The Wakimoto modules considered above correspond to the skyscraper sheaves. What objects of the category of quasi-coherent sheaves correspond to the irreducible $\widehat{\mathfrak{sl}}_2$-modules $\mathbb{V}_0(\chi)$ and $\mathbb{V}_{-2}(\chi)$? The answer is that $\mathbb{V}_0(\chi)$ corresponds to the trivial line bundle $\mathcal{O}$ on $\mathbb{P}^1$ and $\mathbb{V}_{-2}(\chi)$ corresponds to the complex $\mathcal{O}(-1)[1]$, i.e., the line bundle $\mathcal{O}(-1)$ placed in cohomological degree -1. The exact sequences in the category of quasicoherent sheaves on $\mathbb{P}^1$

$$0 \to \mathcal{O}(-1) \to \mathcal{O} \to \mathcal{O}_p \to 0,$$

where $p \in \mathbb{P}^1$ and $\mathcal{O}_p$ is the skyscraper sheaf supported at p, become the exact sequences in the category $\widehat{\mathfrak{sl}}_2\text{-mod}_\chi^{I^0}$

$$0 \to \mathbb{V}_0(\chi) \to W_{u(t)} \to \mathbb{V}_{-2}(\chi) \to 0,$$

where $u(t)$ is the solution of (10.4.19) corresponding to the point p.

Now we observe that in the category $\widehat{\mathfrak{sl}}_2\text{-mod}_\chi^{I^0}$ there is a complete symmetry between the objects $\mathbb{V}_0(\chi)$ and $\mathbb{V}_{-2}(\chi)$, because they are related by an involution of $\widehat{\mathfrak{sl}}_2$ which gives rise to an auto-equivalence of $\widehat{\mathfrak{sl}}_2\text{-mod}_\chi^{I^0}$. In particular, we have $\mathrm{Ext}^1(\mathbb{V}_0(\chi), \mathbb{V}_{-2}(\chi)) \simeq \mathbb{C}^2$. But in the category $D^b(\mathrm{QCoh}(\mathbb{P}^1))$ we have

$$\mathrm{Ext}^1(\mathcal{O}, \mathcal{O}(-1)[1]) = H^2(\mathbb{P}^1, \mathcal{O}(-1)) = 0.$$

This is the reason why the category $D^b(\mathrm{QCoh}(\mathbb{P}^1))$ cannot be equivalent to $\widehat{\mathfrak{sl}}_2\text{-mod}_\chi^{I^0}$. It needs to be replaced with the category $D^b(\mathrm{QCoh}(\Pi T\mathbb{P}^1))$ of quasicoherent sheaves on the DG Springer fiber $\Pi T\mathbb{P}^1$. This is a good illustration of the necessity of using the DG Springer fibers in the equivalence (10.4.14).

10.5 From local to global Langlands correspondence

We now discuss the implications of the local Langlands correspondence studied in this chapter for the global geometric Langlands correspondence.

The setting of the global correspondence that was already outlined in Section 1.1.6. We recall that in the classical picture we start with a smooth projective curve X over $\mathbb{F}_q$. Denote by F the field $\mathbb{F}_q(X)$ of rational functions on X. For any closed point x of X we denote by F_x the completion of F at x and by $\mathcal{O}_x$ its ring of integers. Thus, we now have a local field F_x attached to each point of X. The ring $\mathbb{A} = \mathbb{A}_F$ of **adèles** of F is by definition the restricted product of the fields F_x, where x runs over the set $|X|$ of all closed points of X (the meaning of the word "restricted" is explained in Section 1.1.6). It has a subring $\mathcal{O} = \prod_{x \in X} \mathcal{O}_x$.

Let G be a (split) reductive group over $\mathbb{F}_q$. Then for each $x \in X$ we have a group $G(F_x)$. Their restricted product over all $x \in X$ is the group $G(\mathbb{A}_F)$. Since $F \subset \mathbb{A}_F$, the group $G(F)$ is a subgroup of $G(\mathbb{A}_F)$.

On one side of the global Langlands correspondence we have homomorphisms $\sigma : W_F \to {}^L G$ satisfying some properties (or perhaps, some more refined data, as in [Art]). We expect to be able to attach to each σ an **automorphic representation** π of $GL_n(\mathbb{A}_F)$.† The word "automorphic" means, roughly, that the representation may be realized in a reasonable space of functions on the quotient $GL_n(F)\backslash GL_n(\mathbb{A})$ (on which the group $GL_n(\mathbb{A})$ acts from the right). We will not try to make this precise. In general, we expect not one but several automorphic representations assigned to σ, which are the global analogues of the L-packets discussed above (see [Art]). Another complication is that the multiplicity of a given irreducible automorphic representation in the space of functions on $GL_n(F)\backslash GL_n(\mathbb{A})$ may be greater than one. We will

† In this section, by abuse of notation, we will use the same symbol to denote a representation of a group and the vector space underlying this representation.

mostly ignore all of these issues here, as our main interest is in the geometric theory (note that these issues do not arise if $G = GL_n$).

An irreducible automorphic representation may always be decomposed as the restricted tensor product $\bigotimes'_{x \in X} \pi_x$, where each π_x is an irreducible representation of $G(F_x)$. Moreover, for all by finitely many $x \in X$ the factor π_x is an **unramified** representation of $G(F_x)$: it contains a non-zero vector invariant under the maximal compact subgroup $K_{0,x} = G(\mathcal{O}_x)$ (see Section 10.3.1). Let us choose such a vector $v_x \in \pi_x$ (it is unique up to a scalar). The word "restricted" means that we consider the span of vectors of the form $\otimes_{x \in X} u_x$, where $u_x \in \pi_x$ and $u_x = v_x$ for all but finitely many $x \in X$.

An important property of the global Langlands correspondence is its compatibility with the local one. We can embed the Weil group W_{F_x} of each of the local fields F_x into the global Weil group W_F. Such an embedding is not unique, but it is well-defined up to conjugation in W_F. Therefore an equivalence class of $\sigma : W_F \to {}^LG$ gives rise to a well-defined equivalence class of $\sigma_x : W_{F_x} \to {}^LG$. We will impose the condition on σ that for all but finitely many $x \in X$ the homomorphism σ_x is unramified (see Section 10.3.1).

By the local Langlands correspondence, to σ_x one can attach an equivalence class of irreducible smooth representations π_x of $G(F_x)$.† Moreover, an unramified σ_x will correspond to an unramified irreducible representation π_x. The compatibility between local and global correspondences is the statement that the automorphic representation of $G(\mathbb{A})$ corresponding to σ should be isomorphic to the restricted tensor product $\bigotimes'_{x \in X} \pi_x$. Schematically, this is represented as follows:

$$\sigma \overset{\text{global}}{\longleftrightarrow} \pi = \bigotimes_{x \in X}{}' \pi_x$$

$$\sigma_x \overset{\text{local}}{\longleftrightarrow} \pi_x.$$

In this section we discuss the analogue of this local-to-global principle in the geometric setting and the implications of our local results and conjectures for the global geometric Langlands correspondence. We focus in particular on the unramified and tamely ramified Langlands parameters.

10.5.1 The unramified case

An important special case is when $\sigma : W_F \to {}^LG$ is everywhere unramified. Then for each $x \in X$ the corresponding homomorphism $\sigma_x : W_{F_x} \to {}^LG$ is unramified, and hence corresponds, as explained in Section 10.3.1, to a semi-simple conjugacy class γ_x in ${}^LG$, which is the image of the Frobenius element under σ_x. This conjugacy class in turn gives rise to an unramified irreducible

† Here we are considering ℓ-adic homomorphisms from the Weil group W_{F_x} to ${}^LG$, and therefore we do not need to pass from the Weil group to the Weil–Deligne group.

representation π_x of $G(F_x)$ with a unique, up to a scalar, vector v_x such that $G(\mathcal{O}_x)v_x = v_x$. The spherical Hecke algebra $H(G(F_x), G(\mathcal{O}_x)) \simeq \operatorname{Rep} {}^L G$ acts on this vector according to formula (10.3.5):

$$H_{V,x} \star v_x = \operatorname{Tr}(\gamma_x, V) v_x, \qquad [V] \in \operatorname{Rep} {}^L G. \tag{10.5.1}$$

The tensor product $v = \bigotimes_{x \in X} v_x$ of these vectors is a $G(\mathcal{O})$-invariant vector in $\pi = \bigotimes'_{x \in X} \pi_x$, which, according to the global Langlands conjecture, is automorphic. This means that π is realized in the space of functions on $G(F) \backslash G(\mathbb{A}_F)$. In this realization the vector v corresponds to a right $G(\mathcal{O})$-invariant function on $G(F) \backslash G(\mathbb{A}_F)$, or, equivalently, a function on the double quotient

$$G(F) \backslash G(\mathbb{A}_F) / G(\mathcal{O}). \tag{10.5.2}$$

Thus, an unramified global Langlands parameter σ gives rise to a function on (10.5.2). This function is the **automorphic function** corresponding to σ. We denote it by f_π. Since it corresponds to a vector in an irreducible representation π of $G(\mathbb{A}_F)$, the entire representation π may be reconstructed from this function. Thus, we do not lose any information by passing from π to f_π.

Since $v \in \pi$ is an eigenvector of the Hecke operators, according to formula (10.5.1), we obtain that the function f_π is a **Hecke eigenfunction** on the double quotient (10.5.2). In fact, the local Hecke algebras $H(G(F_x), G(\mathcal{O}_x))$ act naturally (from the right) on the space of functions on (10.5.2), and f_π is an eigenfunction of this action. It satisfies the same property (10.5.1).

To summarize, the unramified global Langlands correspondence in the classical setting may be viewed as a correspondence between unramified homomorphisms $\sigma : W_F \to {}^L G$ and Hecke eigenfunctions on (10.5.2) (some irreducibility condition on σ needs to be added to make this more precise, but we will ignore this).

What should be the geometric analogue of this correspondence when X is a complex algebraic curve?

As explained in Section 1.2.1, the geometric analogue of an unramified homomorphism $W_F \to {}^L G$ is a homomorphism $\pi_1(X) \to {}^L G$, or, equivalently, since X is assumed to be compact, a holomorphic ${}^L G$-bundle on X with a holomorphic connection (it automatically gives rise to a flat connection, see Section 1.2.3). The global geometric Langlands correspondence should therefore associate to a flat holomorphic ${}^L G$-bundle on X a geometric object on a geometric version of the double quotient (10.5.2). As we argued in Section 1.3.2, this should be a $\mathcal{D}$-module on an algebraic variety whose set of points is (10.5.2).

Now, it is known that (10.5.2) is in bijection with the set of isomorphism classes of G-bundles on X. This key result is due to A. Weil (see, e.g., [F7]). This suggests that (10.5.2) is the set of points of the moduli space of G-

bundles on X. Unfortunately, in general this is not an algebraic variety, but an algebraic stack, which locally looks like a quotient of an algebraic variety by an action of an algebraic group. We denote it by Bun_G. The theory of $\mathcal{D}$-modules has been developed in the setting of algebraic stacks like Bun_G in [BD1], and so we can use it for our purposes. Thus, we would like to attach to a flat holomorphic ${}^LG$-bundle E on X a $\mathcal{D}$-module Aut_E on Bun_G. This $\mathcal{D}$-module should satisfy an analogue of the Hecke eigenfunction condition, which makes it into a **Hecke eigensheaf** with eigenvalue E. This notion is spelled out in [F7] (following [BD1]), where we refer the reader for details.

Thus, the unramified global geometric Langlands correspondence may be summarized by the following diagram:

$$\boxed{\text{flat } {}^LG\text{-bundles on } X} \longrightarrow \boxed{\text{Hecke eigensheaves on } \mathrm{Bun}_G}$$

$$E \quad \mapsto \quad \mathrm{Aut}_E$$

(some irreducibility condition on E should be added here).

The unramified global geometric Langlands correspondence has been proved in [FGV2, Ga2] for $G = GL_n$ and an arbitrary irreducible GL_n-local system E, and in [BD1] for an arbitrary simple Lie group G and those ${}^LG$-local systems which admit the structure of an oper. In this section we discuss the second approach.

To motivate it, we ask the following question:

> *How to relate this global correspondence to the local geometric Langlands correspondence discussed above?*

The key element in answering this question is a **localization functor** from $\widehat{\mathfrak{g}}$-modules to (twisted) $\mathcal{D}$-modules on Bun_G. In order to construct this functor we note that for a simple Lie group G the moduli stack Bun_G has another realization as a double quotient. Namely, let x be a point of X. Denote by $\mathcal{O}_x$ the completed local ring and by $\mathcal{K}_x$ its field of fractions. Let $G(\mathcal{K}_x)$ the formal loop group corresponding to the punctured disc $D_x^\times = \operatorname{Spec} \mathcal{K}_x$ around x and by G_{out} the group of algebraic maps $X\backslash x \to G$, which is naturally a subgroup of $G(\mathcal{K}_x)$. Then Bun_G is isomorphic to the double quotient

$$\mathrm{Bun}_G \simeq G_{\mathrm{out}} \backslash G(\mathcal{K}_x)/G(\mathcal{O}_x). \tag{10.5.3}$$

This is a "one-point" version of (10.5.2). This is not difficult to prove at the level of $\mathbb{C}$-points, but the isomorphism is also true at the level of algebraic stacks (see [BL, DrSi]).

The localization functor that we need is a special case of the following general construction. Let $(\mathfrak{g}, K)$ be a Harish-Chandra pair and $\mathfrak{g}\text{-mod}^K$ the category of K-equivariant $\mathfrak{g}$-modules (see Section 10.2.2). For a subgroup

$H \subset G$ let $\mathcal{D}_{H\backslash G/K}$-mod be the category of $\mathcal{D}$-modules on $H\backslash G/K$. Then there is a localization functor [BB, BD1] (see also [F7, FB])

$$\Delta : \mathfrak{g}\text{-mod}^K \to \mathcal{D}_{H\backslash G/K}\text{-mod}.$$

Now let $\widehat{\mathfrak{g}}$ be a one-dimensional central extension of $\mathfrak{g}$, which becomes trivial when restricted to the Lie subalgebras $\operatorname{Lie} K$ and $\operatorname{Lie} H$. Suppose that this central extension can be exponentiated to a central extension $\widehat{G}$ of the corresponding Lie group G. Then we obtain a $\mathbb{C}^\times$-bundle $H\backslash\widehat{G}/K$ over $H\backslash G/K$. Let $\mathcal{L}$ be the corresponding line bundle. Let $\mathcal{D}_{\mathcal{L}}$ be the sheaf of differential operators acting on $\mathcal{L}$. Then we have a functor

$$\Delta_{\mathcal{L}} : \widehat{\mathfrak{g}}\text{-mod}^K \to \mathcal{D}_{\mathcal{L}}\text{-mod}.$$

In our case we take the Lie algebra $\widehat{\mathfrak{g}}_{\kappa_c,x}$, the critical central extension of the loop algebra $\mathfrak{g} \otimes \mathcal{K}_x$ and the subgroups $K = G(\mathcal{O}_x)$ and $H = G_{\text{out}}$ of $G(\mathcal{K}_x)$. It is known (see, e.g., [BD1]) that the corresponding line bundle $\mathcal{L}$ is the square root $K^{1/2}$ of the canonical line bundle on Bun_G.† We denote the corresponding sheaf of twisted differential operators on Bun_G by $\mathcal{D}_{\kappa_c}$. Then we have the localization functor

$$\Delta_{\kappa_c,x} : \widehat{\mathfrak{g}}_{\kappa_c,x}\text{-mod}^{G(\mathcal{O}_x)} \to \mathcal{D}_{\kappa_c}\text{-mod}.$$

We restrict it further to the category $\widehat{\mathfrak{g}}_{\kappa_c,x}\text{-mod}_{\chi_x}^{G(\mathcal{O}_x)}$ for some

$$\chi_x \in \operatorname{Op}_{^LG}(D_x^\times).$$

According to Lemma 10.3.2, this category is non-zero only if $\chi_x \in \operatorname{Op}_{^LG}^\lambda(D_x)$ for some $\lambda \in P^+$. Suppose for simplicity that $\lambda = 0$ and so $\chi_x \in \operatorname{Op}_{^LG}(D_x)$. In this case, by Theorem 10.3.3, the category $\widehat{\mathfrak{g}}_{\kappa_c,x}\text{-mod}_{\chi_x}^{G(\mathcal{O}_x)}$ is equivalent to the category of vector spaces. Its unique, up to an isomorphism, irreducible object is the quotient $\mathbb{V}_0(\chi_x)$ of the vacuum module

$$\mathbb{V}_{0,x} = \operatorname{Ind}_{\mathfrak{g}(\mathcal{O}_x)\oplus\mathbb{C}\mathbf{1}}^{\widehat{\mathfrak{g}}_{\kappa_c,x}} \mathbb{C}$$

by the central character corresponding to χ_x. Hence it is sufficient to determine $\Delta_{\kappa_c,x}(\mathbb{V}_0(\chi_x))$.

The following theorem is due to Beilinson and Drinfeld [BD1].

Theorem 10.5.1 (1) *The $\mathcal{D}_{\kappa_c}$-module $\Delta_{\kappa_c,x}(\mathbb{V}_0(\chi_x))$ is non-zero if and only if there exists a global $^L\mathfrak{g}$-oper on X, $\chi \in \operatorname{Op}_{^LG}(X)$ such that $\chi_x \in \operatorname{Op}_{^LG}(D_x)$ is the restriction of χ to D_x.*

(2) *If this holds, then $\Delta_{\kappa_c,x}(\mathbb{V}_0(\chi_x))$ depends only on χ and is independent of the choice of x in the sense that for any other point $y \in X$, if $\chi_y = \chi|_{D_y}$, then $\Delta_{\kappa_c,x}(\mathbb{V}_0(\chi_x)) \simeq \Delta_{\kappa_c,y}(\mathbb{V}_0(\chi_y))$.*

† Recall that by our assumption G is simply-connected. In this case there is a unique square root.

(3) *For any* $\chi = (\mathcal{F}, \nabla, \mathcal{F}_{^L B}) \in \mathrm{Op}_{^L G}(X)$ *the* $\mathcal{D}_{\kappa_c}$*-module* $\Delta_{\kappa_c,x}(\mathbb{V}_0(\chi_x))$ *is a non-zero Hecke eigensheaf with the eigenvalue* $E_\chi = (F, \nabla)$.

Thus, for any $\chi \in \mathrm{Op}_{^L G}(X)$, the $\mathcal{D}_{\kappa_c}$-module $\Delta_{\kappa_c,x}(\mathbb{V}_0(\chi_x))$ is the sought-after Hecke eigensheaf Aut_{E_χ} corresponding to the $^L G$-local system E_χ under the global geometric Langlands correspondence (10.5.1).† For an outline of the proof of this theorem from [BD1], see [F7]. The key point is that the irreducible modules $\mathbb{V}_0(\chi_y)$, where $\chi_y = \chi|_{D_y}$, corresponding to all points $y \in X$, are Hecke eigenmodules as we saw in Section 10.3.5.

A drawback of this construction is that not all $^L G$-local systems on X admit the structure of an oper. In fact, under our assumption that $^L G$ is a group of the adjoint type, the $^L G$-local systems, or flat bundles $(\mathcal{F}, \nabla)$, on a smooth projective curve X that admit an oper structure correspond to a unique $^L G$-bundle $\mathcal{F}_{\mathrm{oper}}$ on X (described in Section 4.2.4). Moreover, for each such flat bundle $(\mathcal{F}_{\mathrm{oper}}, \nabla)$ there is a unique $^L B$-reduction $\mathcal{F}_{\mathrm{oper}, {^L B}}$ satisfying the oper condition. Therefore $\mathrm{Op}_G(D)$ is the fiber of the forgetful map

$$\mathrm{Loc}_{^L G}(X) \to \mathrm{Bun}_{^L G}, \qquad (\mathcal{F}, \nabla) \mapsto \mathcal{F},$$

over the oper bundle $\mathcal{F}_{\mathrm{oper}}$.

For a $^L G$-local system $E = (\mathcal{F}, \nabla)$ for which $\mathcal{F} \neq \mathcal{F}_{\mathrm{oper}}$, the above construction may be modified as follows (see the discussion in [F7] based on an unpublished work of Beilinson and Drinfeld). Suppose that we can choose an $^L B$-reduction $\mathcal{F}_{^L B}$ satisfying the oper condition away from a finite set of points $y_1, \ldots, y_n$ and such that the restriction χ_{y_i} of the corresponding oper χ on $X \backslash \{y_1, \ldots, y_n\}$ to $D_{y_i}^\times$ belongs to $\mathrm{Op}_{^L G}^{\lambda_i}(D_{y_i}) \subset \mathrm{Op}_{^L G}(D_{y_i}^\times)$ for some $\lambda_i \in P^+$. Then we can construct a Hecke eigensheaf corresponding to E by applying a multi-point version of the localization functor to the tensor product of the quotients $\mathbb{V}_{\lambda_i}(\chi_{y_i})$ of the Weyl modules $\mathbb{V}_{\lambda_i, y_i}$ (see [F7]).

The main lesson of Theorem 10.5.1 is that in the geometric setting the localization functor gives us a powerful tool for converting local Langlands categories, such as $\widehat{\mathfrak{g}}_{\kappa_c,x}\text{-mod}_{\chi_x}^{G(\mathcal{O}_x)}$, into global categories of Hecke eigensheaves. Therefore the emphasis shifts to the study of local categories of $\widehat{\mathfrak{g}}_{\kappa_c,x}$-modules. We can learn a lot about the global categories by studying the local ones. This is a new phenomenon which does not have any obvious analogues in the classical Langlands correspondence.

In the case at hand, the category $\widehat{\mathfrak{g}}_{\kappa_c,x}\text{-mod}_{\chi_x}^{G(\mathcal{O}_x)}$ turns out to be very simple: it has a unique irreducible object, $\mathbb{V}_0(\chi_x)$. That is why it is sufficient to consider its image under the localization functor, which turns out to be the desired Hecke eigensheaf Aut_{E_χ}. For general opers, with ramification, the corresponding local categories are more complicated, as we have seen

† More precisely, Aut_{E_χ} is the $\mathcal{D}$-module $\Delta_{\kappa_c,x}(\mathbb{V}_0(\chi_x)) \otimes K^{-1/2}$, but here and below we will ignore the twist by $K^{1/2}$.

above, and so are the corresponding categories of Hecke eigensheaves. We will consider examples of these categories in the next section.

10.5.2 Global Langlands correspondence with tame ramification

Let us first consider ramified global Langlands correspondence in the classical setting. Suppose that we are given a homomorphism $\sigma : W_F \to {}^L G$ that is ramified at finitely many points $y_1, \ldots, y_n$ of X. Then we expect that to such σ corresponds an automorphic representation $\bigotimes'_{x \in X} \pi_x$ (more precisely, an L-packet of representations). Here π_x is still unramified for all $x \in X \backslash \{y_1, \ldots, y_n\}$, but is *ramified* at $y_1, \ldots, y_n$, i.e., the space of $G(\mathcal{O}_{y_i})$-invariant vectors in π_{y_i} is zero. In particular, consider the special case when each $\sigma_{y_i} : W_{F_{y_i}} \to {}^L G$ is tamely ramified (see Section 10.4.1 for the definition). Then, according to the results presented in Section 10.4.1, the corresponding L-packet of representations of $G(F_{y_i})$ contains an irreducible representation π_{y_i} with non-zero invariant vectors with respect to the Iwahori subgroup I_{y_i}. Let us choose such a representation for each point y_i.

Consider the subspace

$$\bigotimes_{i=1}^{n} \pi_{y_i}^{I_{y_i}} \otimes \bigotimes_{x \neq y_i} v_x \subset \bigotimes_{x \in X}{}' \pi_x, \tag{10.5.4}$$

where v_x is a $G(\mathcal{O}_x)$-vector in $\pi_x, x \neq y_i, i = 1, \ldots, n$. Then, because $\bigotimes'_{x \in X} \pi_x$ is realized in the space of functions on $G(F) \backslash G(\mathbb{A}_F)$, we obtain that the subspace (10.5.4) is realized in the space of functions on the double quotient

$$G(F) \backslash G(\mathbb{A}_F) / \prod_{i=1}^{n} I_{y_i} \times \prod_{x \neq y_i} G(\mathcal{O}_x). \tag{10.5.5}$$

The spherical Hecke algebras $H(G(F_x), G(\mathcal{O}_x)), x \neq y_i$, act on the subspace (10.5.4), and all elements of (10.5.4) are eigenfunctions of these algebras (they satisfy formula (10.5.1)). At the points y_i, instead of the action of the commutative spherical Hecke algebra $H(G(F_{y_i}), G(\mathcal{O}_{y_i}))$, we have the action of the non-commutative affine Hecke algebra $H(G(F_{y_i}), I_{y_i})$. Thus, we obtain a subspace of the space of functions on (10.5.5), which consists of Hecke eigenfunctions with respect to the spherical Hecke algebras $H(G(F_x), G(\mathcal{O}_x)), x \neq y_i$, and which realize a module over $\bigotimes_{i=1}^{n} H(G(F_{y_i}), I_{y_i})$ (which is irreducible, since each π_{y_i} is irreducible).

This subspace encapsulates the automorphic representation $\bigotimes'_{x \in X} \pi_x$ the way the automorphic function f_π encapsulates an unramified automorphic representation. The difference is that in the unramified case the function f_π spans the one-dimensional space of invariants of the maximal compact subgroup $G(\mathcal{O})$ in $\bigotimes'_{x \in X} \pi_x$, whereas in the tamely ramified case the subspace (10.5.4) is in general a multi-dimensional vector space.

Now let us see how this plays out in the geometric setting. As we discussed before, the analogue of a homomorphism $\sigma : W_F \to {}^L G$ tamely ramified at point $y_1, \ldots, y_n \in X$ is now a local system $E = (\mathcal{F}, \nabla)$, where $\mathcal{F}$ a ${}^L G$-bundle $\mathcal{F}$ on X with a connection ∇ that has regular singularities at $y_1, \ldots, y_n$ and unipotent monodromies around these points. We will call such a local system **tamely ramified** at $y_1, \ldots, y_n$. What should the global geometric Langlands correspondence attach to such a local system? It is clear that we need to find a geometric object replacing the finite-dimensional vector space (10.5.4) realized in the space of functions on (10.5.5).

Just as (10.5.2) is the set of points of the moduli stack Bun_G of G-bundles, the double quotient (10.5.5) is the set of points of the moduli stack $\mathrm{Bun}_{G,(y_i)}$ of G-bundles on X with **parabolic structures** at $y_i, i = 1, \ldots, n$. By definition, a parabolic structure of a G-bundle $\mathcal{P}$ at $y \in X$ is a reduction of the fiber $\mathcal{P}_y$ of $\mathcal{P}$ at y to a Borel subgroup $B \subset G$. Therefore, as before, we obtain that a proper replacement for (10.5.4) is a category of $\mathcal{D}$-modules on $\mathrm{Bun}_{G,(y_i)}$. As in the unramified case, we have the notion of a Hecke eigensheaf on $\mathrm{Bun}_{G,(y_i)}$. But because the Hecke functors are now defined using the Hecke correspondences over $X \backslash \{y_1, \ldots, y_n\}$ (and not over X as before), an "eigenvalue" of the Hecke operators is now an ${}^L G$-local system on $X \backslash \{y_1, \ldots, y_n\}$ (rather than on X). Thus, we obtain that the global geometric Langlands correspondence now should assign to a ${}^L G$-local system E on X tamely ramified at the points $y_1, \ldots, y_n$ a *category* $\mathcal{A}ut_E$ of $\mathcal{D}$-modules on $\mathrm{Bun}_{G,(y_i)}$ with the eigenvalue $E|_{X \backslash \{y_1, \ldots, y_n\}}$,

$$E \mapsto \mathcal{A}ut_E.$$

We will construct these categories using a generalization of the localization functor we used in the unramified case (see [FG2]). For the sake of notational simplicity, let us assume that our ${}^L G$-local system $E = (\mathcal{F}, \nabla)$ is tamely ramified at a single point $y \in X$. Suppose that this local system on $X \backslash y$ admits the structure of a ${}^L G$-oper $\chi = (\mathcal{F}, \nabla, \mathcal{F}_{{}^L B})$ whose restriction χ_y to the punctured disc $D_y^\times$ belongs to the subspace $\mathrm{Op}_{{}^L G}^{\mathrm{nilp}}(D_y)$ of nilpotent ${}^L G$-opers.

For a simple Lie group G, the moduli stack $\mathrm{Bun}_{G,y}$ has a realization analogous to (10.5.3):

$$\mathrm{Bun}_{G,y} \simeq G_{\mathrm{out}} \backslash G(\mathcal{K}_y)/I_y.$$

Let $\mathcal{D}_{\kappa_c, I_y}$ be the sheaf of twisted differential operators on $\mathrm{Bun}_{G,y}$ acting on the line bundle corresponding to the critical level (it is the pull-back of the square root of the canonical line bundle $K^{1/2}$ on Bun_G under the natural projection $\mathrm{Bun}_{G,y} \to \mathrm{Bun}_G$). Applying the formalism of the previous section, we obtain a localization functor

$$\Delta_{\kappa_c, I_y} : \widehat{\mathfrak{g}}_{\kappa_c, y}\text{-mod}^{I_y} \to \mathcal{D}_{\kappa_c, I_y}\text{-mod}.$$

However, in order to make contact with the results obtained above we also consider the larger category $\widehat{\mathfrak{g}}_{\kappa_c,y}\text{-mod}^{I_y^0}$ of I_y^0-equivariant modules, where $I_y^0 = [I_y, I_y]$.

Set

$$\operatorname{Bun}'_{G,y} = G_{\text{out}} \backslash G(\mathcal{K}_y)/I_y^0,$$

and let $\mathcal{D}_{\kappa_c, I_y^0}$ be the sheaf of twisted differential operators on $\operatorname{Bun}'_{G,y}$ acting on the pull-back of the line bundle $K^{1/2}$ on Bun_G. Let $\mathcal{D}_{\kappa_c, I_y^0}\text{-mod}$ be the category of $\mathcal{D}_{\kappa_c, I_y^0}$-modules. Applying the general formalism, we obtain a localization functor

$$\Delta_{\kappa_c, I_y^0} : \widehat{\mathfrak{g}}_{\kappa_c,y}\text{-mod}^{I_y^0} \to \mathcal{D}_{\kappa_c, I_y^0}\text{-mod}. \tag{10.5.6}$$

We note that a version of the categorical affine Hecke algebra $\mathcal{H}(G(\mathcal{K}_y), I_y)$ discussed in Section 10.4.5 naturally acts on the derived categories of the above categories, and the functors Δ_{κ_c, I_y} and Δ_{κ_c, I_y^0} intertwine these actions. Equivalently, one can say that this functor intertwines the corresponding actions of the affine braid group associated to ${}^L G$ on the two categories (as in [B]).

We now restrict the functors Δ_{κ_c, I_y} and Δ_{κ_c, I_y^0} to the subcategories $\widehat{\mathfrak{g}}_{\kappa_c,y}\text{-mod}_{\chi_y}^{I_y}$ and $\widehat{\mathfrak{g}}_{\kappa_c,y}\text{-mod}_{\chi_y}^{I_y^0}$, respectively. By using the same argument as in [BD1], we obtain the following analogue of Theorem 10.5.1.

Theorem 10.5.2 *Fix $\chi_y \in \operatorname{Op}_{{}^L G}^{\text{nilp}}(D_y)$ and let M be an object of the category $\widehat{\mathfrak{g}}_{\kappa_c,y}\text{-mod}_{\chi_y}^{I_y}$ (resp. $\widehat{\mathfrak{g}}_{\kappa_c,y}\text{-mod}_{\chi_y}^{I_y^0}$). Then:*

(1) $\Delta_{\kappa_c, I_y}(M) = 0$ (resp., $\Delta_{\kappa_c, I_y^0}(M) = 0$) unless χ_y is the restriction of a regular oper $\chi = (\mathcal{F}, \nabla, \mathcal{F}_{{}^L B})$ on $X \backslash y$ to $D_y^\times$.

(2) In that case $\Delta_{\kappa_c, y}(M)$ (resp., $\Delta_{\kappa_c, I_y^0}(M)$) is a Hecke eigensheaf with the eigenvalue $E_\chi = (\mathcal{F}, \nabla)$.

Thus, we obtain that if $\chi_y = \chi|_{D_y^\times}$, then the image of any object of $\widehat{\mathfrak{g}}_{\kappa_c,y}\text{-mod}_{\chi_y}^{I_y}$ under the functor Δ_{κ_c, I_y} belongs to the category $\mathcal{A}ut_{E_\chi}^{I_y}$ of Hecke eigensheaves on $\operatorname{Bun}_{G,y}$. Now consider the restriction of the functor Δ_{κ_c, I_y^0} to $\widehat{\mathfrak{g}}_{\kappa_c,y}\text{-mod}_{\chi_y}^{I_y^0}$. As discussed in Section 10.4.3, the category $\widehat{\mathfrak{g}}_{\kappa_c,y}\text{-mod}_{\chi_y}^{I_y^0}$ coincides with the corresponding category $\widehat{\mathfrak{g}}_{\kappa_c,y}\text{-mod}_{\chi_y}^{I_y,m}$ of I_y-monodromic modules. Therefore the image of any object of $\widehat{\mathfrak{g}}_{\kappa_c,y}\text{-mod}_{\chi_y}^{I_y^0}$ under the functor Δ_{κ_c, I_y^0} belongs to the subcategory $\mathcal{D}^m_{\kappa_c, I_y^0}\text{-mod}$ of $\mathcal{D}_{\kappa_c, I_y^0}\text{-mod}$ whose objects admit an increasing filtration such that the consecutive quotients are pull-backs of $\mathcal{D}_{\kappa_c, I_y}$-modules from $\operatorname{Bun}_{G,y}$. Such $\mathcal{D}_{\kappa_c, I_y^0}$–modules are called **monodromic**.

Let $\mathcal{A}ut_{E_\chi}^{I_y,m}$ be the subcategory of $\mathcal{D}^m_{\kappa_c, I_y^0}\text{-mod}$ whose objects are Hecke eigensheaves with eigenvalue E_χ.

Thus, we obtain the functors

$$\Delta_{\kappa_c, I_y} : \widehat{\mathfrak{g}}_{\kappa_c,y}\text{-mod}_{\chi_y}^{I_y} \to \mathcal{A}ut_{E_\chi}^{I_y}, \tag{10.5.7}$$

$$\Delta_{\kappa_c, I_y^0} : \widehat{\mathfrak{g}}_{\kappa_c,y}\text{-mod}_{\chi_y}^{I_y^0} \to \mathcal{A}ut_{E_\chi}^{I_y,m}. \tag{10.5.8}$$

It is tempting to conjecture (see [FG2]) that these functors are equivalences of categories, at least for generic χ. Suppose that this is true. Then we may identify the *global* categories $\mathcal{A}ut_{E_\chi}^{I_y}$ and $\mathcal{A}ut_{E_\chi}^{I_y,m}$ of Hecke eigensheaves on Bun_{G,I_y} and Bun'_{G,I_y^0} with the *local* categories $\widehat{\mathfrak{g}}_{\kappa_c,y}\text{-mod}_{\chi_y}^{I_y}$ and $\widehat{\mathfrak{g}}_{\kappa_c,y}\text{-mod}_{\chi_y}^{I_y^0}$, respectively. Therefore we can use our results and conjectures on the local Langlands categories, such as $\widehat{\mathfrak{g}}_{\kappa_c,y}\text{-mod}_{\chi_y}^{I_y^0}$, to describe the global categories of Hecke eigensheaves on the moduli stacks of G-bundles on X with parabolic structures.

Recall that we have the following conjectural description of the derived category of I_y^0-equivariant modules, $D^b(\widehat{\mathfrak{g}}_{\kappa_c,y}\text{-mod}_{\chi_y})^{I_y^0}$ (see formula (10.4.14)):

$$D^b(\widehat{\mathfrak{g}}_{\kappa_c,y}\text{-mod}_{\chi_y})^{I_y^0} \simeq D^b(\mathrm{QCoh}(\widetilde{\mathrm{Sp}}^{\mathrm{DG}}_{\mathrm{Res}(\chi_y)})). \tag{10.5.9}$$

The corresponding I_y-equivariant version is

$$D^b(\widehat{\mathfrak{g}}_{\kappa_c,y}\text{-mod}_{\chi_y})^{I_y} \simeq D^b(\mathrm{QCoh}(\mathrm{Sp}^{\mathrm{DG}}_{\mathrm{Res}(\chi_y)})), \tag{10.5.10}$$

where we replace the non-reduced DG Springer fiber by the reduced one:

$$\mathcal{O}_{\mathrm{Sp}_u^{\mathrm{DG}}} = \mathcal{O}_{\widetilde{\mathcal{N}}} \overset{L}{\underset{\mathcal{O}_{L_{\mathfrak{g}}}}{\otimes}} \mathbb{C}_u.$$

If the functors (10.5.7), (10.5.8) are equivalences, then by combining them with (10.5.9) and (10.5.10), we obtain the following conjectural equivalences of categories:

$$D^b(\mathcal{A}ut_{E_\chi}^{I_y}) \simeq D^b(\mathrm{QCoh}(\mathrm{Sp}^{\mathrm{DG}}_{\mathrm{Res}(\chi_y)})), \tag{10.5.11}$$

$$D^b(\mathcal{A}ut_{E_\chi}^{I_y,m}) \simeq D^b(\mathrm{QCoh}(\widetilde{\mathrm{Sp}}^{\mathrm{DG}}_{\mathrm{Res}(\chi_y)})). \tag{10.5.12}$$

In other words, the derived category of a global Langlands category (monodromic or not) corresponding to a local system tamely ramified at $y \in X$ is equivalent to the derived category of quasicoherent sheaves on the DG Springer fiber of its residue at y (non-reduced or reduced).

Again, these equivalences are supposed to intertwine the natural actions on the above categories of the categorical affine Hecke algebra $\mathcal{H}(G(\mathcal{K}_y), I_y)$ (or, equivalently, the affine braid group associated to ${}^LG$).

The categories appearing in (10.5.11), (10.5.12) actually make sense for an arbitrary ${}^LG$-local system E on X tamely ramified at y. It is therefore tempting to conjecture that these equivalences still hold in general:

$$D^b(\mathcal{A}ut_E^{I_y}) \simeq D^b(\mathrm{QCoh}(\mathrm{Sp}^{\mathrm{DG}}_{\mathrm{Res}(E)})), \tag{10.5.13}$$

$$D^b(\mathcal{A}ut_E^{I_y,m}) \simeq D^b(\mathrm{QCoh}(\widetilde{\mathrm{Sp}}^{\mathrm{DG}}_{\mathrm{Res}(E)})). \tag{10.5.14}$$

The corresponding localization functors may be constructed as follows. Suppose that we can represent a local system E on X with tame ramification at y by an oper χ on the complement of finitely many points $y_1, \ldots, y_n$, whose restriction to $D_{y_i}^\times$ belongs to $\mathrm{Op}^{\lambda_i}_{^LG}(D_{y_i}) \subset \mathrm{Op}_{^LG}(D_{y_i}^\times)$ for some $\lambda_i \in P^+$. Then, in the same way as in the unramified case, we construct localization functors from $\widehat{\mathfrak{g}}_{\kappa_c,y}\text{-mod}_{\chi_y}^{I_y}$ to $\mathcal{A}ut_E^{I_y}$ and from $\widehat{\mathfrak{g}}_{\kappa_c,y}\text{-mod}_{\chi_y}^{I_y^0}$ to $\mathcal{A}ut_E^{I_y,m}$ (here, as before, $\chi_y = \chi|_{D_y^\times}$), and this leads us to the conjectural equivalences (10.5.13), (10.5.14).

These equivalences may be viewed as the geometric Langlands correspondence for tamely ramified local systems.

The equivalences (10.5.13) also have family versions in which we allow E to vary. It is analogous to the family version (10.4.13) of the local equivalences. As in the local case, in a family version we can avoid using DG schemes.

The above construction may be generalized to allow local systems tamely ramified at finitely many points $y_1, \ldots, y_n$. The corresponding Hecke eigensheaves are then $\mathcal{D}$-modules on the moduli stack of G-bundles on X with parabolic structures at $y_1, \ldots, y_n$. Non-trivial examples of these Hecke eigensheaves arise already in genus zero. These sheaves were constructed explicitly in [F3] (see also [F5, F6]), and they are closely related to the Gaudin integrable system.

10.5.3 Connections with regular singularities

So far we have only considered the categories of $\widehat{\mathfrak{g}}_{\kappa_c}$-modules corresponding to ^LG-opers χ on X, which are regular everywhere except at a point $y \in X$ (or perhaps, at several points) and whose restriction $\chi|_{D_y^\times}$ is a nilpotent oper in $\mathrm{Op}^{\mathrm{nilp}}_{^LG}(D_y)$. In other words, $\chi|_{D_y^\times}$ is an oper with regular singularity at y with residue $\varpi(-\rho)$. However, we can easily generalize the localization functor to the categories of $\widehat{\mathfrak{g}}_{\kappa_c}$-modules corresponding to ^LG-opers, which have regular singularity at y with *arbitrary* residue.

Recall that in Section 9.4.3 for each oper $\chi \in \mathrm{Op}^{\mathrm{RS}}_{^LG}(D)_{\varpi(-\lambda-\rho)}$ with regular singularity and residue $\varpi(-\lambda-\rho)$ we have defined the category $\widehat{\mathfrak{g}}_{\kappa_c}\text{-mod}_\chi^{I^0}$ of I^0-equivariant $\widehat{\mathfrak{g}}_{\kappa_c}$-modules with central character χ. The case of $\lambda = 0$ is an "extremal" case when the category $\widehat{\mathfrak{g}}_{\kappa_c}\text{-mod}_\chi^{I^0}$ is most complicated. On the other "extreme" is the case of generic opers χ corresponding to a generic λ. In this case, as we saw in Section 9.4.3, the category $\widehat{\mathfrak{g}}_{\kappa_c}\text{-mod}_\chi^{I^0}$ is quite simple: it contains irreducible objects $\mathbb{M}_{w(\lambda+\rho)-\rho}(\chi)$ labeled by the Weyl group of $\mathfrak{g}$, and each object of $\widehat{\mathfrak{g}}_{\kappa_c}\text{-mod}_\chi^{I^0}$ is a direct sum of these irreducible modules. Here $\mathbb{M}_{w(\lambda+\rho)-\rho}(\chi)$ is the quotient of the Verma module $\mathbb{M}_{w(\lambda+\rho)-\rho}, w \in W$, by the central character corresponding to χ.

For other values of λ the structure of $\widehat{\mathfrak{g}}_{\kappa_c}\text{-mod}_\chi^{I^0}$ is somewhere in-between these two extreme cases.

Recall that we have a localization functor (10.5.6)

$$\Delta^\lambda_{\kappa_c,I^0_y} : \widehat{\mathfrak{g}}_{\kappa_c,y}\text{-mod}^{I^0_y} \to \mathcal{D}_{\kappa_c,I^0_y}\text{-mod}$$

from $\widehat{\mathfrak{g}}_{\kappa_c,y}\text{-mod}_{\chi_y}^{I^0_y}$ to a category of $\mathcal{D}$-modules on Bun'_{G,I_y} twisted by the pull-back of the line bundle $K^{1/2}$ on Bun_G. We now restrict this functor to the subcategory $\widehat{\mathfrak{g}}_{\kappa_c,y}\text{-mod}_{\chi_y}^{I^0_y}$ where χ_y is a ${}^LG$-oper on D_y with regular singularity at y and residue $\varpi(-\lambda-\rho)$.

Consider first the case when $\lambda \in \mathfrak{h}^*$ is generic. Suppose that χ_y extends to a regular oper χ on $X\backslash y$. One then shows in the same way as in Theorem 10.5.2 that for any object M of $\widehat{\mathfrak{g}}_{\kappa_c,y}\text{-mod}_{\chi_y}^{I^0_y}$ the corresponding $\mathcal{D}_{\kappa_c,I^0_y}$-module $\Delta_{\kappa_c,I^0_y}(M)$ is a Hecke eigensheaf with eigenvalue E_χ, which is the ${}^LG$-local system on X with regular singularity at y underlying χ (if χ_y cannot be extended to $X\backslash y$, then $\Delta^\lambda_{\kappa_c,I_y}(M) = 0$, as before). Therefore we obtain a functor

$$\Delta_{\kappa_c,I^0_y} : \widehat{\mathfrak{g}}_{\kappa_c,y}\text{-mod}_{\chi_y}^{I^0_y} \to \mathcal{A}ut_{E_\chi}^{I^0_y},$$

where $\mathcal{A}ut_{E_\chi}^{I^0_y}$ is the category of Hecke eigensheaves on Bun'_{G,I_y} with eigenvalue E_χ.

Since we have assumed that the residue of the oper χ_y is generic, the monodromy of E_χ around y belongs to a regular semi-simple conjugacy class of ${}^LG$ containing $\exp(2\pi i\lambda)$. In this case the category $\widehat{\mathfrak{g}}_{\kappa_c,y}\text{-mod}_{\chi_y}^{I^0_y}$ is particularly simple, as we have discussed above. We expect that the functor Δ_{κ_c,I^0_y} sets up an equivalence between $\widehat{\mathfrak{g}}_{\kappa_c,y}\text{-mod}_{\chi_y}^{I^0_y}$ and $\mathcal{A}ut_{E_\chi}^{I^0_y}$.

We can formulate this more neatly as follows. For $M \in {}^LG$ let $\mathcal{B}_M$ be the variety of Borel subgroups containing M. Observe that if M is regular semi-simple, then $\mathcal{B}_M$ is a set of points which is in bijection with W. Therefore our conjecture is that $\mathcal{A}ut_{E_\chi}^{I^0_y}$ is equivalent to the category $\text{QCoh}(\mathcal{B}_M)$ of quasicoherent sheaves on $\mathcal{B}_M$, where M is a representative of the conjugacy class of the monodromy of E_χ.

Consider now an arbitrary ${}^LG$-local system E on X with regular singularity at $y \in X$ whose monodromy around y is regular semi-simple. It is then tempting to conjecture that, at least if E is generic, this category has the same structure as in the case when E has the structure of an oper, i.e., it is equivalent to the category $\text{QCoh}(\mathcal{B}_M)$, where M is a representative of the conjugacy class of the monodromy of E around y.

On the other hand, if the monodromy around y is unipotent, then $\mathcal{B}_M$ is nothing but the Springer fiber Sp_u, where $M = \exp(2\pi i u)$. The corresponding category $\mathcal{A}ut_E^{I_y^0}$ was discussed in Section 10.5.2 (we expect that it coincides with $\mathcal{A}ut_E^{I_y,m}$). Thus, we see that in both "extreme" cases – unipotent monodromy and regular semi-simple monodromy – our conjectures identify the derived category of $\mathcal{A}ut_E^{I_y^0}$ with the derived category of the category $\mathrm{QCoh}(\mathcal{B}_M)$ (where $\mathcal{B}_M$ should be viewed as a DG scheme $\widetilde{\mathrm{Sp}}_u^{\mathrm{DG}}$ in the unipotent case). One is then led to conjecture, most ambitiously, that for *any* ${}^L G$-local system E on X with regular singularity at $y \in X$ the derived category of $\mathcal{A}ut_E^{I_y^0}$ is equivalent to the derived category of quasicoherent sheaves on a suitable DG version of the scheme $\mathcal{B}_M$, where M is a representative of the conjugacy class of the monodromy of E around y:

$$D^b(\mathcal{A}ut_E^{I_y^0}) \simeq D^b(\mathrm{QCoh}(\mathcal{B}_M^{\mathrm{DG}})).$$

This has an obvious generalization to the case of multiple ramification points, where on the right hand side we take the Cartesian product of the varieties $\mathcal{B}_{M_i}^{\mathrm{DG}}$ corresponding to the monodromies. Thus, we obtain a conjectural realization of the categories of Hecke eigensheaves, whose eigenvalues are local systems with regular singularities, in terms of categories of quasicoherent sheaves.

Let us summarize: by using the representation theory of affine Kac–Moody algebras at the critical level we have constructed the local Langlands categories corresponding to local Langlands parameters; namely, ${}^L G$-local systems on the punctured disc. We then applied the technique of localization functors to produce from these local categories, the global categories of Hecke eigensheaves on the moduli stacks of G-bundles on a curve X with parabolic structures. These global categories correspond to global Langlands parameters: ${}^L G$-local systems on X with ramifications. We have used our results and conjectures on the structure of the local categories to investigate these global categories.

In this chapter we have explained how this works for unramified and tamely ramified local systems. But we expect that analogous methods may also be used to analyze local systems with arbitrary ramification. In this way the representation theory of affine Kac–Moody algebras may one day fulfill the dream of uncovering the mysteries of the geometric Langlands correspondence.

Appendix

A.1 Lie algebras

A **Lie algebra** is a vector space $\mathfrak{g}$ with a bilinear form (the Lie bracket)

$$[\cdot,\cdot] : \mathfrak{g} \otimes \mathfrak{g} \longrightarrow \mathfrak{g},$$

which satisfies two additional conditions:

- $[x, y] = -[y, x]$
- $[x, [y, z]] + [y, [z, x]] + [z, [x, y]] = 0.$

The last of these conditions is called the **Jacobi identity**.

Lie algebras are closely related to Lie groups. The tangent space at the identity element of a Lie group naturally has the structure of a Lie algebra. So, thinking about Lie algebras is a way of linearizing problems coming from the theory of Lie groups.

A **representation** of a Lie algebra $\mathfrak{g}$, or equivalently, a $\mathfrak{g}$**-module**, is a pair (V, ρ) of a vector space V and a map $\rho : \mathfrak{g} \to \mathrm{End}(V)$ such that

$$\rho([x, y]) = \rho(x)\rho(y) - \rho(y)\rho(x).$$

We will often abuse notation and call the representation V rather than specifying the homomorphism ρ; however, ρ will always be implicit.

A.2 Universal enveloping algebras

A standard tool in the study of Lie algebras and their representations is the universal enveloping algebra. This is an associative algebra that can be constructed from any Lie algebra and has the property that any module over the Lie algebra can be regarded as a module over this associative algebra.

As we want to be able to multiply elements of $\mathfrak{g}$ together in the associative algebra it makes sense to look at the **tensor algebra** of $\mathfrak{g}$

$$T(\mathfrak{g}) = \mathbb{C} \oplus \mathfrak{g} \oplus \mathfrak{g}^{\otimes 2} \oplus \mathfrak{g}^{\otimes 3} \oplus \cdots .$$

If we have a representation (V, ρ) of $\mathfrak{g}$, we clearly want want the corresponding

$T(\mathfrak{g})$-module to be such that the action of $T(\mathfrak{g})$ restricts to (V, ρ) on the second summand. Since it should be a module over an associative algebra, we also want

$$\rho(g \otimes h) = \rho(g)\rho(h).$$

However, the Lie algebra modules satisfy one additional relation

$$\rho([g, h]) = \rho(g)\rho(h) - \rho(h)\rho(g).$$

Therefore the elements $[g, h]$ and $g \otimes h - h \otimes g$ of $T(\mathfrak{g})$ should act in the same way. This suggests that we should identify $[g, h]$ and $g \otimes h - h \otimes g$ in the tensor algebra and this leads us to the following definition.

The **universal enveloping algebra** $U(\mathfrak{g})$ is the quotient of the tensor algebra $T(\mathfrak{g})$ by the two sided ideal generated by elements of the form

$$g \otimes h - h \otimes g - [g, h] .$$

Note that we have a natural map $\mathfrak{g} \to U(\mathfrak{g})$.

It is called universal because it has the following universal property: *If A is any associative algebra (regarded naturally as a Lie algebra) and $f : \mathfrak{g} \to A$ is a Lie algebra homomorphism then it may be uniquely completed to the following commutative diagram:*

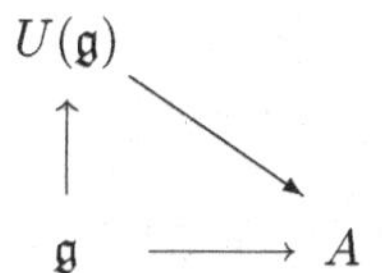

There is a natural **filtration** (sometimes called the PBW filtration) on the universal enveloping algebra coming from the gradation on the tensor algebra. The ith term of the filtration is denoted by $U(\mathfrak{g})_{\leqslant i}$.

The structure of the universal enveloping algebra is very easy to work out in the case that $\mathfrak{g}$ is abelian. In this case the generators for the ideal are simply $g \otimes h - h \otimes g$ (as the bracket is zero). This is exactly the definition of the symmetric algebra generated by $\mathfrak{g}$. Hence we get

$$U(\mathfrak{g}) \cong \mathrm{Sym}(\mathfrak{g}).$$

In the case where $\mathfrak{g}$ is not abelian, the situation is more complicated.

Given any filtered algebra

$$0 = F_{\leqslant -1} \subset F_{\leqslant 0} \subset F_{\leqslant 1} \subset \cdots \subset F$$

we define the **associated graded algebra** to be

$$\mathrm{gr}\, F = \bigoplus_{i \geqslant 0} F_{\leqslant i}/F_{\leqslant i-1}.$$

This is naturally a graded algebra and considered as vector spaces we have $\operatorname{gr} F \cong F$, but non-canonically. Taking the associated graded algebra of a quotient is reasonably easy:

$$\operatorname{gr}(F/I) = \operatorname{gr}(F)/\operatorname{Symb}(I)$$

where Symb is the **symbol map**. Let x be an element of a filtered algebra F. There is an integer i such that $x \in F_{\leqslant i}$ but $x \notin F_{\leqslant i-1}$. The symbol of x is defined to be the image of x in the quotient $F_{\leqslant i}/F_{\leqslant i-1}$. In other words, the symbol map is picking out the leading term of x.

So, for the universal enveloping algebra we have

$$\operatorname{gr} U(\mathfrak{g}) = \operatorname{gr} T(\mathfrak{g})/\operatorname{Symb}(I).$$

As the tensor algebra is already graded, $\operatorname{gr} T(\mathfrak{g}) \cong T(\mathfrak{g})$.

Lemma A.2.1 *The ideal* Symb(I) *is generated by the symbols of the generators for the ideal I.*

The proof of this fact is the heart of most proofs of the Poincaré–Birkhoff–Witt theorem. It fundamentally relies on the fact that the bracket satisfies the Jacobi identity. If we were using an arbitrary bilinear form to define the ideal I (e.g., if we used the ideal generated by $a \otimes b - b \otimes a - B(a, b)$), there would be elements in Symb(I) not obtainable from the symbols of the generators. The symbols of the generators are easy to work out and are just $g \otimes h - h \otimes g$. The corresponding quotient is Sym$(\mathfrak{g})$. Hence we have shown that

$$\operatorname{gr} U(\mathfrak{g}) \cong \operatorname{Sym}(\mathfrak{g})$$

for an arbitrary Lie algebra $\mathfrak{g}$.

In other words, as vector spaces, $U(\mathfrak{g})$ and Sym$(\mathfrak{g})$ are isomorphic, but non-canonically. This is often quoted in the following form.

Theorem A.2.2 (Poincaré–Birkhoff–Witt) *Let $J^a, a = 1, \ldots, \dim \mathfrak{g}$, be an ordered basis for $\mathfrak{g}$ (as a vector space), then the universal enveloping algebra has a basis given by elements of the form $J^{a_1} \ldots J^{a_n}$, where $a_1 \leq \ldots \leq a_n$.*

Monomials of this form are often called **lexicographically ordered**. One of the main reasons for studying the universal enveloping algebras is that the category of $U(\mathfrak{g})$-modules is equivalent to the category of $\mathfrak{g}$-modules (this is just the statement that any representation of $\mathfrak{g}$ can be given a unique structure of module over $U(\mathfrak{g})$).

A.3 Simple Lie algebras

In this book we consider simple finite-dimensional complex Lie algebras. Here "simple" means that the Lie algebra has no non-trivial ideals. These Lie algebras have been completely classified. They are in one-to-one correspondence with the positive definite **Cartan matrices** $A = (a_{ij})_{i,j=1,\dots,\ell}$. These are symmetrizable matrices with integer entries satisfying the following conditions: $a_{ii} = 2$, and for $i \neq j$ we have $a_{ij} \leq 0$ and $a_{ij} = 0$ if and only if $a_{ji} \neq 0$ (see [K2] for more details). Such matrices, and hence simple Lie algebras, fit into four infinite families and six exceptional ones. The families are called A_n, B_n, C_n, and D_n; the exceptional Lie algebras are called E_6, E_7, E_8, F_4 and G_2.

A simple Lie algebra $\mathfrak{g}$ admits a Cartan decomposition

$$\mathfrak{g} = \mathfrak{n}_- \oplus \mathfrak{h} \oplus \mathfrak{n}_+.$$

Here $\mathfrak{h}$ is a Cartan subalgebra, a maximal commutative Lie subalgebra of $\mathfrak{g}$ consisting of semi-simple elements (i.e., those whose adjoint action on $\mathfrak{g}$ is semi-simple). Its dimension is ℓ, which is the rank of the Cartan matrix corresponding to $\mathfrak{g}$; we will also call ℓ the **rank of** $\mathfrak{g}$. The Lie subalgebra $\mathfrak{n}_+$ (resp., $\mathfrak{n}_-$) is called the upper (resp., lower) nilpotent subalgebra. Under the adjoint action of $\mathfrak{h}$ the nilpotent subalgebras decompose into eigenspaces as follows

$$\mathfrak{n}_\pm = \bigoplus_{\alpha \in \Delta_+} \mathfrak{n}_{\pm\alpha},$$

where

$$\mathfrak{n}_{\pm\alpha} = \{x \in \mathfrak{n}_\pm \mid [h, x] = \pm\alpha(h)x\}.$$

Here $\Delta_+ \subset \mathfrak{h}^*$ and $\Delta_- = -\Delta_+$ are the sets of positive and negative **roots** $\pm\alpha$ of $\mathfrak{g}$, respectively, for which $\mathfrak{n}_{\pm\alpha} \neq 0$. In fact, all of these subspaces are one-dimensional, and we will choose generators e_α and f_α of $\mathfrak{n}_\alpha$ and $\mathfrak{n}_{-\alpha}$, respectively.

It is known that Δ_+ contains a subset of **simple roots** $\alpha_i, i = 1, \dots, \ell$, where $\ell = \dim \mathfrak{h}$ is the rank of $\mathfrak{g}$. All other elements of Δ_+ are linear combinations of the simple roots with non-negative integer coefficients. The simple roots form a basis of $\mathfrak{h}^*$. Let $h_i, i = 1, \dots, \ell$, be the basis of $\mathfrak{h}$ that is uniquely determined by the formula

$$\alpha_j(h_i) = a_{ij}.$$

These are the **simple coroots**.

Let us choose non-zero generators $e_i = e_{\alpha_i}$ (resp., $f_i = f_{\alpha_i}$) of the one-dimensional spaces $\mathfrak{n}_{\alpha_i}$ (resp., $\mathfrak{n}_{-\alpha_i}$), $i = 1, \dots, \ell$. Then $\mathfrak{n}_+$ (resp., $\mathfrak{n}_-$) is generated by $e_1, \dots, e_\ell$ (resp., $f_1, \dots, f_\ell$) subject to the following **Serre relations**:

$$(\operatorname{ad} e_i)^{-a_{ij}+1} \cdot e_j = 0, \qquad (\operatorname{ad} f_i)^{-a_{ij}+1} \cdot f_j = 0, \qquad i \neq j.$$

Here the a_{ij}'s are the entries of the Cartan matrix of $\mathfrak{g}$.

The Lie algebra $\mathfrak{g}$ is generated by $h_i, h_i, f_i, i = 1, \ldots, \ell$, subject to the Serre relations together with

$$[h_i, e_j] = a_{ij} e_j, \quad [h_i, f_j] = -a_{ij} f_j, \quad [e_i, f_j] = \delta_{i,j} h_i.$$

To satisfy the last set of relations we might need to rescale the e_i's (or the f_i's) by non-zero numbers.

For example, the Lie algebra A_n is the Lie algebra also called $\mathfrak{sl}_{n+1}$. This is the vector space of $(n+1) \times (n+1)$ matrices with zero trace and bracket given by $[A, B] = AB - BA$. This is the Lie algebra of the special linear group $\mathrm{SL}_{n+1}(\mathbb{C})$, which consists of $(n+1) \times (n+1)$ matrices with determinant 1. To see this, we write an elements of $\mathrm{SL}_{n+1}(\mathbb{C})$ in the form $I_{n+1} + \epsilon M$, where $\epsilon^2 = 0$ and see what restriction the determinant condition puts on M. This is exactly the trace 0 condition.

The Cartan subalgebra consists of diagonal matrices. The Lie algebras $\mathfrak{n}_+$ and $\mathfrak{n}_-$ are the upper and lower nilpotent subalgebras, respectively, with the zeros on the diagonal. The generators e_i, h_i and f_i are the matrices $E_{i,i+1}$, $E_{ii} - E_{i+1,i+1}$, and $E_{i+1,i}$, respectively.

There are four lattices attached to $\mathfrak{g}$: the weight lattice $P \subset \mathfrak{h}^*$ consists of linear functionals $\phi : \mathfrak{h} \to \mathbb{C}$ such that $\phi(h_i) \in \mathbb{Z}$; the root lattice $Q \subset P$ is spanned over $\mathbb{Z}$ by the simple roots $\alpha_i, i = 1, \ldots, \ell$; the coweight lattice $\check{P} \subset \mathfrak{h}$ consists of elements on which the simple roots take integer values; and the coroot lattice $\check{Q}$ is spanned by the simple coroots $h_i, i = 1, \ldots, \ell$.

The lattice P is spanned by elements $\omega_i \in \mathfrak{h}^*$ defined by $\omega_i(h_j) = \delta_{i,j}$. The coroot lattice is spanned by elements $\check{\omega}_i \in \mathfrak{h}$ defined by the formula $\alpha_i(\check{\omega}_i) = \delta_{i,j}$.

We note that P may be identified with the group of characters $H_{\mathrm{sc}} \to \mathbb{C}^\times$ of the Cartan subgroup H_{sc} of the connected simply-connected Lie group G with the Lie algebra $\mathfrak{g}$ (such as SL_n), and $\check{P}$ is identified with the group of cocharacters $\mathbb{C}^\times \to H_{\mathrm{adj}}$, where H_{adj} is the Cartan subgroup of the group of adjoint type associated to G (with the trivial center, such as PGL_n).

Let $s_i, i = 1, \ldots, \ell$, be the linear operator (reflection) on $\mathfrak{h}^*$ defined by the formula

$$s_i(\lambda) = \lambda - \langle \lambda, \check{\alpha}_i \rangle \alpha_i.$$

These operators generate an action of the **Weyl group** W, associated to $\mathfrak{g}$, on $\mathfrak{h}^*$.

A.4 Central extensions

By a central extension of a Lie algebra $\mathfrak{g}$ by an abelian Lie algebra $\mathfrak{a}$ we understand a Lie algebra $\widehat{\mathfrak{g}}$ that fits into an exact sequence of Lie algebras

$$0 \longrightarrow \mathfrak{a} \longrightarrow \widehat{\mathfrak{g}} \longrightarrow \mathfrak{g} \longrightarrow 0. \tag{A.4.1}$$

Here $\mathfrak{a}$ is a Lie subalgebra of $\widehat{\mathfrak{g}}$, which is central in $\widehat{\mathfrak{g}}$, i.e., $[a, x] = 0$ for all $a \in \mathfrak{a}, x \in \mathfrak{g}$.

Suppose that $\mathfrak{a} \simeq \mathbb{C}\mathbf{1}$ is one-dimensional with a generator $\mathbf{1}$. Choose a splitting of $\widehat{\mathfrak{g}}$ as a vector space, $\widehat{\mathfrak{g}} \simeq \mathbb{C}\mathbf{1} \oplus \mathfrak{g}$. Then the Lie bracket on $\widehat{\mathfrak{g}}$ gives rise to a Lie bracket on $\mathbb{C}\mathbf{1} \oplus \mathfrak{g}$ such that $\mathbb{C}\mathbf{1}$ is central and

$$[x, y]_{\text{new}} = [x, y]_{\text{old}} + c(x, y)\mathbf{1},$$

where $x, y \in \mathfrak{g}$ and $c : \mathfrak{g} \otimes \mathfrak{g} \to \mathbb{C}$ is a linear map. The above formula defines a Lie bracket on $\widehat{\mathfrak{g}}$ if and only if c satisfies:

- $c(x, y) = -c(y, x)$.
- $c(x, [y, z]) + c(y, [z, x]) + c(z, [x, y]) = 0$.

Such expressions are called *two-cocycles* on $\mathfrak{g}$.

Now suppose that we are given two extensions $\widehat{\mathfrak{g}}_1$ and $\widehat{\mathfrak{g}}_2$ of the form (A.4.1) with $\mathfrak{a} = \mathbb{C}\mathbf{1}$ and a Lie algebra isomorphism ϕ between them which preserves their subspaces $\mathbb{C}\mathbf{1}$. Choosing splittings $\imath_1 : \mathbb{C}\mathbf{1} \oplus \mathfrak{g} \to \widehat{\mathfrak{g}}_1$ and $\imath_2 : \mathbb{C}\mathbf{1} \oplus \mathfrak{g} \to \widehat{\mathfrak{g}}_2$ for these extensions, we obtain the two-cocycles c_1 and c_2. Define a linear map $f : \mathfrak{g} \to \mathbb{C}$ as the composition of $\imath_1$, ϕ and the projection $\widehat{\mathfrak{g}}_2 \to \mathbb{C}\mathbf{1}$ induced by $\imath_2$. Then it is easy to see that

$$c_2(x, y) = c_1(x, y) + f([x, y]).$$

Therefore the set (actually, a vector space) of isomorphism classes of one-dimensional central extensions of $\mathfrak{g}$ is isomorphic to the quotient of the vector space of the two-cocycles on $\mathfrak{g}$ by the subspace of those two-cocycles c for which

$$c(x, y) = f([x, y]), \qquad \forall x, y \in \mathfrak{g},$$

for some $f : \mathfrak{g} \to \mathbb{C}$ (such cocycles are called coboundaries). This quotient space is the second Lie algebra cohomology group of $\mathfrak{g}$, denoted by $H^2(\mathfrak{g}, \mathbb{C})$.

A.5 Affine Kac–Moody algebras

If we have a complex Lie algebra $\mathfrak{g}$ and a commutative and associative algebra A, then $\mathfrak{g} \underset{\mathbb{C}}{\otimes} A$ is a Lie algebra if we use the bracket

$$[g \otimes a, h \otimes b] = [g, h] \otimes ab.$$

What exactly is A? One can usually think of commutative, associative algebras as functions on some manifold M. Note that

$$\mathfrak{g} \underset{\mathbb{C}}{\otimes} \text{Fun}(M) = \text{Map}(M, \mathfrak{g})$$

is just the Lie algebra of maps from M to $\mathfrak{g}$. This is naturally a Lie algebra with respect to the pointwise Lie bracket (it coincides with the Lie bracket given by the above formula).

Let us choose our manifold to be the circle, so we are thinking about maps from the circle into the Lie algebra. We then have to decide what type of maps we want to deal with. Using smooth maps would require some analysis, whereas we would like to stay in the realm of algebra – so we use algebraic (polynomial) maps. We consider the unit circle embedded into the complex plane with the coordinate t, so that on the circle $t = e^{i\theta}$, θ being the angle. Polynomial functions on this circle are the same as Laurent polynomials in t. They form a vector space denoted by $\mathbb{C}[t, t^{-1}]$. Using this space of functions, we obtain the **polynomial loop algebra**

$$L\mathfrak{g} = \mathfrak{g} \otimes \mathbb{C}[t, t^{-1}]$$

with the bracket given as above.

It turns out that there is much more structure and theory about the centrally extended loop algebras. As we saw above, these are classified by the second cohomology group $H^2(L\mathfrak{g}, \mathbb{C})$, and for a simple Lie algebra $\mathfrak{g}$ this is known to be one-dimensional. The corresponding cocycles are given by formula (1.3.3). They depend on the invariant inner product on $\mathfrak{g}$ denoted by κ. We will denote the central extension by $\widehat{\mathfrak{g}}_{\kappa}^{\mathrm{pol}}$.

There is a formal version of the above construction where we replace the algebra $\mathbb{C}[t, t^{-1}]$ of Laurent polynomials by the algebra $\mathbb{C}((t))$ of formal Laurent series. This gives is the formal loop algebra $\mathfrak{g} \otimes \mathbb{C}((t)) = \mathfrak{g}((t))$. Its central extensions are parametrized by its second cohomology group, which, with the appropriate continuity condition, is also a one-dimensional vector space if $\mathfrak{g}$ is a simple Lie algebra. The central extensions are again given by specifying an invariant inner product κ on $\mathfrak{g}$. The Lie bracket is given by the same formula as before – note that the residue is still well defined. We denote the corresponding Lie algebra $\widehat{\mathfrak{g}}_{\kappa}$. This is the affine Kac-Moody algebra associated to $\mathfrak{g}$ (and κ).

References

[AB] A. Arkhipov and R. Bezrukavnikov, *Perverse sheaves on affine flags and Langlands dual group*, Preprint math.RT/0201073.

[ABG] S. Arkhipov, R. Bezrukavnikov, and V. Ginzburg, *Quantum groups, the loop Grassmannian, and the Springer resolution*, Journal of AMS **17** (2004) 595–678.

[AG] S. Arkhipov and D. Gaitsgory, *Another realization of the category of modules over the small quantum group*, Adv. Math. **173** (2003) 114–143.

[Art] J. Arthur, *Unipotent automorphic representations: conjectures*, Asterisque **171-172** (1989) 13–71.

[ATY] H. Awata, A. Tsuchiya, and Y. Yamada, *Integral formulas for the WZNW correlation functions*, Nucl. Phys. **B 365** (1991) 680–698.

[BV] D.G. Babbitt and V.S. Varadarajan, *Formal reduction theory of meromorphic differential equations: a group theoretic view*, Pacific J. Math. **109** (1983) 1–80.

[BL] A. Beauville and Y. Laszlo, *Un lemme de descente*, C.R. Acad. Sci. Paris, Sér. I Math. **320** (1995) 335–340.

[Bei] A. Beilinson, *Langlands parameters for Heisenberg modules*, Preprint math.QA/0204020.

[BB] A. Beilinson and J. Bernstein, *A proof of Jantzen conjectures*, Advances in Soviet Mathematics **16**, Part 1, pp. 1–50, AMS, 1993.

[BD1] A. Beilinson and V. Drinfeld, *Quantization of Hitchin's integrable system and Hecke eigensheaves*, Preprint, available at www.math.uchicago.edu/~arinkin

[BD2] A. Beilinson and V. Drinfeld, *Chiral algebras*, Colloq. Publ. **51**, AMS, 2004.

[BD3] A. Beilinson and V. Drinfeld, *Opers*, Preprint math.AG/0501398.

[BZ] J. Bernstein and A. Zelevinsky, *Induced representations of reductive p-adic groups*, I, Ann. Sci. ENS **10** (1977) 441–472.

[B] R. Bezrukavnikov, *Noncommutative Counterparts of the Springer Resolution*, Preprint math.RT/0604445.

[Bo] R. Borcherds, *Vertex algebras, Kac–Moody algebras and the monster*, Proc. Natl. Acad. Sci. USA **83** (1986) 3068–3071.

[Bor1] A. Borel, *Admissible representations of a semi-simple group over a local field with vectors fixed under an Iwahori subgroup*, Inv. Math. **35** (1976) 233–259.

[Bor2] A. Borel, e.a., *Algebraic D–modules*, Academic Press, 1987.

[C] H. Carayol, *Preuve de la conjecture de Langlands locale pour* GL_n*: travaux de Harris-Taylor et Henniart*, Séminaire Bourbaki, Exp. No. 857, Astérisque **266** (2000) 191–243.

[CT] A. Chervov and D. Talalaev, *Quantum spectral curves, quantum integrable systems and the geometric Langlands correspondence*, Preprint hep-th/0604128.

[CG] N. Chriss and V. Ginzburg, *Representation theory and complex geometry*, Birkhäuser, 1997.

[CK] I. Ciocan-Fountanine and M. Kapranov, *Derived Quot schemes*, Ann. Sci. ENS **34** (2001) 403–440.

[dBF] J. de Boer and L. Feher, *Wakimoto realizations of current algebras: an explicit construction*, Comm. Math. Phys. **189** (1997) 759–793.

[De1] P. Deligne, *Equations différentielles à points singuliers réguliers*, Lect. Notes in Math. **163**, Springer, 1970.

[De2] P. Deligne, *Les constantes des équations fonctionnelles des fonctions L*, in *Modular Functions one Variable II*, Proc. Internat. Summer School, Univ. Antwerp 1972, Lect. Notes Math. **349**, pp. 501–597, Springer 1973.

[Di] J. Dixmier, *Enveloping Algebras*, North-Holland Publishing Co., 1977.

[Do] V. Dotsenko, *The free field representation of the $su(2)$ conformal field theory*, Nucl. Phys. **B 338** (1990) 747–758.

[Dr1] V.G. Drinfeld, *Langlands conjecture for $GL(2)$ over function field*, Proc. of Int. Congress of Math. (Helsinki, 1978), pp. 565–574.

[Dr2] V.G. Drinfeld, *Two-dimensional ℓ-adic representations of the fundamental group of a curve over a finite field and automorphic forms on $GL(2)$*, Amer. J. Math. **105** (1983) 85–114.

[Dr3] V.G. Drinfeld, *Moduli varieties of F-sheaves*, Funct. Anal. Appl. **21** (1987) 107–122.

[Dr4] V.G. Drinfeld, *The proof of Petersson's conjecture for $GL(2)$ over a global field of characteristic p*, Funct. Anal. Appl. **22** (1988) 28–43.

[DS] V. Drinfeld and V. Sokolov, *Lie algebras and KdV type equations*, J. Sov. Math. **30** (1985) 1975–2036.

[DrSi] V. Drinfeld and C. Simpson, *B-structures on G-bundles and local triviality*, Math. Res. Lett. **2** (1995) 823–829.

[EF] D. Eisenbud and E. Frenkel, Appendix to M. Mustata, *Jet schemes of locally complete intersection canonical singularities*, Invent. Math. **145** (2001) 397–424.

[FF1] B. Feigin and E. Frenkel, *A family of representations of affine Lie algebras*, Russ. Math. Surv. **43**, N 5 (1988) 221–222.

[FF2] B. Feigin and E. Frenkel, *Representations of affine Kac–Moody algebras and bosonization*, in *Physics and mathematics of strings*, pp. 271–316, World Scientific, 1990.

[FF3] B. Feigin and E. Frenkel, *Representations of affine Kac–Moody algebras, bosonization and resolutions*, Lett. Math. Phys. **19** (1990) 307–317.

[FF4] B. Feigin and E. Frenkel, *Affine Kac–Moody algebras and semi-infinite flag manifolds*, Comm. Math. Phys. **128** (1990) 161–189.

[FF5] B. Feigin and E. Frenkel, *Semi-infinite Weil complex and the Virasoro algebra*, Comm. Math. Phys. **137** (1991) 617–639.

[FF6] B. Feigin and E. Frenkel, *Affine Kac–Moody algebras at the critical level and Gelfand–Dikii algebras*, Int. Jour. Mod. Phys. **A7**, Supplement 1A (1992) 197–215.

[FF7] B. Feigin and E. Frenkel, *Integrals of motion and quantum groups*, in Proceedings of the C.I.M.E. School *Integrable Systems and Quantum Groups*, Italy, June 1993, Lect. Notes in Math. **1620**, Springer, 1995.

[FF8] B. Feigin and E. Frenkel, *Integrable hierarchies and Wakimoto modules*, in *Differential Topology, Infinite-Dimensional Lie Algebras, and Applications*, D. B. Fuchs' 60th Anniversary Collection, A. Astashkevich and S. Tabachnikov (eds.), pp. 27–60, AMS, 1999.

[FFR] B. Feigin, E. Frenkel, and N. Reshetikhin, *Gaudin model, Bethe ansatz and critical level*, Comm. Math. Phys. **166** (1994) 27–62.

[FK] E. Freitag and R. Kiehl, *Etale Cohomology and the Weil conjecture*, Springer, 1988.

[F1] E. Frenkel, *Affine Kac–Moody algebras and quantum Drinfeld–Sokolov reduction*, Ph.D. Thesis, Harvard University, 1991.

[F2] E. Frenkel, *Free field realizations in representation theory and conformal field theory*, in Proceedings of the International Congress of Mathematicians (Zürich 1994), pp. 1256–1269, Birkhäuser, 1995.

[F3] E. Frenkel, *Affine algebras, Langlands duality and Bethe Ansatz*, in Proceedings of the International Congress of Mathematical Physics, Paris, 1994, ed. D. Iagolnitzer, pp. 606–642, International Press, 1995; arXiv: q-alg/9506003.

[F4] E. Frenkel, *Wakimoto modules, opers and the center at the critical level*, Advances in Math. **195** (2005) 297–404.

[F5] E. Frenkel, *Gaudin model and opers*, in Infinite Dimensional Algebras and Quantum Integrable Systems, eds. P. Kulish, e.a., Progress in Math. **237**, pp. 1–60, Birkhäuser, 2005.

[F6] E. Frenkel, *Opers on the projective line, flag manifolds and Bethe Ansatz*, Mosc. Math. J. **4** (2004) 655–705.

[F7] E. Frenkel, *Lectures on the Langlands Program and conformal field theory*, Preprint hep-th/0512172.

[FB] E. Frenkel and D. Ben-Zvi, *Vertex algebras and algebraic curves*, Second Edition, Mathematical Surveys and Monographs, vol. 88. AMS 2004.

[FG1] E. Frenkel and D. Gaitsgory, *D-modules on the affine Grassmannian and representations of affine Kac–Moody algebras*, Duke Math. J. **125** (2004) 279–327.

[FG2] E. Frenkel and D. Gaitsgory, *Local geometric Langlands correspondence and affine Kac–Moody algebras*, Preprint math.RT/0508382.

[FG3] E. Frenkel and D. Gaitsgory, *Fusion and convolution: applications to affine Kac–Moody algebras at the critical level*, Preprint math.RT/0511284.

[FG4] E. Frenkel and D. Gaitsgory, *Localization of $\widehat{\mathfrak{g}}$-modules on the affine Grassmannian*, Preprint math.RT/0512562.

[FG5] E. Frenkel and D. Gaitsgory, *Geometric realizations of Wakimoto modules at the critical level*, Preprint math.RT/0603524.

[FG6] E. Frenkel and D. Gaitsgory, *Weyl modules and opers without monodromy*, Preprint.

[FGV1] E. Frenkel, D. Gaitsgory, and K. Vilonen, *Whittaker patterns in the geometry of moduli spaces of bundles on curves*, Annals of Math. **153** (2001) 699–748.

[FGV2] E. Frenkel, D. Gaitsgory, and K. Vilonen, *On the geometric Langlands conjecture*, Journal of AMS **15** (2001) 367–417.

[FKW] E. Frenkel, V. Kac, and M. Wakimoto, *Characters and fusion rules for $\mathcal{W}$-algebras via quantized Drinfeld-Sokolov reduction*, Comm. Math. Phys. **147** (1992) 295–328.

[FT] E. Frenkel and C. Teleman, *Self-extensions of Verma modules and differential forms on opers*, Compositio Math. **142** (2006) 477–500.

[FLM] I. Frenkel, J. Lepowsky, and A. Meurman, *Vertex operator algebras and the Monster*. Academic Press, 1988.

[FHL] I. Frenkel, Y.-Z. Huang, and J. Lepowsky, *On axiomatic approaches to vertex operator algebras and modules*, Mem. Amer. Math. Soc. **104** (1993), no. 494.

[FMS] D. Friedan, E. Martinec, and S. Shenker, *Conformal invariance, supersymmetry and string theory*, Nuclear Phys. **B271** (1986) 93–165.

[Fu] D. Fuchs, *Cohomology of infinite–dimensional Lie algebras*, Consultants Bureau, New York, 1986.

[Ga1] D. Gaitsgory, *Construction of central elements in the Iwahori Hecke algebra via nearby cycles*, Inv. Math. **144** (2001) 253–280.

[Ga2] D. Gaitsgory, *On a vanishing conjecture appearing in the geometric Langlands correspondence*, Ann. Math. **160** (2004) 617–682.

[Ga3] D. Gaitsgory, *The notion of category over an algebraic stack*, Preprint math.AG/0507192.

[GM] S.I. Gelfand, Yu.I. Manin, *Homological algebra*, Encyclopedia of Mathematical Sciences **38**, Springer, 1994.

[G1] V. Ginzburg, *Perverse sheaves on a loop group and Langlands duality*, Preprint alg-geom/9511007.

[G2] V. Ginzburg, *Loop Grassmannian cohomology, the principal nilpotent and Kostant theorem*, Preprint math.AG/9803141.

[GW] R. Goodman and N. Wallach, *Higher-order Sugawara operators for affine Lie algebras*, Trans. Amer. Math. Soc. **315** (1989) 1–55.

[GMS] V. Gorbounov, F. Malikov, and V. Schechtman, *On chiral differential operators over homogeneous spaces*, Int. J. Math. Math. Sci. **26** (2001) 83–106.

[HT] M. Harris and R. Taylor, *The geometry and cohomology of some simple Shimura varieties*, Annals of Mathematics Studies **151**, Princeton University Press, 2001.

[Ha] T. Hayashi, *Sugawara operators and Kac–Kazhdan conjecture*, Invent. Math. **94** (1988) 13–52.

[He] G. Henniart, *Une preuve simple des conjectures de Langlands pour* GL(n) *sur un corps p-adique*, Invent. Math. **139** (2000) 439–455.

[K1] V. Kac, *Laplace operators of infinite-dimensional Lie algebras and theta functions*, Proc. Nat. Acad. Sci. U.S.A. **81** (1984) no. 2, Phys. Sci., 645–647.

[K2] V.G. Kac, *Infinite-dimensional Lie Algebras*, 3rd Edition, Cambridge University Press, 1990.

[K3] V. Kac, *Vertex Algebras for Beginners*, Second Edition. AMS 1998.

[KK] V. Kac and D. Kazhdan, *Structure of representations with highest weight of infinite-dimensional Lie algebras*, Adv. in Math. **34** (1979) 97–108.

[KL] D. Kazhdan and G. Lusztig, *Proof of the Deligne–Langlands conjecture for Hecke algebras*, Inv. Math. **87** (1987) 153–215.

[Ko] B. Kostant, *Lie group representations on polynomial rings*, Amer. J. Math. **85** 1963, 327–402.

[Ku] S. Kudla, *The local Langlands correspondence: the non-Archimedean case*, in *Motives* (Seattle, 1991), pp. 365–391, Proc. Sympos. Pure Math. **55**, Part 2, AMS, 1994.

[Laf] L. Lafforgue, *Chtoucas de Drinfeld et correspondance de Langlands*, Invent. Math. **147** (2002) 1–241.

[L] R.P. Langlands, *Problems in the theory of automorphic forms*, in Lect. Notes in Math. **170**, pp. 18–61, Springer Verlag, 1970.

[La] G. Laumon, *Transformation de Fourier, constantes d'équations fonctionelles et conjecture de Weil*, Publ. IHES **65** (1987) 131–210.

[LRS] G. Laumon, M. Rapoport, and U. Stuhler, *$\mathcal{D}$-elliptic sheaves and the Langlands correspondence*, Invent. Math. **113** (1993) 217–338.

[Lu1] G. Lusztig, *Classification of unipotent representations of simple p-adic groups*, Int. Math. Res. Notices (1995) no. 11, 517–589.

[Lu2] G. Lusztig, *Bases in K-theory*, Represent. Theory **2** (1998) 298–369; **3** (1999) 281–353.

[Mi] J.S. Milne, *Étale cohomology*, Princeton University Press, 1980.

[MV] I. Mirković and K. Vilonen, *Geometric Langlands duality and representations of algebraic groups over commutative rings*, Preprint math.RT/0401222.

[Mu] D. Mumford, *Geometric invariant theory*, Springer, 1965.

[PRY] J.L. Petersen, J. Rasmussen, and M. Yu, *Free field realizations of 2D current algebras, screening currents and primary fields*, Nucl. Phys. **B 502** (1997) 649–670.

[Sa] I. Satake, *Theory of spherical functions on reductive algebraic groups over p–adic fields*, IHES Publ. Math. **18** (1963) 5–69.

[Sz] M. Szczesny, *Wakimoto Modules for Twisted Affine Lie Algebras*, Math. Res. Lett. **9** (2002), no. 4, 433–448.

[V] D.A. Vogan, *The local Langlands conjecture*, Contemporary Math. **145**, pp. 305–379, AMS, 1993.

[W] M. Wakimoto, *Fock representations of affine Lie algebra* $A_1^{(1)}$, Comm. Math. Phys. **104** (1986) 605–609.

Index

adéle, 6
admissible representation, 301
affine flag variety, 346
affine Grassmannian, 21, 296, 317, 318, 340
affine Hecke algebra, *see* Hecke algebra, affine
affine Kac–Moody algebra, 28, 371
annihilation operator, 141
associated graded algebra, 78, 367
automorphic function, 355
automorphic representation, 7, 353, 359

big cell, 129

Cartan decomposition, 108, 369
Cartan matrix, 369
Cartan subalgebra, 369
Casimir element, 32
center
  at the critical level, 77, 117, 234
  of a vertex algebra, 76
  of the enveloping algebra, 120, 124
central character, 297
central charge, 68
central extension, 27, 370
conformal dimension, 40
conformal vector, 68
congruence subgroup, 2, 18, 302
connection, 11, 109
  flat, 11, 109
constructible sheaf, 23
  ℓ-adic, 22
contragredient representation, 21
convolution, 302
coweight, 370
creation operator, 141
critical level, *see* level, critical

$\mathcal{D}$-module, 24, 303, 317, 318, 347, 355
DG scheme, 335
disc, 68, 81, 90, 93
  punctured, 17, 93
Drinfeld–Sokolov reduction, 239
dual Coxeter number, 64

equivariant module, 302
exponent of a Lie algebra, 112

field, 40
flag variety, 129
flat G-bundle, *see* G-bundle, flat
flat connection, *see* connection, flat
flat vector bundle, *see* vector bundle, flat
formal coordinate, 90
formal delta-function, 39
formal loop group, *see* loop group
Frobenius automorphism, 3, 318, 327, 339, 354
fundamental group, 9

G-bundle, 14
  flat, 15
Galois group, 3, 9
gauge transformation, 15, 109
Grothendieck alteration, 194, 335

Harish-Chandra category, 303
Harish-Chandra module, 303
Harish-Chandra pair, 303
Hecke algebra, 302
  affine, 329
  affine, categorical, 346, 361
  categorical, 304
  spherical, 306, 340
  spherical, categorical, 317
Hecke eigenfunction, 355
Hecke eigenmodule, 318, 320, 340
Hecke eigensheaf, 356, 360
Hecke eigenvector, 307, 355
highest weight vector, 185
holomorphic vector bundle, *see* vector bundle, holomorphic
horizontal section, 13

integrable representation, 19
intertwining operator, 197
invariant inner product, 28
Iwahori subgroup, 302, 327

jet scheme, 84, 286

Kac–Kazhdan conjecture, 190
Killing form, 28, 63

L-packet, 6, 328, 353
Langlands correspondence
  global, 6, 8, 353
  global geometric, tamely ramified, 363
  global geometric, unramified, 356
  global, tamely ramified, 359
  global, unramified, 355
  local, 1, 4, 6, 354
  local geometric, 29
  local geometric, tamely ramified, 331, 337
  local geometric, unramified, 307, 310, 322
  local, tamely ramified, 327
  local, unramified, 306
Langlands dual group, 6, 100, 108, 117, 296, 306, 319
Langlands dual Lie algebra, 117, 221
level, 28
  critical, 30, 37, 64, 99, 140, 161, 162, 166
lexicographically ordered monomial, 368
Lie algebra, 366
  simple, 368
local field, 1
local system, 18, 297
  ℓ-adic, 22
  tamely ramified, 332, 342, 360
  trivial, 307, 341
locality, 41
localization functor, 356, 360, 363, 365
loop algebra
  formal, 26, 372
  polynomial, 372
loop group, 9, 18, 20, 23, 26, 30, 295

maximal compact subgroup, 21, 302, 305
Miura oper, 193, 226
  generic, 226
  nilpotent, 258
Miura transformation, 229, 241, 349
module over a Lie algebra, 366
monodromic module, 334
monodromy, 13, 17, 327, 332, 333, 339, 364

nilpotent cone, 86, 260, 330
nilpotent Miura oper, *see* Miura oper, nilpotent
nilpotent oper, *see* oper, nilpotent
normal ordering, 37, 49

oper, 109, 297
  nilpotent, 248, 333
  regular, 310
  with regular singularity, 244
operator product expansion, 56

PBW filtration, *see* universal enveloping algebra, filtration
Poincaré–Birkhoff–Witt theorem, 368
principal G-bundle, *see* G-bundle
principal gradation, 112
projective connection, 96, 101, 107
projective structure, 102
punctured disc, *see* disc, punctured

reconstruction theorem, 51, 52
reduction of a principal bundle, 104
regular singularity, 16, 244, 360, 363
relative position, 261
  generic, 194
representation of a Lie algebra, 366
residue, 28
  of a nilpotent oper, 253
  of an oper with regular singularity, 244
Riemann–Hilbert correspondence, 24
root, 369
  simple, 369

Satake isomorphism, 306
  categorical, 317
Schwarzian derivative, 98, 114
screening operator, 192
  of $\mathcal{W}$-algebra, 219
  of the first kind, 200, 206
  of the second kind, 203, 208
Segal–Sugawara operator, 37, 62, 74, 87, 98, 119, 171
smooth representation, 2, 19, 29, 178, 183
space of states, 40
spherical Hecke algebra, *see* Hecke algebra, spherical
Springer fiber, 261, 329, 348
Springer variety, 260
state-field correspondence, 41
Steinberg variety, 330, 347
symbol, 368

Tannakian formalism, 15, 317, 340
torsor, 90
transition function, 11
translation operator, 40

universal enveloping algebra, 367
  completed, 34
  filtration, 367
unramified representation, 305

vacuum vector, 40
vacuum Verma module, *see* Verma module, vacuum
vector bundle, 11
  flat, 11
  holomorphic, 13
Verma module, 132, 185, 273
  contragredient, 133
  vacuum, 45
vertex algebra, 40
  commutative, 42
  conformal, 68
  quasi-conformal, 173

vertex algebra homomorphism, 68
vertex operator, 41
vertex Poisson algebra, 221
Virasoro algebra, 66

$\mathcal{W}$-algebra, 220
Wakimoto module, 169
  generalized, 183
  of critical level, 166, 343, 351
weight, 370
Weil group, 4, 7, 354
Weil–Deligne group, 4, 306, 326, 339
Weyl algebra, 138
Weyl group, 370
Weyl module, 279
Wick formula, 147

Printed in the United States
by Baker & Taylor Publisher Services